Lecture Notes in Computer Science 8936

Commenced Publication in 1973
Founding and Former Series Editors:
Gerhard Goos, Juris Hartmanis, and Jan van Leeu

Xiangjian He   Suhuai Luo   Dacheng Tao
Changsheng Xu   Jie Yang
Muhammad Abul Hasan (Eds.)

MultiMedia Modeling

21st International Conference, MMM 2015
Sydney, NSW, Australia, January 5-7, 2015
Proceedings, Part II

 Springer

Volume Editors

Xiangjian He
Dacheng Tao
Muhammad Abul Hasan
University of Technology, Sydney, NSW, Australia
E-mail:{xiangjian.he, dacheng.tao, muhammad.hasan}@uts.edu.au

Suhuai Luo
University of Newcastle, NSW, Australia
E-mail: suhuai.luo@newcastle.edu.au

Changsheng Xu
Chinese Academy of Sciences, Institute of Automation
National Lab of Pattern Recognition, Beijing, China
E-mail: csxu@nlpr.ia.ac.cn

Jie Yang
Shanghai Jitotong University, Shanghai, China
E-mail: jieyang@sjtu.edu.cn

ISSN 0302-9743   e-ISSN 1611-3349
ISBN 978-3-319-14441-2   e-ISBN 978-3-319-14442-9
DOI 10.1007/978-3-319-14442-9
Springer Cham Heidelberg New York Dordrecht London

Library of Congress Control Number: 2014957756

LNCS Sublibrary: SL 3 – Information Systems and Application, incl. Internet/Web
and HCI

Typesetting: Camera-ready by author, data conversion by Scientific Publishing Services, Chennai, India

Printed on acid-free paper

Springer is part of Springer Science+Business Media (www.springer.com)

Preface

These proceedings contain the papers presented at MMM 2015, the 21st International Conference on MultiMedia Modeling. The conference was organized by University of Technology, Sydney, and was held during January 5–7, 2015, at the Aerial UTS Function Centre, Sydney Australia.

We were delighted to welcome all attendees to MMM 2015. We believe that they had a wonderful stay in Australia and that their visit was both enjoyable and rewarding. We were very proud to welcome visitors from both Australia and abroad and are delighted to be able to include in the proceedings such high-quality papers for oral presentation, poster presentation, special sessions, demonstrations, and video search showcase.

MMM 2015 received 189 submissions across four categories, consisting of 136 main conference full-paper submissions, 24 special session full-paper submissions, 18 demonstration submissions, and 11 video search showcase submissions. Of these submissions, there are 27 authors (6%) from Australia, 130 (28%) from Europe, 287 (62%) from Asia, and 19 (4%) from the Americas. All main conference submissions were reviewed by at least three members of the Program Committee, to whom we owe a debt of gratitude for providing their valuable time to MMM 2015. All papers submitted to the special sessions, demonstration sessions, and video search showcase sessions were also reviewed by at least three reviewers.

Of the 136 main conference full-paper submissions, 49 were selected for oral presentation, which equates to a 36% acceptance rate. A further 24 papers were chosen for poster presentation. The accepted contributions represent the state of the art in multimedia modeling research and cover a diverse range of topics including: image and video processing, multimedia encoding and streaming, applications of multimedia modeling, and 3D and augmented reality. For the three special sessions, a total of 18 papers were accepted for MMM 2015. The three special sessions were "Personal (Big) Data Modeling for Information Access and Retrieval," "Social Geo-Media Analytics and Retrieval," and "Image or Video Processing, Semantic Analysis, and Understanding." In addition, nine demonstrations and nine video showcase papers were accepted for MMM 2015.

We would like to thank our invited keynote speakers for their stimulating contributions to the conference. Special thanks go to the Organizing Committee for their contributions and great efforts toward the success of this event, and Steering Committee for constant support and timely advice.

In addition, we wish to thank all authors who spent their time and effort to submit their work to MMM 2015, and all of the participants and student volunteers for their contributions and valuable support.

Our gratitude also goes to the MMM 2015 Program Committee members and the other invited reviewers for the large number of reviews required for MMM 2015.

We are grateful to the sponsors for providing support to the conference, including the University of Technology, Sydney, Business Events Sydney, Business Events Australia, Australia Government AusAID, and Qantas.

January 2015

Xiangjian He
Changsheng Xu
Dacheng Tao
Suhuai Luo
Jie Yang
Muhammad Abul Hasan

Organization

Steering Committee

Phoebe Chen	La Trobe University, Australia
Tat-Seng Chua	National University of Singapore, Singapore
Yang Shiqiang	Tsinghua University, China
Kiyoharu Aizawa	University of Tokyo, Japan
Noel E. O'Connor	Dublin City University, Ireland
Cess G.M. Snoek	University of Amsterdam, The Netherlands
Meng Wang	Hefei University of Technology, China
R. Manmatha	University of Massachusetts, USA
Cathal Gurrin	Dublin City University, Ireland
Klaus Schoeffmann	Klagenfurt University, Austria
Benoit Huet	Eurecom, France

Organizing Committee

Honorary Co-chairs

Massimo Piccardi	University of Technology Sydney, Australia
Phoebe Chen	La Trobe University, Australia
Tat-Seng Chua	National University of Singapore, Singapore

General Co-chairs

Xiangjian He	University of Technology Sydney, Australia
Changsheng Xu	Chinese Academy of Science, China

Program Co-chairs

Dacheng Tao	University of Technology Sydney, Australia
Suhuai Luo	University of Newcastle, Australia
Jie Yang	Shanghai Jiaotong University, China

Organizing Co-chairs

Qiang Wu	University of Technology Sydney, Australia
Tao Mei	Microsoft Research Asia, China

Local Chair

Jian Zhang	University of Technology Sydney, Australia

Special Session Co-chairs

Min Xu	University of Technology Sydney, Australia
Lexing Xie	Australian National University, Australia

Demos Co-chairs

Cathal Gurrin   Dublin City University, Ireland
Björn Þór Jónsson   Reykjavík University, Iceland

Financial Chair

Wenjing Jia   University of Technology Sydney, Australia

Tutorial Co-chairs

Richard Xu   University of Technology Sydney, Australia
Shuicheng Yan   National University of Singapore, Singapore

Publication Chair

Muhammad Abul Hasan   University of Technology Sydney, Australia

Publicity Co-chairs

Zhengjun Zha   Chinese Academy of Sciences, China
Jitao Sang   Chinese Academy of Sciences, China
Yinjie Lei   Sichuan University, China

Video Search Showcase Co-chairs

Werner Bailer   Joanneum Research, Austria
Klaus Schoeffmann   Klagenfurt University, Austria

Web Masters

Vera Chung   University of Sydney, Australia
Angus Ma   University of Technology Sydney, Australia
David Kim   University of Sydney, Australia
Feng Sha   University of Sydney, Australia
Benedict Goh   University of Sydney, Australia

Program Committee

Xiangjian He   University of Technology, Sydney, Australia
Klaus Schöffmann   University of Klagenfurt, Austria
Suhuai Luo   The University of Newcastle, Australia
Dong Liu   Columbia University, USA
Jingdong Wang   Microsoft Research Asia, China
Wenjing Jia   University of Technology Sydney, Australia
Cha Zhang   Microsoft Research, USA
Wolfgang Huerst   Utrecht University, The Netherlands

Table of Contents – Part II

Applications

A Proxemic Multimedia Interaction over the Internet of Things   1
  Ali Danesh, Mukesh Saini, and Abdulmotaleb El Saddik

Outdoor Air Quality Inference from Single Image   13
  Zheng Zhang, Huadong Ma, Huiyuan Fu, and Xinpeng Wang

Multimodal Music Mood Classification by Fusion of Audio
and Lyrics   26
  Hao Xue, Like Xue, and Feng Su

Multidimensional Context Awareness in Mobile Devices   38
  *Zhuo Wei, Robert H. Deng, Jialie Shen, Jixiang Zhu, Kun Ouyang,
  and Yongdong Wu*

AttRel: An Approach to Person Re-Identification by Exploiting
Attribute Relationships   50
  *Ngoc-Bao Nguyen, Vu-Hoang Nguyen, Thanh Ngo Duc,
  Duy-Dinh Le, and Duc Anh Duong*

Sparsity-Based Occlusion Handling Method for Person
Re-identification   61
  *Bingyue Huang, Jun Chen, Yimin Wang, Chao Liang, Zheng Wang,
  and Kaimin Sun*

Visual Attention Driven by Auditory Cues: Selecting Visual Features
in Synchronization with Attracting Auditory Events   74
  *Jiro Nakajima, Akisato Kimura, Akihiro Sugimoto,
  and Kunio Kashino*

A Synchronization Ground Truth for the Jiku Mobile Video Dataset   87
  Mario Guggenberger, Mathias Lux, and Laszlo Böszörmenyi

Mobile Image Analysis: Android vs. iOS   99
  *Claudiu Cobârzan, Marco A. Hudelist, Klaus Schoeffmann,
  and Manfred Jürgen Primus*

Dynamic User Authentication Based on Mouse Movements Curves   111
  Zaher Hinbarji, Rami Albatal, and Cathal Gurrin

Sliders Versus Storyboards – Investigating Interaction Design for
Mobile Video Browsing   123
  Wolfgang Hürst and Miklas Hoet

Performance Evaluation of Students Using Multimodal Learning
Systems 135
    Subhasree Basu, Roger Zimmermann, Kay L. O'Halloran,
    Sabine Tan, and Marissa K.L.E.

Is Your First Impression Reliable? Trustworthy Analysis Using Facial
Traits in Portraits.. 148
    Yan Yan, Jie Nie, Lei Huang, Zhen Li, Qinglei Cao,
    and Zhiqiang Wei

Wifbs: A Web-Based Image Feature Benchmark System.............. 159
    Marcel Spehr, Sebastian Grottel, and Stefan Gumhold

Personality Modeling Based Image Recommendation 171
    Sharath Chandra Guntuku, Sujoy Roy, and Lin Weisi

Aesthetic QR Codes Based on Two-Stage Image Blending............ 183
    Yongtai Zhang, Shihong Deng, Zhihong Liu, and Yongtao Wang

Person Re-identification Using Data-Driven Metric Adaptation 195
    Zheng Wang, Ruimin Hu, Chao Liang, Junjun Jiang, Kaimin Sun,
    Qingming Leng, and Bingyue Huang

A Novel Optimized Watermark Embedding Scheme for Digital
Images 208
    Feng Sha, Felix Lo, Yuk Ying Chung, Xiaoming Chen, and
    Wei-Chang Yeh

User-Centred Evaluation to Interface Design of E-Books 220
    Yang-Cheng Lin

A New Image Decomposition and Reconstruction Approach – Adaptive
Fourier Decomposition.. 227
    Can He, Liming Zhang, Xiangjian He, and Wenjing Jia

Video Showcase

Graph-Based Browsing for Large Video Collections.................... 237
    Kai Uwe Barthel, Nico Hezel, and Radek Mackowiak

Enhanced Signature-Based Video Browser.......................... 243
    Adam Blažek, Jakub Lokoč, Filip Matzner, and Tomáš Skopal

VERGE: A Multimodal Interactive Video Search Engine 249
    Anastasia Moumtzidou, Konstantinos Avgerinakis,
    Evlampios Apostolidis, Fotini Markatopoulou,
    Konstantinos Apostolidis, Theodoros Mironidis,
    Stefanos Vrochidis, Vasileios Mezaris, Ioannis Kompatsiaris,
    and Ioannis Patras

IMOTION — A Content-Based Video Retrieval Engine 255
    Luca Rossetto, Ivan Giangreco, Heiko Schuldt, Stéphane Dupont,
    Omar Seddati, Metin Sezgin, and Yusuf Sahillioğlu

A Storyboard-Based Interface for Mobile Video Browsing 261
    Wolfgang Hürst, Rob van de Werken, and Miklas Hoet

Collaborative Browsing and Search in Video Archives
with Mobile Clients 266
    Claudiu Cobârzan, Manfred Del Fabro, and Klaus Schoeffmann

The Multi-stripe Video Browser for Tablets 272
    Marco A. Hudelist and Qing Xu

NII-UIT Browser: A Multimodal Video Search System 278
    Thanh Duc Ngo, Vinh-Tiep Nguyen, Vu Hoang Nguyen,
    Duy-Dinh Le, Duc Anh Duong, and Shin'ichi Satoh

Interactive Known-Item Search Using Semantic Textual and Colour
Modalities 282
    Zhenxing Zhang, Rami Albatal, Cathal Gurrin, and Alan F. Smeaton

Demonstration

ImageMap - Visually Browsing Millions of Images 287
    Kai Uwe Barthel, Nico Hezel, and Radek Mackowiak

Dynamic Hierarchical Visualization of Keyframes
in Endoscopic Video 291
    Jakub Lokoč, Klaus Schoeffmann, and Manfred Del Fabro

Facial Aging Simulator by Data-Driven Component-Based Texture
Cloning 295
    Daiki Kuwahara, Akinobu Maejima, and Shigeo Morishima

Affective Music Recommendation System Based on the Mood of Input
Video 299
    Shoto Sasaki, Tatsunori Hirai, Hayato Ohya, and Shigeo Morishima

MemLog, an Enhanced Lifelog Annotation and Search Tool 303
    Lijuan Marissa Zhou, Brian Moynagh, Liting Zhou, TengQi Ye,
    and Cathal Gurrin

Software Solution for HEVC Encoding and Decoding 307
    Shengbin Meng, Jun Sun, and Zongming Guo

A Surveillance Video Index and Browsing System Based on Object
Flags and Video Synopsis 311
    Gensheng Ye, Wenjuan Liao, Jichao Dong, Dingheng Zeng,
    and Huicai Zhong

A Web Portal for Effective Multi-model Exploration 315
    Tomáš Grošup, Přemysl Čech, Jakub Lokoč, and Tomáš Skopal

Wearable Cameras for Real-Time Activity Annotation 319
    Jiang Zhou, Aaron Duane, Rami Albatal, Cathal Gurrin,
    and Dag Johansen

Personal (Big) Data Modeling for Information Access & Retrieval

Making Lifelogging Usable: Design Guidelines for Activity Trackers 323
    Jochen Meyer, Jutta Fortmann, Merlin Wasmann, and Wilko Heuten

Towards Consent-Based Lifelogging in Sport Analytic 335
    Håvard Johansen, Cathal Gurrin, and Dag Johansen

A Multi-Dimensional Data Model for Personal Photo Browsing 345
    Björn Þór Jónsson, Grímur Tómasson, Hlynur Sigurþórsson,
    Áslaug Eiríksdóttir, Laurent Amsaleg, and Marta Kristín Lárusdóttir

Discriminative Regions: A Substrate for Analyzing Life-Logging Image
Sequences 357
    Mohammad Moghimi, Jacqueline Kerr, Eileen Johnson,
    Suneeta Godbole, and Serge Belongie

Fast Human Activity Recognition in Lifelogging 369
    Stefan Terziyski, Rami Albatal, and Cathal Gurrin

Social Geo-Media Analytics and Retrieval

Iron Maiden While Jogging, Debussy for Dinner? - An Analysis of
Music Listening Behavior in Context 380
    Michael Gillhofer and Markus Schedl

Travel Recommendation via Author Topic Model Based Collaborative
Filtering 392
    Shuhui Jiang, Xueming Qian, Jialie Shen, and Tao Mei

Robust User Community-Aware Landmark Photo Retrieval 403
    Lin Wu, John Shepherd, Xiaodi Huang, and Chunzhi Hu

Cross-Domain Concept Detection with Dictionary Coherence by
Leveraging Web Images 415
    Yongqing Sun, Kyoko Sudo, and Yukinobu Taniguchi

Semantic Correlation Mining between Images and Texts with Global
Semantics and Local Mapping 427
    Jiao Xue, Youtian Du, and Hanbing Shui

Image Taken Place Estimation via Geometric Constrained Spatial
Layer Matching   436
    Yisi Zhao, Xueming Qian, and Tingting Mu

Image or Video Processing, Semantic Analysis, and Understanding

Recognition of Meaningful Human Actions for Video Annotation Using
EEG Based User Responses   447
    Jinyoung Moon, Yongjin Kwon, Kyuchang Kang, Changseok Bae,
    and Wan Chul Yoon

Challenging Issues in Visual Information Understanding Researches   458
    Kyuchang Kang, Yongjin Kwon, Jinyoung Moon, and Changseok Bae

Emotional Tone-Based Audio Continuous Emotion Recognition   470
    Mengmeng Liu, Hui Chen, Yang Li, and Fengjun Zhang

A Computationally Efficient Algorithm for Large Scale Near-Duplicate
Video Detection   481
    Dawei Liu and Zhihua Yu

SLOREV: Using Classical CAD Techniques for 3D Object Extraction
from Single Photo   491
    Pan Hu, Hongming Cai, and Fenglin Bu

Hessian Regularized Sparse Coding for Human Action Recognition   502
    Weifeng Liu, Zhen Wang, Dapeng Tao, and Jun Yu

Robust Multi-label Image Classification with Semi-Supervised Learning
and Active Learning   512
    Fuming Sun, Meixiang Xu, and Xiaojun Jiang

Photo Quality Assessment with DCNN that Understands Image Well . . .   524
    Zhe Dong, Xu Shen, Houqiang Li, and Xinmei Tian

Non-negative Low-Rank and Group-Sparse Matrix Factorization   536
    Shuyi Wu, Xiang Zhang, Naiyang Guan, Dacheng Tao,
    Xuhui Huang, and Zhigang Luo

Two-Dimensional Euler PCA for Face Recognition   548
    Huibin Tan, Xiang Zhang, Naiyang Guan, Dacheng Tao,
    Xuhui Huang, and Zhigang Luo

Multiclass Boosting Framework for Multimodal Data Analysis   560
    Shixun Wang, Peng Pan, Yansheng Lu, and Sheng Jiang

Author Index   573

Table of Contents – Part I

Image and Video Processing

An Efficient Hybrid Steganography Method Based on Edge Adaptive
and Tree Based Parity Check 1
Hayat Al-Dmour, Noman Ali, and Ahmed Al-Ani

Secure Client Side Watermarking with Limited Key Size 13
Jia-Hao Sun, Yu-Hsun Lin, and Ja-Ling Wu

Orderless and Blurred Visual Tracking via Spatio-temporal Context 25
*Manna Dai, Peijie Lin, Lijun Wu, Zhicong Chen, Songlin Lai,
Jie Zhang, Shuying Cheng, and Xiangjian He*

Coupled Discriminant Multi-Manifold Analysis with Application to
Low-Resolution Face Recognition 37
Junjun Jiang, Ruimin Hu, Zhen Han, Liang Chen, and Jun Chen

Text Detection in Natural Images Using Localized Stroke Width
Transform 49
Wenyan Dong, Zhouhui Lian, Yingmin Tang, and Jianguo Xiao

Moving Object Tracking with Structure Complexity Coefficients 59
Yuan Yuan, Yuming Fang, and Lin Weisi

Real-Time People Counting across Spatially Adjacent Non-overlapping
Camera Views 71
Ryota Akai, Naoko Nitta, and Noboru Babaguchi

Binary Code Learning via Iterative Distance Adjustment.............. 83
Zhen-fei Ju, Xiao-jiao Mao, Ning Li, and Yu-bin Yang

What Image Classifiers Really See – Visualizing Bag-of-Visual Words
Models 95
Christian Hentschel and Harald Sack

Coupled-View Based Ranking Optimization for Person
Re-identification 105
*Mang Ye, Jun Chen, Qingming Leng, Chao Liang, Zheng Wang,
and Kaimin Sun*

Wireless Video Surveillance System Based on Incremental Learning
Face Detection.. 118
    Wenjuan Liao, Dingheng Zeng, Liguo Zhou, Shizheng Wang,
    and Huicai Zhong

An Automatic Rib Segmentation Method on X-Ray Radiographs....... 128
    Xuechen Li, Suhuai Luo, and Qingmao Hu

Content-Based Discovery of Multiple Structures from Episodes of
Recurrent TV Programs Based on Grammatical Inference 140
    Bingqing Qu, Félicien Vallet, Jean Carrive, and Guillaume Gravier

FOCUSING PATCH: Automatic Photorealistic Deblurring for Facial
Images by Patch-Based Color Transfer 155
    Masahide Kawai and Shigeo Morishima

Efficient Compression of Hyperspectral Images Using Optimal
Compression Cube and Image Plane................................ 167
    Rui Xiao and Manoranjan Paul

Automatic Chinese Personality Recognition Based on Prosodic
Features 180
    Huan Zhao, Zeying Yang, Zuo Chen, and Xixiang Zhang

Robust Attribute-Based Visual Recognition Using Discriminative
Latent Representation 191
    Yuqi Wang, Yunfei Gong, and Qiang Liu

An Analysis of Time Drift in Hand-Held Recording Devices 203
    Mario Guggenberger, Mathias Lux, and Laszlo Böszörmenyi

A Real-Time People Counting Approach in Indoor Environment 214
    Jun Luo, Jinqiao Wang, Huazhong Xu, and Hanqing Lu

Multi-instance Feature Learning Based on Sparse Representation for
Facial Expression Recognition 224
    Yuchun Fang and Lu Chang

Object Detection in Low-Resolution Image via Sparse
Representation 234
    Wenhua Fang, Jun Chen, Chao Liang, Xiao Wang, Yuanyuan Nan,
    and Ruimin Hu

A Novel Fast Full Frame Video Stabilization via Three-Layer Model 246
    Wei Long, Jie Yang, Dacheng Song, Xiaogang Chen,
    and Xiangjian He

Multimedia Mining and Retireval

Cross-Modal Self-Taught Learning for Image Retrieval 257
  Liang Xie, Peng Pan, Yansheng Lu, and Sheng Jiang

Multimedia Social Event Detection in Microblog 269
  Yue Gao, Sicheng Zhao, Yang Yang, and Tat-Seng Chua

A Study on the Use of a Binary Local Descriptor and Color Extensions
of Local Descriptors for Video Concept Detection 282
  Foteini Markatopoulou, Nikiforos Pittaras, Olga Papadopoulou,
  Vasileios Mezaris, and Ioannis Patras

Content-Based Image Retrieval with Gaussian Mixture Models 294
  Christian Beecks, Merih Seran Uysal, and Thomas Seidl

Improving Interactive Known-Item Search in Video with the Keyframe
Navigation Tree 306
  Marco A. Hudelist, Klaus Schöffmann, and Qing Xu

Large-Scale Image Mining with Flickr Groups 318
  Alexandru Lucian Ginsca, Adrian Popescu, Hervé Le Borgne,
  Nicolas Ballas, Phong Vo, and Ioannis Kanellos

FISIR: A Flexible Framework for Interactive Search in Image Retrieval
Systems 335
  Sheila M. Pinto-Cáceres, Jurandy Almeida,
  M. Cecília C. Baranauskas, and Ricardo da S. Torres

Auditory Scene Classification with Deep Belief Network 348
  Like Xue and Feng Su

An Improved Content-Based Music Recommending Method with
Weighted Tags 360
  Lu Ding, Ning Zheng, Jiang Xu, and Ming Xu

A Unified Model for Socially Interconnected Multimedia-Enriched
Objects 372
  Theodora Tsikrika, Katerina Andreadou, Anastasia Moumtzidou,
  Emmanouil Schinas, Symeon Papadopoulos, Stefanos Vrochidis,
  and Ioannis Kompatsiaris

Concept-Based Multimodal Learning for Topic Generation 385
  Cheng Wang, Haojin Yang, Xiaoyin Che, and Christoph Meinel

Audio Secret Management Scheme Using Shamir's Secret Sharing 396
  M. Abukari Yakubu, Namunu C. Maddage, and Pradeep K. Atrey

Live Version Identification with Audio Scene Detection 408
  Kazumasa Ishikura, Aiko Uemura, and Jiro Katto

Community Detection Based on Links and Node Features in Social
Networks 418
   Fengli Zhang, Jun Li, Feng Li, Min Xu, Richard Xu,
   and Xiangjian He

Multimedia Encoding and Streaming

Scaling and Cropping of Wavelet-Based Compressed Images in Hidden
Domain.. 430
   Kshitij Kansal, Manoranjan Mohanty, and Pradeep K. Atrey

MAP: Microblogging Assisted Profiling of TV Shows 442
   Xiahong Lin, Zhi Wang, and Lifeng Sun

Improved Rate-Distortion Optimization Algorithms for HEVC Lossless
Coding 454
   Fangdong Chen and Houqiang Li

A Novel Error Concealment Algorithm for H.264/AVC................ 466
   Jinlei Zhang and Houqiang Li

Edge Direction-Based Fast Coding Unit Partition for HEVC Screen
Content Coding.. 477
   Mengmeng Zhang and Yangxiao Ou

Signal-Aware Parametric Quality Model for Audio and Speech over IP
Networks 487
   SongBo Xie, Yuhong Yang, Ruimin Hu, Yanye Wang,
   Hongjiang Yu, ShaoLong Dong, Li Gao, and Cheng Yang

3D and Augmented Reality

Patch-Based Disparity Remapping for Stereoscopic Images 498
   Dawei Lu, Huadong Ma, Liang Liu, and Huiyuan Fu

3D Depth Perception from Single Monocular Images.................. 510
   Hang Xu, Kan Li, FuYu Lv, and JianMeng Pei

Muscular Movement Model Based Automatic 3D Facial Expression
Recognition 522
   Qingkai Zhen, Di Huang, Yunhong Wang, and Liming Chen

Azimuthal Perceptual Resolution Model Based Adaptive 3D Spatial
Parameter Coding 534
   Li Gao, Ruimin Hu, Yuhong Yang, Xiaocheng Wang, Weiping Tu,
   and Tingzhao Wu

Flat3D: Browsing Stereo Images on a Conventional Screen 546
  Wenjing Geng, Ran Ju, Xiangyang Xu, Tongwei Ren,
  and Gangshan Wu

Online 3D Shape Segmentation by Blended Learning 559
  Feiqian Zhang, Zhengxing Sun, Mofei Song, and Xufeng Lang

Factorizing Time-Aware Multi-way Tensors for Enhancing Semantic
Wearable Sensing 571
  Peng Wang, Alan F. Smeaton, and Cathal Gurrin

Author Index 583

A Proxemic Multimedia Interaction over the Internet of Things

Ali Danesh, Mukesh Saini, and Abdulmotaleb El Saddik

MCRLab, University of Ottawa, Ottawa ON K1N 6N5, Canada
Division of Engineering, New York University Abu Dhabi, UAE
{aahma078,msain2,elsaddik}@uottawa.ca

Abstract. With the rapid growth of online devices, a new concept of *Internet of Things* (IoT) is emerging in which everyday devices will be connected to the Internet. As the number of devices in IoT is increasing, so is the complexity of the interactions between the user and the devices. There is a need to design intelligent user interfaces that could assist users in interactions with multiple devices. The present study proposes a proximity-based user interface for multimedia devices over IoT. The proposed method employs a cloud-based decision engine to support user to choose and interact with the most appropriate device, reliving the user from the burden of enumerating available devices manually. The decision engine observes the multimedia content and device properties, learns user preferences adaptively, and automatically recommends the most appropriate device to interact. The system evaluation shows that the users agree with the proposed interaction 70% of the times.

Keywords: Proxemic interaction, multimedia interaction, user interface, elicitation study.

1 Introduction

The Internet has been expanding very rapidly with time. Now, more than 2.7 billion people (almost 39% of world's population) have access to it and use it in their daily life [1]. The number of devices that are connected to the Internet has been growing dramatically. Therefore, a new concept of Internet of Things (IoT) is emerging [2]. More than 11.2 billion devices were connected to the Internet in 2013 and it is predicted that there will be around 50 billion devices online by 2020 [3]. These uniquely identifiable objects and devices move and interact with each other to accomplish various tasks. In other words, IoT is like a big and dynamic society of objects and people. But we are far from the ubiquitous computing vision of Weiser [4] due to two main reasons: the lack of an appropriate task-centered User Interface (UI) design approach and the lack of reliable support for distributable user interfaces in ubiquitous environments [5]. Thus, there is a need for a new generation of specifically designed user interfaces to interact with multiple devices IoT.

X. He et al. (Eds.): MMM 2015, Part II, LNCS 8936, pp. 1–12, 2015.

More than 7 billion people constantly use verbal and nonverbal communication techniques to interact with their society all across the world. Verbal communication is considered as the main channel since it is used explicitly. Still, nonverbal communication such as Proxemics, Haptics, Body Language, etc. also plays an important role in the human society because it uses implicit interactions. IoT is like human society, except its size is few times bigger than human society. Thus, nonverbal techniques can be used in order to build effective user interfaces for IoT.

The present study proposes a distributed user interface that provides a suitable environment for multimedia interactions over the IoT. It consists of a cloud-based decision engine that manages the proxemic interactions. This engine is called Proxemic Interaction Unit (PIU). PIU uses multimedia devices within the IoT as elements of a universal UI in order to assist users to make use of their surrounding devices. We collected the effectual multimedia interaction variables such as device properties, multimedia content attributes, user's preferences, etc. to propose a scoring mechanism for the PIU. A group of these variables are obtained by conducting a user survey and elicitation study while the rest are extracted from the literature. To further personalize user preferences, we update the variables after every interaction. The evaluation shows that 70% of the times the user interface chooses the same device a user would. We reckon that the user preference learning mechanism will further increase the accuracy of the user interface.

The rest of this paper firstly presents a review of related works in Section 2. Next, Section 3 describes the architecture and details of proposed UI. The results of an evaluation user study are given in Section 4, which is followed by brief discussion in Section 5. This study ends with the conclusion and future work in Section 6.

2   Related Work

Researchers have been trying to develop new UIs over the IoT for the last few years. However, the idea of using proxemic interactions has been around for a while. Hall highlighted the influence of proxemic behavior on interpersonal communication in 1966 [6]. He divided proxemic interactions into two levels: micro-level, which studies the way people interact with each other in daily life and macro-level, which reviews the space organization of houses, buildings and ultimately towns. More recently, Greenberg et al. proposed a practical version of proxemic interaction that considers people, digital devices, and non-digital objects [7]. They defined five dimensions for proxemic interaction: distance, orientation, movement, identity, and location. Every measured change in any dimension can trigger an interaction. Marquardt et al. used this terminology to develop a proximity toolkit in order to aid fast prototyping of proxemic interactions [8]. Despite the fact that researchers have been trying to define a more precise structure and terminology for proxemic interactions recently, proxemics have been involved in applications for a long time.

Researchers study the interactions between smart objects in the IoT (e.g. [9]). However, the research community has paid more attention to the human-device interactions (e.g. [10,11]). Ju et al. proposed the Range as a public interactive whiteboard, which supports co-located, ad-hoc meetings [12]. It uses the proximity sensing to proactively manage the transitions between display and authors (e.g. to clear the space for writing). Wang et al. introduced Proxemic Peddler, which is a public display [13]. This public display can capture and preserve the attention of pedestrians. Both of the aforementioned studies are examples of human-device proxemic interactions in micro-level where the proximity space is very small (e.g. one room). Researchers also have developed some proximity aware applications at macro-level (e.g. multiple rooms). The active badge, which was introduced by Want et al., is one of them [14]. They embedded an infrared beacon into their badge in order to connect it to a sensory network. Using this architecture, they could track employees in the office and redirect their phone calls to their current stations. Active badge was successfully implemented and used in large scale.

Moreover, there are some studies that focus on possible human-device proxemic interactions in households. The EasyLiving is one of the earliest projects in this group [15]. It emphasizes on an architecture, which can connect different devices together in order to enhance the user experience. Ramani et al. proposed a new location tracking system for media appliances in the wireless home networks, which can be used to build a proxemic media redirection system [16]. Recently, Sørensen et al. introduced and evaluated the AirPlayer, which is a multi-room music system that uses the proxemic interaction [17].

In conclusion, there is no custom designed proxemic interface for multimedia interactions. The proposed UI is designed and optimized for proxemic multimedia interaction, which makes it a task-specific UI. Furthermore, we believe that a completely distributed solution (e.g. [17]) cannot meet the growing requirements of the IoT users. Therefore, the proposed method employs a cloud-based decision engine (PIU), which considers additional information such as multimedia content properties, user's preferences, devise capabilities etc. in order to improve multimedia interactions. Moreover, the current trend is that a cloud-base database keeps the shared resources including multimedia data. Hence, our design of a cloud-based decision engine is in accordance with the current technological trends.

3 Proposed System

The main purpose of the proposed UI is to provide an environment within the IoT that supports the proxemic interactions for multimedia devices. The proposed system has three main components: proxemic interaction unit (PIU), media redirection unit, and tracking engine. PIU is the central part of the system. Figure 1 presents the abstract algorithm of PIU. As we can see in the figure, the algorithm begins by receiving the location information of the devices. PIU uses the tracking engine to find the locations and update it on the location database.

Next, PIU checks for possible proxemic interaction. Then PIU calculates a score between 0 and 1 for each device based on content properties, user preferences, and device capabilities. Consequently, the algorithm finds the device with maximum score. If it is not the same device as the one in use, it notifies the user and recommends a media redirection. If user accepts the recommendation, PIU asks the media redirection unit to handle this process. After each interaction, PIU updates the scoring coefficients to learn user's preferences adaptively.

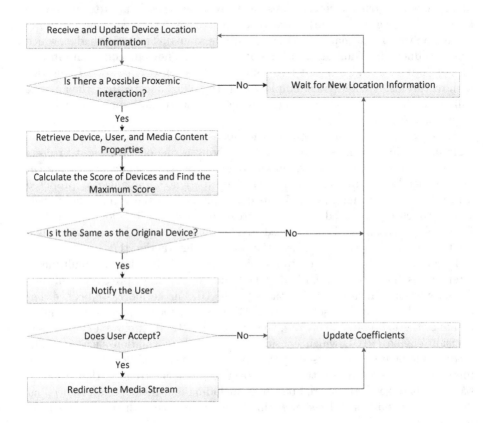

Fig. 1. The abstract algorithm of proxemic interaction unit

The device scoring and recommendation step in the aforementioned algorithm is the most important step since it can assist users to engage with a larger number of devices in the IoT. To recommend the most appropriate device in the given scenario, we should find and imply different variables in the scoring mechanism. Therefore, we start by introducing these variables and then present the scoring mechanism. There are three groups involved in the proposed solution: devices, users, and multimedia content. Below we study these groups in order to find effective variables for the scoring mechanism.

3.1 Devices

Devices are the first and most important objects in our system. Since the proposed system provides an environment for proxemic interactions, we begin by explaining device's proximity behavior. Hall defined four perimeters around each person: intimate space, personal space, social space, and public space (Table 1) [6]. We are using same categorization for devices according to their effective quality in each space: intimate devices, personal devices, social devices, and public devices. For example, smartphones have small screens, so they are usually used by individuals separately. Moreover, they have the best visual quality when they are close to the user (e.g. 20 cm which is in the intimate space). Users can still see the smartphone screen when they are farther away, but their effective quality decreases. Similarly, users enjoy big screen TVs most when they are at an acceptable distance, i.e., 3.5 m to 6 m. Hence, they are placed in the category of public devices. Let $T(x)$ be a function that returns an integer between 0 to 3 depending on the type of device x. Table 1 provides examples of all 4 groups of device spaces and function $T()$ value for each device type. We will use function $T()$ later while calculating scored for each device.

Table 1. Hall's personal space definitions and device examples for each space

Space Name	Space Area	Device Examples	$T()$
Intimate Space	$distance \leq 0.45\ m$	smartphones	0
Personal Space	$0.45\ m \leq distance \leq 1.2\ m$	tablets, laptops	1
Social Space	$1.2\ m \leq distance \leq 3.6\ m$	PCs, digital displays	2
Public Space	$3.6\ m \leq distance \leq 7.6\ m$	TVs, home stereos	3

3.2 User Survey

It is important to characterize user priorities in order to design effective user interface. We conducted a user survey to study user's attitudes and preferences. Participants had to answer 17 questions regarding the frequency of using multimedia contents, their device preferences, and satisfaction levels. We prepared questions such that they do not require any specific domain knowledge. Due to lack of space, we could not present the designed questionnaire here. The website we used for the study and the questionnaire details can be found at[1]. Totally, 149 people of an average age of 28.62 years participated in our study with the following gender distribution: 53.7% male and 46.3% female. They were engineers, employees, physicians, university students and professors, etc. with different background levels in IT; 53% of them were involved in a profession that requires a high level of IT knowledge while the other 47% were not involved in those kinds of jobs.

Results of this user survey revealed some interesting points. First, we could not find any correlation between the gender and device preferences. However, when

---

[1] https://docs.google.com/forms/d/
1Od28XvMpmZ1Kq5p5xKa8aQ5xAwpaLgbspMmqagrSHP8/viewform

we grouped our participants into young ($\leq 30\ years$) and old ($\gtrless 30\ years$) users, we found that old participants are more interested in using TVs and PCs than tablets and smartphones. In addition to that, profession had an effect on the device preferences. Participants who were involved in jobs that need a high level of IT knowledge were keener to use PCs and TVs. We also found that frequent users (who spend more than 1 hour per day) prefer TVs and PCs more than non-frequent users (who spend up to 1 hour per day).

To summarize, based on the results of this user survey, we were convinced to include three characteristics of the engaged user, u, in our scoring mechanism: participant's age (U^a), profession type (U^p), and multimedia usage habits (U^h). We used the user ratings to initialize preference coefficients corresponding to these characteristics, which are updated over time according to user interactions.

3.3 Multimedia Content

Multimedia content properties can influence the user's decision regarding the playback device. For example, a video content cannot be streamed to a home stereo, which only supports audio inputs. So, the type of the content should be considered. In addition, there is a user study that shows the length of multimedia content can affect the user's choice for playback device [18]. It categorizes videos based on their length: very short videos (up to 2 minutes, e.g. social network clips), short videos (up to 7 minutes, e.g. music videos), medium videos (up to 22 minutes, e.g. soap operas and animations), long videos (up to 45 minutes, e.g. TV series), and very long videos (longer than 45 minutes). Results of this study indicates that the length of the first half of videos that are played on the smartphones over the WiFi connection is around 50 seconds while it is almost 100 seconds for tablets. This is due to several factors such as screen dimensions and resolutions, battery capacity, etc. Hence, we decided to use three properties of the given multimedia content, c, in our scoring mechanism: audio of content (C^a), video of content (C^v), and duration of the content (C^d). We have grouped long and very long videos together, as a result, C^d can take four values depending on the length: 0 - very short, 1 - short, 2 - medium, and 4 - long.

3.4 Scoring Mechanism

When the PIU finds out that there is a possible proxemic interaction, it starts the scoring mechanism. Let $\mathcal{D} = \{D_i | 1 \leq i \leq n\}$ be the set of available devices. For the given content c and a user u, the the score of ith device, s_i, is calculated as follows:

$$s_i = f_{ic} \times m_{ic} \times h_{iu} \times a_{iu} \times p_{iu} \times d_{iu} \tag{1}$$

where f_{ic} is a flag that indicates capability of device D_i to play content c, m_{ic} is device appropriateness coefficient for the given content, h_{iu}, a_{iu}, and p_{iu} are user's habit, age, and profession factors for the device D_i. The last element is d_{iu} is the device effectiveness for the distance between device D_i and user u. The value of all of elements is defined between 0 and 1. We have chosen to multiply individual factor to obtain s_i because for optimal results we want all factor to be unity; on the other hand, if even one factor is zero, the device is useless. For example, if a device cannot play the given content, the value of other elements do not matter at all and that device should get 0 score. In the following text, we explain each element in details.

Playback Capability (f_{ic}). This element represents the playback capability of D_i for the given multimedia content c. It is a binary variable, so it is either 0 (i.e. i^{th} device cannot play c) or 1 (i.e. i^{th} device can play c). Equation 2 shows the Boolean expression that is used to compute f_{ic}:

$$f_{ic} = (D_i^a \odot C_c^a) \wedge (D_i^v \odot C_c^v) \tag{2}$$

where D_i^a and D_i^v logical variables which are **true** if device D_i is capable of playing audio and video respectively, otherwise **false**. Similarly, C_c^a and C_c^v are **true** if the given multimedia content c has audio and video playback requirements, otherwise **false**.

Device Appropriateness (m_{ic}). As we explained in Subsection 3.3, the length of given content (C_c^d) can affect the user's playback device preferences. The value of m_{ic} is selected from the 4×4 matrix (M), which its rows represent the length of the given content and its rows show the device space types:

$$m_{ic} = M(C_c^d, T(D_i)) \tag{3}$$

The first row of M is dedicated to the content with length ≤ 2 min, second row for the content with length between 2 min and 7 min, third row for the content with between length 7 min and 22 min, and the last row covers the rest (length ≥ 22 min). The columns follow this order: intimate devices, personal devices, social devices, and public devices. The same order is followed in the other defined matrices that have the device space type as column indicator. We defined these thresholds and matrix entries using the results of Li et al.'s survey [18] as follows:

$$M = \begin{bmatrix} 1 & 0.8 & 0.6 & 0.4 \\ 0.8 & 1 & 0.8 & 0.6 \\ 0.6 & 0.8 & 1 & 0.8 \\ 0.4 & 0.6 & 0.8 & 1 \end{bmatrix} \tag{4}$$

Habit (h_{iu}), Age (a_{iu}), Profession (p_{iu}) Factors. To determine the factors of user's multimedia usage history (h_{iu}), age (a_{iu}), and profession (p_{iu}); we define three binary functions as follows:

$$U_u^h = \begin{cases} 0 \text{ if } u \text{ is a frequent user} \\ 1 \text{ if } u \text{ is a non-frequent user} \end{cases}$$

$$U_u^a = \begin{cases} 0 \text{ if age of } u \text{ is less than 30} \\ 1 \text{ if age } u \text{ is greater than or equal to 30} \end{cases} \tag{5}$$

$$U_u^p = \begin{cases} 0 \text{ if } u \text{ does a high IT knowledge required job} \\ 1 \text{ otherwise} \end{cases}$$

The factor h_{iu} depends on the given user's multimedia usage habits. As we explained earlier (Subsection 3.2), user's multimedia playing time can influence the preferred playback device. The habit factor h_{iu} is calculated as follows:

$$h_{iu} = H(U_u^h, T(D_i)) \qquad (6)$$

where H is defined as a 2×4 matrix where rows represent usage frequency (frequent, non-frequent) and columns represent device space type. The age and profession factors are also calculated in the similar way as follows:

$$a_{iu} = A(U_u^a, T(D_i)) \qquad (7)$$

$$p_{iu} = P(U_u^p, T(D_i)) \qquad (8)$$

where A and P are again 2×4 matrices. The first row of A keeps a_{iu} values for users who are younger ($Ua \leq 30years$) and the second row presents a_{iu} values for the older users ($30years \lesssim Ua$). Similarly, first row of P represents users who work in a high IT knowledge required job and the second row presents rest of the users. The initial values of matrix elements of H, A, and P are determined based on user rating in the survey as follows:

$$H = \begin{bmatrix} 0.8 & 0.7 & 1 & 0.9 \\ 0.8 & 0.8 & 0.9 & 0.9 \end{bmatrix} \quad A = \begin{bmatrix} 0.7 & 0.8 & 1 & 0.9 \\ 0.8 & 0.7 & 0.9 & 1 \end{bmatrix} \quad P = \begin{bmatrix} 0.8 & 0.8 & 0.9 & 1 \\ 0.8 & 0.8 & 1 & 0.9 \end{bmatrix} \qquad (9)$$

Distance Effectiveness (d_{iu}). The distance effectiveness d_{iu} is calculated using the distance between the given user u and device D_i. Let x_{iu} be the distance between user and device. We defined a unique device space effectiveness function for each type of device (where type is defined in terms of space). Figure 2 shows the plot of the defined functions. These functions are extracted from the observations on device manuals and user's actions. For example, a big screen TV is mostly effective when $3.5\ m \leq x_{iu} \leq 6\ m$. On the other hand, it is not convenient for the user to watch the TV when the distance is around $0.5\ m$. You can see this behavior on the *Public function* in Figure 2. Similarly, user cannot clearly see a smartphone screen when too close ($x_{iu} \approx 0\ m$). However, the smartphones are very effective when x_{iu} is around $0.25\ m$. Afterwards, when x_{iu} increases, their effectiveness (d_{iu}) decreases due to their small screen size. Generally, we can argue that the effectiveness function is asymmetric around its peak.

Finally, we could fit Weibull II with four parameters on the observed curves in order to define effectiveness function. It is restricted between 0 and 1. Equation 10 shows the general form of this function where x_{iu} is the distance between the user and device; and x_0, a, b, and c are four parameters, which are different for each device spaces.

$$d_{iu} = a \times \left(\frac{c-1}{c}\right)^{\frac{1-c}{c}} \times \left|\frac{x_{iu}-x_0}{b} + \left(\frac{c-1}{c}\right)^{\frac{1}{c}}\right|^{c-1} \times e^{-\left|\frac{x_{iu}-x_0}{b}+\left(\frac{c-1}{c}\right)^{\frac{1}{c}}\right|^c} + \frac{c-1}{c} \qquad (10)$$

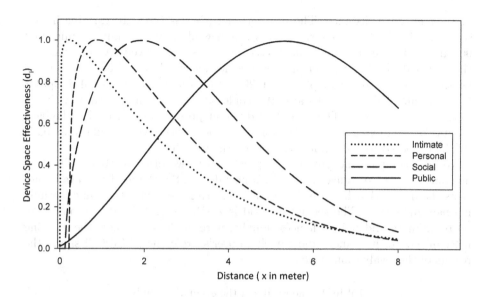

Fig. 2. The plot of all versions of d_{iu} functions

3.5 Adaptation Mechanism

The proposed solution considers different variables to calculate the score for each device (Equation 1) in order to find the best possible device for media redirection. Although we used the results of different surveys in order to define the values of the coefficients in the scoring mechanism, each person may have different preferences. Therefore, we designed an adaptation mechanism in the proposed solution, which updates the coefficient values based on the responses of individuals after each interaction. The PIU selects the corresponding values for coefficients and calculates the scores. For example, imagine PIU uses user's age coefficient a_{iu} to calculate the score for device D_i, and D_i is recommended to the user since it has the highest score. So, when the user responds to the recommendation, the value of a_{iu} is updated using the following equation:

$$a_{iu,new} = a_{iu,old} \times \left(\frac{1}{1 + \frac{n_r}{n_t + 1}}\right) \tag{11}$$

where $a_{iu,new}$ is the new coefficient's value, $a_{iu,old}$ is the old coefficient, n_r is number of rejected recommendations for this type of device space, and n_t is the total number of recommendations for this type of device space.

4 Evaluation

To evaluate the proposed system, we defined 4 scenarios and conducted a user survey. In these scenarios, participants were given access to all types of devices in their effective distance. The content's length was the only variable in our scenarios. In the first scenario, it was 2 min while it was 7 min, 22 min, and 90 min in the second, third and fourth scenarios respectively. As an example, the user had to select his/her preferred device for watching a 7 min video while he/she has access to a smartphone in 0.25 m, tablet in 0.75 m, personal computer in 2.5 m, and TV in 5 m. Totally, 10 people participated in our study with age ranging from 20 to 40. The users had different professions and media playback habits. We compared the user responses to the device recommendations made by the proposed user interface to measure the usability.

We also compare the proposed approach with two other methods. The first one only uses the distance to suggest a new device. So, the method always suggests the device that has user in its most effective area, with the same learning mechanism as ours. The second method is AirPlayer [17]. AirPlayer only supports intimate and public devices. Furthermore, it does not have the learning mechanism. So, it cannot adapt itself to user's preferences. Table 2 shows the results of this evaluation study.

Table 2. The results of the evaluation study

Method		scenario 1	scenario 2	scenario 3	scenario 4	Average
Our	1st suggestion	90%	30%	70%	90%	70%
Method	$1^{st}/2^{nd}$ suggestion	100%	80%	80%	100%	90%
Distance	1st suggestion	90%	30%	60%	90%	67.5%
Only	$1^{st}/2^{nd}$ suggestion	100%	70%	70%	100%	85%
AirPlayer [17]	–	100%	30%	30%	90%	62.5%

In this study, the total accuracy of the first suggestion for our system is equal to 70% while it was 67.5% for the distance only method and 62.5% for the AirPlayer. But, since the proposed method can learn from the user's responses and adapt itself to his/her preferences, we decided to study the accuracy of these methods after two interactions. This is not applicable to the AirPlayer due to its lack of learning mechanism. The total accuracy after two interactions for our system was 90%, however it was 85% for the distance only method. To summarize, we can say that the proposed method had a higher accuracy in these scenarios. As a result, it could provide an environment that has more successful proxemic multimedia interactions.

5   Discussion

There are three points that should be discussed regarding the proposed system. First, the proposed solution checks the identity of users by tracking their intimate devices. Then, it uses their personal information to recommend a device to them and adaptively updates itself for each user. Also, each device or user updates the location information when it has a movement. The devices are recommended based on the distance between them. Therefore, we can conclude that the proposed solution directly uses 3 out of 5 proxemic interaction dimensions that were introduced earlier by Greenberg et al [7]. The remaining two dimensions can be involved in this system as well. The orientation of the user and location of devices may influence the user's decision on playback device.

Second, we should compare the proposed UI with the definition of proxemic interaction by Hall [6]. The proposed UI is designed to work over the IoT. So, it can handle proxemic interactions in multiple rooms (i.e. macro-level). But it provides an environment for micro-level proxemic interactions as well. Devices within a single room can interact with each other when a measured change occurs in any of the 3 mentioned proxemic dimensions.

The last point is regarding the engagement mechanism. Since the number of online devices is growing very fast, most of the times users are surrounded by a number of devices. But, since it is not easy to migrate from one device to the other, users may not change the device that they are using. However, the proposed UI considers the user's preferences and proxemic information to suggest a new device, which can increase the number of accepted recommendations. Also, it can assist users through the migration process by handling redirection process in the background. Hence, the proposed UI facilitates the engagement process for the users and provides them more options to use.

6   Conclusion and Future Work

This paper presented a new UI, which is designed for multimedia devices within the IoT. The proposed UI provides a proxemic interaction experience to the users. The proposed solution involves user's preferences and media content properties in addition to the proxemic information in its scoring mechanism. Then, it recommends a new device to the user in order to motivate him/her to engage with the new device. The proposed algorithm adaptively trains itself towards user's attitude over the time based on the feedback from the user. The scoring mechanism is validated in 4 scenarios by a user study and it has the acceptable average accuracy of 70%. Using the feedback from the rejection ratio, system trains itself towards user's attitude over the time and reaches average accuracy of 90%. In the future, we want to include more devices and interaction modalities to build a generic, distributed user interface for IoT.

References

1. Sanou, B.: ICT Facts and Figures. International Telecommunications Union, Geneva (2013)
2. Mattern, F., Floerkemeier, C.: From the Internet of Computers to the Internet of Things. In: Sachs, K., Petrov, I., Guerrero, P. (eds.) Buchmann Festschrift. LNCS, vol. 6462, pp. 242–259. Springer, Heidelberg (2010)
3. Connections Counter: The Internet of Everything in Motion, http://newsroom.cisco.com/feature-content?type=webcontent&articleId=1208342

4. Weiser, M.: The computer for the 21st century. Scientific American, 94–104 (1991)
5. Luyten, K., Coninx, K.: Distributed user interface elements to support smart interaction spaces. In: 7th IEEE International Symposium on Multimedia, pp. 277–286. IEEE Press, New York (2005)
6. Hall, E.T.: The hidden dimension. Anchor Books, New York (1969)
7. Greenberg, S., Marquardt, N., Ballendat, T., Diaz-Marino, R., Wang, M.: Proxemic interactions: The new ubicomp? Interactions 18(1), 42–50 (2011)
8. Marquardt, N., Diaz-Marino, R., Boring, S., Greenberg, S.: The proximity toolkit: Prototyping proxemic interactions in ubiquitous computing ecologies. In: 24th Annual ACM Symposium on User Interface Software and Technology, pp. 315–326. ACM, New York (2011)
9. Kortuem, G., Kawsar, F., Fitton, D., Sundramoorthy, V.: Smart objects as building blocks for the internet of things. IEEE Internet Computing 14(1), 44–51 (2010)
10. Broll, G., Rukzio, E., Paolucci, M., Wagner, M., Schmidt, A., Hussmann, H.: Perci: Pervasive service interaction with the internet of things. IEEE Internet Computing 13(6), 74–81 (2009)
11. Guinard, D., Trifa, V.: Towards the web of things: Web mashups for embedded devices. In: MEM 2009 in Proceedings of WWW 2009. ACM, New York (2009)
12. Ju, W., Lee, B.A., Klemmer, S.R.: Range: Exploring implicit interaction through electronic whiteboard design. In: The 2008 ACM Conference on Computer Supported Cooperative Work, pp. 17–26. ACM, New York (2008)
13. Wang, M., Boring, S., Greenberg, S.: Proxemic peddler: A public advertising display that captures and preserves the attention of a passerby. In: The 2012 International Symposium on Pervasive Displays, pp. 3:1–3:6. ACM, New York (2012)
14. Want, R., Hopper, A., Falcao, V., Gibbons, J.: The active badge location system. ACM J. Transactions on Information Systems 10(1), 91–102 (1992)
15. Brumitt, B., Meyers, B., Krumm, J., Kern, A., Shafer, S.: EasyLiving: Technologies for intelligent environments. In: Thomas, P., Gellersen, H.-W. (eds.) HUC 2000. LNCS, vol. 1927, pp. 12–29. Springer, Heidelberg (2000)
16. Ramani, I., Bharadwaja, R., Rangan, P.V.: Location tracking for media appliances in wireless home networks. In: IEEE International Conference on Multimedia and Expo, pp. 769–772. IEEE Press, New York (2003)
17. Sørensen, H., Kristensen, M.G., Kjeldskov, J., Skov, M.B.: Proxemic interaction in a multi-room music system. In: 25th Australian Computer-Human Interaction Conference: Augmentation, Application, Innovation, Collaboration, pp. 153–162. ACM, New York (2013)
18. Li, Z., Lin, J., Akodjenou, M.I., Xie, G., Kaafar, M.A., Jin, Y., Peng, G.: Watching Videos from Everywhere: A Study of the PPTV Mobile VoD System. In: The 2012 ACM Conference on Internet Measurement Conference, pp. 185–198. ACM, New York (2012)

Outdoor Air Quality Inference
from Single Image

Zheng Zhang, Huadong Ma, Huiyuan Fu, and Xinpeng Wang

Beijing Key Lab of Intelligent Telecomm. Software and Multimedia
Beijing University of Posts and Telecomm., Beijing 100876, P.R.China
{zhangzheng,mhd,fhy}@bupt.edu.cn, xpsmiler@gmail.com

Abstract. Along with rapid urbanization and industrialization processes, many developing countries are suffering from air pollution. Air
quality varies non-linearly, the effective range of an air quality monitoring station is limited. While there are seldom air quality monitoring stations in cities, it is difficult to know the exact air quality of everywhere.
How to obtain the air quality fast and conveniently will attract much
attention. In this paper, we present an air quality inference approach
based on air quality index(AQI) decision tree from a single image. We
first extract several corresponding features such as medium transmission,
power spectrum slope, contrast, and saturation from the single image.
Then we construct a decision tree of AQI values, in accordance with
the distance between the features we extract previously. For each none-
leaf node of the decision tree, we use five classifiers to choose the next
node respectively. We collect a dataset of high quality registered and
calibrated images named Outdoor Air Quality Image Set(OAQIS). The
dataset covers a wide range of daylight illumination and air pollution
conditions. We evaluate our approach on the dataset, the results show
the effective of our method.

Keywords: Outdoor air auality inference, outdoor air quality image set,
AQI decision tree.

1 Introduction

Over the last few decades, many developing countries have experienced urbanization and industrialization processes on an unprecedented scale. In China, more
than 120 cities have populations of more than one million. This rapid growth in
such a short period of time has not only led to a remarkable increase in material
wealth and a higher standard of living, but has also caused serious complicated
air pollution, that various types of air pollution (e.g. coal smog, photochemical
smog, dust storm and haze) are presented together in China [1]. The government
has built some air quality monitoring stations in metropolis, figure 1 shows the
distribution of air quality monitoring stations in Beijing (6336 square miles).
While air quality varies non-linearly, so that the effective range of an air quality
monitoring station is limited. It is impossible to know the exact air quality of

X. He et al. (Eds.): MMM 2015, Part II, LNCS 8936, pp. 13–25, 2015.

Fig. 1. Air quality monitoring station in Beijing

everywhere in the city, so how to obtain the air quality fast and conveniently will attract much attention.

Many existing methods are based on satellite remote sensing technologies that only can reflect the air quality of atmosphere which is far from the ground air quality[2][3][4][5][6]. The authors of [2] estimated the ground-level concentration of fine particulate mass (PM2.5) for January 2001 to October 2002 using space-based measurements from the Moderate Resolution Imaging Spectroradiometer (MODIS) and the Multiangle Imaging Spectroradiometer (MISR) satellite instruments, and additional information from a global chemical transport model (GEOS-CHEM). The authors of [3] used a generalized linear regression model to examine the relationship between ground-level PM2.5 measurements and aerosol optical thickness from Multiangle Imaging Spectroradiometer (MISR) measurements in the eastern United States. The authors of [11] proposed a semi-supervised learning framework for estimating atmospheric visibility using off-the-shelf cameras.

Recently, some works focused on air quality inference via massive sensing data. The authors of [7] inferred the air quality information based on the historical and real-time air quality data reported by existing monitor stations, and a variety of data sources such as meteorology, traffic flow, human mobility, structure of road networks, and point of interests. The authors of [8] proposed a big spatio-temporal data framework for the analysis of China Severe Smog. They collected about 35,000,000 detailed historical and real-time air quality records (contain the concentrations of PM2.5 and air pollutants including SO_2, CO, NO_2, $O3$ and PM10) and 30,000,000 meteorological records in 77 major cities of China through air quality and weather stations. It conducts scalable correlation analysis to find

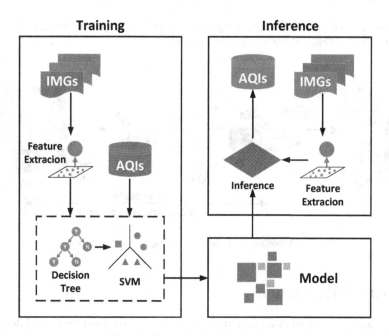

Fig. 2. Framework of our system

the possible short-term and long-term factors to PM2.5. There were some works for collecting air quality data via mobile equipments. For example, the authors of [9] presented a vehicular-based mobile approach for measuring fine-grained air quality in real-time, the authors of [10] designed a wireless distributed mobile air pollution monitoring system which utilized city buses to collect pollutant gases.

However, these works require massive sensing data which is difficult to obtain. Moreover, these works need complex algorithms to deal with the massive sensing data. In this paper, we present an air quality inference approach based on an AQI decision tree from a single image. We first extract several features such as medium transmission, power spectrum slope, contrast, and saturation from single images. Then we construct a decision tree of AQI values, which is in accordance with the distance between the features we extract previously. For each none-leaf node of the decision tree, the next node selection based on SVM classifiers. Figure 2 shows the framework of our system. We collect a dataset of high quality registered and calibrated images named OAQIS. The dataset covers a wide range of daylight illumination and air pollution conditions.

To summarize, we develop an approach for air quality inference from single image with the following contributions:

- We propose an air quality inference approach from single image based on decision tree and SVM.
- We construct a dataset of high quality registered and calibrated images, which covers a wide range of daylight illumination and air quality conditions.

It has the potential value for image processing, atmospheric sciences and can be used as a testbed for many algorithms.

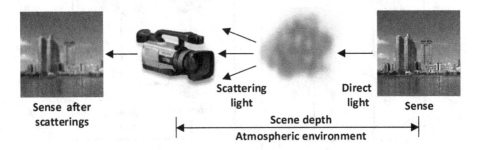

Fig. 3. Atmospheric light scattering and attenuation

2   Feature Extraction

For inferring the air quality from images, we extract some features first. There are four different features, i.e. medium transmission, power spectrum slope, contrast, and saturation, developed and combined in our approach. These features are derived by analyzing the visual and spectral clues from images.

2.1   Medium Transmission

Particulate matter is one of the main air pollution sources. In the process of transmission, light intensity attenuated because of the particulate matter scattering. Figure 3 demonstrates the process of atmospheric light scattering and attenuation. In order to analyze the impact of particulate matter scattering of an image, we extract the medium transmission.

In computer vision and computer graphics, the model widely used [12][13][14] to describe the formation of a fog/haze image is as follows:

$$I(x) = t(x)J(x) + (1 - t(x))A, \tag{1}$$

where I denotes the observed intensity, J denotes the scene radiance, A denotes the atmospheric light, and t denotes the medium transmission describing the portion of the light that is not scattered and reaches the camera. The authors of [14] found that, in most of the local regions which do not cover the sky, it is very often that some pixels have very low intensity in at least one color (rgb) channel, called dark channel prior. They estimated the medium transmission by

$$\tilde{t}(x) = 1 - \min_{c}(\min_{y \in \Omega(x)}(\frac{J^c(y)}{A^c})). \tag{2}$$

where J^c is a color channel of J, $\Omega(x)$ is a local patch centered at x, and A^c is always positive.

2.2   Power Spectrum Slope

With the decreasing of the air quality, the captured image becomes low-resolution even blur. Due to the low-pass-filtering characteristic of a blurred region, some high frequency components are lost. So the amplitude spectrum slope of a blurred region tends to be steeper than that of an unblurred region. In order to analyze the impact of low-resolution of an image, we extract the power spectrum slope [15]. First, we compute the power spectrum of an image I with size $N \times N$ by taking the squared magnitude after Discrete Fourier transform (DFT)

$$S(u, v) = \frac{1}{N^2} | I(u, v) |^2, \tag{3}$$

where $I(u, v)$ denotes the Fourier transformed image. We represent the two-dimensional frequency in polar coordinates, thus $u = f \cos \theta$ and $v = f \sin \theta$, f denotes the radius of power spectrum image, and θ denotes the angle of the polar coordinates, and we construct $S = (f, \theta)$. By summing the power spectra S over all directions θ, and using polar coordinates, $S(f)$ can be approximated by

$$S(f) = \sum_{\theta} S(f, \theta). \tag{4}$$

The authors of [16] find that $S(f)$ approximate to an exponential function of f, so slope of power spectrum α can be calculated as

$$\alpha \approx \ln B - \frac{\ln (S(f))}{\ln (f)}, \tag{5}$$

where B denotes a constant.

2.3   Contrast

There are various particulate matters in the atmosphere, light intensity attenuated during the transmission because of these particulate matters. Therefore, the same scenes in different air pollution conditions have different contrast. The Michelson contrast is commonly used for patterns where both bright and dark features are equivalent and take up similar fractions of the area. However, the Michelson contrast does not consider the error due to noise points. We compute the contrast according to Root Mean Square(RMS) [17]

$$C = \left(\frac{N_I \sum L_{(x,y)}^2 - (\sum L_{(x,y)})^2}{N_I^2} \right)^{\frac{1}{2}}, \tag{6}$$

where $L_{(x,y)}$ denotes the luminance of the pixel (x, y) of the image, N_I denotes the number of pixels in the image.

AQI Values	Levels of Health Concern	Colors	Health Implications
0 - 50	Good	Green	No health implications.
51 - 100	Moderate	Yellow	Few hypersensitive individuals should reduce outdoor exercise.
101 - 150	Unhealthy for Sensitive Groups	Orange	Slight irritations may occur,individuals with breathing or heart problems should reduce outdoor exercise.
151 - 200	Unhealthy	Red	Slight irritations may occur,individuals with breathing or heart problems should reduce outdoor exercise.
201 - 300	Very Unhealthy	Purple	Healthy people will be noticeably affected.People with breathing or heart problems will experience reduced endurance in activities.These individuals and elders should remain indoors and restrict activities.
301 - 500	Hazardous	Maroon	Healthy people will experience reduced endurance in activities.There may be strong irritations and symptoms and may trigger other illnesses.Elders and the sick should remain indoors and avoid exercise.Healthy individuals should avoid out door activities.

Fig. 4. AQI values, descriptors, colors, and health implications

2.4   Normalized Saturation

We also consider the color information of images for air pollution inference. As the saturation is independent of illumination, it can represent different images under various illumination conditions. For an image I we calculate the normalized saturation for each pixel by

$$S_{(x,y)} = \frac{S_{x,y} - min(S_I)}{max(S_I) - min(S_I)},\qquad(7)$$

where $max(S_I)$ is the maximum saturation value and $min(S_I)$ is the minimum saturation value of image I. For convenience of calculations in the following steps, we compute the histogram of the normalized saturation of the images.

3   Air Quality Inference

Using the above air pollution image features, our approach consists of two steps. In the first step, we construct a decision tree about AQI categories. Then, for each none-leaf node of the tree, we choose the next node by using SVM, and inference the air pollution level of the image.

3.1   AQI Decision Tree

AQI is a number used by government agencies to communicate to the public how polluted the air is. AQI values are divided into categories, and each category is assigned a descriptor and a color code. In many countries, AQI is divided into six categories indicating increasing levels of health concern as shown in Figure 4. An AQI value over 300 represents hazardous air quality whereas if it is below 50 the air quality is good.

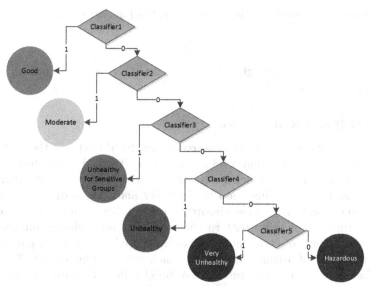

Fig. 5. The AQI decision tree

In this paper, we construct a decision tree of AQI levels as shown in Figure 5. First, we divide all categories into two subclasses, one is an air quality level, the other one is the category of remaining air quality levels. Then, we divide subclasses into the next level class, and repeat this procedure until all classes can no longer be divided. The authors of [18] proposed four algorithms of decision tree construct, we use one of them in our approach. We calculate the Euclidean distances between the class centers, and recursively separate the farthest class from the remaining classes. First, we calculate the class centers $c_i (i = 1, 2, \cdots, N)$ by

$$c_i = \frac{1}{|X_i|} \sum_{x \in X_i} x, \tag{8}$$

where X_i denotes a set of training data including in class i, and $|X_i|$ denotes the number of elements included in X_i. Then, we calculate the normalized distances by

$$\bar{c} = \frac{1}{N} \sum_{i=1}^{N} c_i, \tag{9}$$

and we compute the distance between class i and class j by

$$d_{i,j} = d_{j,i} = \|\frac{c_i - c_j}{\bar{c}}\|. \tag{10}$$

Then, we compute the smallest value of $d_{i,j}$ for each class

$$l_i = \min_{j=1,2,\cdots,N, j \neq i} d_{i,j}, \tag{11}$$

and set the class k which has the largest l_i as the farthest class

$$k = arg \max_{i=1,2,\cdots,N} l_i, \tag{12}$$

and separate this class from the others. Repeat the procedure until all classes can no longer be separated.

3.2 SVM Based Node Choice

For each none-leaf node of the decision tree, we use SVM to choose the next node, and inference the air pollution level of the image. These different features can be organized to form a 22-dimensional feature vector. However, the importance or effect of each feature dimension in similarity matching is different. Feature weighting and selection become important issues when aiming to find out an optimal feature combination pattern that describes air pollution images best. We calculate weights of each feature dimension based on the whole database to increase the effect of feature dimensions which are more important. The main idea is that the better the feature dimension classifies the database, the more effective the dimension is and thus deserves a higher weight. The higher inter-class variance and lower intra-class variance the feature dimension has, the better it is to distinguish different categories.

Similarity matching is performed between feature vectors based on weighted distance as

$$d(c_i, c_j) = \sqrt{\sum_{i=1}^{n} (c_i - c_j)^2 W}, \tag{13}$$

in which c_i and c_j could be calculated by Equation (8), $W = \{w_1, w_2, \cdots, w_n\}$ are normalized weights according to the ratio of inter-variance to intra-variance.

4 Experiments and Analysis

4.1 Dataset

We evaluate our approach on the dataset we construct named OAQIS(Outdoor Air Quality Image Set). As shown in Figure 6, the dataset contains a lot of high quality registered and calibrated images. It covers a wide range of daylight illumination conditions and air pollution conditions. The images are captured automatically every 15 minutes for 10 hours each day in Beijing, where suffers serious air pollution. The spatial resolution of each image is 1280 × 720 pixels. The images are tagged with a variety of ground truth data such as weather and illumination conditions actual scene depths and PM2.5 values. Our dataset is collected under a wide variety of seasons, weather and illumination conditions. The main purpose of this dataset is to provide an extensive testbed for the evaluation of existing appearance models, and provide insight needed to develop new appearance models. The data collection is ongoing and we plan to acquire images for one year.

Fig. 6. Exemplar images from OAQIS. The concentration of PM2.5 in the images changes from low to high.

4.2 Ground Truth

We have tagged our images with a variety of ground truth information. Most important categories of the ground truth we collected are the concentration of particles in the air and weather information. We collect the concentration of particles in the air by a Dylos air quality monitor and a Hexin air quality monitor. We gather data for an hour continuously, set the average value as the hour value. We automatically collect standard weather information from Weather China web site [19]. This includes information about weather condition (sunny, cloudy, overcast, mist, fog, haze, rain, snow etc), temperature, humidity and wind, as shown in Figure 7.

4.3 Experimental Results

We evaluate our approach on APIS, and performance on the dataset demonstrates the effectiveness of the proposed approach. First, we divide 2000 images into 10 parts, while choosing 9 parts as training set, leaving one as testing part. Then we use 10-fold cross-validation to test 10 times, and use the average result as the final result. We justify the performance of features described in Section 2. F_m, F_p, F_c, F_s are the notations of medium transmission, power spectrum slope, contrast, and saturation. As shown in Table 1, the accuracy of these features are improved a lot by using all of the features, and the saturation feature has the best performance of the four features, meanwhile the contrast feature has the worst. So that, the weight are different, in accordance with the performance of these features.

Conditions at 2014.05.12 10:33am	
PM2.5	34µg/m^3
Weather	Sunny
Temperature	73.4 F /23.0 C
Humidity	44%
Wind	NW 1.6 - 3.3m/s

Fig. 7. Sample ground truth data. The image was captured at May 12, 2014 in Beijing. The concentration of PM2.5 acquired by our air quality monitor, and weather data obtained from Weather China websites.

Table 1. Results related to features

PM2.5	Features				
	F_m	F_p	F_c	F_s	$F_m + F_p + F_c + F_s$
Accuracy	0.57	0.63	0.42	0.67	0.77

In order to evaluate the proposed approach, we randomly select five days in OAQIS. Figure 8 shows the measured values of PM2.5 and the inference result of PM2.5 by the proposed approach on $2014.5.26 - 2014.5.30$ at $9 \sim 11 : 00$am and $2 \sim 4 : 00$pm. The measured values were collected every 5 minutes by a Hexin air quality monitor during the experiment period. We calculate the mean value of each hour, and set it as the PM2.5 label of the images captured in the same period. As shown in Figure 8(a) and (b), some measured values of PM2.5 on May 29 change a lot within a short time, but the mean value of that period is not so big, so the inference result is consistent with the correct label, the proposed approach is effective.

4.4 Efficiency

The experiments were evaluated on a 64-bit PC with a Dual-Core CPU @3.20GHz and 4GB memory. We use Matlab R2013b to extract features from images, and use Visual Studio 2013 for the inference part. In our first experiment, we compare the time needed to extract different features from images, and the procedure of inference. As shown in table 2, on average, the consumption of feature extraction are 3.0136 seconds, 0.3661 seconds, 8.9181 seconds, 0.9068 seconds, and the inference procedure need 0.029 seconds. So that, the total time cost of the whole procedure is 13.2336s, our approach is effective. We could simply inference the air quality from an image without spending much more time.

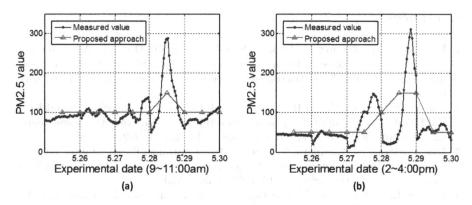

Fig. 8. Experimental result of the proposed approach on 2014.5.26 − 2014.5.30 at 9 ∼ 11 : 00am and 2 ∼ 4 : 00pm.

Table 2. Efficiency study

Procedures	Features				Inference	Total
	F_m	F_p	F_c	F_s		
Time(s)	3.0136	0.3661	8.9181	0.9068	0.029	13.2336

5   Conclusion

We have presented an approach for air quality inference from a single image. Our approach is based on AQI decision tree and five SVM classifiers, developed and combined four different features, i.e. medium transmission, power spectrum slope, contrast, and saturation. We first extract these features from single images. Then we construct a decision tree of AQI values, in accordance with the distance between the features. For each none-leaf node of the decision tree, the next node selection is based on SVM classifiers. We collect a dataset of high quality registered and calibrated images named OAQIS. The dataset covers a wide range of daylight illumination and air pollution conditions, and has potential implications for image processing, atmospheric sciences and can be used as a testbed for many algorithms. We evaluate our approach on the dataset, the results show the effective of our method. In the future, we will focus on fine-grained air quality inference from unrestricted images and consider the influnce of wind, humidity, weather on air quality inference. We also plan to keep on collecting images by using various devices to enlarge our dataset and release it.

Acknowledgments. The research reported in this paper is supported by the National Natural Science Foundation of China under Grant No. 61332005; The Funds for Creative Research Groups of China under Grant No. 61121001; Jiangsu Provincial Natural Science Foundation of China under Grant No. BK2011170; The Key Technologies R&D Program of China under Grant No. 2013BAK01B02, the Cosponsored Project of Beijing Committee of Education.

References

1. Li, J., Wang, Z., Zhuang, G., Luo, G., Sun, Y., Wang, Q.: Mixing of Asian mineral dust with anthropogenic pollutants over East Asia: A model case study of a super-duststorm in March 2010. Atmospheric Chemistry and Physics 12(16), 7591–7607 (2012)
2. Van Donkelaar, A., Martin Randall, V., Park Rokjin, J.: Estimating ground-level PM2.5 using aerosol optical depth determined from satellite remote sensing. Journal of Geophysical Research: Atmospheres(1984-2012) 111(D21) (2006)
3. Liu, Y., Sarnat, J.A., Kilaru, V., Jacob, D.J., Koutrakis, P.: Estimating ground-level PM2. 5 in the eastern United States using satellite remote sensing. Environmental Science & Technology 39(9), 3269–3278 (2005)
4. Lamsal, L.N., Martin, R.V., Van Donkelaar, A., Steinbacher, M., Celarier, E.A., Bucsela, E., Dunlea, E.J., Pinto, J.P.: Ground-level nitrogen dioxide concentrations inferred from the satellite-borne Ozone Monitoring Instrument. Journal of Geophysical Research: Atmospheres(1984-2012), 113(D16) (2008)
5. Martin, R.V.: Satellite remote sensing of surface air quality. Atmospheric Environment 42(34), 7823–7843 (2008)
6. Li, S., Chen, L., Zheng, F., Han, D., Wang, Z.: Design and application of haze optic thickness retrieval model for beijing olympic games. In: IEEE International Geoscience and Remote Sensing Symposium, pp. II-507–II-510. IEEE, Cape Town (2009)
7. Zheng, Y., Liu, F., Hsieh, H.P.: U-Air: When urban air quality inference meets big data. In: The 19th ACM SIGKDD International Conference on Knowledge Discovery and Data Mining, pp. 1436–1444. ACM, Chicago (2013)
8. Chen, J., Chen, H., Pan, J.Z., Wu, M., Zhang, N., Zheng, G.: When big data meets big smog: A big spatio-temporal data framework for China severe smog analysis. In: The 2nd ACM SIGSPATIAL International Workshop on Analytics for Big Geospatial Data, pp. 13–22. ACM, Orlando (2013)
9. Devarakonda, S., Sevusu, P., Liu, H., Liu, R., Iftode, L., Nath, B.: Real-time air quality monitoring through mobile sensing in metropolitan areas. In: The 2nd ACM SIGKDD International Workshop on Urban Computing, p. 15. ACM, Chicago (2013)
10. Al-Ali, A.R., Zualkernan, I., Aloul, F.: A mobile GPRS-sensors array for air pollution monitoring. Sensors Journal 10(10), 1666–1671 (2010)
11. Graves, N., Newsam, S.: Camera-based visibility estimation: Incorporating multiple regions and unlabeled observations. Ecological Informatics 23, 62–68 (2013)
12. Narasimhan, S.G., Nayar, S.K.: Vision and the atmosphere. International Journal of Computer Vision 48(3), 233–254 (2002)
13. Tan, R.T.: Visibility in bad weather from a single image. In: IEEE Conference on Computer Vision and Pattern Recognition, pp. 1–8. IEEE, Anchorage (2008)
14. Kaiming, H., Jian, S., Xiaoou, T.: Single image haze removal using dark channel prior. IEEE Transactions on Pattern Analysis and Machine Intelligence 33(12), 2341–2353 (2011)
15. Renting, L., Zhaorong, L., Jiaya, J.: Image partial blur detection and classification. In: IEEE Conference on Computer Vision and Pattern Recognition, pp. 1–8. IEEE, Anchorage (2008)

16. Burton, G.J., Moorhead, I.R.: Color and spatial structure in natural scenes. Applied Optics 26(1), 157–170 (1987)
17. Peli, E.: Contrast in complex images. JOSA A 7(10), 2032–2040 (1990)
18. Takahashi, F., Abe, S.: Decision-tree-based multiclass support vector machines. In: The 9th International Conference on Neural Information Processing, pp. 1418–1422. IEEE, Singapore (2002)
19. Weather China, http://www.weather.com.cn/

Multimodal Music Mood Classification by Fusion of Audio and Lyrics

Hao Xue, Like Xue, and Feng Su*

State Key Laboratory for Novel Software Technology,
Nanjing University, Nanjing 210023, China
suf@nju.edu.cn

Abstract. Mood analysis from music data attracts both increasing research and application attentions in recent years. In this paper, we propose a novel multimodal approach for music mood classification incorporating audio and lyric information, which consists of three key components: 1) lyric feature extraction with a recursive hierarchical deep learning model, preceded by lyric filtering with discriminative reduction of vocabulary and synonymous lyric expansion; 2) saliency based audio feature extraction; 3) a Hough forest based fusion and classification scheme that fuses two modalities at the more fine-grained sentence level, utilizing the time alignment cross modalities. The effectiveness of the proposed model is verified by the experiments on a real dataset containing more than 3000 minutes of music.

Keywords: music mood, mood classification, multimodal, hough forest, recursive autoencoder.

1   Introduction

With the explosion of huge digital music libraries over the past decade, music moods have been shown that becoming a desirable access point to music databases, and automatic recognition and classification of mood have received increasing attention in recent years. Compare to traditional characteristics of music such as genre or artist, mood recognition is more subjective and difficult to quantify, which make it more challenging [6].

Since music is typically composed of both the audio track and the lyric text, existing methods for music mood classification can be categorized into three groups, according to which modality of data the method exploits.

Audio-Based Music Mood Classification. As the most important part of music, audio content has been studied by most research in mood classification with variant machine learning methods [8,9,19]. As one of the first publications on this task, Li and Ogihara [8] used SVM for the multiple binary classification tasks with three audio features: rhythm, timbre and pitch. Similarly, [19] found

---

* Corresponding author.

X. He et al. (Eds.): MMM 2015, Part II, LNCS 8936, pp. 26–37, 2015.

the popular SVM classifier with spectral features usually give the superior performance. Lu [9] used a variety of audio features including intensity, timbre and rhythm and the Gaussian Mixture model to detect and classify music mood.

Text Sentiment Classification. Lyrics have rich semantic information and significant impact on human perception of music [1]. As one of the first experiments on this topic, Pang [13] applied document sentiment classification on movie reviews, which used simple bag-of-words feature and received the best result by unigram-based model with SVM classifiers. Nakagawa [11] proposed a dependency tree-based method for sentence-level binary sentiment classification, which calculated the polarities for each dependency subtree as hidden variables and obtained the polarity of the whole sentence based on these hidden variables. Socher et al. [15] proposed a method for learning vector representation of sentences and predicting their polarities based on recursive Autoencoders.

Multi-Modal Music Mood Classification Using Lyric and Audio. Recently, several studies [7,18,17,4] have shown that combining audio with lyric will make some improvements in classification accuracy, and generally adopted a document-level bag of early fused multi-modal features representation for the music. Laurier et al. [7] used language model differences (LMD) to select words for BOW features and combined them with acoustic features, yielding the improved accuracy of mood categorization. Yang and Lee [17] converted lyric words into 182 psychological categories, showing that the fused features improved the classification results. Hu [4] compared some lyric features with specific audio and combined features. In [18], Yang et al. demonstrated that lyrics can complement the semantic information of audio signals.

In this paper, we propose a novel multimodal approach for music mood classification incorporating audio and lyric information. To quantify a mood state, we employ the most common Russell's 2-dimensional Valence-Arousal (V-A) definition space [14] for music mood categories, in which the valence value (positive/negative) corresponds to the polarity of mood and the arousal value represents the intensity of mood ranging from low to high. On the basis of the V-A space, we define four categories of mood: *happy, angry, sad and relaxed.* Our algorithm framework consists of three key components, as shown in Fig. 1:

1. Text feature extraction from lyric with a recursive hierarchical deep learning model, preceded by lyric filtering with discriminative reduction of vocabulary and synonymous lyric expansion.
2. Audio feature extraction on the basis of the saliency analysis of audio signals.
3. A Hough forest based multi-modal fusion and classification scheme, which fuses the information of two modalities on a more fine-grained level, i.e. the sentence level, utilizing the time alignment cross modalities.

The motive of our approach is, the many-to-many mapping between the lyric word and its possible acoustic characteristic, considering one lyric word may appear with variant vocal expressions or background sounds at different time of music, generally causes ambiguity in discriminating the music mood at the document-level or based on any single modality, while the temporal correlation

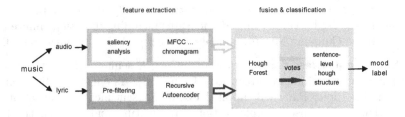

Fig. 1. Block diagram of the proposed algorithm

between the audio and the lyric features in the finer sentence-level context of music should help to reduce this ambiguity.

The rest of the paper is organized as follows. We introduce the lyric-based text feature extraction in Section 2 and the audio feature extraction in section 3. In Section 4 we describe the Hough forest based multi-modal fusion and classification framework. Section 5 describes the dataset used in our experiment, and Section 6 presents the experimental results.

2   Text Features from Lyric

Lyrics provide valuable semantic information about the mood of music. Previous feature extraction methods for lyrics mostly employ a BOW representation at the document-level, which ignores the finer grained relationship between words and sentences. Different from them, we propose to extract text features from the lyrics at the sentence-level based on one deep learning model [15]. We exploit the time information about every sentence encoded in specific lyric files (e.g. the LRC format) to divide the lyric into sentences for text feature extraction as well as aligning them with the audio features. Moreover, we propose a pre-filtering of lyric based on discriminability and synonymy for better learning of text features.

2.1   Distributed Representation of Word

The concept of distributed representation of word was firstly proposed by Hinton [3] in 1986, and Bengio[2] first presented a neural probabilistic language model for learning the representation for text, in which words are represented by continuous vectors of parameters. The semantic relationship between words can then be learned in the vector space, in which, besides other favorable properties, we can measure the semantic similarity between words with e.g. Euclidean distance metric. For example, the distance between happy and glad will be far smaller than that between happy and angry in this representation.

We use the 50-dimension dataset of word representation induced by Turian [16], which is learned by a semi-supervised neural language model for several weeks. Based on this dataset, we convert the discrete lyric words into the continuous vector space for further feature extraction.

2.2 Feature Extraction with Semi-supervised Recursive Autoencoder

Autoencoder is an unsupervised multi-layer deep learning algorithm aiming to learn an output similar to its input, while its hidden layer can be used as a reduced vector representation of the input. In our work, to convert one lyric sentence composed of varying number of words to one single feature vector with fixed dimensionality, one specific variant of the Autoencoder model, the *recursive Autoencoder (RAE)*, is exploited, which has a recursive tree structure of nodes/cells, as shown in Fig. 2. At each layer of RAE, a three-layer Autoencoder maps the two input vectors to two output vectors, while the single hidden node is used as the newly reduced representation for the two input vectors, and is then fed into the Autoencoder at the higher layer of RAE. Repeating this process recursively, finally we obtain a vector representation for the entire set of input vectors (i.e. sentence) at the root of the tree. Moreover, to enable semi-supervised training of RAE, a softmax layer is added to the parent nodes in every RAE layer of the tree. Now the objective of the whole RAE is to minimize the reconstruction error plus the cross-entropy error of softmax layer.

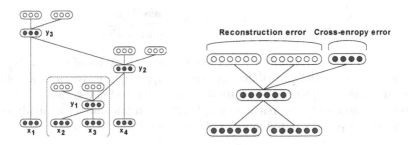

Fig. 2. Example of the tree structure of recursive Autoencoder (left) and the structure of one semi-supervised RAE layer with the softmax layer denoted by red (right) [15]

For example, for one sentence represented as a sequence of 4 word vectors $s = (x_1, x_2, x_3, x_4)$ in Fig. 2, which are initially used as the leaf nodes of the tree, the representation vector at the nonterminal node y_1 is computed from the vector of words $(x2, x3)$, and then y_2 is computed from $(y1, x4) = (x2, x3, x4)$. The whole sentence is finally represented by the vector y_3, which has the same dimensionality as the word vector and is used as the text feature for the sentence.

2.3 Lyric Pre-filtering

One significant difference between lyric and other ordinary text data like articles is the *sparsity* of the words appearing in the lyric, which has two implications. First, the words constituting the lyric of one music usually belong to a very limited subset of the whole vocabulary. Second, different words have large variances of the occurrence frequencies in and across lyrics. Such sparsity of word-level features usually results in drops of the efficiency of the word-based representation model, therefore we propose certain filtering of the lyric before text feature extraction, by constructing both the more discriminative vocabulary and the denser representation of lyric itself.

Discriminability-Guided Vocabulary Reduction. Since not all words in the lyric vocabulary have equivalent discriminabilities for the music mood, we propose to reconstruct a reduced vocabulary that comprises only the most discriminative words, and correspondingly filter out those discarded words from the lyric of music.

The basic idea for selecting discriminative words for the reduced vocabulary is that words appearing frequently in one language model (i.e. mood) while rarely appearing in another model should get higher scores in the ranking of the discriminability of word. Different from Laurier's work [7] that considers just the binary positive/negative classification for one category at one time, we want to extract discriminative words across all mood categories. Thereby, the ranking score $LMD(w)$ for each word w is computed as:

$$LMD(w) = \sum_{c=1}^{K} LMD(w,c), \quad LMD(w,c) = \frac{abs(P(w|c) - P(w| \neq c))}{\sqrt{\max(P(w|c), P(w| \neq c))}} \quad (1)$$

where, $P(w|c)$ is the probability of word w appearing in lyrics of mood category c and $P(w| \neq c)$ is the probability of word w appearing in lyrics of all other mood categories except c, K is the total number of mood categories.

Based on the ranking scores of words, the reduced vocabulary is finally constructed with a predesignated number of words with the highest scores.

Lyric Expansion by Synonymy. Due to the sparsity of the lyric words, there could be words that are semantically discriminative for specific mood but are not selected into the vocabulary due to their low frequency of occurrence $P(w|c)$ in lyric according to Eqn.(1), which are consequently filtered out from the lyric. To include such word (more precisely, its synonymous peer) in mood classification for better accuracy rather than discarding it, we propose an expansion method for lyrics based on synonymy of words.

Specifically, for one word in the lyric but not listed in the vocabulary, we look for its synonymous peers in the vocabulary. If such peer exists, the word is replaced with the peer in the lyric. Otherwise, the word is discarded as described previously. To look for synonyms, we exploit WordNet [10], an ontology library for English provided by University of Princetion, which is organized as groups of synonyms called synset and describes relations between these synsets. Thereby, words are substituted in lyric if they are in the same synset or some adjacent synsets that have similar semantic and are related to the original synset. Note that, for higher efficiency, we can equivalently expand the vocabulary (instead of lyric) based on synonyms, which is then used to filter the lyric as aforementioned.

3 Audio Features

Three common audio features, MFCC, spectral contrast and chromagram, are exploited in our work for capturing acoustic characteristics of music for mood categorization. To join three features together, which are extracted from signal windows of variant length respectively, the shorter-term features such as MFCC and spectral contrast falling in the longer window of chromagram are averaged to align with the latter. Then, the mean and variance of the composite features within the span of one lyric sentence are used as its feature of audio modality.

Furthermore, considering the dynamic regions of music could be perceptionally more important in characterizing the mood than relatively stationary audio regions, we propose a *saliency*-based subsampling/filtering step for the audio features extracted. Specifically, given the power spectrum output of short-term discrete Fourier transform on the audio track of music and followed by Mel-scale spectral filtering, we take the local spectral peaks (maxima) in the spectrogram as the *salient points*, whose energy is strictly bigger than all energies in immediately adjacent frequency (Mel band) and time locations. Since such local temporal-spectral peaks could be very dense and with redundancy, we further perform a temporal filtering on the candidate peaks to preserve only the representative ones whose energy stays above a temporal masking threshold as defined by Eqn.(2):

$$Thr[n] = \alpha^{\Delta t} E[n-1] \exp -\frac{(\Delta t)^2}{2 * \sigma^2} \qquad (2)$$

where, Δt is the temporal distance between the considered peak n and the previous peak chosen in the same Mel band whose energy is denoted by $E[n-1]$, and α and σ are two parameters controlling the temporal falling speed of the threshold and thus the final density of salient points.

Finally, given the salient points detected, we use only the audio features at the temporal locations of the salient points to construct the audio modality feature for each lyric sentence as aforementioned.

4 Fusion and Classification

Given the sequence of sentence-level features of the audio and text modalities extracted from each sentence of music of certain mood, in this section, we propose a Hough forest-based ensemble recognition model to fuse the two modalities and classify it to one of the predefined music mood categories. The main differences of the proposed method from other existing ones are in two aspects: 1) We employ a *bag of sentences* representation for music data, and thereby the recognition of the music's mood relies the ensemble of individual votes casted by every music segment. Many other methods usually characterize one music by the single holistic description. 2) We integrate the fusion of cues about the potential mood of the music from two modalities into the Hough voting mechanism of the Hough forest model, as one form of model-level fusion, instead of simply fusing the classification outputs from individual modalities. On the other hand, the feature-level fusion can also be easily integrated into the model.

4.1   Hough Forest Classification Model

Hough forest is basically a random forest consisting of a collection of decision tree predictors, each classifying the input feature vector and outputting its vote on the possible class label, and the final output of the forest is then the integration of the individual votes. The Hough forest, which is built on basis of random forest, further exploit a generalized Hough transform to aggregate votes on any parameters of interest casted by individual samples, which endows Hough forest great flexibility as well as superior generalization for many tasks.

In our work, the Hough forest model is used to learn the mapping between the multimodal features of the individual sentence of music and their probabilistic votes on potential categories of the music mood. Given the set of sentence-level features of training musics in different moods, we associate the mood label r_i ($r_i \in [1..K]$, K is the number of mood categories) of the training music and the music id n_i and the sentence id t_i to each sentence-level feature v_i, and then train the Hough forest with all $\{< v_i, r_i, n_i, t_i >\}$.

Depending on the form of fusion, we employ two different voting schemes in the Hough forest.

4.2   Feature-Level Fusion

In this configuration, the features from two different modalities for the same one sentence are concatenated as one composite feature. Given an unknown testing music s, we send each composite feature s_j corresponding to one of sentence segments of s down the forest and collect the set R_j of training samples contained in the predicted leaf nodes of the forest that are reached by s_j.

Then, the probabilistic votes on the mood category casted by all segments $\{s_j\}$ are then aggregated with the generalized Hough transform. Specifically, we initialize a 1-D Hough matrix $H_c(k)$ ($k = 1..K$) with each element of the matrix containing the accumulated votes for the corresponding mood category k. Since each training sample $< v_i, r_i, n_i, t_i >$ in R_j of segment s_j makes a contribution $u(j)$ to the hypothesis that the music belongs to the mood category r_i, we increment the corresponding bin of the Hough matrix $H_c(r_i)$ as:

$$H_c(r_i) = H_c(r_i) + u(j), \quad u(j) = \frac{w(j)}{|R_j|} \tag{3}$$

where, $w(j)$ is the weight associated with the segment s_j, denoting the importance of its votes. We fix $w(j) = 1.0$ for simplicity in this work.

After all the segments $\{s_j\}$ are processed, the maxima over all $H_c(k)$ ($k = 1..K$) indicates the MAP estimate for the potential mood category.

4.3   Model-Level Fusion

In this scheme, we make explicit use of the time alignment constraint of two modalities at the sentence-level, and exploit two separate Hough forests, one for the audio modality and the other for the text modality, to more efficiently capture the characteristics of music segments in each modality. Correspondingly, during either the training or the testing phase, the audio and text features extracted for the sentences of music are fed into the respective Hough forest. Then, for one input unknown music, a different voting scheme is employed to fuse the two modalities in a hierarchical Hough space. The basic idea is, besides accumulating the categorical votes casted by training samples in both modalities, the *temporal correlation* of the votes in two modalities at the sentence level also gives cues or constraints for the potential mood category, and thus can be exploited in specific Hough space for improved prediction accuracy.

Specifically, on the basis of the sentence-based hierarchical representation of musics, we define a *two-layer* Hough structure, in which the top layer is the same 1-D Hough matrix $H_c(k)$ for mood category as in the feature-level fusion scheme, to accumulate the categorical votes from the bottom layer for each sentence segment. In the bottom layer of the Hough structure, for each training music n, we maintain a *sentence-level* 2-D Hough matrix:

$$H_s^n(m, t)_{[2 \times T_n]} \tag{4}$$

where, in the first modality dimension, $m = 1$ for audio modality and $m = 2$ for text modality. For the second temporal dimension in terms of sentence, $t \in [1..T_n]$, T_n is the number of sentences of the music n.

For one unknown music, we employ a bottom-up two-phase voting scheme. We first initialize the Hough matrix $H_c(k)$ to zero. Then, for each sentence segment j of the input music, we proceed with following steps:

1. Initialize the Hough matrix $H_s^n(m, t)$ to zero.
2. Send the audio feature s_j^A and text feature s_j^T of the input segment down the corresponding audio-specific and text-specific Hough forests, and collect the sets of training samples R_j^A (for audio modality) and R_j^T (for text modality) at the predicted leaf nodes of the forests.
3. For each training sample $< v_i, r_i, n_i, t_i > \in R_j^A$ in the audio modality (the process for the text modality is same with $m = 2$),

$$H_s^{n_i}(m, t_i) = H_s^{n_i}(i, t_i) + u_A(j), \quad m = 1, \quad u_A(j) = \frac{w_A(j)}{|R_j^A|} \tag{5}$$

where, $w_A(j)$ is similar to that described in previous section.
4. After all training samples in both R_j^A and R_j^T have casted their votes on the sentence-level Hough matrix $H_s^n(m, t)$, we compute the categorical vote by current sentence segment of the input music by fusing two modalities in $H_s^n(m, t)$ as follows:
   → For each training music n comprising segment samples in R_j^A and R_j^T:
   (a) Compute the *correlation coefficient* **Cor$_n$** between the two modalities $H_s^n(1, \cdot)$ and $H_s^n(2, \cdot)$, taking the vote number in each modality of H_s^n as a random variable and the actual vote numbers at each sentence as different samples of it.

(b) Sum up modality-specific votes across sentences as

$$\bar{H}_s^n(m) = \sum_{t=1}^{T_n} H_s^n(m,t), \quad m = 1,2 \tag{6}$$

(c) Compute the categorical vote $\bar{H}_s^n$ by the training music n by combining two modalities:

$$\bar{H}_s^n = (1.0 + \mathbf{Cor_n})(\alpha \bar{H}_s^n(1) + (1.0 - \alpha)\bar{H}_s^n(2)) \tag{7}$$

The combination factor α is chosen to 0.3 by experiments in our work.

5. For every training music n, increment the corresponding bin of the top layer Hough matrix $H_c(r_n)$ as $H_c(r_n) = H_c(r_n) + \frac{1}{|r_n|}\bar{H}_s^n$, supposing music n belongs to the mood category r_n, and $|r_n|$ denotes the number of training musics in mood category r_n.

After all segments of the input music are processed, the maxima over all $H_c(k)$ ($k = 1..K$) indicates the MAP estimate for the potential mood category.

5   Dataset

We build the experimental dataset of musics based on their mood tags found on the website last.fm, which is a music recommendation website with large amount music associated with social tags. We used the tag API provided by the website to generate the list of similar tags for each mood category and then looked for the musics mostly tagged with corresponding tags. On the other hand, as our method exploits the time alignment between lyrics and audio tracks at the sentence level, we acquire such time information from the LRC files found on Internet, which provide the beginning time stamp for each sentence of lyric.

Specifically, our dataset is composed of 781 musics at 44.1kHz sampling rate with total length more than 3000 minutes, which are nearly evenly distributed over 4 mood categories. The corresponding LRC files consist of over 30000 sentences of lyric after some manually editing to remove errors and noises.

6   Results

To evaluate the proposed music mood classification framework, a set of experiments are conducted on the aforementioned dataset. The training set consists of 400 randomly chosen musics, 100 per mood category, and the rest musics are used as testing samples. A 23ms analysis window with 50% overlapping is applied on the audio track of music to extract MFCC and spectral contrast features. For chromagram features, the window length is 368ms with step size of 46ms. The Hough forest has 30 decision trees with the maximum depth set to 15.

Table. 1 shows the average classification accuracy with different configurations of the proposed framework. As the result shows, the fusion of multiple modalities

Table 1. Comparisons of average accuracy (%) with different configurations of the proposed framework: each modality solely, multimodality fusion at either feature-level or model-level, and further with vocabulary reduction and/or lyric expansion

Model Configuration	Mood Category				Avg.
	happy	angry	sad	relaxed	
text modality	51.4	83.8	42.4	22.8	50.1
audio modality	43.0	77.0	36.4	55.4	53.0
text+audio **feature** fusion	49.5	86.4	39.4	46.5	55.4
text(*reduced*)+audio **feature** fusion	56.0	86.4	37.4	51.5	57.8
text(*expansion*)+audio **feature** fusion	53.3	86.4	38.4	53.5	57.9
text+audio **model** fusion	49.5	87.9	44.4	38.6	55.1
text(*reduced*)+audio **model** fusion	56.1	90.5	48.5	38.6	58.4
text(*expansion*)+audio **model** fusion	54.2	89.2	51.5	44.6	59.9

outperforms the unimodal methods (either audio or text), as the latter are limited in exploiting full information about music mood from all modalities data. Also, the proposed model-level fusion scheme is shown averagely superior to the feature-level fusion. Moreover, the proposed discriminability-guided vocabulary reduction and lyric expansion mechanism is shown capable of promoting the accuracy further. Note the 'expansion' in the table means lyric expansion after vocabulary reduction, which yields the averagely higher accuracy than using the 'reduced' vocabulary solely.

Fig. 3 shows the change of average accuracy with different number of words in the vocabulary by the text modality only, the feature-level and model-level fusion of both modalities by the Hough forest model, all with the proposed lyric pre-filterings. Note that the proposed multimodal classification model achieves the peak performance around the vocabulary size of 1000 words out of the total

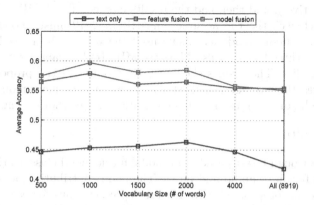

Fig. 3. Comparisons of accuracy with different vocabulary sizes by the text modality only, the feature-level and model-level fusion of two modalities in Hough forest

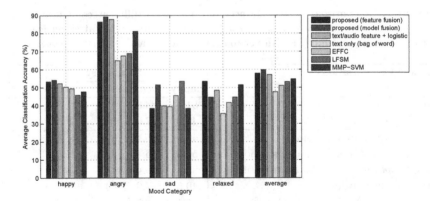

Fig. 4. Comparisons of average accuracy over 4 mood classes by: 1) the proposed method with feature-level fusion (57.9%), 2) the proposed method with model-level fusion (59.9%), 3) the proposed text/audio features with logistic classifier (57.1%), 4) a bag of text features model [5] (47.6%), 5) EFFC [18] (51.1%), 6) LFSM [18] (53.2%) and 7) MMP-SVM [12] (54.7%)

8919 words, while it's around 2000 words for the single text modality scheme. The difference shows the effect of the correlation (or some form of constraints) of representations cross modalities, which contributes essentially to the overall superiority of the multimodal scheme to the unimodal one.

In Fig. 4, we compare the overall classification accuracy for the 4 mood categories by different methods. The bag-of-word model of text features was used in [5], which consists of a vector of terms weighed by tf-idf and reduced to 1500 dimension by the CHI-based feature selection. The EFFC (Early Fusion by Feature Concatenation) method [18] concatenated the text and audio (MFCC) features with the SVM classifier, while the LFSM (Late Fusion by Subtask Merging) method [18] used separately the text and audio features to classify valence and arousal respectively, and then had the results merged. The MMP-SVM method [12], which achieved the top accuracy in MIREX 2012 music mood classification test by the audio modality solely, used the total 312-dim audio features extracted from three frameworks - Marsyas, MIR Toolbox and PsySound3, with the SVM classifier. The results show the effectiveness of the proposed model, which achieves the averagely highest classification accuracy among both the unimodal (audio or text) and multimodal state-of-art methods being compared.

7   Conclusion

We propose a Hough forest based multimodal music mood classification method based on the fusion of audio and lyrics. The main contribution of the paper is twofold: First, we propose effective feature representations for each of the text and audio modality. Second, we propose a hierarchical Hough voting scheme to fuse the two modalities efficiently. The experimental results demonstrate the effectiveness of the proposed method.

Acknowledgments. Research supported by the National Science Foundation of China under Grant Nos. 61003113, 61272218 and 61321491.

References

1. Ali, Omar, S., Zehra, Peynircioglu, F.: Songs and emotions: Are lyrics and melodies equal partners? Psychology of Music 34(4), 511–534 (2006)
2. Bengio, Y., Schwenk, H., Senscal, J.S., Morin, F., Gauvain, J.L.: Neural probabilistic language models. JMLR 3, 1137–1155 (2003)
3. Hinton, G.E.: Learning distributed representations of concepts. In: 8th Annual Conference of the Cognitive Science Society, pp. 1–12 (1986)
4. Hu, X., Downie, J.S.: When lyrics outperform audio for music mood classification: A feature analysis. In: ISMIR 2010, pp. 619–624 (2010)
5. Hu, X., Downie, J.S., Ehmann, A.F.: Lyric text mining in music mood classification. In: ISMIR 2009, pp. 411–416 (2009)
6. Kim, Schmidt, E.M., Migneco, R., Youngmoo, E.: Music emotion recognition: A state of the art review. In: ISMIR 2010, pp. 255–266 (2010)
7. Laurier, C., Grivolla, J., Herrera, P.: Multimodal music mood classification using audio and lyrics. In: ICMLA 2008, pp. 688–693 (2008)
8. Li, T., Ogihara, M.: Detecting emotion in music. In: ISMIR 2003, pp. 239–240 (2003)
9. Lu, L., Liu, D., Zhang, H.J.: Automatic mood detection and tracking of music audio signals. IEEE TASLP 14(1), 5–18 (2006)
10. Miller, G.A.: Wordnet: A lexical database for english. Communications of the ACM 38(11), 39–41 (1995)
11. Nakagawa, T., Iuni, K., Kurohashi, S.: Dependency tree-based sentiment classification using crfs with hidden variables. In: Human Language Technologies: The 2010 Annual Conference of the North American Chapter of the Association for Computational Linguistics, pp. 786–794 (2010)
12. Panda, R., Paiva, R.P.: Mirex 2012: Mood classification tasks submission (2012)
13. Pang, B., Lee, L., Vaithyanathan, S.: Thumbs up? sentiment classification using machine learning techniques. In: EMNLP 2002, pp. 78–86 (2002)
14. Russell, J.A.: A circumplex model of affect. Journal of Personality and Social Psychology, 1161–1178 (1980)
15. Socher, R., Pennington, J., Huang, E.H., Ng, A.Y., Manning, C.D.: Semi-supervised recursive autoencoders for predicting sentiment distributions. In: EMNLP 2011, pp. 151–161 (2011)
16. Turian, J., Ratinov, L., Bengio, Y.: Word representations: A simple and general method for semi-supervised learning. In: 48th Annual Meeting of the Association for Computational Linguistics, pp. 384–394 (2010)
17. Yang, D., Lee, W.S.: Disambiguating music emotion using software agents. In: ISMIR 2004, pp. 218–223 (2004)
18. Yang, Y.-H., Lin, Y.-C., Cheng, H.-T., Liao, I.-B., Ho, Y.-C., Chen, H.H.: Toward multi-modal music emotion classification. In: Huang, Y.-M.R., Xu, C., Cheng, K.-S., Yang, J.-F.K., Swamy, M.N.S., Li, S., Ding, J.-W. (eds.) PCM 2008. LNCS, vol. 5353, pp. 70–79. Springer, Heidelberg (2008)
19. Yang, Y.-H., Chen, H.-H.: Machine recognition of music emotion: A review. ACM Transactions on Intelligent Systems and Technology 3(3) (May 2012)

Multidimensional Context Awareness in Mobile Devices

Zhuo Wei[1], Robert H. Deng[1], Jialie Shen[1],
Jixiang Zhu[2], Kun Ouyang[2], and Yongdong Wu[3]

[1] Singapore Management University, Singapore
[2] Wuhan University, China
[3] Institute for Infocomm Research, Singapore
{zhuowei,robertdeng,jlshen}@smu.edu.sg
{jixiang.jason.zhu,oyk1115}@gmail.com
wydong@i2r.a-star.edu.sg

Abstract. With the increase of mobile computation ability and the development of wireless network transmission technology, mobile devices not only are the important tools of personal life (e.g., education and entertainment), but also emerge as indispensable "secretary" of business activities (e.g., email and phone call). However, since mobile devices could work under complex and dynamic local and network conditions, they are vulnerable to local and remote security attacks. In real applications, different kinds of data protection are required by various local contexts. To provide appropriate protection, we propose a multidimensional context (**MContext**) scheme to comprehensively model and characterize the scene and activity of mobile users. Further, based on the scheme and RBAC, we also develop a novel access control system. Our experimental results indicate that it achieves promising performance comparing to traditional RBAC (Role-based Access Control).

Keywords: Mobile security, context awareness, access control.

1 Introduction

The growing pervasiveness of the mobile device has changed the way we go about our daily lives. With increasing computing and storage capabilities, mobile device emerges as dominant computing platform for end-users to access the Internet services and connect the people. Meanwhile, most mobile devices are equipped with a wide range of advanced sensors, such as cameras, speakers, microphone, and accelerometer. They can be used for various purposes and open up a wide range of opportunities to develop new applications for business. It comes as no surprise that more and more employees bring their own mobile devices to workplace and frequently access sensitive company information (named BYOD - Bring Your Own Device). While it has been proven that BYOD can improve productivity of employee and make the company look like a flexible and attractive

X. He et al. (Eds.): MMM 2015, Part II, LNCS 8936, pp. 38–49, 2015.

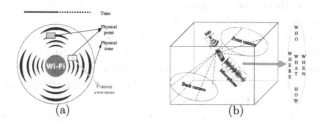

Fig. 1. (a) Existing access control model; (b) Multidimensional context aware model

employer, the practice could create various kinds of data security risks. For example, when users read confidential documents or have confidential discussion via phone lines, this can create high risk of information leak or loss.

To provide appropriate data security protection under different real scenarios, accurate context modelling and detection becomes more and more important. Motivated by the observation, this research aims to design, develop and evaluate an automatic multidimensional context aware security techniques tailor-made for mobile device users. To support an effective user context modeling, we propose a novel and efficient scheme called *MContext* to characterize contextual information based on mobile users' scene and behaviours. By using raw signal from various sensors in mobile device, *MContext* can effectively model and infer mobile users' context including: *Who, When , Where, What* and *How*.

The contextual information can be further applied to design security strategy of mobile device users when they use different applications. In this paper, based on the scheme and RBAC, we have developed and fully implemented a novel access control system. To validate the proposed scheme, we carry out large scale experimental study and our results show that *MContext* model is highly effective and can dynamically detect current situation. Meanwhile, *MContext* aware RBAC system demonstrates high efficiency and effectiveness while deployed in ubiquitous devices.

The rest of this paper is organized as follows. We introduce the related works in Section 2. Section 3 and Section 4 describe the multidimensional context aware model and access control system, respectively. Finally, we give experimental results and performance analysis in Section 5. Section 6 concludes the paper.

2   Related Works

Recently mobile device security attracts a lot of research attentions and the core research challenge is closely associated with the end node problem, wherein a device is used to access both sensitive and risky networks/services. Because of Internet-based risks, the very risk-averse organizations provide devices specifically for Internet use. Thus, it is very important for company to understand the behaviour of its own employees while using mobile devices. Previous literatures proposed several access control techniques for mobile device users, which are classified into three major categories: temporal, physical zone and physical point domains as shown in Figure 1(a).

The first category is related to temporal domain. In many practical scenarios, users may be restricted to assume roles only at predefined time periods. Furthermore, the roles may only be invoked on pre-specified intervals of time depending upon when certain actions are permitted. For example, Bertino et al. presented TRBAC (Temporal Role Based Access Control)) [1] and GTRBAC (Generalized Temporal Role Based Access Control) models [2]. Both proposed models can be used for the access control for mobile device users inside time interval.

Secondly, the information about where mobile user presents can be applied to characterize objects, user positions, and geographically bounded roles. Roles are activated based on the position of the user. Besides a physical position, obtained from a given mobile terminal or a cellular phone, users are also assigned a logical and device independent position, representing the feature (e.g., room, floor, region) in which they are located. For example, with GPS (Global Positioning System) and GIS (Geographic Information System) information, Bertino et al. propose the geometric-based RBAC [3,4]. Similarly, by taking use of WIFI information. In [5], Ray et al. design a location-aware RBAC. All of information can be used for the access control for mobile device users inside physical zones.

The third access control management of mobile device users depends on extra devices and can be used inside physical places (points). A user provides a service provider with role and location tokens, e.g., Radio-frequency identification, along with a request. The service provider consults with a role authority and a location authority to verify the tokens and evaluate the policy. The typical example of system in this category is the smartphone-based system developed by Gey [6]. Using the system, user can exercise her authority to gain access to rooms inside university building, and by which she can delegate that authority to other users.

More recently, a few more comprehensive access control schemes have been developed for mobile device users by combining temporal, physical zone/point analysis techniques, e.g., STARBAC [7]. However, the existing techniques are not good enough for BYOD. On the one hand, they are unable to identify the higher level contexts (e.g., meeting, leisure, travelling), which a user involves. Mobile device access control requires high-level visual information to analyse scene of employee or environment context and then invoke the appropriate security policy and measure. On the other hand, mobile device application is getting ubiquitous now. It cannot depend on extra devices for its access control, e.g., you cannot setup radio frequency identification devices at public places.

3 Multidimensional Context Aware Model

It is not hard find that a conflict always exists between employee's flexibility and employer's management: employees can flexibly use mobile devices to access sensitive company data or important government documents at anytime and anywhere while employers need to protect data/documents in case of security risks. A possible way to help employers to perform the access control of mobile device users is to understand employees' context information (e.g., location and behaviours) and perform different access control policies.

Our proposed model is a systematic and comprehensive approach to achieve effective context awareness for mobile device users. It can effectively exploit both static and dynamic scene and event recognition to classify users' location and activity. The goal of proposed model is to tell a *who* (user identification), *when* (current time), *where*, *what* and *how* story of mobile device users to employers. Figure 1(b) illustrates the multidimensional context awareness for mobile device users. *Where* and *What* are the location, scene and behaviours of mobile device users.

3.1 *Who, When* and *How* Verification

When a user wants to access a remote server database with his/her mobile device, it is necessary to supply user' profile (*When, Who*, and *How*) to the server in order to verify his/her validity. For example, when a businessman requires checking email outside company by own mobile, he must provide name and password, i.e., *Who*, to the server. Meanwhile, his login/logout time log, i.e., *When*, also are recorded by the server. In addition, user's network and device type, i.e., *How*, can further be analysed by servers in order to verify the user authorities.

3.2 *Where*: Physical Location Recognition

Physical location are generally classified into two categories: indoor (e.g., bedroom or office) and outdoor (e.g., bus or square), which can be inferred from mobile devices sensors, e.g., GPS (out-door), 3G and WIFI (indoor). With present technology, we are able to recognize users location precisely and efficiently, thereby determining the IOR (Indoor or Outdoor) information of mobile users. In this paper, we exploit GPS signal information as well as hybrid location technology provided by third-party map API to determine IOR.

GPS: Nowadays, GPS (Global Positioning System) is widely applied for localization in smart device related application. Its signal differentiates obviously from indoor and outdoor, e.g., the acquisition speed of indoor is less than 10s with the accuracy of 10m to 20m; while the acquisition speed of outdoor is expected to be larger than 20s with the accuracy of more than 500m. Hence, by using the acquisition time T and the accuracy A, we can coarsely infer users' location. Assuming N_1 is the first impact factor for location decision, if $T < 10$ and $A < 50$, $N_1 = 1$, otherwise, $N_1 = 0$.

Hybrid Position: Most of the users are inside the architectural complexes in cities, where the GPS accuracy can not be guaranteed. It also takes a long time to locate the position, which can affect the accuracy as well. In this study, a third-party map SDK, the GOOGLEMAP SDK[1], is adopted for the hybrid

---

[1] https://developers.google.com/maps/

positioning solution which integrates WIFI, GPS, and base stations, in order to enhance the accuracy when users connect to WIFI or GPRS network. By using the relevant API, we can achieve more accurate locations and translate users coordinates into specific locations. We collect a set of building coordinates at advance, and set proximity alert based on the users current coordinates and the collected ones. When users are approaching the building, the proximity alert will be activated. The alert signal is recorded as the second impact factor N_2. When the proximity alert is activated, $N_2 = 1$, otherwise, $N_2 = 0$.

Algorithm: The algorithm to determine whether the user indoor or outdoor is as follows.

$$\mu = \begin{cases} (1 - \alpha)N_1 + \alpha N_2, & if\ WIFI\ is\ accessable \\ N_1, & otherwise \end{cases} \tag{1}$$

Where μ is IOR decision. Under perfect circumstance, when $\mu = 1$, the user is in the indoor. However, in reality, since the accuracy of exiting positioning technology cannot reach 100, we can adjust the parameter α. By doing so, the accuracy is expected to exceed 80% when μ is larger than the threshold value, e.g., $\mu = 0.7$.

3.3   *What*

Although users' location (***Where***) is verified and recognized, it cannot decide if mobile users stay at a secure context. Hence, it is necessary to further know following context: **scene, situation** (e.g., are users' environment crowd, noise?), **behaviour** (e.g., are users static or moving?). By taking use of microphone, camera, accelerometer of mobile devices, robust temporal features (audio and accelerometer) and spatial features (images) are extracted and send to classifiers for sensing users' context.

What - Scene Recognition. In the scenario of scene recognition, mobile device is able to provide lots of information such as lighting, acceleration, temperature, audio and videos. The solution of context recognition based on those scenarios has been studied widely. Scene recognition based on audio has advantages like simplicity of data acquisition, abundant information amount and maturity of existing processing approaches. Considering the limitation of computing capability of mobile device and realizability of the model, our recognition scheme mainly takes use of audio as well as location information. The scheme consists of off-line training and real-time classifying. During the training process, classifiers are trained and built, which includes feature extraction, classifier construction, machine learning and model evaluation. In the real-time classifying process, the best performance classifier at last process is utilized and environment information such as audio and IOR from mobile device are extracted and sent as input to trained classifier to get predict values.

Recording Procedure: In order to reflect the characteristics of the scene, we have collected plenty of audio data[2]. Totally, there are 1350 audio records with no less than 10 seconds each in nine scenes: Indoor (Classroom, Quiet room and Market hall), Outdoor (Square, Park and Street), and Transportation (Bus, Metro, and Train). Audio are recorded in PCM standard with sampling rate of 41.1 KHz, 16 bit mono.

Feature Extraction: Currently, literatures proposed a variety of approaches for audio feature extraction, e.g., zero-crossing rate (ZRC), Mel Frequency Cepstral Coefficients (MFCC), spectral centroid and linear prediction coefficients (LPC), etc. Among these approaches, MFCC is often exploited in audio processing due to its accuracy and robustness properties, such as speech recognition, music genre and instrument categorization. In this paper, we also utilize MFCC as features for subsequent classification and prediction. Regardless of the accuracy of classification, the 13 dimensional MFCC vectors extracted from audio can be used as input data to train and build the classifier. To improve the performance, the IOR information obtained by location recognition scheme is added to feature vectors as the 14th dimension of them.

Classification Method: Plenty of approaches designed for audio retrieval and classification have been proposed. Our main focus is on finding methods that are suitable for implementation on mobile devices. In this paper, we build and train our Bayesian network model as a classifier on PC, then download it to mobile devices for the inference procedure of Bayesian network which is much simpler than its building procedure.

Description of the Model: Let $U = \{x_1, x_2, ..., x_n\}$ be a set of variables. A Bayesian network consists of a network structure B_s, which is a directed acyclic graph over U, and a set of probability tables $B_p = p(u|pa(u)|u \varepsilon U)$, where $pa(u)$ is the set of parents of u in B_s. A Bayesian network represents a probability distributions:

$$P(U) = \prod_{u \varepsilon U} p(u|pa(u)) \tag{2}$$

The task of a Bayesian network classifier is to classify $y = \hat{y}$ called the class variable, given a $X = \{x_1, x_2, ..., x_n\}$ set of input variables called attribute variables. A classifier is actually a mapping function $h = X \to y$.

Learning Procedure: A classifier is built in this period given a training data set. The learning procedure consists of two stages: firstly learning a network structure B_s, then learning the probability tables B_p. There are various approaches to structure learning, such as *local score metrics*, *global score metrics*, *conditional independence tests*, and so on. Each kind of approach has corresponding search algorithms, e.g., hill climbing, simulated annealing, and tabu search. Considering the limitation of the calculating capacity of mobile devices, *local score*

---

[2] It is available at `http://1drv.ms/1sQ9Pxj`.

metrics and $K2$ algorithm [8] are exploited as the structure learning approach and search algorithm, respectively. For probability tables learning, we estimate the conditional probability tables by averaging all sub-structures of the network structures we have learned before. This is achieved by estimating the conditional probability table of a node x_i as a weighted average of all conditional probability tables of x_i given subsets of its parents $pa(x_i)$.

Inference: Assuming we have built a classifier through the learning procedure, $P(y|X)$ is calculated simply using the distribution $P(U)$ represented by the Bayesian network:

$$P(y|X) = P(U)/P(X)$$
$$\propto P(U)$$
$$= \prod_{u \varepsilon U} p(u|pa(u)) \tag{3}$$

Since all variables in X is known, no more extra complicated inference algorithms are needed. The output of the classifier is $\hat{y}$ where $P(\hat{y}|X) = argmax_y P(y|X)$.

What: Environment Recognition. When a user uses the phone with confidential operations, such as checking confidential document or talking on a confidential lines, there is information leak risk due to spy and eavesdropping. Therefore, estimating security level of surrounding environment becomes very important. In this paper, environment estimation refers to the number of people around and behind users. Generally, the more people around the user, the more congestion will be, the more insecure the environment will become, hence the risk of being watched is emerging. We exploits both front camera and microphone to analyze environments. Firstly, algorithm SC (Speaker Count) is used to determine number of people surrounding users, e.g., Crowd++ method [9]. SC algorithm splits audio file to equal length of speck slices, and extracts sound information and features in human voice frequency range. Eventually, SC algorithm will calculate the number people speaking based on unsupervised algorithm. Secondly, with the front camera, **MContext** model captures images as users access important/privacy data, and exploits face detection schemes (e.g., OpenCV[3]) to recognize human faces behind users.

What: Activity Recognition. As mobile users access important company or personal privacy data, above context awareness modules, e.g., **Where, What (scene)**, will be activated once. However, context of users may be changed due to user activities (e.g., walking), hence the new location, scene or environment may require different security policies. In this paper, we use mobile sensors' signals, e.g., accelerometer and rotation sensors, to dynamically recognize users behaviours, such as walking speed and direction. Once those activities which

---

[3] http://opencv.org/

may hint that users' location, scene or environment are changed are recognized, *MContext* model will require processing static modules again, i.e., **Where** or **What**, in order to apply the right security policies for new user context.

4   Context Aware Access Control

Most of the present access control models still resting on assigning permissions to users based on **Who**, i.e., users identity, may cannot guarantee the security of sensitive data when it comes to mobile devices. The mobility of the device leads to the uncertainty of location, environment and other contexts. That is to say, even the same user should be assigned different permissions under difference circumstance. In this section, we propose a new access control model named *MContext* aware RBAC (MCARBAC), a modified model of RBAC[11]. An example of our model was implemented after that.

4.1   *MContext* Aware RBAC Model

The MCARBAC model consists of: a set of basic element sets; a set of RBAC relations involving those element sets (containing subsets of Cartesian products denoting valid assignments); and a set of mapping function that yield instances of members from one element set for a given instance from another element set.

Figure 2 shows basic architecture of *MContext* aware RBAC, which includes five basic data elements: MContexts (MCS), roles (ROLES), objects (OBS), operations (OPS) and permissions (PMS). The model is defined in terms of contexts assigned to roles and data access rights assigned to roles. Moreover, a set of sessions (SESSIONS) are also included in the model where each session is a mapping between a *MContext* and an activated subset of roles assigned to the *MContext*. The most significant difference between RBAC and MCARBAC is that the element users (USERES) in the former is replaced by MContexts (MCS) in the latter.

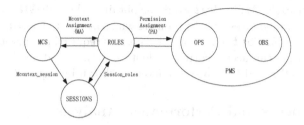

Fig. 2. *MContext* aware RBAC

4.2   Prototype Implementation of *MCRBAC*

A prototype of MCRBAC has been fully implemented using Android operating system. In our prototype, we define the element sets as ROLES $\{R1, R2, R3\}$ and PMS $\{P1, P2, P3\}$. We also assume that the elements in ROLES as well as PMS are hierarchical, which suggests all privileges of elements at low hierarchy are also privileges of elements at higher hierarchy. Here $R_1 < R_2 < R_3$ and $P_1 < P_2 < P_3$. Then we assign permissions to roles (PA) and *MContext* to ROLES (MA). The mapping function of PA is defined as $assigned\_permissions(R_i) = P_i$, where $R_i \in$ ROLES, $P_i \in$ PMS; and the mapping of MA is defined as Table 1. Less safe the *MContext* was, the lower its assigned role in the hierarchy would be.

Table 1. The mapping of MContext Assignment

Location		High crowd	Low crowd
	Classroom	R_2	R_3
Indoor	Quiet Room	R_3	R_3
	Market Hall	R_1	R_2
	Square	R_2	R_3
Outdoor	Park	R_2	R_3
	Street	R_1	R_2
Transpo	Bus	R_1	R_2
-rtation	Metro	R_1	R_2
	Train	R_1	R_3

Table 2. Categorization of OPS & OBS and the PMS each type requires

Source		Confid -ential	General
	Email	P_3	P_2
Files	Message	P_2	P_1
	Contact	P_2	P_1
Commu	Phone Call	P_3	P_1
nication	Wechat	P_2	P_1
	Facetime	P_2	P_1
Money	Bank account	P_3	\
Apps	Amazon account	P_3	\
	Paypal account	P_3	\

"\": resources are not considered as general.

Since the accessible operations (OPS) and objects (OBS) on the mobile device are so complex and diverse, it is difficult to design an access control matrix (ACM) that contains all of them. Thus, for the sake of simplicity, OPS and OBS of mobile devices are categorized into three hierarchical types, and each type of OPS and OBS requires correspondent level of PMS, as shown in Table 2.

When the user attempts to execute some operations or access some objects of the mobile device, the *MContext* would be obtained through *MContext* model and a SESSION would be established during which a subset of activated roles will be assigned to the *MContext*. If the roles possess required permissions assigned to them, then they can access correspondent operations and objects. Otherwise, access to those OPS and OBS are restricted as shown in Figure 3(d).

5   Experiments and Performance Analysis

5.1   Experiments

Our system is developed on Android 4.0 platform and Figure 3 shows its main interface. To gain reliable test result, we run test 100 times in each of 9 scenes to

Fig. 3. The interface of *MContext* aware access control system. (a) Interface; (b) Classroom example; (c) Face detection; (d) File control.

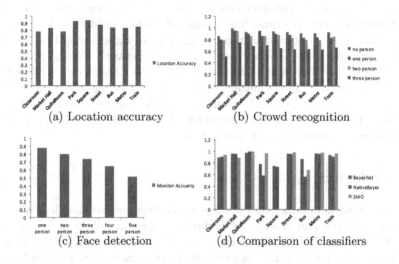

Fig. 4. Experiments results

obtain the accuracy of location recognition scheme. 1350 recordings of 9 scenes are split into two data sets at the proportion of 66% to train and test the classifiers. Random division of recordings into the training and tests sets is done 100 times. As for comparison, two classification models SMO[12] and Naive Bayes[13] are implemented as the counterparts of our Bayes Net Model. Also, environment recognition scheme is also evaluated 100 times each scene on the condition that different number of persons (ranging from 0 to 3) speak respectively. Figure 4 illustrates experimental results and demonstrates a set of promising performance. Figure 4(a) shows that the recognition rate of location is about 85.1% for nine scenes; Figure 4(b) illustrates the recognition rates of crowd with SC methods; Figure 4(c) shows that accuracy rate of the faces detection is about

71.8%; Figure 4(d) illustrates the accuracy rate comparison among proposed scheme (90.2%), Native Bayesian (84.4%) and SMO (82.7%).

5.2 Performance Analysis

As mobile device users take their devices to access sensitive data, presented model can automatically and friendly perform context awareness. That is, all operations are transparent to mobile device users, e.g., image/audio capturing, features extraction. *MContext* aware model consists of static and dynamic context aware models. Static context aware model integrates scene and environment categorizations once; while dynamic context aware model takes use of temporal information (e.g., accelerometer) to percept users' context changing.

Good recognition that can be used for supporting real applications, largely depends on a major factors: 1) size of training dataset and 2) classification scheme. There is a trade-off between accuracy and efficiency. When more learning examples are consider, classifier can achieve better accuracy but needs more hours to complete training time. Since proposed model considers both spatial and temporal features, it guarantees nice accuracy rate. Experimental results show that the recognition the accuracy rate of Bayesian network is about 90.2%. It is better than Native Bayesian 84.4% and SMO 82.7%. The delay of *MContext* is about 2.358 second. In the future, we may try to leverage the computational and storage capability of cloud computing which improves the efficiency, i.e., performing light-weight feature extraction at mobile device while performing computational expensive classification in the cloud.

6   Conclusions

With increasing computing and storage capabilities, smart mobile devices are changing our lives and emerge as dominant computing platform for different kinds of end-users. BYOD users can access important company or government data at anytime and anywhere, however, they are vulnerable to local and remote attacks. In this paper, *MContext* aware access control system was proposed in order to protect local/remote data security. Experimental results indicate that *MContext* model enjoy high accuracy and good robustness. In addition, proposed access control system perfectly matches the requirement provided by various real applications (such as BYOD). In future, we plan to develop more advanced access control algorithm to improve effectiveness of data protection.

Acknowledgements. This work was partly supported by National Natural Science Funds of China (Grant No. 61402199). Jialie Shen was supported by Singapore Ministry of Education Academic Research Fund Tier 2 (MOE2013-T2-2-156), Singapore.

References

1. Bertino, E., Bonatti, P.A., Ferrari, E.: TRBAC: A Temporal Role-based Access Control Model. ACM Transactions on Information and System Security 4(3), 191–233 (2001)
2. Joshi, J.B.D., Bertino, E., Ghafoor, A.: Temporal Hierarchies and Inheritance Semantics for GTRBAC. In: Proceedings of the Seventh ACM Symposium on Access Control Models and Technologies, pp. 74–83 (2002)
3. Bertino, E., Catania, B., Damiani, M.L., Perlasca, P.: Geo-rbac: A spatially aware rbac. In: Proceedings of the Tenth ACM Symposium on Access Control Models and Technologies, pp. 29–37. ACM (2005)
4. Damiani, M.L., Bertino, E., Catania, B., Perlasca, P.: Geo-rbac: A Spatially Aware RBAC. ACM Transactions on Information and System Security 10(1) (2007)
5. Ray, I., Kumar, M., Yu, L.: LRBAC: A Location-aware Role-based Access Control Model. In: Bagchi, A., Atluri, V. (eds.) ICISS 2006. LNCS, vol. 4332, pp. 147–161. Springer, Heidelberg (2006)
6. Bauer, L., Cranor, L.F., Reiter, M.K., Vaniea, K.: Lessons Learned from the Deployment of a Smartphone-based Access-Control System. In: Proceedings of the 3rd Symposium on Usable Privacy and Security, pp. 64–75 (2007)
7. Aich, S., Sural, S., Majumdar, A.: STARBAC: Spatiotemporal Role Based Access Control. In: Meersman, R. (ed.) OTM 2007, Part II. LNCS, vol. 4804, pp. 1567–1582. Springer, Heidelberg (2007)
8. Cooper, G., Herskovits, E.: A Bayesian Method for the Induction of Proba-bilistic Networks from Data. Machine Learning 9, 309–347 (1992)
9. Xu, C., Li, S., Liu, G., Zhang, Y.: Crowd++: Unsupervised Speaker Count with Smartphones. In: Proceedings of the 2013 ACM International Joint Conference on Pervasive and Ubiquitous Computing, pp. 43–52 (2013)
10. Cheveigne, A.D., Kawahara, H.: YIN, a Fundamental Frequency Estimator for Speech and Music. The Journal of the Acoustical Society of America 111(4), 1917–1930 (2002)
11. Ferraiolo, D.F., Sandhu, R., Gavrila, S.: Proposed NIST Standard for Role-based Access Control. ACM Transactions on Information and System Security 4(3), 224–274 (2001)
12. Platt, J.: Sequetial minimal optimization: A Fast Algorithm for Training Support Vector Machines, Technical Report MST-TR-98-14, Microsoft Research (1998)
13. Langley, P., Iba, W., Thompson, K.: An Analysis of Bayesian Classifiers. In: The Tenth National Conference on Artificial Intelligence, pp. 223–228. AAAI Press and MIT Press (1992)

AttRel: An Approach to Person Re-Identification by Exploiting Attribute Relationships

Ngoc-Bao Nguyen, Vu-Hoang Nguyen, Thanh Ngo Duc, Duy-Dinh Le,
and Duc Anh Duong

Multimedia Communications Laboratory
University of Information Technology, VNU-HCM
Ho Chi Minh City, Vietnam
10520228@gm.uit.edu.vn, {vunh,thanhnd,ledduy,ducda}@uit.edu.vn

Abstract. Person Re-Identification refers to recognizing people across cameras with non-overlapping capture areas. To recognize people, their images must be represented by feature vectors for matching. Recent state-of-the-art approaches employ semantic features, also known as attributes (e.g. wearing-bags, jeans, skirt), for presentation. However, such presentations are sensitive to attribute detection results which can be irrelevant due to noise. In this paper, we propose an approach to exploit relationships between attributes for refining attribute detection results. Experimental results on benchmark datasets (VIPeR and PRID) demonstrate the effectiveness of our proposed approach.

Keywords: Person Re-Identification, Attribute Relationships, Re-Score, Learning Relationships.

1 Introduction

Surveillance camera systems are very widespread in public places such as banks, supermarkets, museums, etc, for the purpose of security. Person re-identification is the problem of matching persons across cameras which helps ease job of operators in camera systems. Because of the quality of surveillance cameras, we are challenged by the low resolution of human images. Furthermore, different cameras in different places can lead to change in viewpoints, illumination, and qualities of images. Finally, cluttered background and occlusion are also challenges in person re-identification.

Visual feature extraction plays an important role in a person re-identification system. Beside low level features, attributes are used recently and demonstrate good performance. Attributes are concepts which are understandable to human such as hair-style, clothing-style, etc.

Because attributes are used to represent images, they affect directly to the accuracy of a person re-identification system. In existing works, different attributes

X. He et al. (Eds.): MMM 2015, Part II, LNCS 8936, pp. 50–60, 2015.

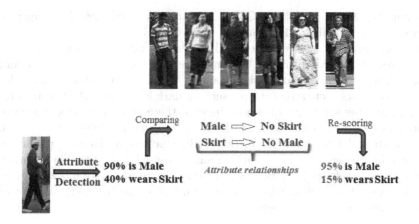

Fig. 1. Using attribute relationships to re-score attribute detection results

are extracted independently. However, there may exist relationships between attributes. For instance, a person who is a man does not likely wear a skirt (which is for women).

If such relationships are utilized, we could refine attribute detection results to more accurate results. For example, in the previous example, the confidence for Skirt Attribute detection result can be decreased to get a more reasonable result (see Figure 1). In this paper, we propose a method for using relationships between attributes to refine attribute detection results to promote person re-identification's performance. Our method's effectiveness is shown via experiments on VIPeR and PRID, two of standard datasets for person re-identification. The experimental results demonstrate the improvement in performance when our proposed method is applied.

The remaining of this paper is organized as follows: Sec. 2 will review related work, Sec. 3 will discuss details of our proposed method, in Sec. 4 we describe the experiments on the VIPeR and the PRID datasets, and Sec. 5 is conclusions of the paper.

2   Related Work

Based on the two main components of a person re-identification system: feature extraction and evaluating similarity, researchers focus on the two corresponding approaches. In [1], the authors use histograms of HSV color systems weighted by the distance of pixels to the asymmetric axis of persons. In [2, 3], the authors histogram of color from three systems RGB, YCbCr, and HSV. Texture features are used in [2] when Gabor filters and Schmid filters are applied to convolve with images in luminance channel. In [4], Schwartz and Davis use co-occurrence matrices for texture feature extraction. In addition, Biologically Inspired Feature is applied recently in person re-identification but shows outperformed accuracy.

In particular, Ma et al. [5] use Gabor filter for low-level biologically inspired features (BIF) and BiCov descriptor to vectorize BIF.

Semantic features have emerged and applied to person re-identification recently. Using concept perceivable by human beings to describe images, semantic features help boost performance significantly. [6, 7] use texture feature and SVM classifiers to detect semantic features such as dark hair, green shirt, male, etc.

For evaluating similarities between images, there are many works proposing different ways to compute distances between images. In addition to fixed distances such as L1 distance, L2 distance, Bhattacharyya distance, researchers also propose various ways of computing dataset-based distances. This approach engage attentions from [2–4]. Besides, learning the transformation between cameras is proposed in [8, 9] in which the assumption is that the cameras and their environments must be unchanged.

3   AttRel: Exploiting Attribute Relationships

When a set of attributes $\{A_1, A_2, \ldots, A_n\}$ is chosen as features for representing an image, the feature vector is then $\vec{f} = (s(A_1), s(A_2), \ldots, (A_n))$ where $s(A_i)$ is score which represents the confidence of A_i on the image.

Existing works extract each attribute independently to each others. In this paper, we propose a technique (illustrated by Figure 1) for utilizing the information by relationships between attributes to improve the accuracy of attribute detection and person re-identification performance.

This approach includes three main points: (i) How to define an attribute relationship (ii) How to learn useful relationships between attributes and (iii) How to use attribute relationships. We will show how to solve all three problems above in this paper.

3.1   Attribute Relationship

In this section, we define attribute relationships which are used to present co-occurrence or non-concurrence of attribute pairs.

Let A $= \{A_1, A_2, \ldots, A_n\}$ be the set of n attributes used to represent a person, (A_i, st_{A_i}) be an item which comprises an attribute and its status in an image ($st_{A_i} = 0$ means attribute disappears and $st_{A_i} = 1$ means attribute appears). For convenience, in this paper, we also write attribute name for $(A_i, 1)$ and No +attribute name for $(A_i, 0)$. An image then can be represented by a set of items $P = \{(A_1, st_{A_1}), (A_2, st_{A_2}), \ldots, (A_n, st_{A_n})\}$

An attribute relationship exists in a dataset D is a constraint between two sets of items X and Y $(X \longrightarrow Y)$ which means that

For an arbitrary image in D, if X happens then Y likely happens.

More specifically, for an attribute representation P of an image in D, if X $\subset$ P then P is likely $\supset$ Y.

For instance, in Vietnam, the skirt is usually seen as typical for females only. This means in Vietnam, there should be relationships between Male attribute and Skirt attribute:

$$(\text{Male},1) \longrightarrow (\text{Skirt},0) \text{ or Male} \longrightarrow \text{No Skirt}$$
$$(\text{Skirt},1) \longrightarrow (\text{Male},0) \text{ or Skirt} \longrightarrow \text{No Male}$$

An attribute representation P is called **contradicted** to a relationship $X \longrightarrow Y$ when $X \subset P$ and $P \not\supset Y$.

3.2   Learning Attribute Relationships

In this section, we introduce a technique for mining attribute relationships from a dataset. Given a dataset of persons with their accompanying attributes, our method would learn a set of relationships between attributes. We use Support and Confidence, measurements in data mining techniques, to evaluate qualities and filter candidate relationships. Support represents the frequency of persons from a dataset that contains both X and Y. Confidence assesses the degree of certainty of each candidate relationship.

$$Support(X \longrightarrow Y) = P(X \cup Y) \tag{1}$$

$$Confidence(X \longrightarrow Y) = P(Y|X) \tag{2}$$

Fig. 2. Defining Attribute Relationships

To filter qualified candidate relationships, thresholds of support and confidence should be set. Candidate relationships with support and confidence over the thresholds are considered as attribute relationships and will be used for re-scoring attribute detection results (mentioned in Section 3.3).

The Learning Algorithm
Input:

- $\{P_i$ where P_i contains attributes accompanying with i^{th} person$\}$
- Min_support, Min_confidence

Output: R is the set of learned attribute relationships.

__Algorithm 1__: Learning Attribute Relationships

   __Step 1:__ Build a set I of all possible $item = (A_i, st_{A_i})$ where A_i is an attribute and st_{A_i} is its status.

   __Step 2:__ Build a set CR all candidate relationships X $\longrightarrow$ Y where:

- X and Y are subsets of I
- $X \cap Y = \emptyset$

__Step 3: Filtering__
   R = $\emptyset$
   Foreach r in CR
      supp = Support(r)
      conf = Confidence(r)
      If (supp $\geq$ Min_support) and (conf $\geq$ Min_confidence)
         $R \longleftarrow R \cup \{r\}$
      End
   End
__Step 4:__ Output R

3.3 Using Attribute Relationships

In this subsection, we will show how to use attribute relationships to make attribute detection results more accurate. First, we discuss the terms of using relationships in order to adjust attribute detection results. Second, we define an adjustment function, and finally, we will show an algorithm uses a combination of multiple attribute relationships.

__Terms of Using Relationships:__ Assuming that we have an attribute relationship r: X $\longrightarrow$ Y. Thus, an attribute detection result which contradicts an attribute relationship (mentioned in Section 3.1) is possibly wrong. Therefore we can use r to adjust the attribute detection result.

__The Adjustment Function:__ Let r: X $\longrightarrow$ Y be an attribute relationship; (A_i, st_{A_i}) be an item of r; p be an attribute detection result which contradicts with r. Let $s(A_i)$ be the A_i's score of p before re-scoring and $s'(A_i) = F(s(A_i), st_{A_i})$ be A_i's score of p after re-scoring. There are two possible cases.

   __Case 1:__ $st_{A_i} = 1$ (A_i __appears__) . The attribute detection result should be adjusted toward the relationship r. Therefore $s'(A_i)$ should be higher than $s(A_i)$ and adjusted to be nearer to 1 (score of attribute which appears). Moreover, the higher $s(A_i)$ is, the higher $s'(A_i)$ should be. Specifically:

$$\begin{cases} 1 \geq F(s(A_i), 1) \geq s(A_i) \\ 0 \leq F(s_1(A_i), 1) < F(s_2(A_i), 1) \leq 1 \quad for\ all\ s_1(A_i) < s_2(A_i) \end{cases} \quad (3)$$

Case 2: $st_{A_i} = 0$ (A_i **Disappears.**) In contrast, $s'(A_i)$ should be lower than $s(A_i)$ and adjusted to be nearer to 0 (score of attribute which disappears). The lower $s(A_i)$ is, the lower $s'(A_i)$ should be. Specifically:

$$\begin{cases} 0 \le F(s(A_i), 0) \le s(A_i) \\ 0 \le F(s_1(A_i), 0) < F(s_2(A_i), 0) \le 1 \quad for\ all\ s_1(A_i) < s_2(A_i) \end{cases} \quad (4)$$

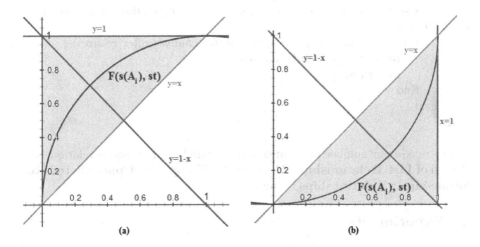

(a)　　　　　　　　　　　　　(b)

Fig. 3. Re-scoring curve. (a) $F(s(A_i), 1)$ is the result of $s(A_i)$ after re-scoring when $st_{A_i} = 1$ and (b) $F(s(A_i), 0)$ is the result of $s(A_i)$ after re-scoring when $st_{A_i} = 0$.

There are many functions $F(s(A_i), st_{A_i})$ satisfying (3) and (4). In this paper, we choose a simple form of $F(s(A_i), st_{A_i})$ for processing efficiency. In particular, our method's function is a part of the circle above (case 1) and below (case 2) the line y = x (Figure 3). Specifically:

$$F(s(A_i), st_{A_i}) = \begin{cases} 1 - m + \sqrt{m^2 + (m-1)^2 - (s(A_i) - m)^2} & if\ st_{A_i} = 1 \\ m - \sqrt{m^2 + (m-1)^2 - (s(A_i) - 1 + m)^2} & if\ st_{A_i} = 0 \end{cases}$$

$$(5)$$

where m $\ge$ 1. m is the impact coefficient. The higher m is, the lower the change of score is and vice-versa. This means $|s'(A_i) - s(A_i)|$ is decreased.

Using Multiple Attribute Relationships: In the case of multiple attribute relationships, the contradicted relationships are ordered by the distances between them and the attribute detection results. The distance implies the probability of changing statuses after re-scoring. The attribute relationships are then applied to re-score in that order. The detail of our algorithm is described below:

Algorithm 2. Re-scoring attribute detection results
Create **List_AttributeRelationship**.
For each person **p**
 Relationships = **List_AttributeRelationship**.
 Step 1: Search **Contradicted_Relationships** of **p** in **Relationships**.
 Step 2: Compute distance between **p** and each status of relationship in
Contradicted_Relationships.
 Step 3: Sorted ascending distance.
 Step 4: For each relationship **r** in **Contradicted_Relationships**
  Using **r** to re-score **p**. *//using adjustment function to re-score each
attribute score*
  If (status changes) *//attribute's status changed after re-scoring*
   Remove **r** in **Relationships**.
   Return **Step 1**.
  End
 End
End

The maximum number of iterations is the number of attribute relationships (length of **List_Relationship**). Algorithm 2 will stop when **Contradicted Relationship** or **Relationships** is empty.

4   Experiments

4.1   Implementation Details

Datasets: We conduct our experiments on VIPeR[10] and PRID[11] datasets. VIPeR contains 1264 images of 632 pedestrians. All images are taken from two non-overlapping viewpoints under varying illumination conditions. The images are normalized to 128x48 pixels. The images of PRID dataset were captured from two outdoor cameras. We use 400 shots of the first 200 person from each view of the single-shot version to carry out experiments. The images are scaled to 128x64 pixels.

For experiments, we bisect each dataset. The first part, called training set,is used for training classifiers, the second part, called test set, is utilized for person re-identification in which camera B is the gallery set and camera A is the probe set. The experiments are carried out by matching each test image in probe set against the gallery set.

Attribute: We use attributes and annotation provided in [12] (include *Redshirt, Blueshirt, Lightshirt, Darkshirt, Greenshirt, Nocoats, Notlightdarkjeans colour, Darkbottoms, Lightbottoms, Hassatchel, Barelegs, Shorts, Jeans, Male, Skirt, Patterned, Midhair, Darkhair, Bald, Hashandbag carrierbag, Hasbackpack*). To detect attributes for image, visual features of images are extracted. We then use classifiers as detectors to detect attributes based on the extracted features. Specifically:

- Feature: We follow the method in [3] to extract feature. Each image is divided into 6 equal horizontal parts. Each part is extracted 8 color channels (RGB, HS, and YCbCr) and 21 texture filters (Gabor and Schmid) from the bright channel.
- Classifier: We use SVM classifier with RBF kernel to detect attributes. The output is changed to scores by Sigmoid function.

Table 1. Learned attribute relationships for VIPeR and PRID

VIPeR	PRID
1. *Darkbottoms* $\longrightarrow$ *No Notlightdarkjeanscolour*	1. *Lightshirt* $\longrightarrow$ *No Darkshirt*
2. *Darkbottoms* $\longrightarrow$ *No Lightbottoms*	2. *Darkshirt* $\longrightarrow$ *No Lightshirt*
3. *Male* $\longrightarrow$ *No Skirt*	3. *Darkbottoms* $\longrightarrow$ *No Lightbottoms*
4. *Darkhair* $\longrightarrow$ *No Midhair*	4. *Lightbottoms* $\longrightarrow$ *No Darkbottoms*
5. *Darkhair* $\longrightarrow$ *No Bald*	5. *Jeans* $\longrightarrow$ *No Barelegs*
	6. *Jeans* $\longrightarrow$ *No Skirt*
	7. *Jeans* $\longrightarrow$ *No Bald*
	8. *Male* $\longrightarrow$ *No Skirt*
	9. *Darkhair* $\longrightarrow$ *No Midhair*
	10. *Darkhair* $\longrightarrow$ *No Bald*

Attribute Relationships: For experiments on VIPeR dataset, we combine PRID and labels provided for GRID dataset, a human annotation of attribute labels [12] and the training set of VIPeR to form the training dataset for learning attribute relationships. Similarly, VIPeR, GRID, and the training set of PRID will be the training set for learning attribute relationships in experiments on PRID. The learned attribute relationships are shown in Table 1 with the confidence and support threshold for VIPeR and PRID are (0.985, 0.48) and (0.99, 0.38) respectively.

Evaluation Metrics: To compare performances of various methods, similar to [1] and [12] we use Rank i, normalized area under the CMC curve (nAUC), and Expected Rank (ER). Rank i is the percentage of people in the probe set whom the groundtruth was found in first i ranked people in the gallery set (The higher value of Rank i implies better performance). CMC measures the presentation of all Rank i. ER is the mean rank of true matches (The best ER is 1 and the higher ER implies worse performance).

4.2 Results and Discussions

Attribute Detection Results: Accuracy of attribute detection in VIPeR and PRID is shown in Table 2. PRID has four attributes without enough data for training classifier (Greenshirt, Notlightdarkjeanscolour, Shorts, Patterned).

Table 2. Attribute detection result

	VIPeR	PRID		VIPeR	PRID
Redshirt	53.15	23.5	Shorts	51.74	-
Blueshirt	53.31	54.4	Jeans	65.14	41
Lightshirt	80.60	70	Male	56.15	46.5
Darkshirt	82.17	82	Skirt	61.99	36.5
Greenshirt	37.54	-	Patterned	35.96	-
Nocoats	65.46	68.5	Midhair	60.88	59
Notlightdarkjeanscolour	47.48	-	Darkhair	62.61	74.5
Darkbottoms	68.14	73.5	Bald	73.82	37.5
Lightbottoms	72.24	58.5	Hashandbagcarrierbag	35.49	48.5
Hassatchel	51.42	46	Hasbackpack	46.85	36
Barelegs	58.36	50	**Mean**	**58.12**	**53.29**

Person Re-identification Result: The re-identification performance is summarized in Table 3 and Table 4. There are two results in each dataset. The first is the result before applying our proposed method (Raw attr). The second is the result after applying learned attribute relationship set to re-score attribute score (AttRel).

Table 3. Person re-identification result in VIPeR dataset

	Raw attr	AttRel
Mean attribute detection	58.12	60.16 (**+2.04%**)
Rank 5	12.30	12.30 (+0%)
Rank 10	23.02	23.97(**+0.95%**)
Rank 25	38.81	38.81 (+0%)
ER	61.96	61.20 (**-0.76 rank**)
nAUC	0.8061	0.8085(**+0.24%**)

From the result, we can see that, the person re-identification is improved when our method is applied. In VIPeR dataset, with AttRel, attribute detection results enhance 2.04% and lead to the improvement at 0.76 rank in expected rank. In PRID, the enhancement of attribute detection accuracy is boosted up to 4.29% and expected rank is 1.32 higher after AttRel is applied. This shows the efficiency of our proposed method.

Besides, we also use sets of attribute relationships which are carefully selected by human being to carry out experiments (Table 5). The results (Table 6 and Table 7) show that person re-identification performances are boosted considerably. This demonstrates the potential of utilizing attribute relationships in person re-identification. The gap between performances in learned attribute relationships and manually defined attribute relationships is because of the limitation of training datasets. Accordingly, learning attribute relationships carefully from diverse datasets is an promising future direction.

Table 4. Person re-identification result in PRID dataset

	Raw attr	AttRel
Mean attribute detection	53.29	57.58 (**+4.29%**)
Rank 5	29	29 (+0%)
Rank 10	42	42 (+0%)
Rank 25	64	65(**+1%**)
ER	23.86	22.54(**-1.32 rank**)
nAUC	0.7659	0.7792(**+1.33%**)

Table 5. Manually defined attribute relationships for VIPeR and PRID

VIPeR	PRID
1. *Blueshirt* ⟶ *No Redshirt*	1. *Midhair* ⟶ *No Bald*
2. *Shorts* ⟶ *Barelegs*	2. *Lightbottoms* ⟶ *No Darkbottoms*
3. *Skirt* ⟶ *Barelegs*	3. *Skirt* ⟶ *No Jeans*
4. *Greenshirt* ⟶ *No Redshirt*	4. *Skirt* ⟶ *No Male*
5. *Shorts* ⟶ *No Skirt*	5. *Blueshirt* ⟶ *No Redshirt*
6. *Notlightdarkjeanscolour* ⟶ *No Jeans*	6. *Lightshirt* ⟶ *No Darkshirt*
	7. *Darkshirt* ⟶ *No Lightshirt*

Table 6. Person re-identification result in VIPeR dataset

	Raw attr	AttRel
Mean attribute detection	58.12	62.52 (**+4.4%**)
Rank 5	12.30	14.51(**+2.21%**)
Rank 10	23.02	23.34(**+0.32%**)
Rank 25	38.81	40.70(**+1.89%**)
ER	61.96	58.71(**-3.25 rank**)
nAUC	0.8061	0.8163(**+1.02%**)

Table 7. Person re-identification result in PRID dataset

	Raw attr	AttRel
Mean attribute detection	53.29	59.94 (**+6.65%**)
Rank 5	29	32 (**+3%**)
Rank 10	42	47 (**+5%**)
Rank 25	64	67 (**+3%**)
ER	23.86	20.1 (**-3.76 rank**)
nAUC	0.7659	0.8034 (**+3.75%**)

5 Conclusions

In this paper, we proposed AttRel, a method for utilizing the information of relationship between attributes in the problem of person re-identification. More specifically, AttRel contains a procedure for learning relationships between attributes and a re-scoring method for adjusting attribute detection results based on the learned attribute relationships. The improved performances in datasets VIPeR and PRID prove the benefit when AttRel is applied. Exploiting larger and more diverse datasets is an open and potential direction for the future.

Acknowledgement. This research is funded by Vietnam National University HoChiMinh City (VNU-HCM) under grant number B2013-26-01.

References

1. Farenzena, M., Bazzani, L., Perina, A., Murino, V., Cristani, M.: Person re-identification by symmetry-driven accumulation of local features. In: 2010 IEEE Conference on omputer Vision and Pattern Recognition (CVPR), pp. 2360–2367. IEEE (2010)
2. Gray, D., Tao, H.: Viewpoint invariant pedestrian recognition with an ensemble of localized features. In: Forsyth, D., Torr, P., Zisserman, A. (eds.) ECCV 2008, Part I. LNCS, vol. 5302, pp. 262–275. Springer, Heidelberg (2008)
3. Prosser, B., Zheng, W.S., Gong, S., Xiang, T., Mary, Q.: Person re-identification by support vector ranking. In: BMVC, vol. 1, p. 5 (2010)
4. Schwartz, W.R., Davis, L.S.: Learning discriminative appearance-based models using partial least squares. In: 2009 XXII Brazilian Symposium on omputer Graphics and Image Processing (SIBGRAPI), pp. 322–329. IEEE (2009)
5. Ma, B., Su, Y., Jurie, F., et al.: Bicov: A novel image representation for person re-identification and face verification. In: British Machive Vision Conference (2012)
6. Layne, R., Hospedales, T.M., Gong, S., et al.: Person re-identification by attributes. In: BMVC, vol. 2, p. 3 (2012)
7. Liu, C., Gong, S., Loy, C.C., Lin, X.: Person re-identification: What features are important? In: Fusiello, A., Murino, V., Cucchiara, R. (eds.) ECCV 2012 Ws/Demos, Part I. LNCS, vol. 7583, pp. 391–401. Springer, Heidelberg (2012)
8. Avraham, T., Gurvich, I., Lindenbaum, M., Markovitch, S.: Learning implicit transfer for person re-identification. In: Fusiello, A., Murino, V., Cucchiara, R. (eds.) ECCV 2012 Ws/Demos, Part I. LNCS, vol. 7583, pp. 381–390. Springer, Heidelberg (2012)
9. Brand, Y., Avraham, T., Lindenbaum, M.: Transitive re-identification. In: Proceedings of the British Machine Vision Conference. BMVA Press (2013)
10. Gray, D., Brennan, S., Tao, H.: Evaluating appearance models for recognition, reacquisition, and tracking. In: IEEE International Workshop on Performance Evaluation of Tracking and Surveillance. Citeseer (2007)
11. Hirzer, M., Beleznai, C., Roth, P.M., Bischof, H.: Person re-identification by descriptive and discriminative classification. In: Heyden, A., Kahl, F. (eds.) SCIA 2011. LNCS, vol. 6688, pp. 91–102. Springer, Heidelberg (2011)
12. Layne, R., Hospedales, T.M., Gong, S.: Attributes-based re-identification. In: Person Re-Identification, pp. 93–117. Springer, Heidelberg (2014)

Sparsity-Based Occlusion Handling Method for Person Re-identification

Bingyue Huang[1], Jun Chen[1,2], Yimin Wang[1], Chao Liang[1,2], Zheng Wang[1], and Kaimin Sun[3]

[1] National Engineering Research Center for Multimedia Software,
School of Computer, Wuhan University, Wuhan, 430072, China
[2] Research Institute of Wuhan University in Shenzhen, China
[3] State Key Laboratory of Information Engineering in Surveying, Mapping, and Remote Sensing, Wuhan University, China
byHuang@whu.edu.cn

Abstract. Person re-identification has recently attracted a lot of research interests, it refers to recognizing people across non-overlapping surveillance cameras. However, person re-identification is essentially a very challenging task due to variations in illumination, viewpoints and occlusions. Existing methods address these difficulties through designing robust feature representation or learning proper distance metric. Although these methods have achieved satisfactory performance in the case of illumination and viewpoint changes, seldom of they can genuinely handle the occlusion problem that frequently happens in the real scene. This paper proposes a sparsity-based patch matching method to handle the occlusion problem in the person re-identification. Its core idea is using a sparse representation model to determine the occlusion state of each image patch, which is further utilized to adjust the weight of patch pairs in the feature matching process. Extensive comparative experiments conducted on two widely used datasets have shown the effectiveness of the proposed method.

Keywords: Person re-identification, non-overlapping camera tracking, occlusion problem, sparse representation.

1 Introduction

Person re-identification, i.e., matching persons across non-overlapping camera views, is gradually becoming a hot research spot in the computer vision and multimedia fields [1–5]. Given a query pedestrian image, the algorithm searches for images of the same person from different cameras, and then ranks all the images based on their visually similarities to the query pedestrian image. However, although much progress has been made in recent years [6–11], person re-identification is still a very challenging task due to difficulties caused by variations in illumination, viewpoints and occlusions (see Fig. 1). These interference factors make appearance difference of the same person under two cameras even

X. He et al. (Eds.): MMM 2015, Part II, LNCS 8936, pp. 61–73, 2015.
© Springer International Publishing Switzerland 2015

(1)   (2)   (3)   (4)   (5)   (6)   (7)   (8)   (9)   (10)

Fig. 1. Examples of ten human image pairs from the VIPeR dataset [30] and the ETHZ dataset [31]. Each column has two images of the same person taken from two different camera views. People suffer significant appearance variations, one of the most problems is occlusion, people may be partly occluded by themselves, other persons or other objects.

larger than that of different persons, and hence lead to erroneous matching results.

A lot of research effort has been devoted to address the above difficulties. Existing approaches on person re-identification generally fall into two categories [2]: the feature based methods [12–17] and the distance learning based methods [6, 7, 18–22].

The former, i.e., feature based methods, focuses on designing a discriminative and robust feature representation for recognizing different people in various camera views [12–17]. Farenzena et al. [12] exploited symmetry and asymmetry perceptual principles to segment a human image into five regions, and then extracted multiple color features and texture features to represent the person's appearance. Gray et al. [13] combined many different kinds of simple features together, such as color, texture and so on. Ma et al. [14] proposed a BiCov method by combining biologically inspired features with covariance descriptors. Cheng et al. [16] utilized the Pictorial Structures to estimate human body configuration and also computed visual features based on different human body parts to cope with pose variations. Liu et al. [17] learned a bottom-up feature importance to adaptively weight features of different individuals rather than using global weights. Nevertheless, constructing and selecting a set of features that are both distinctive and stable is extremely difficult in itself, let alone under conditions where viewpoint changes usually cause significant appearance variations.

The latter kind, i.e., distance learning based methods, searches for the optimal distance metric under which images of the same person are more similar, while images of different persons are more dissimilar [6, 7, 18–22]. Weinberger et al. [18] proposed a large margin nearest neighbor (LMNN) approach to solve k-nearest neighbor classification problem by gathering instances of the same label together and separating instances of different label apart. Kostinger et al. [6] introduced a KISSME method, applied a simple but effective strategy to learn a distance metric from equivalence constraints, based on a statistical inference perspective. Mignon et al. [7] introduced pairwise constrained component analysis (PCCA) for learning distance metric from sparse pairwise similarity/dissimilarity constraints in high dimensional input space. Li et al. [20] developed a locally-adaptive decision function (LADF) that jointly models a distance metric and a locally adaptive thresholding rule, and achieved good results. Liu et al. [21] presented a man-in-the-loop method to allow user quickly refine ranking performance, and achieved significant improvement when comparing with other metric learning methods. However, most of these methods simply match images by directly computing the feature distance with learned metric, seldom consider the variation of different viewpoints, causing misalignment problems between images. Due to the fact that viewpoint variations are quite common in non-overlapping surveillance cameras (see Fig. 1) and they usually lead to large matching errors, solving the misalignment problem caused by large viewpoint variations is of great importance.

In recent years, Zhao et al. [10] considered the misalignment problem by adopting an adjacency constrained patch matching method to build dense correspondence between image pairs. They represented each image with overlapping patches, and for each patch of one image, they found the most similar patch from another image and measured the distance between them, then the summation of distances of all patch pairs was defined as the distance of the image pair. This method showed effectiveness in finding misaligned patches, but ignored the occlusion situation. However, in practice applications, occlusion frequently happens across non-overlapping camera views. Take Fig. 1 as an example, the red backpack of (1) in the second row can not be seen in the first row because of self-occlusion; some parts of the person of (6) in the first row can not be seen in the second row because of occluded by some objects; and the person of (7) in the second row is partly occluded by another person comparing to that in the first row. In this condition, people may be partly occluded by themselves, by other persons or other objects, resulting in some patches of people can be seen in one view while can not be seen in another view. Therefore, the occluded patch cannot find very similar patches in another image and thus naturally leading to the matching error.

In this paper, we propose a sparsity-based occlusion handling (SOH) method for person re-identification. Through assigning each image patch an occlusion value, which indicates the occlusion degree of the patch, we adjust the weight of patch pairs in the feature matching process. More precisely, given a probe image P and a gallery image Q, we first divide two images into a grid of local patches by overlapped sliding windows. Then, for each probe patch P_{ij} of image P, a sparse representation is computed based on a dictionary composed of all gallery patches of image Q. Intuitively, the smaller the representation error is, the more similar the probe patch and gallery patches are. Hence, we use the representation error as an indication to measure the occlusion degree of the probe image in the pair-wise matching process. Finally, an occlusion-weighted patch matching method is adopted to generate more robust person re-identification result. Experiments conducted on two challenging datasets show the effectiveness and robust of the proposed method.

2 Approach

In this section, we present our sparsity-based occlusion handling (SOH) approach for person re-identification. First, the patch-based matching method [10] is introduced. Then, we make a particular description of our sparsity-based occlusion handling (SOH) method, which is the core work of this paper. Finally, we show how to apply the SOH method to the patch-based matching method.

2.1 Patch-Based Matching Method

For easy to discuss, we consider to measure the distance of two images denoted as P and Q. In order to solve viewpoint variations and misalignment problems, a patch-based matching method is adopted. A brief introduction is given as follows.

First, both images are represented as a set of patches. Each patch is extracted features and all the features of all patches are consisted together to denote the image. Specifically, image P is divided into a grid of local patches by overlapped sliding windows, assuming that image P is divided into n rows and m columns, the image can be represented as:

$$P = \{P_{ij} | i = 1, 2, \ldots, n; j = 1, 2, \ldots, m\}, \tag{1}$$

where P_{ij} denotes the features of the patch of image P centered at the i-th row and the j-th column. In the same way, image Q is also represented as:

$$Q = \{Q_{ij} | i = 1, 2, \ldots, n; j = 1, 2, \ldots, m\}. \tag{2}$$

Then, for a patch P_{ab} of P, we search the most similar patch in Q.

As discussed in section 1, viewpoint variations generally happen in horizontal direction. Searching the most similar patch in all patches of Q is unreasonable and time-consuming. Therefore, a horizontal constraint is exploited. As shown in Fig. 2, in form, the search set $S_{P_{ab}}^Q$ is defined as:

$$S_{P_{ab}}^Q = \{Q_{ij} | i = a - l, \ldots, a, \ldots, a + l; j = 1, 2, \ldots, m\}, \tag{3}$$

where l is a constraint parameter to control the size of the search set. $l = 1$ is chosen in our experiments.

The most similar patch of P_{ab} in the search set $S_{P_{ab}}^Q$ is denoted as $Q_{P_{ab}}$ and represented by:

$$Q_{P_{ab}} = \underset{Q_{ij} \in S_{P_{ab}}^Q}{\arg\min} \, d(P_{ab}, Q_{ij}), \tag{4}$$

where $d(P_{ab}, Q_{ij})$ represents the Euclidean distance between P_{ab} and Q_{ij}.

And the distance between image P and Q is denoted as:

$$d(P, Q) = \sum_{m,n} d(P_{ab}, Q_{P_{ab}}). \tag{5}$$

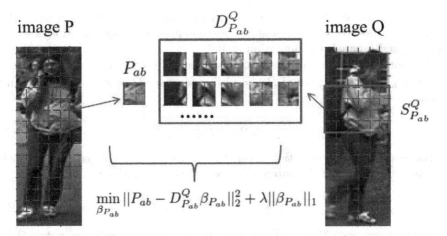

Fig. 2. Diagram of our sparsity-based occlusion handling scheme. Patch P_{ab} of image P is matched with image Q by the sparsity-based representation model. The dictionary $D^Q_{P_{ab}}$ of P_{ab} is obtained from Q.

2.2   Sparsity-Based Occlusion Handling

It is easy to see that the patch-based matching method [10] does not consider the occlusion problem, and therefore causing matching errors. In order to handle the occlusion problem, motivated by the widely successful applications of sparse representation in many tasks [23–26] and the effectiveness of it in handling occlusions in object tracking [26, 27], we apply the sparse representation to person re-identification. As the sparse representation model in object tracking can not handle misalignment problem, we design a new sparsity-based occlusion handing method for person re-identification.

As is known to all, if a collection of representative samples is found for one sample such as audio and image, then this sample can be sparsely represented by these representative ones [25]. In the same way, in our work, each image patch is sparsely represented by its neighboring patches. If the patch is occluded, it can no longer find its true neighboring patches and thus causing large representation error. Based on the above, the representation error can be used to evaluate the occlusion state of each image patch [25, 26]. We handle each patch of an image by the sparsity-based representation model, and the diagram of our sparsity-based occlusion handling (SOH) method is shown in Fig. 2. More details are as follows.

Considering patch P_{ab} of image P, the sparsity-based representation model is denoted as:

$$\min_{\beta_{P_{ab}}} ||P_{ab} - D^Q_{P_{ab}}\beta_{P_{ab}}||^2_2 + \lambda||\beta_{P_{ab}}||_1, \tag{6}$$

where P_{ab} is the feature vector of the patch of image P centered at the a-th row and the b-th column. $D^Q_{P_{ab}}$ is the dictionary of P_{ab} which is made up of several special selected patches from image Q (see Fig. 2), and can be denoted as $D^Q_{P_{ab}} = \{Q_{ij}|i = a-l, \ldots, a, \ldots, a+l; j = 1, 2, \ldots, m\}$, l is a constraint parameter in horizontal direction and $l = 1$ is chosen in our setting. $\beta_{P_{ab}}$ represents the sparse coefficient vector of P_{ab} and can be computed by the optimization function of the sparsity-based representation model.

As done in [26], the representation error can be used to check the occlusion. So the representation error of P_{ab} according to image Q is denoted as $E^Q_{P_{ab}}$ and defined as:

$$E^Q_{P_{ab}} = ||P_{ab} - D^Q_{P_{ab}}\beta_{P_{ab}}||^2_2. \tag{7}$$

Based on the representation error, the occlusion value of patch P_{ab} which indicates the degree of the patch being occluded is defined as:

$$P(P_{ab}, Q) = exp((E^Q_{P_{ab}})^2). \tag{8}$$

In the same way, $P(Q_{P_{ab}}, P)$ represents the occlusion value of patch $Q_{P_{ab}}$ according to image P.

From the above equation, we can see that the occlusion value becomes small when the representation error of the patch is getting smaller. In this way, we handle each patch of image P and get all the occlusion values of image P.

2.3 Occlusion-Weighted Patch Matching

In this subsection, the occlusion value is applied to the patch-based matching method [10] to measure the similarity between image P and Q.

For easy to apply the occlusion value into the patch-based matching method, we first transform the distance to similarity by a Gaussian function as follows:

$$sim(P_{ab}, Q_{P_{ab}}) = exp(\frac{-d(P_{ab}, Q_{P_{ab}})^2}{2\sigma^2}), \tag{9}$$

where σ is the bandwidth of the Gaussian function.

Then, considering image P and Q, we design a bidirectional matching function to compute the similarity score. The similarity score is defined by:

$$sim(P, Q) = \sum_{m,n} \frac{(1 - P(P_{ab}, Q)) \cdot sim(P_{ab}, Q_{P_{ab}}) \cdot (1 - P(Q_{P_{ab}}, P))}{|P(P_{ab}, Q) - P(Q_{P_{ab}}, P)| + \alpha}, \tag{10}$$

where α is a balance factor for controlling the correction. $sim(P_{ab}, Q_{P_{ab}})$ is the similarity score composed by the similarity scores of patch-based matching method and salience learning in [10].

Based on the above equation, we can see that patches more likely being occluded will count less than patches with low occlusion value. Therefore, the similarity between image P and Q will be more reliable. In addition, we combine the occlusion value of image P according to image Q with the occlusion value of image Q according to image P to obtain a more reliable similarity score of image P and Q.

3   Experiments

In order to evaluate the performance of our work, we conduct experiments on two publicly available datasets, the VIPeR dataset [30] and the ETHZ dataset [31]. The reasons why we select these two datasets are that: 1) these two datasets are public available and the most widely used datasets for person re-identification; 2) most of images in both datasets have occlusion problems.

3.1   Datasets and Experimental Settings

VIPeR Dataset [30]. The VIPeR dataset is collected by Gray et al. [30] from two cameras in outdoor academic environment. It contains 1,264 images of 632 pedestrian pairs. Each pedestrian pair consists of two images of the same pedestrian taken from two different cameras, under different viewpoints, poses and light conditions. Most of the pedestrian pairs contain a viewpoint change larger than 90 degrees. All images are normalized to 128×48 for experiments.

ETHZ Dataset [31]. The ETHZ dataset [31] is collected by Schwartz, et al. in [29] from moving cameras. The dataset is made up of three video sequences, i.e., SEQ. #1 contains 83 pedestrians, for a total of 4,857 images; SEQ. #2 contains 35 pedestrians, for a total of 1,936 images; SEQ. #3 contains 28 pedestrians, for a total of 1,762 images. Images in this dataset are suffered from a wide range of variations in human appearance and illumination because of captured from moving cameras in uncontrolled conditions. One of the most challenging aspects is occlusion. All images are normalized to 64×32 for experiments.

Experimental Settings. In order to make a fair comparison with [10], the same procedure as described in [10] is exploited. To be specific, first, each image is divided into patches of size 10×10 pixels with a grid step size 4. And each image patch is represented by a feature descriptor vector. Dense color histograms feature and dense SIFT feature are extracted and concatenated as a vector with length 672. For detecting the occlusion patches, we simply adopt the gray-scale pixel feature, i.e, concatenating the gray pixel to a vector with length 100. Then, the test images are divided into a probe set and a gallery set. Each image in the probe set is matched with all images in the gallery set, and the rank of the real match is recorded. The parameters in our experiments are set the same as [10]. Specifically, the searching set size parameter $l = 1$, the sparse representation model parameter $\lambda = 0.01$, and the similarity parameter $\sigma = 2.8$ for VIPeR dataset and $\sigma = 1.6$ for ETHZ dataset. We repeat evaluation for 10 times, and take the average as the final result to get a stable statistics.

All the experiment results are evaluated by the Cumulated Matching Characteristics (CMC) curve [15] as well as many papers for person re-identification [1, 2, 6, 7, 9, 12]. The Cumulative Matching Characteristic (CMC) curve represents the percentage of finding the true matching in the top k ranks.

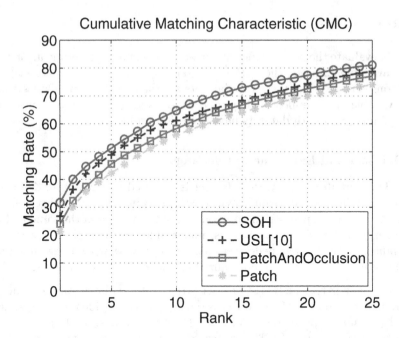

Fig. 3. Comparing results on VIPeR dataset

Table 1. Comparing of used components in each method

Method	used components
SOH	patch + salience + occlusion
USL [10]	patch + salience
PatchAndOcclusion	patch + occlusion
Patch	patch

3.2   Results and Analysis

We conduct experiments mainly on two aspects: first, we validate the effectiveness of our work for handling the occlusion problem in person re-identification; second, we show comparing results of our method with several state-of-the-art approaches.

We first demonstrate the effectiveness of our work for occlusion handling. In the VIPeR dataset, we randomly select half of the dataset, i.e., 316 image pairs, for testing [10, 12]. Images from one camera view are used as probe set and those from the other camera view as gallery set. The matching result of our SOH method is compared with results of other three methods, i.e., the result of the Unsupervised Salience Learning (USL) [10], the result of using both the patch-based matching method (see Section 2.1) and our sparsity-based occlusion

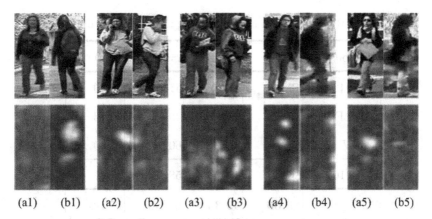

(a1)   (b1)   (a2)   (b2)   (a3)   (b3)   (a4)   (b4)   (a5)   (b5)

Fig. 4. Examples of five image pairs and their occlusion maps. Each person has two images taken from two different views. In the occlusion map, brighter region means higher chances the region being occluded.

handling method (see Section 2.2), and the result of only using the patch-based matching method. All the comparison results are shown in Fig. 3. The used components in each method are also compared in Table 1. It is obvious that our SOH method has better performance than Unsupervised Salience Learning (USL) [10], and the matching result of using both the patch-based matching method and the sparsity-based occlusion handling method outperforms that of only using the patch-based matching method. The results demonstrate the effectiveness of our work and indicate that incorporating the occlusion handling scheme is the main contribution of the improvement even without the salience learning in [10]. The reason of the good performance is that we exploit a sparisity-based occlusion handling model to determine the occlusion state of each image patch, which is further utilized to adjust the weight of patch pairs in the feature matching process and thus reducing the noise introduced by occlusion patches. As Fig.4 shows, brighter region means higher chances the region being occluded, so they are given smaller weight in the feature matching process.

We also compare the performance of our method against state-of-the-art approaches, as our method does not use a learned metric at measuring the similarity of a pair images, we only compare it with some representative feature based methods. The experiments are performed on the VIPeR dataset and the ETHZ dataset.

In the VIPeR dataset, following [10, 12], we randomly select half of the dataset, i.e., 316 image pairs, for testing. Images from one camera view are used as probe set and those from the other camera view as gallery set. We compare our SOH method with Ensemble of Localized Features (ELF) [13], Symmetry-Driven Accumulation of Local Features (SDALF) [12], Local Descriptors encoded by Fisher Vectors (LDFV) [28] and Unsupervised Salience Learning (USL) [10]. For a fair comparison, we directly use the results published on their original paper, and the same testing set in these approaches are used in our experiments.

Table 2. VIPeR dataset: top ranked matching rates in [%] with 316 persons

Method	r=1	r=5	r=10	r=20
ELF [13]	12.00	31.00	41.00	58.00
SDALF [12]	19.87	38.89	49.37	65.73
LDFV [28]	22.34	47.00	60.04	71.00
USL [10]	26.31	46.61	58.86	72.77
SOH	**30.06**	**49.34**	**62.50**	**77.53**

The comparison results are shown in Table 2. It is obvious that our methods outperforms all comparative methods. Specially, the matching rate at rank 1 of our method is about 30.06%, versus 26.31% for USL, 19.87% for SDALF and 12.0% for ELF, and the matching rate at rank 10 of the proposed method is about 62.50%, versus 58.86% for USL, 49.37% for SDALF and 41.00% for ELF. The improvement is due to two aspects: first, we use a patch-based matching method which effectively addresses the misalignment problem. Second, a sparsity-based occlusion handling model is exploited to reduce the noise introduced by occlusion patches.

In the ETHZ dataset, one person has more than one images. For the purpose of fair comparisons, we use the same settings as described in [12, 29], and the same single-shot evaluation strategy is applied in our work. For each person, we randomly select one image for gallery and the remained for probe. The comparison results of the three sequences are shown in Fig. 5, our SOH method is compared to USL [10] and other three state-of-the-art approaches, i.e., Partial Least Squares (PLS) [29], SDALF [12] and Covariance Descriptor based on Bio-inspired Features (BiCov) [14]. It is obvious that our SOH method outperforms

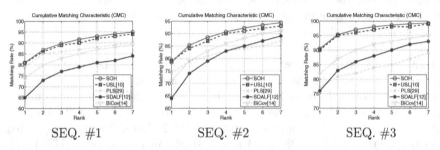

SEQ. #1   SEQ. #2   SEQ. #3

Fig. 5. ETHZ dataset: performances comparison using CMC curves on SEQ. #1, SEQ. #2, and SEQ. #3.

most state-of-the-art approaches and has competitive performance with USL [10]. All the comparison results demonstrate the effectiveness and competitive of our SOH method.

4   Conclusion

In this paper, we propose a sparsity-based occlusion handling (SOH) approach for person re-identification. The patch-based matching method is combined with a sparse representation model to determine whether a patch of an image taken from one camera view is occluded on the other camera view. Extensive experiments show that our method outperforms several state-of-the-art methods for person re-identification and is effective and robust on two different challenging public datasets, VIPeR and ETHZ.

Acknowledgement. The research was supported by the National Nature Science Foundation of China (61303114, 61231015, 61170023), the Specialized Research Fund for the Doctoral Program of Higher Education (20130141120024), the Technology Research Project of Ministry of Public Security (2014JSYJA016), the Fundamental Research Funds for the Central Universities (2042014kf0250), the China Postdoctoral Science Foundation funded project (2013M530350), the major Science and Technology Innovation Plan of Hubei Province (2013AAA020), the Key Technology R&D Program of Wuhan (2013030409020109), the Guangdong-Hongkong Key Domain Break-through Project of China (2012A090200007), and the Special Project on the Integration of Industry, Education and Research of Guangdong Province (2011B090400601).

References

[1] Prosser, B., Zheng, W.S., Gong, S., Xiang, T.: Person Re-Identification by Support Vector Ranking. In: British Machive Vision Conference (BMVC), p. 5 (2010)

[2] Zheng, W.S., Gong, S., Xiang, T.: Person re-identification by probabilistic relative distance comparison. In: Computer Vision and Pattern Recognition (CVPR), pp. 649–656 (2011)

[3] Figueira, D., Bazzani, L., Minh, H.Q., Cristani, M., Bernardino, A., Murino, V.: Semi-supervised multi-feature learning for person re-identification. In: Advanced Video and Signal Based Surveillance (AVSS), pp. 111–116 (2013)

[4] Salvagnini, P., Bazzani, L., Cristani, M., Murino, V.: Person re-identification with a ptz camera: An introductory study. In: Conference on Image Processing (ICIP) (2013)

[5] Barbosa, I.B., Cristani, M., Del Bue, A., Bazzani, L., Murino, V.: Re-identification with RGB-D sensors. In: Fusiello, A., Murino, V., Cucchiara, R. (eds.) ECCV 2012 Ws/Demos, Part I. LNCS, vol. 7583, pp. 433–442. Springer, Heidelberg (2012)

[6] Kostinger, M., Hirzer, M., Wohlhart, P., Roth, P., Bischof, H.: Large scale metric learning from equivalence constraints. In: Computer Vision and Pattern Recognition (CVPR), pp. 2288–2295 (2012)

[7] Mignon, A., Jurie, F.: PCCA: A new approach for distance learning from sparse pairwise constraints. In: Computer Vision and Pattern Recognition (CVPR), pp. 2666–2672 (2012)

[8] Zheng, W.S., Gong, S., Xiang, T.: Re-identification by relative distance comparison. In: Pattern Analysis and Machine Intelligence (PAMI), pp. 653–668 (2013)

[9] Tao, D., Jin, L., Wang, Y., Yuan, Y., Li, X.: Person Re-Identification by Regularized Smoothing KISS Metric Learning. In: Circuits and Systems for Video Technology (CSVT), pp. 1675–1685 (2013)

[10] Zhao, R., Ouyang, W., Wang, X.: Unsupervised Salience Learning for Person Re-identification. In: Computer Vision and Pattern Recognition (CVPR), pp. 3586–3593 (2013)

[11] Pedagadi, S., Orwell, J., Velastin, S., et al.: Local fisher discriminant analysis for pedestrian re-identification. In: Computer Vision and Pattern Recognition (CVPR), pp. 3318–3325 (2013)

[12] Farenzena, M., Bazzani, L., Perina, A., Murino, V., Cristani, M.: Person re-identification by symmetry-driven accumulation of local features. In: Computer Vision and Pattern Recognition (CVPR), pp. 2360–2367 (2010)

[13] Gray, D., Tao, H.: Viewpoint invariant pedestrian recognition with an ensemble of localized features. In: Forsyth, D., Torr, P., Zisserman, A. (eds.) ECCV 2008, Part I. LNCS, vol. 5302, pp. 262–275. Springer, Heidelberg (2008)

[14] Ma, B.P., Su, Y., Jurie, F., et al.: BiCov: A novel image representation for person re-identification and face verification. In: British Machive Vision Conference (BMVC) (2012)

[15] Wang, X., Doretto, G., Sebastian, T., Rittscher, J., Tu, P.: Shape and appearance context modeling. In: International Conference on Computer Vision (ICCV), pp. 1–8 (2007)

[16] Cheng, D.S., Cristani, M., Stoppa, M., Bazzani, L., Murino, V.: Custom pictorial structures for re-identification. In: British Machive Vision Conference (BMVC), p. 6 (2011)

[17] Liu, C., Gong, S., Loy, C.C., Lin, X.: Person re-identification: What features are important? In: Fusiello, A., Murino, V., Cucchiara, R. (eds.) ECCV 2012 Ws/Demos, Part I. LNCS, vol. 7583, pp. 391–401. Springer, Heidelberg (2012)

[18] Weinberger, K.Q., Blitzer, J., Saul, L.K.: Distance metric learning for large margin nearest neighbor classification. In: Advances in Neural Information Processing Systems (NIPS), pp. 1473–1480 (2005)

[19] Dikmen, M., Akbas, E., Huang, T.S., Ahuja, N.: Pedestrian recognition with a learned metric. In: Kimmel, R., Klette, R., Sugimoto, A. (eds.) ACCV 2010, Part IV. LNCS, vol. 6495, pp. 501–512. Springer, Heidelberg (2011)

[20] Li, Z., Chang, S., Liang, F., Huang, T.S., Cao, L., Smith, J.R.: Learning locally-adaptive decision functions for person verification. In: Computer Vision and Pattern Recognition (CVPR), pp. 3610–3617 (2013)

[21] Liu, C., Loy, C.C., Gong, S., Wang, G.: Pop: Person reidentification post-rank optimisation. In: International Conference on Computer Vision (ICCV), pp. 441–448 (2013)

[22] Hirzer, M., Roth, P.M., Köstinger, M., Bischof, H.: Relaxed pairwise learned metric for person re-identification. In: Fitzgibbon, A., Lazebnik, S., Perona, P., Sato, Y., Schmid, C. (eds.) ECCV 2012, Part VI. LNCS, vol. 7577, pp. 780–793. Springer, Heidelberg (2012)

[23] Jianchao, Y., Kai, Y., Yihong, G., Thomas, H.: Linear spatial pyramid matching using sparse coding for image classification. In: Computer Vision and Pattern Recognition (CVPR), pp. 1794–1801 (2009)

[24] Wright, J., Yang, A.Y., Ganesh, A., Sastry, S.S., Ma, Y.: Robust face recognition via sparse representation. In: Pattern Analysis and Machine Intelligence (PAMI), pp. 210–227 (2009)

[25] Wright, J., Ma, Y., Mairal, J., Sapiro, G., Huang, T.S.: Shuicheng Yan.: Sparse representation for computer vision and pattern recognition. Proceedings of the IEEE, 1031–1044 (2010)

[26] Zhong, W., Lu, H.C., Yang, M.H.: Robust object tracking via sparsity-based collaborative model. In: Computer Vision and Pattern Recognition (CVPR), pp. 1838–1845 (2012)

[27] Jia, X., Lu, H., Yang, M.-H.: Visual tracking via adaptive structural local sparse appearance model. In: Computer Vision and Pattern Recognition (CVPR), pp. 1822–1829 (2012)

[28] Ma, B., Su, Y., Jurie, F.: Local descriptors encoded by fisher vectors for person re-identification. In: Fusiello, A., Murino, V., Cucchiara, R. (eds.) ECCV 2012 Ws/Demos, Part I. LNCS, vol. 7583, pp. 413–422. Springer, Heidelberg (2012)

[29] Schwartz, W.R., Davis, L.S.: Learning discriminative appearance-based models using partial least squares. In: Computer Graphics and Image Processing (SIBGRAPI), pp. 322–329 (2009)

[30] Gray, D., Brennan, S., Tao, H.: Evaluating appearance models for recognition, reacquisition, and tracking. In: IEEE International Workshop on Performance Evaluation of Tracking and Surveillance (PETS) (2007)

[31] Ess, A., Leibe, B., Van Gool, L.: Depth and appearance for mobile scene analysis. In: International Conference on Computer Vision (ICCV), pp. 1–8 (2007)

Visual Attention Driven by Auditory Cues

Selecting Visual Features in Synchronization with Attracting Auditory Events

Jiro Nakajima[1], Akisato Kimura[2], Akihiro Sugimoto[3], and Kunio Kashino[2]

[1] Chiba University
nakajima13@chiba-u.jp
[2] Communication Science Laboratories, NTT Corporation
akisato@ieee.org, kashino.kunio@lab.ntt.co.jp
[3] National Institute of Informatics
sugimoto@nii.ac.jp

Abstract. Human visual attention can be modulated not only by visual stimuli but also by ones from other modalities such as audition. Hence, incorporating auditory information into a human visual attention model would be a key issue for building more sophisticated models. However, the way of integrating multiple pieces of information arising from audio-visual domains still remains a challenging problem. This paper proposes a novel computational model of human visual attention driven by auditory cues. Founded on the Bayesian surprise model that is considered to be promising in the literature, our model uses surprising auditory events to serve as a clue for selecting synchronized visual features and then emphasizes the selected features to form the final surprise map. Our approach to audio-visual integration focuses on using effective visual features alone but not all available features for simulating visual attention with the help of auditory information. Experiments using several video clips show that our proposed model can better simulate eye movements of human subjects than other existing models in spite that our model uses a smaller number of visual features.

Keywords: Visual attention, auditory cues, Bayesian surprise, synchronization, feature selection.

1 Introduction

Human beings have capability of detecting various kinds of objects without any thought or effort. *Visual attention* is considered to play a significant role in achieving this function. In fact, visual attention is one of the built-in mechanisms of the human visual system that quickly selects regions most likely to attract human interest in a visual scene. Such a pre-selection mechanism focusing only on relevant data would be essential in enabling computers to undertake subsequent processing such as generic object recognition or scene understanding.

With this background, many researches have been reported to simulate visual attention in several research fields including psychophysics, neuroscience and computer vision (see extensive survey papers, e.g. [2,3,11] for details). These researches usually

X. He et al. (Eds.): MMM 2015, Part II, LNCS 8936, pp. 74–86, 2015.

take a *bottom-up* approach, meaning that a given video signal is the only resource for simulating visual attention. Nevertheless, they have enabled us to investigate in detail the process of visual search and simulate its performance.

A large amount of effort for developing computational models of human visual attention has ever been devoted to only *visual* processing. Human visual attention, however, can be easily modulated by other modalities. As an intuitive example, when we hear something interest or strange we tend to look at the direction of sounds even if that direction is not so visually salient. As such, sounds are often strongly related to events that draw human visual attention. We will be able to further augment computational models of human visual attention if we incorporate auditory information into them. However, the way of integrating information arising from both audio and visual domains still remains a challenging problem.

This paper proposes a novel model of human visual attention driven by auditory cues. In our model, auditory information plays a supportive role in simulating visual attention, in contrast to standard multi-modal fusion approaches [21,18,19,5,14]. More concretely, we take an approach that detects visual features in synchronization with surprising auditory events. Our strategy is built on two recent psychophysical studies:

1. Audio-visual temporal alignment leads to benefits in visual attention if changes in the component signals are both *synchronized* and *transient* [4].
2. Auditory attention *modulates* visual attention in a *feature-specific* manner [1].

Following these findings, our model first detects *transient* events using the Bayesian surprise model in visual [8] and auditory [20] domains separately, and then looks for visual features in *synchronization* with detected auditory events. Surprise maps are then *modulated* by the selected features, in a similar manner to the guided search [23], one of well-founded psychophysical models that explicitly implements characteristics of target stimuli.

2 Related Work

Building computational models of human visual attention has attracted much attention especially in the last decade. Here we briefly review just a couple of related studies due to the space limitation. Extensive surveys can be found in e.g. [2,3,11] which include the history, detailed taxonomies and related psychophysical findings.

A seminal work as regards bottom-up models of human visual attention is the *saliency map* model proposed by Itti, Koch and Niebur [9]. In this model, the concept of *saliency* as a measure of attractiveness of human visual attention was first introduced into a computational model. Since it is simple, easy to implement and produces reasonable output for various kinds of images, it has had a considerable impact on broader research areas such as image processing, pattern recognition, computer vision, robotics and neuroscience [3].

The saliency map model has been further extended by Itti and Baldi to develop the *Bayesian surprise* model [8] that incorporates temporal dynamics of the human visual system. In this model, saliency is formulated by the Kullback-Leibler divergence between probabilistic density functions (PDFs) of expected and obtained visual features.

Therefore, continuously similar visual features give low saliency values, while unexpected visual features such as sudden changes provide high saliency values. Bayesian inference methods have been introduced also in several other computational models [24,16,6,13]. Such a probabilistic model enables us to handle various types of features with different characteristics into a unified framework. This is why we adopt the Bayesian surprise model as the basis of our new model.

Meanwhile, several mechanisms developed in visual attention models have also been introduced into auditory attention models, for example, the saliency map model [10], SUN (Saliency Using Natural statistics) [22] and Bayesian surprise [20].

However, human visual attention models with the help of auditory information has not been well studied. This might be because solid psychophysical findings about characteristics of audio information in human visual perception have been recently developed. In turn, most existing methods took multi-modal fusion approaches and concentrated on improving application performances, and thus the compatibility with the human visual perception is rather out of focus. Video summarization [12,15,5] is one of the popular applications of audio-visual saliency. Robotics [21,18,19] has also been an attracting application for last several years.

To the best of our knowledge, this is the first work that explicitly incorporates solid psychophysical findings of auditory-based attention modulation into a computational model of human visual attention.

3   Proposed Model

3.1   Framework

Figure 1 depicts the framework of the proposed model. As shown in this figure, our proposed model consists of four main steps.

(1) Bayesian surprise. The first step extracts surprising events in visual and auditory domains individually where image and audio signals are separately applied to the Bayesian surprise model. For a given input video, 360 visual surprise maps with different types of features and a single auditory surprise signal are extracted. The details will be described in Section 3.2.

(2) Synchronization detector. The second step evaluates synchronization of each visual surprise map with the auditory surprise signal. For this purpose, synchronization detectors are attached to every location in each of the 360 visual surprise maps and the auditory surprise signals, resulting in 360 maps. Every map is averaged over pixels to create a sequence describing how synchronized the corresponding visual surprise map is with the auditory surprise. The details will be described in Section 3.3.

(3) Features selection. The third step is devoted to selecting visual features that well synchronize with the auditory surprise. Counting the number of samples with a sufficient level of synchronization for every sequence, we obtain a histogram representing the degree of synchronization for every visual surprise map with the auditory surprise. Remembering that every visual surprise map corresponds to a specific type of features, feature selection based on audio-visual synchronization can be implemented by binarizing the histogram. The details will be described in Section 3.4.

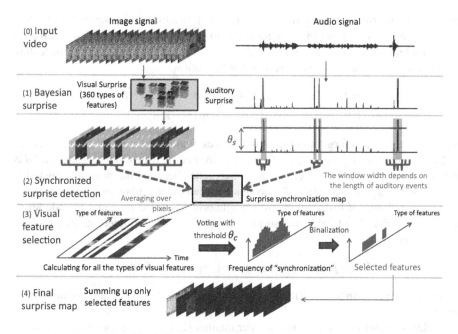

Fig. 1. Framework of the proposed model

(4) Final surprise map. The last step is for forming the final surprise map composed of visual surprise maps with the selected visual features. The details will be described in Section 3.4.

Our proposed model detects transient auditory events, and then selects visual features in synchronization with detected auditory events to modulate the final saliency maps. Note that the proposed model is built on a two-pass algorithm, where the first 3 steps are devoted to selecting visual features that describe major audio-visual events in the input video to produce the final map in the last step.

3.2 Bayesian Surprise Model

Here, we briefly review the Bayesian surprise model proposed by Itti and Baldi [8]. We introduce it to obtain visual surprise maps.

Center-surround feature maps are first generated. They are extracted in parallel over 12 feature channels (intensity, 2 color opponents, 4 orientations, temporal onset and 4 directed motion energies) and 6 spatial scales, yielding $12 \times 6 = 72$ feature maps in total.

Local surprise detectors are then attached to every location in each of the 72 feature maps. Suppose that every pixel value $f(t, x)$ received from feature map f at location x and time t obeys a Poisson distribution and it holds a conjugate Gamma prior $\gamma(\,\cdot\,;\alpha, \beta)$ with parameters (α, β). Once $f(t, x)$ is observed, the posterior $\gamma(\,\cdot\,;\alpha', \beta')$ can be obtained using the Bayes rule. Namely, at each time step t, the posterior $\gamma(f(t-1, x); \alpha_V(t-1, x), \beta_V(t-1, x))$ at the previous step can be used as a prior to obtain the current posterior $\gamma(f(t, \boldsymbol{x}); \alpha_V(t, \boldsymbol{x}), \beta_V(t, \boldsymbol{x}))$. In addition, 5 cascade detectors are implemented at every pixel in every feature map so that the model can detect surprises at several temporal scales. In summary, the update rule of parameters (α, β) at feature map f, time t and cascade level d is described as follows:

$$\alpha_V(t; d) = \xi \alpha_V(t-1; d) + \alpha_V(t; d-1)/\beta_V(t; d-1),$$
$$\alpha_V(t; 0) = f(t), \quad \beta_V(t; d) = \xi \beta_V(t-1; d) + 1,$$

where $0 < \xi < 1$ is a forgetting factor and indices f and $\boldsymbol{x}$ are omitted for simplicity.

Local temporal surprise $S_{V,T}(t; f, d)$ at feature map f, time t and cascade level d is determined as the Kullback-Leibler (KL) divergence between the prior and posterior, while spatial surprise $S_{V,S}(t; f, d)$ is as that between the neighborhood prior (modeled as a weighted sum of distributions over neighborhoods at the previous cascade level) and the posterior. The total visual surprise $S_V(t; f, d)$ is determined according to the original paper [8] as

$$S_V(t; f, d) = (S_{V,T}(t; f, d) + S_{V,S}(t; f, d)/20)^{1/3}.$$

As we see, we have in total 72(feature maps) $\times$5(cascade levels) = 360 visual surprise maps.

Auditory surprise is derived in a similar manner [20], where a spectrogram $F(t, \omega)$ extracted via short-time Fourier transform (STFT) is used as an observation. Following the same update rule as the visual surprise, we can obtain parameters (α, β) of the posterior at time t and frequency ω as

$$\alpha_A(t; \omega) = \xi \alpha_A(t-1; \omega) + F(t, \omega), \quad \beta_A(t; \omega) = \xi \beta_A(t-1; \omega) + 1.$$

Auditory surprise $S_A(t; \omega)$ at time t and frequency ω is determined as the KL divergence between the prior and posterior. The final auditory surprise $S_A(t)$ is obtained as the mean over all the frequencies. As a result, we have a single auditory surprise signal.

3.3 Synchronization Detector

Following the recent psychophysical insight that audio-visual temporal alignment affects visual attention if changes of component signals are synchronized and transient [4], our model detects synchronized audio-visual events in videos from the output of Bayesian surprise models. Since both audio and visual signals have been converted into the "surprise" domain under the same logic, we can adopt a simple approach based on cross correlation.

A synchronization detector comprises the following 3 steps: Detecting surprising auditory events, pixel-wise cross correlations, and averaging over frames.

(1) Segments of surprising auditory events are first extracted from the auditory surprise signal $S_A(t)$. We exploit a simple approach that extracts segments with a surprise value $S_A(t)$ greater than a predefined threshold θ_s, resulting a set of segments $T_{S,i}$ $(i, 1, 2, \dots)$.

(2) For every segment $T_{S,i}$ normalized cross correlation (NCC) is calculated between the auditory surprise signal $S_A(t)$ and visual surprises at every location $\boldsymbol{x}$ in each of the

Table 1. Details of video clips

	Video 1	Video 2	Video 3	Video 4	Video 5	Video 6
clip name	advert_bbc4_bees	advert_bbc4_library	sports_kendo	basketball_of_sports	documentary_adrenaline	BBC_wildlife_eagle
# frames	246	246	101	246	195	107
fps	30	30	30	30	30	30

360 visual surprise maps $S_V(t; f, d)$. A window width for computing NCC depends on the length of an auditory event, namely the length $|T_{S,i}|$ of the segment. Through this process, 360 maps are obtained, each representing how synchronized every pixel in the corresponding visual surprise map is with the auditory surprise signal.

(3) Every synchronization map is finally averaged over pixels to obtain a sequence $c(t; f, d)$ that describes how synchronized the visual surprise map $S_V(t; f, d)$ is with the auditory surprise $S_A(t)$.

3.4 Features Selection

Once we have detected visual events synchronized with auditory events, the next step is to find dominant visual features in the detected events and to emphasize them to compute the final surprise map. This harmonizes our model with the finding that auditory attention modulates visual attention in a feature-specific manner [1].

First, the number of samples greater than a pre-specified threshold θ_c is counted in every sequence $c(t; f, d)$ to create a histogram with 360 (= the number of visual features) bins. Since every visual surprise map corresponds to a specific pair of feature type f and cascade level d (cf. Section 3.2), the histogram represents how dominant each feature is in synchronized audio-visual events. We do not accumulate $c(t; f, d)$ over time t to create a histogram because surprise values (accordingly, cross correlation values as well) at different frames cannot be compared in principle. We instead evaluate at each time t whether $c(t; f, d)$ is greater than a threshold or not, and if it is we vote for the corresponding visual feature.

Feature selection can be achieved by just binarizing the histogram, where a threshold for the binarization is adaptively chosen so that its slight change significantly impacts on the number of selected features. Only the visual surprise maps of the selected features (with active in the binarized histogram) are accumulated to form the final surprise map. In this way, our proposed model uses a smaller number of visual features than 360 for forming the final map.

4   Experiments

We experimentally evaluated our proposed model. We selected 6 video clips (advert bbc4 bees, advert bbc4 library, sports kendo, basketball of sports, documentary adrenaline, BBC wildlife eagle), all of which are provided by the DIEM project[1]. Table 1 illustrates the details of the video clips. We showed them to 15 human subjects. While

---

[1] http://thediemproject.wordpress.com

Table 2. NSS with optimal threshold values (bold letters: highest NSS for each video)

	Video 1	Video 2	Video 3	Video 4	Video 5	Video 6
baseline (Itti2009 [8])	0.299	1.524	0.636	**2.763**	1.275	0.450
proposed (θ_s, θ_c: optimal)	**0.935**	**1.842**	**0.801**	**2.763**	**1.287**	**0.621**
proposed ($\theta_s = 0, \theta_c$: optimal)	0.605	1.543	**0.801**	2.485	1.275	0.589

Table 3. Selected features using the optimal thresholds

	Baseline	Video 1	Video 2	Video 3	Video 4	Video 5	Video 6
Intensity	30	0	8	8	30	0	6
Color	60	4	17	27	60	8	23
Orientation	120	0	46	39	120	0	7
Onset	30	0	0	0	30	1	0
Motion	120	0	0	0	120	14	0
Total	360	4	71	74	360	23	36

the subjects were watching the video clips, their eye movements were recorded using an eye tracker Tobii TX300. Note that we showed all the video clips to the subjects together with audio signals and originally collected their eye movements rather than directly exploiting the accompanying eye traces by the DIEM project. We extracted gaze points from the eye movements by removing micro-saccades. Namely, we removed all the eye movements greater than 2.12 pixels per millisecond, and identified all the remainings as gaze points (ground truths for the evaluation). As a metric to quantify how well a model predicts actual human eye movements, we used the normalized scan-path saliency (NSS) [8] calculated from the gaze points.

We first evaluated how two thresholds, i.e., auditory surprise threshold θ_s and correlation threshold θ_c, have impact on NSS. We changed the two threshold values independently and averaged NSS scores over frames for each video. Fig. 2 visualizes the averaged NSS scores for each video in terms of the heat map. In this visualization, red areas indicate threshold pairs with better performance than the Bayesian surprise model [8] (we call this the baseline model) and blue ones are the opposite. Table 2 shows the optimal values for θ_s and θ_c found in Fig. 2 where the optimal values mean the values that produce best NSS in our model. We remark that we also show the optimal value for θ_c under the condition that[2] $\theta_s = 0$.

From Fig. 2, we see that for Videos 1, 2 and 6, the proposed model is in most cases superior to or compatible with the baseline model and that for Videos 3 and 4 the proposed models is almost compatible with the baseline model for any threshold values. In contrast, for Video 5 our model could not achieve the compatible level with the baseline in most cases. Table 2 shows that our model with optimal threshold values produced higher NSS scores than the baseline model except for Video 4. NSS scores for Video 4 are fairly high themselves and, moreover, the ball is the only moving object and almost all image features are similar over frames in the video; these bring difficulty in selecting

---

[2] $\theta_s = 0$ is equivalent to computing correlation between audio and visual surprises for all the frames in the video.

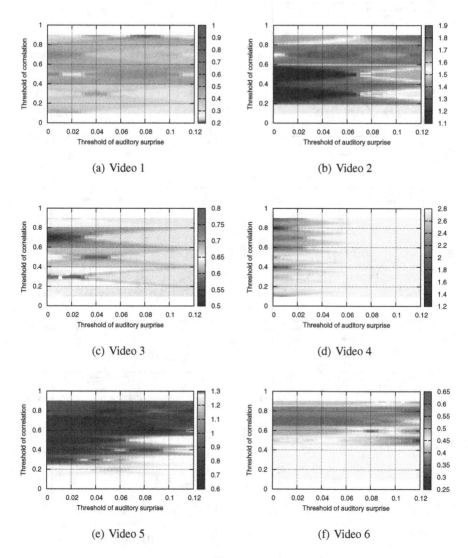

(a) Video 1

(b) Video 2

(c) Video 3

(d) Video 4

(e) Video 5

(f) Video 6

Fig. 2. NSS scores under different threshold values (horizontal axis: auditory surprise threshold θ_s; vertical axis: correlation threshold θ_c). Red areas indicate thresholds with better performance than the baseline and blue ones are the opposite.

features synchronized with auditory events, which results in failure of further improvement of NSS scores. We also observe in Table 2 that we have better NSS scores by restricting the correlation computation only to the frames where auditory events occur than by correlation computation using all the frames in the video.

We see in Fig. 2 that for the same auditory surprise threshold value, a larger correlation threshold value tends to achieve better NSS for several videos while for the same

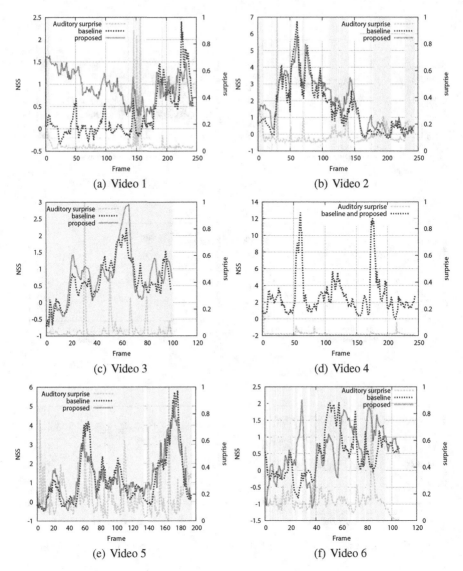

Fig. 3. NSS scores with the optimal thresholds and auditory surprises (horizontal axis: frame; vertical axis: NSS or normalized auditory surprise; light blue frame: surprising auditory events).

correlation threshold value, the auditory surprise threshold does not affect significant difference on NSS. This observation indicates that image features that have stronger correlation with the audio signal contribute to increase NSS more for any level of auditory surprise. This is consistent with the psychophysical finding that audio-visual temporal alignment leads to benefits in visual attention if changes in the component signals are both synchronized and transient [4]. We should note that two thresholds, in particular the correlation threshold, should be carefully chosen depending on the video for better performance.

To each video, we used the optimal threshold values shown in Table 2 to compute NSS scores for frames in the video. We illustrated in Fig. 3 detailed time trends of NSS scores for each video where the blue and red lines represent NSS scores for baseline and proposed models respectively. Note that in Video 4, the NSS score with the optimal threshold values was the same as that by the baseline model; blue and red lines are identical. Time trends of normalized auditory surprise are also shown in Fig. 3 using green lines. We see that in many frames of all the videos, the proposed model outperformed the baseline model. This is remarkable for Video 1. For Video 6, which model has a better NSS score heavily depends on the frame though the averaged NSS for our model is better. This is because Video 6 has several sudden changes in the scene and thus image features that are synchronized with auditory events also easily change.

Table 3 shows the number of selected visual features by the our model with the optimal threshold values. We can see that only a small fraction of 360 types of features were selected. We also observe that categories such as intensity or color of selected features highly depend on each input video. This is reasonable because what image features are closely correlated with auditory events depends on the video. This also justifies our implementation with a two-pass algorithm.

In order to show the effectiveness of our proposed model, we compared performances with the state-of-the-art models in addition to the baseline model [8]. They are the saliency map model [7], and the audio-visual attention model using the sound localization [14]. We also compared our model with the model, hereafter called the random feature selection model, in which we randomly selected a given number of image features among all image features used in [8], where the number of features to be selected was set in accordance with the number of image features determined by the optimal threshold values (see Table 2). We remark that NSS of the random feature selection model was computed as the average of NSS scores over 20 trials in random feature selections because selected image features vary at each iteration and produce a different NSS score.

Figure 4 illustrates averages of NSS scores over frames for each video and for each model. Note that the random feature selection model has error bars representing the standard deviation over 20 trials. We remark that all the image features (360 features) are selected for Video 4 and thus three models (the random feature selection model, the baseline model and our model) produce the same NSS score.

We see in Fig. 4 that our proposed model produced best NSS scores for all the video, outperforming the other models. The audio-visual attention model using the sound localization [14] did not achieve a good level in spite that it additionally uses auditory signals. This will be because it could not detect the sound source location accurately. Interestingly, the random feature selection model tends to outperform the baseline model. This indicates that using all the image features does not necessarily perform better. Using a smaller number of image features may be better. The number of selected features in our model may suggest such numbers though further investigation on required number of features is left for future work.

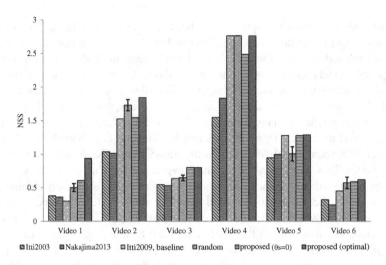

Fig. 4. Comparison of NSS averaged over frames for each video

5   Concluding Remarks

This paper proposed a novel computational model of human visual attention driven by auditory cues. Our model first detects synchronized and transient audio-visual events using a framework of Bayesian surprise and then selects dominant visual features in the detected events to form the final output. Our approach stands on using auditory features as a synchrony cue for selecting visual features. Differently from just fusing audio-visual information, our approach boosts the ability of visual information by selecting visual features synchronized with surprising auditory events. We remark that our approach is in line with recent psychophysical findings as well. The experimental evaluation with human eye movements demonstrated that our model outperformed the state-of-the-art models, in particular, the baseline model [8] in spite that we used a smaller number of visual features.

We used correlation to evaluate synchronization between audio and visual surprises. Mutual information can be also used as a measure for synchronization [17]. We are currently working for using mutual information instead of correlation. Our proposed model provides just one way to incorporate auditory cues into a computational model of human visual attention. We can thus improve our model into several directions in future, e.g. the introduction to adaptive image feature selection depending on the auditory event or the location in the image and machine learning strategies for capturing generic structures of audio-visual events.

References

1. Ahveninen, J., Jaaskelainen, I.P., Belliveau, J.W., Hamalainen, M., Lin, F.H., Raij, T.: Disso-ciable influences of auditory object *vs.* spatial attention on visual system oscillatory activity. PLoS One 7(6), e38511 (2012)
2. Begum, M., Karray, F.: Visual attention for robotic cognition: A survey. IEEE Transactions on Autonomous Mental Development 3(1), 92–105 (2011)
3. Borji, A., Itti, L.: State-of-the-art in visual attention modeling. IEEE Transactions on Pattern Analysis and Machine Intelligence 35(1), 185–207 (2013)
4. Van der Burg, E., Cass, J., Olivers, C.N.L., Theeuwes, J., Alais, D.: Efficient visual search from synchronized auditory signals requires transient audiovisual events. PLoS One 5(5), e10664 (2010)
5. Evangelopoulos, G., Zlatintsi, A., Potamianos, A., Maragos, P., Rapantzikos, K., Skoumas, G., Avrithis, Y.: Multimodal saliency and fusion for movie summarization based on aural, visual, and textual attention. IEEE Transactions on Multimedia 15(7), 1553–1568 (2013)
6. Gao, D., Han, S., Vasconcelos, N.: Discriminant saliency, the detection of suspicious coin-cidences, and applications to visual recognition. IEEE Transactions on Pattern Analysis and Machine Intelligence 31(6), 989–1005 (2009)
7. Itti, L., Dhavale, N., Pighin, F.: Realistic avatar eye and head animation using a neurobio-logical model of visual attention. In: Proc. SPIE 48th Annual International Symposium on Optical Science and Technology, vol. 5200, pp. 64–78. SPIE Press, Bellingham (2003)
8. Itti, L., Baldi, P.: Bayesian surprise attracts human attention. Vision Research 49(10), 1295–1306 (2009)
9. Itti, L., Koch, C., Niebur, E.: A model of saliency-based visual attention for rapid scene analysis. IEEE Transactions on Pattern Analysis and Machine Intelligence 20(11), 1254–1259 (1998)
10. Kayser, C., Petkov, C., Lippert, M., Logothesis, N.: Mechanisms for allocating auditory at-tention: An auditory saliency map. Current Biology 15, 1943–1947 (2005)
11. Kimura, A., Yonetani, R., Hirayama, T.: Computational models of human visual attention and their implementations: A survey. IEICE Transactions 96-D(3), 562–578 (2013)
12. Ma, Y.F., Hua, X.S., Lu, L., Zhang, H.J.: A generic framework of user attention model and its application in video summarization. IEEE Transactions on Multimedia 7(5), 907–919 (2005)
13. Miyazato, K., Kimura, A., Takagi, S., Yamato, J.: Real-time estimation of human vi-sual attention with dynamic Bayesian network and MCMC-based particle filter. In: ICME, pp. 250–257. IEEE (2009)
14. Nakajima, J., Sugimoto, A., Kawamoto, K.: Incorporating audio signals into constructing a visual saliency map. In: Klette, R., Rivera, M., Satoh, S. (eds.) PSIVT 2013. LNCS, vol. 8333, pp. 468–480. Springer, Heidelberg (2014)
15. Ngo, C.W., Ma, Y.F., Zhang, H.J.: Video summarization and scene detection by graph mod-eling. IEEE Transactions on Circuits and Systems for Video Technology 15(2), 296–305 (2005)
16. Pang, D., Kimura, A., Takeuchi, T., Yamato, J., Kashino, K.: A stochastic model of selective visual attention with a dynamic Bayesian network. In: Proc. IEEE International Conference on Multimedia and Expo. (ICME), pp. 1073–1076. IEEE (2008)
17. Rolf, M., Asada, M.: Visual attention by audiovisual signal-level synchrony. In: Proc. 9th ACM/IEEE International Conference on Human-Robot Interaction Workshop on Attention Models in Robotics: Visual Systems for Better HRI (2014)
18. Ruesch, J., Lopes, M., Bernardino, A., Hornstein, J., Santos-Victor, J., Pfeifer, R.: Multi-modal saliency-based bottom-up attention a framework for the humanoid robot iCub. In: IEEE International Conference on Robotics and Automation (ICRA), pp. 962–967 (2008)

19. Schauerte, B., Kühn, B., Kroschel, K., Stiefelhagen, R.: Multimodal saliency-based attention for object-based scene analysis. In: Proc. 24th International Conference on Intelligent Robots and Systems (IROS). IEEE/RSJ (2011)
20. Schauerte, B., Stiefelhagen, R.: Wow! Bayesian surprise for salient acoustic event detection. In: Proc. 38th International Conference on Acoustics, Speech, and Signal Processing, (ICASSP) (2013)
21. Spexard, T., Hanheide, M., Sagerer, G.: Human-oriented interaction with an anthropomorphic robot. IEEE Transactions on Robotics 23(5), 852–862 (2007)
22. Tsuchida, T., Cottrell, G.: Auditory saliency using natural statistics. In: Proc. Annual Meeting of the Cognitive Science (CogSci), pp. 1048–1053 (2012)
23. Wolfe, J., Cave, K., Franzel, S.: Guided search: an alternative to the feature integration model for visual search. Journal of Experimental Psychology: Human Perception and Performance 15(3), 419–433 (1989)
24. Zhang, L., Tong, M.H., Marks, T.K., Shan, H., Cottrell, G.W.: SUN: A Bayesian framework for saliency using natural statistics. Journal of Vision 8(7) (2008)

A Synchronization Ground Truth for the Jiku Mobile Video Dataset

Mario Guggenberger, Mathias Lux, and Laszlo Böszörmenyi

Institute of Information Technology,
Alpen-Adria-Universität Klagenfurt,
9020 Klagenfurt am Wörthersee, Austria
{mg,mlux,lb}@itec.aau.at

Abstract. This paper introduces and describes a manually generated synchronization ground truth, accurate to the level of the audio sample, for the Jiku Mobile Video Dataset, a dataset containing hundreds of videos recorded by mobile users at different events with drama, dancing and singing performances. It aims at encouraging researchers to evaluate the performance of their audio, video, or multimodal synchronization methods on a publicly available dataset, to facilitate easy benchmarking, and to ease the development of mobile video processing methods like audio and video quality enhancement, analytics and summary generation that depend on an accurately synchronized dataset.

Keywords: Audio, video, multimedia, crowd, events, synchronization, time drift.

1   Introduction

With the incredibly fast proliferation of mobile devices capable of video recording, it is now easier than ever for people to quickly record interesting moments at the press of a button. For the research community, this opens up a lot of new and interesting opportunities. As an example, if you have recently been to a concert you might have noticed that people are constantly taking pictures and recording video clips. By the end, a huge dataset distributed over many devices has been generated by the crowd. Supposed there is a way to access this dataset, many interesting post processing methods can be applied to it. To name a few, there is the possibility to detect highlights and key moments by looking at the frequency of concurrent recordings, since people tend to capture what they consider to be most interesting to them or to friends that they want to show the capture. Recordings can be temporally stitched together to get a complete and continuous coverage of the whole event. Even better, vivid videos can be created by switching between different perspectives or showing different shots side-by-side. Quality can be improved by picking the best audio and video tracks from parallel recordings. 3D scenes can be reconstructed from recordings of different angles. It can even help in forensics, e.g. by reconstructing a crime scene and calculating where a gunshot came from.

X. He et al. (Eds.): MMM 2015, Part II, LNCS 8936, pp. 87–98, 2015.

The key to all these applications is precise automatic synchronization, a topic extensively researched in recent years, aiming to replace the tedious and very time-consuming manual work [16]. While an experienced user can synchronize a pair of recordings in a matter of minutes, it still costs him many hours to synchronize a large dataset. The difficulty of the problem is determined by multiple dimensions and grows with increasing clip amounts, decreasing clip lengths, decreasing perceived clip quality, and wider time frames where the clips are scattered in. To synchronize automatically, algorithms usually look at the audio or video content of the recordings and try to find unique events occurring in multiple recordings, which are then taken as reference points for aligning the recordings on a timeline. There are many published methods and algorithms for automatic synchronization to choose from, but authors usually evaluate them on their own custom datasets. This makes it impossible to compare them in terms of computational complexity, spacial complexity, synchronization rate, and synchronization accuracy.

To mitigate this situation, we contribute an accurate synchronization ground truth for a large publicly available mobile video dataset, and even consider the effect of time drift between the recording devices. It can be used to evaluate current and future synchronization methods, and serve as a foundation for methods that build upon synchronized audio and video tracks.

2   Related Work

There are many methods for audio and video synchronization, and a recent overview of synchronization methods is presented in [10]. Mathematical formulations of the synchronization problem can be found in [16,19,10]. There is no publicly available dataset with a precise synchronization ground truth, and individual methods are usually evaluated on custom datasets. Shrestha et al. [17] created a custom dataset captured at two different events by two video cameras, a wedding in a church and a dance event inside a hall, with a total runtime of 3 hours and 45 minutes. In follow-up works, they first extended the dataset with three additional events [18], and later extended it with two concert events [16] covered by 9, respectively 10 cameras. Both extensions consisted of short clips of 20 seconds to 5 minutes length, their total runtime is unknown. Kennedy and Naaman [9] evaluated their work on a reasonably big dataset sourced from YouTube from three big music concerts with about 200 videos each and runtimes between 1 and 10 minutes. Shankar et al. [14] used a custom dataset with videos recorded with mobile and handheld devices at cricket, baseball and football matches, but they did not describe it more detailed. The most recent work was conducted by Casanovas and Cavallaro [10], who again extended the dataset from [16] with additional events. All of these datasets are either too small, not distributable due to copyright restrictions, out-dated and not available any more, or do not capture the real-world characteristics of our use-case. If datasets are too small, they might (un)intentionally mask problems of complexity. If clips are too short or taken from homogeneous sources, they might mask drift. If the perceptual quality of clips is too high or they are recorded in lab settings, they might mask low robustness.

Time drift has been mostly ignored in the multimedia community. The problem itself is well known and has been covered in network delay measurements [12] or to identify physical network devices through fingerprinting [15]. In multimedia, [10] is the first paper presenting a synchronization method that, to our knowledge, identifies and acknowledges the time drift problem. We have also already presented a demo application for media synchronization that can semi-automatically handle drift [4], and we described a measurement method in [5].

3   Jiku Mobile Video Dataset

The Jiku Mobile Video Dataset [13] is a collection of crowdsourced videos captured at 5 different events across Singapore by 4 to 15 recording devices in parallel, mostly in HD resolution. The events feature drama, dancing and singing performances. It aims at providing a publicly available collection of videos that (i) captures the unique characteristics of mobile video, (ii) supports researchers in working on solutions instead of spending time gathering test data, and (iii) enables benchmarking by leading to comparability of related methods and algorithms. It is to our knowledge the only currently and publicly available dataset of this kind, and by far the largest (Table 1) and most recent dataset available for event synchronization in general. An additional feature is the complementary metadata of each video recording comprised of compass and accelerometer readings. Potential applications suggested by the authors are (i) *video quality enhancement* by complementing information from multiple concurrent recordings from different viewpoints, (ii) *audio quality enhancement* by improving the audio track of a video with audio data from other concurrent audio tracks, (iii) *virtual directing* by automatically presenting the best shot out of a number of concurrent recordings to the viewer, and switching between them to create vivid multi-camera presentations, (iv) *occlusion detection* to support the selection of recordings that present the intended view of a scene, (v) *video sharing* by simulating events with a multitude of users transmitting their recordings over a network, and (vi) *mobile video analytics* including face detection, tracking, segmentation and de-shaking. Almost all of these suggestions rely on concurrent recordings, which implicates the need of an exact time-based synchronization. The clips are organized by a naming scheme consisting of an event ID, the date of the event, the ID of the recording device, and the recording start timestamp. By looking at the filename, they can be split into the five different event sets, and further divided into subsets by the recording device ID. The timestamps are too inaccurate and cannot be used for synchronization, as described in [17].

4   Methodology

This section describes the process of generating the ground truth. The goal was, for each set of event recordings in the dataset, to (i) lay out all recordings on a

Table 1. Breakdown of the Jiku Mobile Video Dataset. Additional detailed characteristics can be found in the original paper [13].

Event	GT_090912	NAF_160312	NAF_230312	RAF_100812	SAF_290512
Cameras	4	8	15	7	8
Recordings	50	66	117	97	143
Total Length	3h 37m	6h 00m	8h 23m	6h 40m	5h 57m

common timeline and (ii) extract the offset of each recording from the start of the timeline as the synchronization ground truth. The timeline begins at zero which equals to the moment the first recording was started, ends at the moment the last recording was stopped, and covers the whole interval in between. All recordings are placed such that all moments from the real event captured on recordings are placed at the same point on the timeline. We chose to synchronize the recordings by their audio tracks, because (i) it allows higher alignment precision due to the much higher audio sampling rate compared to the video frame rate, (ii) it provides humans a compact overview of the time dimension in the form of audio waveform envelopes which facilitates easy spotting and validation of matching points, and (iii) most currently existing synchronization algorithms work on audio data. The omnidirectionality of audio makes it also much easier to detect overlaps in the time domain than the strict unidirectionality of video, where cameras could be looking at totally different excerpts of the event scene.

While synchronization on audio tracks automatically leads to synchronized video tracks, they will not be as accurately synchronized due to the difference between the speed of sound and speed of light, and the fact that people in a crowd usually record from different positions with different distances from the target scene. Given the sound traveling at 340 m/s and neglecting the much higher speed of light, a difference of 10 meters distance yields a skew of $\approx$ 30 ms or $\approx$ 1 video frame at 30 fps. Luckily, time shifts between video tracks are less likely to be detected by humans, and offsets below the frame rate cannot be detected at all. In contrast, an audio offset of 30 ms is usually very noticeable. According to ITU, subjective research has shown that acceptability thresholds are at about +90 ms to −185 ms [7]. The ATSC found this numbers inadequate and recommends to stay within +15 ms and −45 ms [1]. In either case, switching between video streams that are out of sync will not always go undetected.

To generate the ground truth and lay out all recordings on the timeline, synchronization points between overlapping recordings had to be found, where a synchronization point is a quadruple consisting of two recordings and two time points that specify where the content in one recording equals the content in another recording. Given such a point, one recording can be adjusted to the other on the timeline such that the two time points are placed on the same time instant, which can be seen as a direct synchronization. An indirect synchronization involves intermediate recordings, such that two non-overlapping recordings A and B can be synchronized when recording A overlaps X and X overlaps B, resulting in a synchronization of A, X, and B. It is not necessary to find synchronization points between all pairs of overlapping recordings, just between as many as are needed for a minimum spanning tree to be built from synchronization points interpreted as edges and recordings as nodes. One such tree then represents a cluster of directly and indirectly overlapping recordings. In the case of coverage gaps where an event is not continuously captured on recordings, multiple unconnected trees are formed. Care must be taken that a synchronization point between two tracks does not automatically lead to the tracks being synchronized over time, it only assures that the content of the two tracks conforms at the exact time points. To synchronize them over time and thus facilitate flawless parallel playback, the drift between the recordings must be detected and eliminated.

4.1 Time Drift Correction

To get our ground truth as precise as possible, we determined the absolute drifts in the Jiku dataset. We did this with the help of the Jiku authors [13] who provided us a mapping of device IDs to recording devices. We gathered devices of the same models and measured their absolute drift at a room temperature of $\approx 25\,°C$ with the same method that we described in [5]. Table 2 lists the recording devices, their dataset IDs and the measured drifts in milliseconds per minute. A positive drift indicates that the real sampling rate of a device is higher than the nominal sampling rate, making the playback time longer than the captured real-time event when played back at the nominal sampling rate. Knowing these drifts, it is now sufficient to synchronize two overlapping recordings at one single point to get them synchronized over their whole overlapping interval. There is still a small fraction of drift error left, resulting from the fact that we did not measure the exact same devices that were used for recording and we do not know the temperatures at which the recordings took place. Series of measurement in our laboratory have shown a standard drift deviation of $\approx 0.1\,ms/min$ between multiple devices of the same model, and temperature changes between $-20\,°C$ and $+50\,°C$ have shown a variance of $\approx 1\,ms/min$ [5], which we assume to also be true for the ones used in the dataset. In our opinion, both of these errors left in the measurements do not have a reasonable impact on our ground truth because (i) the temperature difference between our laboratory and the actual air temperature at recording time in Singapore is presumably much lower than between the extreme bounds in our laboratory measurements and (ii) the recordings in the dataset are short enough to minimize its impact. Out of the 481 recordings in the dataset, only 19 are longer than 15 minutes, and more than 75% stay below 5 minutes runtime.

4.2 Manual Synchronization

The manual synchronization was done by an author of this paper who has a lot of experience in multi-track recording and post-production of audio and video data and has had the pleasure to synchronize tracks on many occasions. Doing this manually, especially when many tracks need to be synchronized, takes

Table 2. Measured absolute drifts in ms/min of the recording devices used to create the Jiku Mobile Video Dataset

Device	IDs	Drift
Samsung GT-i9023 Nexus S	15, 16, 19, 20	−0.37
Samsung GT-i9000 Galaxy S	5	+0.26
Samsung GT-i9100 Galaxy S II	2, 3, 4, 11, 12, 13, 14, 17, 18, 21, 23	+15.95
Samsung GT-i9250 Galaxy Nexus	0, 1, 6, 7, 8, 9, 10	+4.78
Samsung GT-i9300 Galaxy S III	22	+0.34

a lot of time and effort. This is why automatic methods are sought after, but both available in the research domain and on the commercial market have not been used by intention since it would contradict the intended purpose of the ground truth. To give the manual process a starting boost, we still applied two automatic approaches to get a rough timeline pre-alignment to start with, but every synchronization point in the final result has been set and verified by hand. The first approach was generating an approximate timeline alignment from the metadata timestamps for all 481 recorded clips. This helped to get a very rough overview of the alignment of recordings and to spot extreme outliers. At this point, almost all recordings were off of their final alignment. The second approach was the application of an audio fingerprinting algorithm [6] that helped to obtain approximate synchronization points for about 50% of all recordings, which specifically helped in those cases where the timestamps were off by a huge amount. The manual work began with the validation and correction of wrong pre-alignments by looking at the waveform amplitude envelopes, trying to find visually matching patterns and listening to the recordings to semantically match them by their content, until all recordings were approximately synchronized. At this stage, the synchronization between recordings was accurate to a few seconds only. Then followed a time consuming manual refinement process, where 397 exact synchronization points were determined by visually looking at the waveforms, aurally listening to the audio data, and fine adjusting their relative offsets until the alignments were as precise as possible, often at sample or even subsample level. It was always followed by a validation step where the overlapping interval was proof-listened. The difficulty of determining a synchronization point varied from easy cases where the signals could be visually matched very clearly to hard cases with extremely distorted signals where only aural matching by repeated careful listening and readjusting was possible. All of this work was done in a custom software specifically developed for synchronization purposes. It took about 20 hours and was approximately cut in half by the automatic pre-alignment. The final result was a list of synchronization points which we transformed into a list of time offsets resulting in the manual ground truth. The timestamps were used as reference to order unconnected clusters of overlapping track groups in time, because this information cannot be inferred from the synchronization points alone.

5 Synchronization Ground Truth

The synchronization ground truth contains, for each of the five events, the start times of all recordings ordered on a timeline, the drift correction factors, and all manually generated synchronization points. Laying out all recordings on a timeline with the specified offsets and changing their runtime by the drift factor results in a synchronized event. The start times are relative to the start time of the first recording at the corresponding event, which is assumed w.l.o.g. as zero, and are calculated from the synchronization points. All specified times are given to a fractional seconds precision of 10^{-7} to enable subsample accuracy. Since all synchronization points have been generated and validated manually, they are very precise on one hand, probably more precise than current algorithms are able to achieve, but on the other hand this means that their precision cannot be measured in numbers. It is guaranteed though, that almost all synchronization points are inexact to at most 10 ms, where most are more precise and only a very small part of very hard to determine synchronization points are off by more. These are cases where humans and also computer algorithms probably reach their current limits. All synchronization points are guaranteed to be exact enough for artifacts of nonsynchronous playback, like echoes, to be unperceivable. It is not guaranteed that video frames of concurrent recordings are in sync, because of the already mentioned difference between the speed of sound and speed of light. We had to exclude all recordings from device 5 in the NAF_230312 set because they were not correctly cut, resulting in multiple noncontinuous shots inside its files that rendered them unsynchronizable. The data is available for download on our website[1] in structured XML files.

5.1 Accuracy

To evaluate the accuracy of our manually generated synchronization points, we chose to cross-correlate short intervals of audio samples that surround the points. The idea was that a low cross-correlation offset with a high correlation coefficient would confirm a synchronization point valid, while a high offset would be an indicator that the manually set synchronization point is inaccurate and can even be improved by the offset. Cross-correlation in general is a computationally expensive operation, but a 1-second interval sufficed because we knew for sure that all potential manual synchronization errors are much smaller, since e.g. an error of 50 ms would stand out heavily and cannot go undetected during validation. It turned out that the correlation results could not be used to automatically classify the manual synchronization points into true and false positives because we were unable to set a reasonable threshold. A problem is that we do not know the maximum achievable correlation coefficient between pairs of recordings, due to noise, the different frequency pickup patterns of the recording devices, and the time drift error. Upon inspection of the results, we found a lot of cases where the correlation offset was rated with a high coefficient but was

---

[1] http://www-itec.aau.at/~maguggen/jikusync/

actually too far off the optimal synchronization point, leading to audible echoes when listened to carefully. In contrast, we had many cases of valid offsets with much lower coefficients. Experiments with different interval lengths, sampling rates, and frequency filtering did not have any significant impact on the results. We could still learn a lot about the ground truth by manually analyzing the results. Looking at Figure 1, we can see that 200 of the 397 synchronization points result in a cross-correlation offset within ±5 ms, and 274 are within ±10 ms. This means that in all these cases, our manually generated synchronization points correlate highly with those calculated by the cross-correlation, confirming the accuracy of our manually generated data. All other cross-correlation results were manually double-checked and found to be more inaccurate compared to the manually identified points. The extreme cases where the cross-correlation offsets lied within the three-digit range happed in very noisy audio tracks where the correlation series are flat and the maximum correlation coefficients not located at distinct peaks, leading to ambiguous results.

5.2   Comparison

To show that the timestamps of the dataset are not reliable enough to be used for synchronization, we compared our ground truth with the timestamps. We measured the time difference of each recording as the error between the ideal position in the event timeline from the ground truth and the position from the timestamp-based synchronization. The distribution of the offsets is shown in Figure 2, which clearly indicates that a timestamp-based synchronization approach is not suitable to be taken as a ground truth because even half a second offset between two concurrent recordings causes a heavily noticeable lag in the audio and video tracks, and larger lags make it often even impossible to perceive two recordings as concurrent. The majority of offsets is greater than one second, and the manually generated ground truth is therefore essential for the development and evaluation of synchronization-dependent methods. A few clips had enormous offsets because the clocks of the recording devices were not set correctly, resulting in timestamps years behind (around January 2000).

5.3   Evaluation

To demonstrate the usefulness of our ground truth, we chose to evaluate the synchronization performance of the well known audio fingerprinting algorithm by Haitsma and Kalker [6] by measuring the preciseness of the calculated synchronization points. This method has been shown to be a promising method for media synchronization in [16] and [3], and we had it already implemented in our own synchronization tool. We applied it with the default parameters as described in the original paper on each of the five events in the Jiku dataset, which yielded 2020 synchronization points in total. Figure 3 shows a histogram distribution of their offsets from the ground truth, binned in steps of 5 milliseconds. Most of the synchronization points are within the range of ±50 ms; 140 are outside the 100 ms range of which most are false positives that are off by many minutes and connect

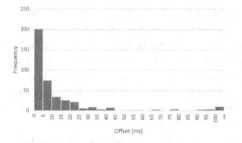

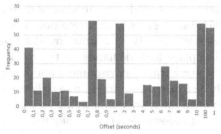

Fig. 1. Distribution of the calculated cross-correlation offsets from the manually generated synchronization points

Fig. 2. Distribution of the error offsets between the timestamp synchronization and the synchronization ground truth

completely unrelated clips. The 95% confidence interval of the mean is between 21.2 ms and 22.8 ms. To test our hypothesis that cross-correlation might improve synchronization results, we applied it on all synchronization points by correlating 1-second audio signal excerpts centered around the positions they point to. This post-processing step improved the fingerprinting results significantly by shifting them towards smaller offsets and almost tripling the synchronization points in the range of ±5 ms. The 95% confidence interval of the mean moved down to 10.2 ms-11.4 ms. The improved results are also shown in Figure 3 for comparison.

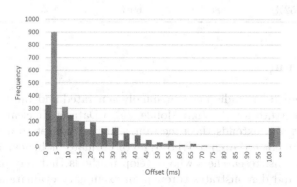

Fig. 3. Histogram distribution of the offsets to the ground truth of all synchronization points as found by the fingerprinting approach (blue), and additionally post-processed by cross-correlation (green)

The overall synchronization rate of the algorithm, which is the number of clips that are covered by the calculated synchronization points, can also be determined with the help of the ground truth. For this, we compared the optimal minimum spanning trees of the overlapping event recordings generated from the ground truth with the minimum spanning trees generated from the computed synchronization points. Table 3 contains for each dataset the number of edges in the optimal MST, the number of determined MST edges by fingerprinting, and the resulting synchronization rate. It shows that this fingerprinting method does not yield satisfying results, owed to the real-life characteristics of the dataset that place high demands on the robustness of synchronization methods due to the uncontrolled environment and heterogeneous sources. There are many heavily distorted audio tracks due to background noise, heavy compression, and poor built-in microphones or analog-to-digital converters that cannot cope with high sound pressure levels like they usually occur at such live events.

Just like we demonstrated the determination of the overall synchronization rate and the individual improvements gained by cross-correlation, our ground truth can be used for the evaluation and comparison of all methods presented in Section 2, where some are expected to perform better. For the fingerprinting method that we evaluated, there are also a few iterative improvements proposed in [8], [2] and [11], which could also be objectively evaluated.

Table 3. Synchronization rate of the fingerprinting method on the Jiku events showing the optimal number of MST edges in the ground truth (MST_{GT}), the achieved number through fingerprinting (MST_{FP}), and the rate in percent

Event	GT_090912	NAF_160312	NAF_230312	RAF_100812	SAF_290512
MST_{GT}	44	63	106	82	102
MST_{FP}	23	54	73	15	78
Rate	52%	86%	69%	18%	76%

6   Conclusion

This paper presents an audio based manually generated and validated synchronization ground truth for the Jiku Mobile Video Dataset. It cleans the dataset from time drift and extends the timestamps in the dataset to a much higher precision. It aims at researchers who want to evaluate or benchmark synchronization algorithms, researchers who develop methods that rely on a synchronized dataset, and demonstrates through an exemplary evaluation experiment how helpful the ground truth can be.

To further improve the dataset, interesting future work could be the determination of the audio to video track offsets to make audio and video data perfectly synchronized at the same time. User studies to determine detectability and acceptability thresholds of offsets between parallel audio tracks are needed to assess the maximum acceptable error offset. Other interesting future work could include the evaluation of different synchronization algorithms on this ground truth to determine the best fit for the evergrowing use-case of crowd sourced mobile video.

Acknowledgments. This work was supported by Lakeside Labs GmbH, Klagenfurt, Austria, and funding from the European Regional Development Fund (ERDF) and the Carinthian Economic Promotion Fund (KWF) under grant 20214/22573/33955. Special thanks go to the authors of the Jiku Mobile Video Dataset for creating and providing it to the community.

References

1. ATSC. Relative Timing of Sound and Vision for Broadcast Operations (IS-191). Advanced Television Systems Committee (June 2003)
2. Baluja, S., Covell, M.: Content fingerprinting using wavelets. In: 3rd European Conference on Visual Media Production, CVMP 2006, pp. 198–207 (November 2006)
3. Duong, N., Howson, C., Legallais, Y.: Fast second screen tv synchronization combining audio fingerprint technique and generalized cross correlation. In: 2012 IEEE International Conference on Consumer Electronics - Berlin (ICCE-Berlin), pp. 241–244 (September 2012)
4. Guggenberger, M., Lux, M., Boszormenyi, L.: Audioalign - synchronization of A/V-streams based on audio data. In: 2012 IEEE International Symposium on Multimedia (ISM), pp. 382–383 (December 2012)
5. Guggenberger, M., Lux, M., Böszörmenyi, L.: An analysis of time drift in handheld recording devices. In: He, X., Xu, C., Tao, D., Luo, S., Yang, J., Hasan, M.A. (eds.) MMM 2015, Part II. LNCS, vol. 8935, pp. 199–209. Springer, Heidelberg (2015)
6. Haitsma, J., Kalker, T.: A highly robust audio fingerprinting system. In: Proceedings of the 3rd International Conference on Music Information Retrieval (ISMIR), Paris, France (2002)
7. ITU. Relative timing of sound and vision for broadcasting (ITU-R BT.1359-1). International Telecommunication Union (November 1998)
8. Ke, Y., Hoiem, D., Sukthankar, R.: Computer vision for music identification. In: IEEE Computer Society Conference on Computer Vision and Pattern Recognition, CVPR 2005, vol. 1, pp. 597–604 (June 2005)
9. Kennedy, L., Naaman, M.: Less talk, more rock: Automated organization of community-contributed collections of concert videos. In: Proceedings of the 18th International Conference on World Wide Web, WWW 2009, pp. 311–320. ACM, New York (2009)
10. Llagostera Casanovas, A., Cavallaro, A.: Audio-visual events for multi-camera synchronization. Multimedia Tools and Applications, 1–24 (2014)
11. Mansoo, P., Hoi-Rin, K., Ro, Y.M., Munchurl, K.: Frequency filtering for a highly robust audio fingerprinting scheme in a real-noise environment. IEICE Transactions on Information and Systems 89(7), 2324–2327 (2006)
12. Moon, S., Skelly, P., Towsley, D.: Estimation and removal of clock skew from network delay measurements. In: Proceedings of the Eighteenth Annual Joint Conference of the IEEE Computer and Communications Societies, INFOCOM 1999, vol. 1, pp. 227–234. IEEE (March 1999)
13. Saini, M., Venkatagiri, S.P., Ooi, W.T., Chan, M.C.: The jiku mobile video dataset. In: Proceedings of the 4th ACM Multimedia Systems Conference, MMSys 2013, pp. 108–113. ACM, New York (2013)

14. Shankar, S., Lasenby, J., Kokaram, A.: Warping trajectories for video synchronization. In: Proceedings of the 4th ACM/IEEE International Workshop on Analysis and Retrieval of Tracked Events and Motion in Imagery Stream, ARTEMIS 2013, pp. 41–48. ACM, New York (2013)

15. Sharma, S., Hussain, A., Saran, H.: Experience with heterogenous clock-skew based device fingerprinting. In: Proceedings of the 2012 Workshop on Learning from Authoritative Security Experiment Results, LASER 2012, pp. 9–18. ACM, New York (2012)

16. Shrestha, P., Barbieri, M., Weda, H., Sekulovski, D.: Synchronization of multiple camera videos using audio-visual features. IEEE Transactions on Multimedia 12(1), 79–92 (2010)

17. Shrestha, P., Weda, H., Barbieri, M., Sekulovski, D.: Synchronization of multiple video recordings based on still camera flashes. In: Proceedings of the 14th Annual ACM International Conference on Multimedia, MULTIMEDIA 2006, pp. 137–140. ACM, New York (2006)

18. Shrstha, P., Barbieri, M., Weda, H.: Synchronization of multi-camera video recordings based on audio. In: Proceedings of the 15th International Conference on Multimedia, MULTIMEDIA 2007, pp. 545–548. ACM, New York (2007)

19. Whitehead, A., Laganiere, R., Bose, P.: Temporal synchronization of video sequences in theory and in practice. In: Seventh IEEE Workshops on Application of Computer Vision, WACV/MOTIONS 2005, vol. 1, 2, pp. 132–137 (January 2005)

Mobile Image Analysis: Android vs. iOS

Claudiu Cobârzan, Marco A. Hudelist, Klaus Schoeffmann,
and Manfred Jürgen Primus

Alpen-Adria-Universität Klagenfurt
9020 Klagenfurt, Austria
{claudiu,marco,ks,mprimus}@itec.aau.at

Abstract. Currently, computer vision applications are becoming more
common on mobile devices due to the constant increase in raw processing
power coupled with extended battery life. The OpenCV framework is a
popular choice when developing such applications on desktop computers
as well as on mobile devices, but there are few comparative performance
studies available. We know of only one such study that evaluates a set of
typical OpenCV operations on iOS devices. In this paper we look at the
same operations, spanning from simple image manipulation like grayscaling and blurring to keypoint detection and descriptor extraction but on
flagship Android devices as well as on iOS devices and with different image resolutions. We compare the results of the same tests running on the
two platforms on the same datasets and provide extended measurements
on completion time and battery usage.

Keywords: mobile devices, OpenCV, performance evaluation, Android,
iOS.

1 Introduction

When it comes to taking photos or shooting short videos, smartphones and
tablet computers are nowadays often preferred over point-and-shoot cameras or
camcorders. Also, they represent a popular alternative to dedicated devices like
mp3 players, mobile games consoles, or even traditional laptops and desktop
PCs when engaged in entertainment activities like listening to music, watching videos, browsing through photos, playing games, etc. This has been made
possible, among other factors, by the increase in processing power and improved
battery life. Multi-core CPUs and GPUs are no longer uncommon both on Apple
and Android devices and installments that provide 64-Bit computing architectures have recently entered the market.

Due to their popularity among consumers as well as their high availability and
versatility, smartphones and tablets tend to become primary computing devices
and, as a result, the interest of computer vision research on such devices is on the
rise. Approaches for performing visual search using mobile devices have already
been proposed [6]. Recently, collaborative video browsing on mobile devices [5]
has been explored for known-item search tasks, as opposed to more standard
solutions which still concentrate on desktop applications [12]. For such mobile

X. He et al. (Eds.): MMM 2015, Part II, LNCS 8936, pp. 99–110, 2015.
© Springer International Publishing Switzerland 2015

approaches, performing feature extraction directly on the device might prove beneficial in various scenarios.

2   Related Work

In [4] the performance and computational cost of various keypoint detection, feature extraction and encoding algorithms are evaluated on a Samsung Galaxy S3 device. To the best of our knowledge, the work in [7] is the only one that deals with OpenCV measurements on iOS mobile devices (other platforms are not considered). Different generations of Apple's iPads and iPhones are used with ground truth measurements being performed on a MacBookPro. The tested OpenCV operations are grouped into three measurement phases as follows: (phase 1) typical computer vision operations: *grayscaling, image blurring* using Gaussian blur, *face* and *edge detection*; (phase 2) *keypoint detection* and *descriptor extraction* operations using SIFT [9] and SURF [2]; (phase 3) *matching descriptors* extracted from two subsequent frames of a video. For the first two phases, a dataset of 5000 images randomly drawn from the MIRFLICKR25000 dataset is used.

The reported results show that high end mobile devices are often only two times slower than common desktop computers. Some devices could match the power of the PC up to 60 percent for certain scenarios concerning face detection and up to 80 percent in terms of grayscaling images. Generation of HSV histograms and Canny edge detection could be performed with about 30 percent of the processing performance of a traditional PC. In terms of keypoint detection and descriptor extraction, the top devices could provide up to 50 percent the performance of the traditional PC in some of the considered cases. Descriptor matching showed a rather good performance in some cases (iPad Air and iPhone 5S), providing about 20 to 65 percent of the performance of a normal PC, depending on the considered descriptors. Those results show that mobile devices have already remarkable performance which will further increase in fast pace over the years, making multimedia content analysis on these devices affordable.

3   Goals and Measurements Setup

The main goal of the current work is to provide a reference as well as a comparison for common OpenCV operations on both Android and iOS mobile devices in terms of completion time and battery usage. We concentrate on those two platforms because they have the largest shares on the smartphone and tablet markets. We consider this important since it gives a clear image about which operations can be performed directly on such mobile devices and which cannot. Previous work [7] has concentrated only on the iOS platform and considered some of the OpenCV operations on a coarser level of granularity (e.g. no details on keypoint detection vs. descriptor extraction completion times). Moreover, to the best of out knowledge, there is no work about battery consumption of commonly used OpenCV operations on mobile devices.

We use the same OpenCV operations as in [7]:

- Phase 1: typical computer vision operations: grayscaling, image blurring using Gaussian blur, face and edge detection;
- Phase 2: keypoint detection and descriptor extraction operations using SIFT [9] and SURF [2].

For both phase 1 & phase 2 we have used 250 pictures drawn from the INRIA Holiday dataset (https://lear.inrialpes.fr/ jegou/data.php where is freely available). The file list is the one at http://ngvb.net/?page_id=158. The pictures all surpass 3 mega pixel and were organized in 3 setups: *(a)* the original 3MP images (noted as the 3MP dataset), *(b)* the images scaled to 50% of the original size (noted as the 3MP_50 dataset) and *(c)* the images scaled to 25% of the original size (noted as the 3MP_25 dataset). The original images vary in size from a minimum of 1536×20148 to a maximum of 2592×3888 (an average of 2547 horizontal by 2016 vertical pixels).

The main motivation in doing so was to perform the tests on data that better reflects the variety of image and video acquisition capabilities of the used devices. Within each set there is minor variance in image resolution since it is expected that future computer vision applications will use heterogeneous data from multiple sources, including data captured by the device as well as data coming from other sources.

The approach was the same as in [7], meaning that for phase 1 & 2 we have tested the corresponding function calls with each image five times in a raw and then we averaged those five measurement times in order to even out possible interventions by the OS. Those averaged times for each image were again averaged to get the overall performance measure for each specific OpenCV function.

The devices used for the experiment include flagship Android devices, namely the Nexus and Galaxy product lines as well as different generations of Apple's iPads and iPhones. The brief specifications of the used devices are presented in Table 1. The CPUs on the iPad Air and the iPhone 5S have a 64 Bit architecture while all the other devices have a 32 Bit CPU architecture. The System On Chip information is presented in Table 2. All Android devices use Android version 4.4.2 except for the GalaxyNote 10.1 which uses Android version 4.3. The iOS devices used version 7.1.2. All devices had OpenCV Release 2.4.9 installed via Google Play and Apple App Store respectively .

4   Results and Discussion

In the following we provide an overview of the obtained results with respect to both completion time and battery usage. It is important to mention that the measured values reflect the performance on a single core of the devices' CPUs since we did not employed custom parallelization or GPU supported calls but just used the default OpenCV functions, which are not parallelized to best of our knowledge. The keypoint detection and descriptor extraction operations

Table 1. Android and iOS devices specification breakdown

Device	CPU		Memory	Screen Size	Screen Resolution	Battery Capacity
	clock	cores				
Galaxy Note 10.1 (2014 Edition)	1.95 GHz	4+4	3 GB	10.1-inch	2560×1600	8220 mAh
Galaxy Note 3	2.3 GHz	4	3 GB LPDDR3	5.7-inch	1920×1080	3200 mAh
Galaxy S4	1.9 GHz	4	2 GB LPDDR3	5.0-inch	1920×1080	2600 mAh
Nexus 7 (2013)	1.5 GHz	4	2 GB DDR3L	7.0-inch	1920×1200	3950 mAh
Nexus 7	1.2 GHz	4	1 GB DDR3L	7.0-inch	1280×800	4325 mAh
Nexus 5	2.3 GHz	4	2 GB LPDDR3	4.95-inch	1920×1080	2300 mAh
iPad Air	1.4 GHz	2	1 GB	9.7-inch	2048×1536	8820 mAh
iPad 4	1.4 GHz	2	1 GB DDR2	9.7-inch	2048×1536	11560 mAh
iPad 3	1 GHz	2	1 GB LP DDR2	9.7-inch	2048×1536	11560 mAh
iPad Mini 1	1 GHz	2	512 MB DDR2	7.9-inch	1024×768	4382 mAh
iPad 2	1 GHz	2	512 MB DDR2	9.7-inch	1024×768	6930 mAh
iPhone 5S	1.3 GHz	2	1 GB LPDDR3	4.0-inch	1136×640	1560 mAh
iPhone 5	1.3 GHz	2	1 GB LPDDR2	4.0-inch	1136×640	1440 mAh
iPhone 4s	800 MHz	2	512 MB DDR2	3.5-inch	960×640	1432 mAh

Table 2. Android and iOS System on Chip information

Device	CPU SoC
Galaxy Note 10.1 (2014 Edition)	Samsung Exynos 5420
Galaxy Note 3	Qualcomm Snapdragon 800
Galaxy S4	Qualcomm Snapdragon 600
Nexus 7 (2013)	Qualcomm Snapdragon S4 Pro
Nexus 7	Nvidia Tegra 3 T30L
Nexus 5	Qualcomm Snapdragon 800
iPad Air	Apple A7
iPad 4	Apple A6X
iPad 3	Apple A5X
iPad Mini 1	Apple A5 (2nd Generation)
iPad 2	Apple A5
iPhone 5S	Apple A7 / Apple M7
iPhone 5	Apple A6

are measured independently. The battery usage measurements were performed with 1% granularity steps for the Android devices and with 5% granularity steps for the Apple devices in accordance with the support provided by the corresponding APIs. The tests were started on all devices with the battery fully charged. This means that there is a ±1% error margin for the Android devices and a ±5% error margin for the iOS devices since it is possible that a previous test has ended just under the threshold and part of it's battery consumption gets counted for the following test.

4.1   Common OpenCV Operations

The common OpenCV operation we have measured (corresponding to the phase 1 measurements we mentioned above) are:

- Grayscaling an image
- Blurring an image using Gaussian Blur
- Face detection in an image
- RGB and HSV histogram calculations
- Edge detection using the Canny algorithm

Blurring was performed using the OpenCV GaussianBlur-function with a kernel size of 21×21 and sigma set to 8.0 in order to produce recognizable result images. For face detection we used a trained cascade classifier and its *detectMultiScale* function. Before the actual detection, the images were grayscaled. The grayscaling was not included in the time measurement. The other used parameters were the scaleFactor set to 1.1, the minNeighbors set to 2 and the minimum size set to 30×30. The RGB histograms used 256 bins within the range of 0 to 256, had uniform set to true and accumulate set to false. The HSV histograms considered 30 hue levels and 32 saturation levels with the standard ranges and used the default values for uniformity (true) and accumulation (false). When detecting edges with the Canny algorithm we first grayscaled the image and then blurred it using a kernel size of 5×5 and a sigma of 1.2 (both operations were not included in the time measurement). After completing those two precursor operations, we measured the Canny function of OpenCV with the thresholds one and two set to 0 and 50 respectively. The measured values are presented in Table 3 and Table 4 for the 3MP and 3MP_50 datasets.

For a visual aid when analyzing the data in Table 3 and Table 4, please refer to Figure 1 and Figure 2. Both use a logarithmic scale for representing the measured values for OpenCV common operations in the case of the 3MP and 3MP_50 datasets. It can be seen that all the Android devices are faster then the fastest iOS device which is for the considered set of operations the iPhone 5S. Among the Android devices the best performances is achieved by the Galaxy Note 10.1. As already noted in [7], the iPad 2 outmatches its successor, the iPad 3 in all measurements. The explanation lies in the fact that since the iPad 3 is equipped with a Retina Display, it processes 4 times more pixels then it's predecessor and this takes its toll on the overall performance despite the newer processor.

Among the considered operation, the most demanding are the face detection, Gaussian blur and Canny edge detection while the least demanding are the HSV histogram and the grayscaling.

The corresponding battery drop levels for completing the considered common operations for the 3MP and 3MP_50 datasets are presented in Table 5. As expected, there is a strong correlation between battery consumption and the time needed to complete the tests.

Table 3. Common Operations (values in ms) - 3MP images

Device	Grayscale	Gaussian Blur	Face Detection	RGB Histogram	HSV Histogram	Canny Edge Detection
Galaxy Note 10.1	8.35	592.62	2471.53	11.64	6.00	145.01
Galaxy Note 3	10.05	680.84	3297.80	19.33	6.40	153.25
Galaxy S4	14.55	886.78	4261.88	28.08	9.96	205.38
Nexus 7 (2013)	14.60	1014.20	3439.15	30.03	9.22	240.46
Nexus 7	11.43	1501.77	3273.17	27.55	20.56	193.18
Nexus 5	7.94	840.87	3494.73	25.19	7.65	144.22
iPad Air	12.95	1260.53	7930.91	69.84	25.06	282.01
iPad 4	21.28	2464.97	10400.87	61.26	33.38	384.51
iPad 3	51.97	4604.35	16556.63	81.00	87.37	797.74
iPad Mini 1	31.26	2769.81	9954.92	47.69	52.92	493.89
iPad Mini 2	13.62	1368.43	8771.00	79.85	27.76	328.01
iPad 2	30.51	2868.00	9842.43	47.09	52.20	488.23
iPhone 5S	8.12	1106.79	6338.24	42.84	15.98	179.60
iPhone 5	13.27	1705.18	6615.14	39.27	21.05	245.03

Table 4. Common Operations (values in ms) - 3MP_50 images

Device	Grayscale	Gaussian Blur	Face Detection	RGB Histogram	HSV Histogram	Canny Edge Detection
Galaxy Note 10.1	2.36	135.85	556.21	4.08	1.82	37.20
Galaxy Note 3	3.33	151.62	734.67	7.90	2.80	44.07
Galaxy S4	3.99	181.88	933.57	9.89	4.06	61.18
Nexus 7 (2013)	4.63	213.22	780.29	8.84	4.30	61.81
Nexus 7	3.33	358.98	798.09	7.11	5.36	46.60
Nexus 5	2.55	220.05	867.93	8.04	3.17	65.85
iPad Air	3.32	270.05	1815.28	19.26	6.17	69.23
iPad 4	6.20	534.86	2666.15	16.02	9.13	98.01
iPad 3	13.35	1077.73	3936.46	20.48	21.95	196.59
iPad Mini 1	13.23	1074.42	3971.17	19.84	21.88	199.14
iPad Mini 2	3.43	289.37	2376.85	22.43	6.88	79.66
iPad 2	12.84	1076.47	3930.68	19.64	21.59	196.88
iPhone 5S	3.43	288.90	2858.57	16.82	6.60	73.12
iPhone 5	6.66	573.81	2951.05	16.91	9.60	102.293

By far, the biggest battery drain can be observed for the face detection operation. This is true for all the devices in our tests. Moreover, in the case of the iPad 3 and the iPhone 5 a full battery charge is not sufficient for completing the face detection test for all the images in the 3MP dataset.

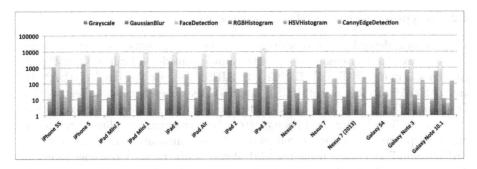

Fig. 1. Measured results of common operations in the case of the 3MP dataset

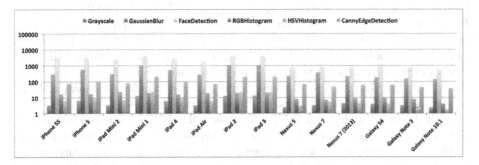

Fig. 2. Measured results of common operations in the case of the 3MP_50 dataset

Table 5. Battery drop levels for common operations (values in %) - 3MP (D1) and 3MP_50 (D2) datasets

Device	Grayscale		Gaussian Blur		Face Detection		RGB Histogram		HSV Histogram		Canny Edge Detection	
	D1	D2	D1	D2	D1	D2	D1	D2	D1	D2	D1	D2
Galaxy Note 10.1	1	0	2	2	40	9	1	0	0	0	3	1
Galaxy Note 3	0	0	13	4	63	14	2	0	1	1	5	1
Galaxy S4	1	0	17	3	86	19	3	0	0	1	5	1
Nexus 7 (2013)	0	0	13	2	86	20	2	0	1	0	5	1
Nexus 7	0	0	17	0	83	18	3	1	2	0	5	1
Nexus 5	1	0	15	5	89	21	2	1	1	0	4	1
iPad Air	0	0	10	0	65	15	0	0	0	0	5	0
iPad 4	0	0	20	5	75	20	5	0	0	0	5	0
iPad 3	5	0	30	5	>100	35	5	0	0	0	10	0
iPad Mini 1	5	0	20	5	80	40	0	0	5	0	5	0
iPad Mini 2	0	0	10	5	70	15	0	0	0	0	5	0
iPad 2	0	0	20	5	75	35	0	5	0	0	5	0
iPhone 5S	0	0	10	5	90	35	0	5	0	0	5	0
iPhone 5	0	0	15	10	>100	45	0	0	5	0	5	0

4.2 Keypoint Detection/Descriptor Extraction

The keypoint detection and descriptor extraction measurements (corresponding to the phase 2 measurements mentioned above) considered built-in OpenCV algorithms, namely ORB [11], BRIEF [3], BRISK [8], SIFT, SURF, FREAK [1] and FAST [10]. In the case of BRIEF and FREAK we used the GoodFeaturesTo-Track algorithm [13] for keypoint detection followed by the descriptor extraction.

In Figure 3 to Figure 6 we have plotted the measurement values corresponding to keypoints extraction as well as descriptor computation for ORB, BRISK, FREAK and FAST while using the 3MP dataset (due to space considerations we do not provide the detailed measurements). The values on the Y axis are in milliseconds. It can be seen that in this case, the measured values for the considered tests on all the devices are similar, except for the iPad Mini 1, iPad 2 and iPad 3. For those 3 devices, the measured values are much higher. Within the Android family, the keypoint detection in the case of ORB clearly outperforms the iOS devices. BRISK shows similar results across all devices with the exception of the iOS devices mentioned earlier. The same is true in the case of FREAK with the observation that the Galaxy Note 10.1 and Nexus 5 provide the fastest results. FAST keypoints detection also takes less on all the Android devices.

In the following we focus on SIFT, SURF as well as BRIEF. The measurements in those cases, also considered the time needed to detect the keypoints as well as the time needed to extract the descriptors based on the detected keypoints. It is important to note that due to patent issues, SIFT and SURF are not included in the official release package of OpenCV for the Android platform, being categorized into the nonfree module. In order to perform the measurements we had to build this nonfree module for Android native projects. One option is to rebuild the whole OpenCV library. We have opted instead for rebuilding only the missing nonfree module[1]. For this we have used the Android NDK, Revision 9c (December 2013).

The exact measured values for the 3MP, 3MP_50 and 3MP_25 datasets are listed in Table 6, Table 7 and Table 8. Although in the case of BRIEF the Android devices achieve better measurement results, for both SIFT and SURF the measured values are much larger than in the case of iOS devices.

For the 3MP dataset, the SIFT measurements could not be performed on any of the devices because the test applications crashed on both Android and iOS with an insufficient memory error message (therefore they are omitted from Table 6). This also happened in the case of the 3MP_50 dataset for the iPad Mini 1 and iPad2. The corresponding values are marked as − in Table 7 and Table 9.

The battery drop information in the case of BRIEF, SIFT and SURF for the 3 considered datasets are provided in Table 9. Please note that for the 3MP dataset, the SURF measurements needed more than one full charge to complete on all tested devices.

---

[1] We took as reference the instructions from
http://web.guohuiwang.com/technical-notes/sift_surf_opencv_android

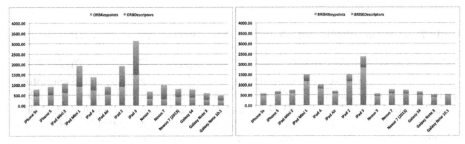

Fig. 3. ORB keypoints and descriptors   **Fig. 4.** BRISK keypoints and descriptors

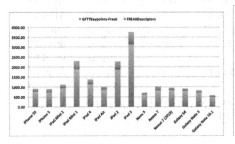

Fig. 5. FREAK keypoints and descrip-   **Fig. 6.** FAST keypoints (no descriptors
tors   computed)

Table 6. Keypoint detection and descriptor extraction (values in ms) - 3MP images

Device	BRIEF		SURF	
	Keypoints	Descriptors	Keypoints	Descriptors
Galaxy Note 10.1	488.78	42.97	82662.90	184723
Galaxy Note 3	677.24	52.20	58601.10	134267.00
Galaxy S4	801.14	75.76	74455.90	173836
Nexus 7 (2013)	824.41	73.78	58248.60	136172.00
Nexus 7	811.50	71.41	45270.00	97972.20
Nexus 5	582.38	48.87	63328.40	148495.00
iPad Air	901.05	54.38	7115.26	20084.18
iPad 4	1134.90	79.62	6552.49	17761.25
iPad 3	3116.00	177.17	12447.22	35201.64
iPad Mini 1	1879.90	126.21	8810.79	22985.91
iPad Mini 2	952.29	56.97	7735.29	21879.85
iPad 2	1866.40	123.42	8586.47	22650.45
iPhone 5S	569.57	36.58	6631.87	18658.79
iPhone 5	729.66	56.50	5120.15	13744.48

Table 7. Keypoint detection and descriptor extraction (values in ms) - 3MP_50 images

Device	BRIEF		SIFT		SURF	
	Keypoints	Descriptors	Keypoints	Descriptors	Keypoints	Descriptors
Galaxy Note 10.1	126.26	16.82	18553.60	20168.90	24220.60	51071.20
Galaxy Note 3	157.51	21.18	21520.80	27542.70	15096.30	33957.30
Galaxy S4	211.50	30.13	26907.70	33548.60	20023.70	46063.10
Nexus 7 (2013)	225.88	33.10	28082.10	33697.40	15718.30	34687.50
Nexus 7	220.16	33.97	38399.30	47780.40	12494.10	27438.70
Nexus 5	180.35	25.91	27057.40	34179.10	18131.50	41822.00
iPad Air	226.52	16.49	3276.46	4780.03	1890.82	5106.70
iPad 4	286.16	25.47	4883.35	6847.57	1846.23	4902.68
iPad 3	756.95	61.63	11560.26	15891.07	3449.07	9811.72
iPad Mini 1	756.18	62.07	−	−	3492.48	9879.97
iPad Mini 2	240.21	17.35	3548.15	5186.26	2236.38	6059.69
iPad 2	751.41	60.96	−	−	3436.40	9795.54
iPhone 5S	218.44	15.75	4209.18	6161.39	2670.63	7242.44
iPhone 5	304.46	26.66	5191.81	7302.17	2085.50	5567.79

Table 8. Keypoint detection and descriptor extraction (values in ms) - 3MP_25 images

Device	BRIEF		SIFT		SURF	
	Keypoints	Descriptors	Keypoints	Descriptors	Keypoints	Descriptors
Galaxy Note 10.1	33.59	7.89	4833.04	5371.13	6201.35	12579.20
Galaxy Note 3	32.55	8.00	4757.97	6533.47	4156.58	8940.18
Galaxy S4	47.74	11.77	6283.44	8221.01	5253.89	11394.90
Nexus 7 (2013)	50.72	12.12	7233.23	9301.98	4043.44	8614.13
Nexus 7	50.33	16.41	9445.54	12471.60	3250.60	6820.61
Nexus 5	39.55	9.75	5940.75	8033.79	4888.02	10580.00
iPad Air	58.52	5.76	914.51	1407.81	587.21	1447.07
iPad 4	76.20	9.38	1185.85	1767.18	564.72	1360.12
iPad 3	187.33	26.09	2874.33	4191.65	905.31	2406.50
iPad Mini 1	186.45	26.56	2955.26	4310.72	919.02	2428.89
iPad Mini 2	61.80	6.09	956.89	1475.02	623.83	1540.25
iPad 2	186.10	25.96	2888.90	4224.87	899.57	2397.98
iPhone 5S	55.33	5.43	1127.66	1742.89	677.56	1674.53
iPhone 5	80.94	9.82	1300.93	1933.76	514.59	1267.40

Table 9. Battery drop levels (values in %) for the 3MP (D1), 3MP_50 (D2) and 3MP_25 (D3) datasets for keypoint detection and descriptor extraction

Device	BRIEF			SIFT			SURF		
	D1	D2	D3	D1	D2	D3	D1	D2	D3
Galaxy Note 10.1	9	2	1	–	55	15	>100	>100	30
Galaxy Note 3	15	3	1	–	93	19	>100	98	24
Galaxy S4	18	5	2	–	>100	25	>100	>100	33
Nexus 7 (2013)	15	4	1	–	94	18	>100	>100	33
Nexus 7	20	6	1	–	>100	35	>100	>100	26
Nexus 5	15	4	1	–	>100	25	>100	>100	34
iPad Air	5	5	0	–	60	15	>100	55	15
iPad 4	10	5	0	–	80	20	>100	45	15
iPad 3	25	10	0	–	>100	45	>100	100	25
iPad Mini 1	10	5	5	–	–	55	>100	>100	25
iPad Mini 2	10	0	0	–	70	15	>100	60	15
iPad 2	15	5	0	–	–	50	>100	100	25
iPhone 5S	10	5	0	–	>100	35	>100	>100	35
iPhone 5	10	5	0	–	25	40	>100	>100	30

5 Conclusion

In this paper we showed results for performance measurements of common OpenCV functions executed on Android and iOS mobile devices. Our measurements were grouped into common operations, keypoint detection and descriptor extraction. The obtained results show that most modern mobile devices can achieve good and in some cases excellent results for the considered operations, meaning that powerful computer vision applications will be possible on such devices in the near future. Overall, better performance was achieved with Android devices. We speculate that this is mainly due to the more powerful hardware. We did not further investigate this issue since this does not constitute the main focus of our work. Another aspect that might have impacted the results, but on a smaller scale, is the difference in compiler implementations used for building the OpenCV library for the considered architectures. Future work will concentrate on parallel and GPU assisted computations for the OpenCV operations on Android, iOS and Windows platforms.

Acknowledgments. This work was funded by the Federal Ministry for Transport, Innovation and Technology (bmvit) and the Austrian Science Fund (FWF): TRP 273-N15 and by Lakeside Labs GmbH, Klagenfurt, Austria, and funding from the European Regional Development Fund (ERDF) and the Carinthian Economic Promotion Fund (KWF) under grant KWF-20214/22573/33955.

References

1. Alahi, A., Ortiz, R., Vandergheynst, P.: Freak: Fast retina keypoint. In: IEEE Conference on Computer Vision and Pattern Recognition (CVPR), pp. 510–517 (2012)
2. Bay, H., Tuytelaars, T., Van Gool, L.: SURF: Speeded up robust features. In: Leonardis, A., Bischof, H., Pinz, A. (eds.) ECCV 2006, Part I. LNCS, vol. 3951, pp. 404–417. Springer, Heidelberg (2006)
3. Calonder, M., Lepetit, V., Strecha, C., Fua, P.: BRIEF: Binary robust independent elementary features. In: Daniilidis, K., Maragos, P., Paragios, N. (eds.) ECCV 2010, Part IV. LNCS, vol. 6314, pp. 778–792. Springer, Heidelberg (2010)
4. Chatzilari, E., Liaros, G., Nikolopoulos, S., Kompatsiaris, Y.: A comparative study on mobile visual recognition. In: Perner, P. (ed.) MLDM 2013. LNCS, vol. 7988, pp. 442–457. Springer, Heidelberg (2013)
5. Cobârzan, C., Hudelist, M.A., Del Fabro, M.: Content-based video browsing with collaborating mobile clients. In: Gurrin, C., Hopfgartner, F., Hurst, W., Johansen, H., Lee, H., O'Connor, N. (eds.) MMM 2014, Part II. LNCS, vol. 8326, pp. 402–406. Springer, Heidelberg (2014)
6. Girod, B., Chandrasekhar, V., Chen, D.M., Cheung, N.-M., Grzeszczuk, R., Reznik, Y.A., Takacs, G., Tsai, S.S., Vedantham, R.: Mobile visual search. IEEE Signal Processing Magazine (2011)
7. Hudelist, M.A., Cobârzan, C., Schoeffmann, K.: Opencv performance measurements on mobile devices. In: Proc. of Int. Conf. on Multimedia Retrieval ICMR 2014, Glasgow, United Kingdom, April 01-04, pp. 479–482 (2014)
8. Leutenegger, S., Chli, M., Siegwart, R.: Brisk: Binary robust invariant scalable keypoints. In: IEEE Int. Conf. on Computer Vision (ICCV), pp. 2548–2555 (2011)
9. Lowe, D.: Distinctive image features from scale-invariant keypoints. International Journal of Computer Vision 60(2), 91–110 (2004)
10. Rosten, E., Drummond, T.: Machine learning for high-speed corner detection. In: Leonardis, A., Bischof, H., Pinz, A. (eds.) ECCV 2006, Part I. LNCS, vol. 3951, pp. 430–443. Springer, Heidelberg (2006)
11. Rublee, E., Rabaud, V., Konolige, K., Bradski, G.: Orb: An efficient alternative to sift or surf. In: IEEE International Conference on Computer Vision (ICCV), pp. 2564–2571 (2011)
12. Schoeffmann, K., Ahlström, D., Bailer, W., Cobârzan, C., Hopfgartner, F., McGuinness, K., Gurrin, C., Frisson, C., Le, D.-D., Del Fabro, M., Bai, H., Weiss, W.: The video browser showdown: A live evaluation of interactive video search tools. International Journal of Multimedia Information Retrieval (2014)
13. Shi, J., Tomasi, C.: Good features to track. In: IEEE Computer Society Conference on Computer Vision and Pattern Recognition, pp. 593–600 (1994)

Dynamic User Authentication Based on Mouse Movements Curves

Zaher Hinbarji, Rami Albatal, and Cathal Gurrin

Insight Centre for Data Analytics,
Dublin City University
zaher.hinbarji@insight-centre.org
https://www.insight-centre.org

Abstract. In this paper we describe a behavioural biometric approach to authenticate users dynamically based on mouse movements only and using regular mouse devices. Unlike most of the previous approaches in this domain, we focus here on the properties of the curves generated from the consecutive mouse positions during typical mouse movements. Our underlying hypothesis is that these curves have enough discriminative information to recognize users. We conducted an experiment to test and validate our model in which ten participants are involved. A back propagation neural network is used as a classifier. Our experimental results show that behavioural information with discriminating features is revealed during normal mouse usage, which can be employed for user modeling for various reasons, such as information asset protection.

Keywords: multimedia security, user modeling, human-computer interaction, mouse dynamics.

1 Introduction

We take care to encrypt or password protect our computers so that we can restrict unauthorised access. However, we don't typically consider the situation of 'friendly' unauthorised access to our private personal archives. Consider the people that have access to your physical computing devices and can theoretically access your content. This work addresses this aspect of multimedia content storage and protection by proposing a user authentication method based on mouse movements alone. One could easily imagine the computer immediately locking out a user who accesses the private data of the data owner. It is common to have three security procedures working together to protect information assets and to ensure the controlled access. These procedures are: authentication, authorization and auditing [1].

We focus here on the first component of security systems, personal identification and authentication, which can be done by something the user knows (e.g., password, PIN code, pattern), something the user has (e.g., card, access token, wrist band) or something the user is or does (e.g., a fingerprint, signature, face, voice, which are known as biometrics) [2]. The need for special hardware devices

X. He et al. (Eds.): MMM 2015, Part II, LNCS 8936, pp. 111–122, 2015.

for data capture is a big limitation of most biometric systems. The advantage of a mouse-based authentication system is that it can be implemented using a regular mouse [2], [3], [4], [5]. User authentication can be achieved statically or dynamically. In the static approach, the system checks the identity of the user once, usually at the beginning of the session so any change of user after that will be unnoticeable to the system. In contrast, dynamic verification checks the user continuously over the session which can effectively prevent session hijacking, however that should be done passively without interrupting the user [6]. In this work, we present a dynamic user authentication based on mouse movements only and using normal mouse devices.

2   Related Work

Several mouse dynamics approaches for dynamic authentication have been proposed in the literature, presenting different types of features. Hayashi et al. [7] presented one of the earliest research in this domain; users were requested to use the mouse for drawing circles or other figures, and then analytics algorithms were applied on features based on the distances between the mouse coordinates and the centre of the shapes. In [2], Pusara and Brodley used the distance, angle and speed between pairs of data points as raw features which then used to produce their mean, standard deviation and the third moment values (distance, angle and speed) over a window of N data points. In Ahmed and Traore's work [3], raw mouse events are aggregated and then classified by action type. Consecutive actions are grouped into sessions, from which features related to movement speed, movement direction, traveled distance are computed producing user signature. Schulz in [5] presented a model in which raw data are broken into mouse curves; length, curvature and inflection points of the curve are used as main features, and a reference signature is built by generating histograms from the curve characteristics of multiple curves. The verification is implemented then by computing the Euclidean distance between the reference signature and the mouse activity observed during authentication time. Jorgensen and Yu [8] evaluated the approaches done by [2] and [3] and discussed some of the limitations in their works. According to [8] it is not clear whether the system detects the differences in mouse behavior or differences among the working environment including the performed task.

To overcome the above mentioned drawback of the state-of-the-art methods, our main challenge is to extract features that can reflect the user behavior patterns regardless of both the task that she/he is performing and the environment she/he is working in. Our underlying hypothesis is that mouse curves have such user-specific information that can be used to model users behavior. Our approach is similar to the one followed by Schulz [5] in using histograms of features extracted from the mouse curves. However, the two approaches differ in two major aspects. First, we use different features and emphasise on using ones that satisfy certain mathematical properties related to task-independence. Second, [5] uses a single 'reference' signature per user. The Euclidean distance between the evaluated signature and reference signature is then used to validate the user. Our method, on the other hand, uses instead a neural network per user which is trained using multiple signatures. The neural network has the potential of better recognition by generalising from different signature training samples.

3 Behavior Modeling

In order to make our approach practical enough to be used in typical working environment, our model uses raw mouse coordinates collected from users during their normal workday activities without any restrictions. The consecutive mouse coordinates are grouped into curves that correspond to the typical performed mouse actions (point-and-click, move, drag-drop). A single curve does not have enough information by itself to refer to its user, that is why we group the curves into sessions in order to study the statistical behavior characteristics observed during each session. A session is a number of consecutive curves belonging to the same user. To show how results are affected by the length of the session we will present the accuracy of the system in terms of 3 different values of session length: 100, 200 and 300 curves.

The objective of our method is to verify the identity of the user based on features of the curves followed when moving the mouse from one point to another. The exact values of those features may vary considerably even for curves belonging to the same user. However, we assume that each feature follows a probability distribution that is unique to each user and can serve as a signature of his/her mouse movements. The probability distribution of each feature is approximated by a normalized histogram computed using a large number of curves belonging to a certain user (session). The histograms of different features belonging to a certain user form the signature of that user, which is the input of our detection algorithm.

4 Mouse Curve Features

In this section, we describe the nine different features used to characterize a single mouse curve. Each feature gives a single value as a descriptor of the curve, except for inflection profile, sharpness profile and central moments that generate five, five and three values respectively. As a result, we have in total 19 values describing each curve. To the best of our knowledge we are the first who introduced these features in this domain except for straightness and inflection profile which are previously used before by [5]. However, we introduce here our own implementation for these features.

A mouse curve is defined as a tuple (ordered list) of two or more 2D points:

$$C = (p_1, p_2, ..., p_n) : n \geq 2, \ p_i = (x_i, y_i) \in \mathbb{R}^2 , \tag{1}$$

where n is the number of points.
A feature of this curve is simply a function of the coordinates of its points:

$$F(C) : \mathbb{R}^{2n} \to \mathbb{S} \subseteq \mathbb{R} , \tag{2}$$

where $\mathbb{S}$ is a subset of the real numbers.

Since those features will be used to characterize a user, they should be task independent, which implies that any feature should be independent of the position, size and orientation of the curve. This means that feature functions should satisfy the following mathematical properties:

– **Translational Invariance**

$$F(p_1 + q, p_2 + q, ..., p_n + q) = F(p_1, p_2, ..., p_n) , \qquad (3)$$

where $q \in \mathbb{R}^2$ is a 2D translation vector.
– **Scale Invariance**

$$F(\alpha p_1, \alpha p_2, ..., \alpha p_n) = F(p_1, p_2, ..., p_n) , \qquad (4)$$

where $\alpha \in \mathbb{R}$ is a scaling factor.
– **Rotational Invariance**

$$F(\mathbf{M} \, p_1, \mathbf{M} \, p_2, ..., \mathbf{M} \, p_n) = F(p_1, p_2, ..., p_n) , \qquad (5)$$

where $\mathbf{M} = \begin{bmatrix} \cos\theta & -\sin\theta \\ \sin\theta & \cos\theta \end{bmatrix}$ is a 2D rotation matrix that rotates points counter-clockwise through an angle θ about the origin.

4.1 Efficiency

The goal of a single mouse movement is to go from the first point p_1 to the last one p_n. The shortest path between those two points is a straight line while any other curve will be longer. Efficiency is defined as the ratio of the length of the shortest path over the length of the curve

$$E = \frac{\sqrt{(x_n - x_1)^2 + (y_n - y_1)^2}}{\sum_{i=1}^{n-1} \sqrt{(x_{i+1} - x_i)^2 + (y_{i+1} - y_i)^2}} \in [0, 1] . \qquad (6)$$

This feature measures how *efficient* a curve is, in achieving its goal. The value of efficiency rages between 0 and 1. A curve with value 1 is the shortest path and the most efficient (It should be a straight line but the converse is not necessarily true), while a curve with value near 0 is a one that moves a lot without going too far. A user whose curves have a high efficiency value tends to move the mouse directly between target positions without making many unnecessary movements.

4.2 Straightness

This feature measure how much a curve resembles a straight line. This is done by studying the correlation between the curve points. First, we compute the covariance matrix of the points coordinates:

$$\Sigma = \begin{pmatrix} \sigma_x^2 & \sigma_{xy} \\ \sigma_{yx} & \sigma_y^2 \end{pmatrix} \qquad (7)$$

σ_x^2 and σ_y^2 are the variances along the x and y axes, respectively. σ_{xy} is the covariance of x and y.

Then we compute the eigenvalue decomposition of the covaraince matrix. The largest eigenvalue (let it be λ_1) represents the largest variance of the points along any direction while the second eigenvalue (let it be λ_2) represents the variance along the direction orthogonal to the previous one. As a straightness measure, we use the following relation:

$$S = \frac{\lambda_1 - \lambda_2}{\lambda_1} \in [0, 1] . \tag{8}$$

Straightness value range between 0 and 1. When the relation between x and y is almost linear, λ_1 is much larger than λ_2 and the straightness is near 1. When the relation is non-linear, λ_1 and λ_2 are close and the straightness is near 0 (see Fig. 1). Our measure of straightness is meant to replace the one given by [5]. Their measure is a yes/no measure that does not account for that fact that some curves appear more straight than others while our measure gives a one value when the curve is exactly a straight line and a positive value proportional to how well the curve resembles a straight line.

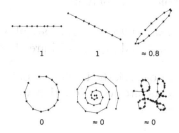

Fig. 1. Different sets of points with their straightness values

4.3   Regularity

This feature measures how regular a curve is by looking at the distances of its points to its geometrical centre.

$$\bar{x} = \frac{1}{n} \sum_{i=1}^{n} x_i, \quad \bar{y} = \frac{1}{n} \sum_{i=1}^{n} y_i$$

$$d_i = \sqrt{(x_i - \bar{x})^2 + (y_i - \bar{y})^2}$$

Regularity is then defined in terms of the mean and standard deviation of those distances as:

$$R = \frac{\mu_d}{\mu_d + \sigma_d} \in [0, 1] , \tag{9}$$

$$\text{where} \quad \mu_d = \frac{1}{n} \sum_{i=1}^{n} d_i, \quad \sigma_d^2 = \frac{1}{n} \sum_{i=1}^{n} (d_i - \mu_d)^2 .$$

Note that curves forming regular polygons, like equilateral triangle or squares, have all their corners at the same distance from their center. This implies that the variance of the distances is zero and thus regularity is one.

4.4   Self-intersection

This feature counts the number of times a curve intersects with itself. To find this number, we test all pairs of line segments for intersection. This approach takes $\mathcal{O}(n^2)$ which is fine because n is small in our case.

Two line segments p_1p_2 and p_3p_4 intersect, if and only if both:

- p_1 and p_2 are on different sides of p_3p_4,
- p_3 and p_4 are on different sides of p_1p_2.

These are respectively true, if and only if :

- Angles $\angle p_1p_3p_4$ and $\angle p_2p_3p_4$ are of opposite signs.
- Angles $\angle p_3p_1p_2$ and $\angle p_4p_1p_2$ are of opposite signs.

These are respectively true, if and only if :

- Cross products $\overrightarrow{p_3p_1} \times \overrightarrow{p_3p_4}$ and $\overrightarrow{p_3p_2} \times \overrightarrow{p_3p_4}$ are of opposite signs.
- Cross products $\overrightarrow{p_1p_3} \times \overrightarrow{p_1p_1}$ and $\overrightarrow{p_1p_4} \times \overrightarrow{p_1p_2}$ are of opposite signs.

As a result, it is sufficient to check the fulfillment of the last two conditions to test whether two line segments intersect. Note that we do not consider touching line segments as intersecting.

4.5   Curvature-Based Features

Mathematically, the *curvature* of a continuously differentiable curve is defined as the rate of change of the tangential angle with respect to the arc length: $\kappa = d\phi/ds$.

Handling curvature of mouse curves is problematic because they are piecewise linear, so the curvature is zero inside the line segments and ill-defined on the corners. One approach to solve this problem is approximating the rough mouse curve by another smooth curve like B-spline (as done in [5]) or Bezier curves and then studying the curvature of the resulting smooth curves. We follow here a different approach.

First let us look at the the line integral of the curvature between two points of the curve

$$\Phi(a,b) = \int_a^b \kappa \, ds = \int_a^b d\phi = \phi(b) - \phi(a) \, . \tag{10}$$

It is called the *total curvature* and it is the total change in the tangential angle. For mouse curves, the total curvature is well-defined and it is a step function that is constant along the line segments and jumps at the corners; The value of the jump at a corner equals to the change in the angle (see Fig. 2).

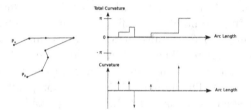

Fig. 2. Curvature and total curvature of a mouse curve (piecewise linear curve)

Since the curvature is the derivative of the total curvature, it is ill-defined at the jumps because of the discontinuity. However, it can still be defined in a distributional sense as a Dirac delta function. The behavior of delta function is well-defined inside integrals and most curvature-based features are defined using integrals. The only feature that goes beyond integration, is the inflection profile but the problem is resolved then by replacing delta functions with Gaussian function.

Formally, the curvature of a mouse curve is then defined as a linear combination of delta functions positioned at the corners and weighted by the angles

$$\kappa(s) = \sum_{i=1}^{n-2} \theta_i \, \delta\left(s - \frac{s_i}{s_{n-1}}\right) \quad \text{where:} \tag{11}$$

$$s_i = \sum_{j=1}^{j=i} \sqrt{(x_j - x_{j+1})^2 + (y_j - y_{j+1})^2}$$

$$\theta_i = \text{sign}(\mathbf{v}_i \times \mathbf{v}_{i+1}) \arccos\left(\frac{\mathbf{v}_i \cdot \mathbf{v}_{i+1}}{\|\mathbf{v}_i\| \|\mathbf{v}_{i+1}\|}\right)$$

$$\mathbf{v}_i = \overrightarrow{p_i p_{i+1}} \, .$$

Notice that in Eq. 11 the positions are rescaled by the length of the curve (s_{n-1}) in order to make the curvature scale-invariant.

Total Angle. This feature measures the total change in angle between the first and last points of the curve. It equals to the integral of the curvature

$$\text{TA} = \int ds \, \kappa(s) = \int ds \sum_{i=1}^{n-2} \theta_i \, \delta(s - s_i) = \sum_{i=1}^{n-2} \theta_i \int ds \, \delta(s - s_i) = \sum_{i=1}^{n-2} \theta_i \, . \tag{12}$$

The last equality holds because the integral of a Dirac delta function is unity.

Bending Energy. This feature measures the total change of angles regardless of the sign. It equals to the L^1-norm of the curvature

$$\text{BE} = \int ds \, |\kappa(s)| = \int ds |\sum_{i=1}^{n-2} \theta_i \, \delta(s - s_i)| = \sum_{i=1}^{n-2} |\theta_i| \int ds \, \delta(s - s_i) = \sum_{i=1}^{n-2} |\theta_i| \, . \tag{13}$$

Inflection Profile. Inflection point is where the curve changes the sign of its curvatures. Computing the number of inflection points naively, by counting the number of sign changes of θ_i, results many spurious inflection points because of the noise on the curvature. Therefore, we need to smooth the curvature by replacing the delta functions with Gaussian functions of the same weight and certain width (see Fig. 3). The number of sign inflection points is the number of sign changes in the smoothed curvature. In order to avoid biasing the result to a certain smoothing level (certain width of the Gaussian), we use the different number of inflection points at five different smoothing levels as features. Taken together, we call these features *inflection profile*.

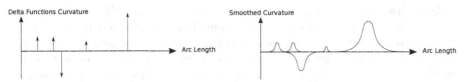

Fig. 3. Curvature can be smoothed by replacing delta functions with Gaussian functions of the same weight and at the same postion

Sharpness Profile. We define a sharp bend as an absolute change in the angle by more than $\pi/2$. A sharp bend may happen on a single corner or may take several corners. To find the number of sharp bends requiring m corners, we sum each m consecutive angles θ_i and then count the values that are greater than $\pi/2$ in absolute value.

We compute the number of sharp bends requiring five different number of corners. Taken together, these features are called *sharpness profile*.

Center and Central Moments. The n-th moment of a normalized positive function around point c is defined as

$$\mu'_n(c) = \int ds \, (s - c)^n \, f(s) \,. \tag{14}$$

The first moment around 0 is called the mean $\mu := \mu'_1(0)$ and the moments around the mean $\mu_n := \mu'_n(\mu)$ are called central moments. Knowing the mean and the first few central moments of a function gives a rough idea about its shape.

We use the moments of the normalized absolute value of the curvature $f(s) = |\kappa(s)|/\int ds \, |\kappa(s)|$ as features. They are computed using the following relations which are obtained by substituting $f(s)$ in Eq. 14:

$$\mu = \frac{\sum_{i=1}^{n-2} s_i \, |\theta_i|}{\sum_{i=1}^{n-2} |\theta_i|} \tag{15}$$

$$\mu_n = \frac{\sum_{i=1}^{n-2} (s_i - \mu)^n \, |\theta_i|}{\sum_{i=1}^{n-2} |\theta_i|} \tag{16}$$

We used the values of the first 3 moments in our feature vectors.

5   Behavior Comparison

As described earlier, we do not use the features themselves for classification but rather their probability distribution. This probability distribution can be approximated using a normalized histogram of the feature values. We determine the binning of the histogram using the equal-frequency discretization algorithm [9]. Given the feature values of all the users together, we sort all the values in ascending order. If the number of required bins is n, we use every nth value of the sorted data as a bin marker, plus the first and last values. Regarding the number of bins, we use eight bins per feature; this has heuristically given good results in approximating the underlying distributions of the features. As a result, our final signature vector has 152 values, which is an 8-value histogram for each one of our 19 features concatenated together.

Authenticating users using a signature is done via artificial neural networks. Each user has his/her own neural network which is trained to recognize his/her signature alone. A neural network automatically selects the most discriminating factors in the features vector by over-weighting the most discriminating features and ignoring the less important ones. The neural network used in our approach is a feed-forward multilayer perceptron network of three layers. The input layer consists of 152 nodes, corresponding to the length of our feature vectors. The hidden layer consists of 50 nodes whereas the network has only one node in the output layer. The output of the network is in the range [0,1]. During training, positive examples are assigned value 1 and negative examples are assigned value 0. However, due to the interpolating nature of the neural network, the output for new examples will lie in between. Therefore, a threshold limit is used to authorize the claimed identity.

To train a neural network to recognise sessions for a targeted user, we consider all the sessions that belong to that user in the training set as positive examples and all other sessions as negative examples. This was carried out in a similar manner for all users. In real life scenario, the first sessions of the user are used passively to build his/her signature. User authentication is the final step in the process; in order to verify the claimed identity of a session we first extract the features vector of that session and use it as an input for the network specifically trained to recognize that identity. Then, the output of the network is compared to an authentication threshold to ensure that it is sufficient enough to authenticate that captured behavior.

6   Evaluation

In order to test our model in normal working environment, our data set is collected from users during their regular workday activities without any kind of pre-defined tasks or restrictions. We developed a background app that intercepts raw mouse events passively and save them for later analysis. Ten users participated in this experiment using Mac OS without changing their mouse settings. The collected data is only mouse coordinates with an indicator of the current performed action (point-and-click, move, drag-drop). Around 16,500 actions are collected for each user during about 24 working hours. Our evaluation is done using three measures: false acceptance rate (FAR), false rejection rate (FRR) and equal error rate (EER). FAR measures the likelihood that the system incorrectly accepts an access attempt by an intruder. Whereas, FRR measures the likelihood that the system incorrectly rejects an access attempt by an authorized user. EER is the value at which both acceptance and rejection errors are equal due to tuning the authentication threshold of the system. The sessions of each user are divided into two equal subsets: training set and testing set. After training the system on the training set only, we test the system by counting the number of misclassifications against all the sessions in the testing set. FRR is computed by counting the number of misclassifications when both the testing session and the trained neural network belong to the same user. On the other hand, FAR is computed by counting the number of misclassifications when the testing session and the trained neural network belong to different users. To make sure that FAR also covers the case of an attacker who has not been seen before by the system, the previous process is repeated ten times (according to the number of users). In each time, one of the users is considered as an outsider; his sessions are excluded during the training phase and are added to the testing set. The total number of misclassifications over the ten times is used to calculate FAR and FRR. To find the value of EER, we conduct the testing by varying the authentication threshold between 0.1 and 0.9. Figure 4 shows how the values of FAR decrease and the FRR increases as the authentication threshold increases for sessions of length 100 curves. The higher the threshold, the lower the probability the system incorrectly accepts an intruder and the higher the probability the system incorrectly rejects an authorized user. The shape of ROC curves for the two other values of session length (200,300) are similar to Figure 4. As expected, the EER decreases when we increase the length of the session. Increasing the session length allow the system to detect more behavioral patterns. However, long sessions give the attacker more time to finish his/her attack before the system detects him/her. Since the session length measured by the number of actions (curves), the actual needed time to authenticate a user may vary considerably even for sessions of the same user depending on how much time the user needs to generate the sufficient mouse actions. Table 1 shows the EER and the corresponding total average time our subjects took to generate the needed curves.

As we introduced before, [5] presented a similar model based on mouse curves too. EER values reported by [5] are 20.6%, 16.6% and 11.1% for sessions of 120, 300 and 3600 curves respectively. By comparing the previous values to our reported error rates (Table 1) we can see that our model achieves good results toward a reliable task-independence mouse based authentication system.

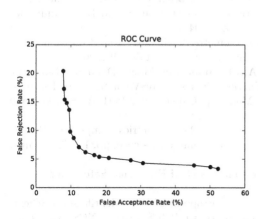

Threshold	FRR(%)	FAR(%)
10%	3.3	52.3
15%	3.6	50.1
20%	3.9	45.4
25%	4.3	30.7
30%	4.8	27.1
35%	5.2	20.8
40%	5.4	18.1
45%	5.7	16.7
50%	6.2	14.2
55%	7.1	12.3
60%	8.7	10.8
65%	9.8	9.8
70%	13.6	9.5
75%	14.8	8.8
80%	15.4	8.2
85%	17.3	8.0
90%	20.4	7.8

Fig. 4. FAR and FRR values according to different authentication thresholds for sessions of length 100 curves and the corresponding ROC curve

Table 1. EER values and the corresponding session length

Session Length (#Curves)	Session Length (Time)	EER	Threshold
100	5.6 min	9.8%	65%
200	14.3 min	7.2%	50%
300	18.7 min	5.3%	55%

7   Conclusions

In this work we have proposed an approach to model the normal human-computer interaction via mouse device that can be used to control access to information asset, multimedia archives and personal devices by continuously authenticating the current user of the system. Our main contribution is introducing mouse-curve based features that satisfy certain mathematical properties related to task-independence. Experimental results of ten subjects achieved an EER of 5.3 percent. Future research will extend the subjects involved in the evaluation process and test consistency and invariance of the model over time. Besides, a statistical analysis will be useful to show the contribution of each feature in the final decision, which can also help us to know how comprehensive the features set is.

Acknowledgments. This publication has emanated from research conducted with the financial support of Science Foundation Ireland (SFI) under grant number SFI/12/RC/2289. We would like to express our deep gratitude to Dr. Ammar Joukhadar and Mr. Khaldoon Ghanem for their valuable support and constructive recommendations on this project.

References

[1] Todorov, D. (ed.): Mechanics of User Identification and Authentication:Fundamentals of Identity Management. Auerbach Publications (2007)

[2] Pusara, M., Brodley, C.E.: User Re-authentication via Mouse Movements. In: Proceedings of the 2004 ACM Workshop on Visualization and Data Mining for Computer Security, pp. 1–8. ACM, USA (2004)

[3] Ahmed, A.A.E., Traore, I.: A New Biometric Technology Based on Mouse Dynamics. IEEE Trans. Dependable Sec. Comput. 4(3), 165–179 (2007)

[4] Nakkabi, Y., Traore, I., Ahmed, A.A.E.: Improving Mouse Dynamics Biometric Performance Using Variance Reduction via ExtractorsWith Separate Features. IEEE Transactions on Systems, Man and Cybernetics, Part A: Systems and Humans 40(6), 1345–1353 (2010)

[5] Schulz, D.A.: Mouse Curve Biometrics. In: 2006 Biometrics Symposium: Special Session on Research at the Biometric Consortium Conference, pp. 1–6 (September 2006)

[6] Denning, D.E.: An Intrusion-Detection Model. IEEE Trans. Softw. Eng. 13(2), 222–232 (1987)

[7] Hayashi, K., Okamoto, E., Mambo, M.: Proposal of User Identification Scheme Using Mouse. In: Han, Y., Quing, S. (eds.) ICICS 1997. LNCS, vol. 1334, pp. 144–148. Springer, Heidelberg (1997)

[8] Jorgensen, Z., Yu, T.: On Mouse Dynamics As a Behavioral Biometric for Authentication. In: Proceedings of the 6th ACM Symposium on Information, Computer and Communications Security, ASIACCS 2011, pp. 476–482. ACM, New York (2011)

[9] Kotsiantis, S., Kanellopoulos, D.: Discretization Techniques: A recent survey (2006)

Sliders Versus Storyboards – Investigating
Interaction Design for Mobile Video Browsing

Wolfgang Hürst and Miklas Hoet

Department Information & Computing Sciences, Utrecht University, The Netherlands
huerst@uu.nl

Abstract. We present a comparative study of two different interfaces for mobile video browsing on tablet devices following two basic concepts - storyboard
designs representing a video's content in a grid-like arrangement of static images extracted from the file, and slider interfaces enabling users to interactively
skim a video's content at random speed and direction along the timeline. Our
results confirm the usefulness and usability of both designs but do not suggest a
clear benefit of either of them in the direct comparison, recommending – among
other identified design issues – an interface integrating both concepts.

Keywords: Mobile interfaces, Mobile video browsing, Interactive multimedia.

1   Introduction

The increasing availability of videos on mobile devices also results in a growing need
for better search and browsing functionality – not just in video archives but within
single files as well. Confirming predictions of earlier studies [1], users these days
want to interactively skim video content, for example, to find parts of particular relevance or skip ones of minor interest. Especially in situations when the searched
information is unknown or hard to describe by queries and when watching videos for
fun, users often rely on personal, interactive skimming of video content in addition to
query-based retrieval. Yet, browsing videos is traditionally a difficult problem due to
the transient nature of the continuous signal. In contrast to static information, such as
text and images, videos are time-dependent, i.e., only a fraction of their content is
displayed at a time. Most common approaches for video browsing therefore usually
follow one of two principles: they either give people more interactive control over the
timeline or dissolve a video's transient nature by representing the dynamic signal via
a static one. Maybe the most omnipresent example for the latter approach are so-
called storyboards, where thumbnails of still images extracted from the video are used
for a time-sorted, grid-like static representation of the content that is easy to browse
by humans. Examples enabling users to manipulate a video's timeline in order to
flexibly browse its content include simple fast forward or slow motion (e.g., to quickly skim a video and to investigate a particular area of interest in detail, respectively)
and, maybe most important, slider interfaces that represent a video's timeline and
allow users to interactively and flexibly "scroll" though its content by providing

X. He et al. (Eds.): MMM 2015, Part II, LNCS 8936, pp. 123–134, 2015.
© Springer International Publishing Switzerland 2015

real-time feedback when dragging the slider's thumb. Applying such approaches to mobile devices, such as smartphones and tablets, introduces a whole set of additional problems, mostly due to their form factor and resulting limited screen estate. Motivated by the ultimate goal to create better browsing interfaces for such mobiles, we present a comparative study between two interface designs for video browsing on tablets – one featuring a storyboard, and one with extended and optimized timeline sliders. Our research aims at identifying advantages and disadvantages of each concept, verifying their usefulness for mobile interface design, and gaining knowledge about how to best integrate them into common video players, considering search and browsing performance as well as usability and personal user preferences. After discussing related work (Section 2), we introduce two interface designs (Section 3) that are evaluated in a comparative study providing answers to aforementioned questions (Section 4) and identifying general design issues and alleys for potential future research (Section 5).

2   Related Work

Video browsing is an important and relevant topic that has gained lots of attention for desktop computing. In the following, we will mostly discuss related work for mobile computing and focus on browsing via interactive exploration, in particular via storyboards and slider interfaces. For excellent general reviews we refer to [2] and [3].

Interfaces using static representations of video content (i.e., thumbnails created from extracted video frames) have been introduced and researched in various incarnations. Examples include different arrangements of such thumbnails, such as film stripes [4] – nowadays often used when browsing video archives – and storyboards [5], i.e., matrix-like structures with thumbnails temporally sorted line-by-line from top left to bottom right – nowadays often used to represent the content of long, single video files. Special cases such as comic-book-style visualizations [6] and 3D representations [7] exist as well. Other researchers investigated the benefit of replacing still images with "moving" thumbnails, i.e., small clips extracted from the video [4], [8]. Related work on mobile devices mostly focused on investigating optimum thumbnail sizes for small screens [9], still versus moving thumbnails [10], and interaction techniques such as paged versus continuous scrolling [11]. Related results, for example, with respect to optimum thumbnail sizes have been considered in our interface designs presented in the next section.

To the best of our knowledge, there is no comparative evaluation of storyboards with interactive timeline-based video browsing using slider interfaces yet. Considering the latter, we can however observe a trend of integrating storyboard-like visualizations into the browsing process, for example, by visualizing related thumbnails on top of the slider's thumb while dragging – as done in the popular YouTube online video platform (cf. Fig. 1). However, especially with small screen sizes on mobile devices, moving the slider even just one pixel can result in a relatively large jump in the file, thus not enabling users to explore content in detail. Several researchers have addressed this *granularity problem* with different interface designs. Examples from mobile video browsing include circular and elastic interfaces [12] as well as the

ZoomSlider [13] where skimming a video is done by left/right gestures on the screen and browsing granularity depends on the gesture's vertical location. While this enables users to flexibly navigate a file at different levels of granularity, it introduces another problem, namely that content is often covered by the user's fingers. [14] addresses this issue with an interface design optimized for smartphone-sized screens that can be operated with one hand. Accidental content hiding is reduced to a minimum by placing the thumbnails to the left of the screen and browsing them via gestures made on the right side. Yet, the solution does not account for the aforementioned granularity problem. [15] addresses both issues with a video browsing interface for tablets that can be operated with one's thumbs when holding the device in both hands. The right side features a thumbnail-enhanced slider for video browsing, while the left side of the screen enables users to activate different functions, including an option to modify the right thumbnail-slider's scale. Their idea of using such "vertical" sliders to make them easier to reach and limit accidental information hiding served as inspiration for the enhances slider interface we introduce in the next section.

3   Interface Designs

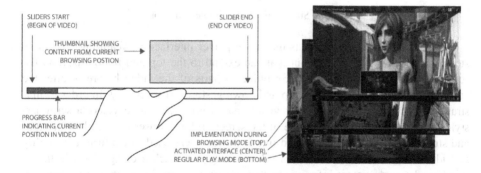

Fig. 1. State of the art video browsing interface implementation

Motivated by the state of the art and as starting point for our research, we implemented an ***initial player interface*** based on the popular YouTube design (cf. Fig. 1). It features a slider at the bottom, representing the length of the video, with a thumb that enables users to quickly access all positions along this timeline. The interface is hidden during normal playback but becomes visible as soon as a user touches the screen. To provide real-time feedback during browsing, a thumbnail image representing the corresponding content is displayed on top of the slider's thumb while it is dragged. Informal testing confirmed obvious advantages such as flexibility, familiarity, and the fact that location feedback with respect to the whole video file is provided. Yet, we also saw that the design suffers from the granularity problem mentioned in Section 2, and the single thumbnail providing only limited content information. In order to deal with these two issues, we introduce two alternative designs that we evaluate in a comparative study in the next section – one storyboard-based, in order to deal with the latter problem, and a modified, enhanced slider interface addressing the granularity issue while also providing more content information during browsing.

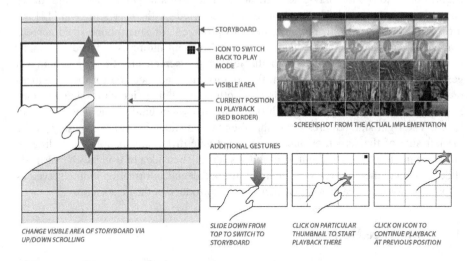

Fig. 2. Storyboard design implementation

Our ***storyboard design*** extends the initial player interface by featuring a separate storyboard with a 5x5 grid layout that can extend to the top and bottom beyond the screen, as illustrated in Figure 2. Scrolling to parts of the video before or after the currently visible area is done via up and down gestures, respectively. In order to illustrate the location of the currently visible part within the whole video, a scrollbar-style icon is added to the right side of the screen. Informal user tests with colleagues and students otherwise not involved in this project were used to confirm its usability and identify potential issues. *Potential advantages* of such a design include that it provides detailed content information that is easy to skim due to its static character. The layout is usually considered intuitive and easy to understand following common interaction designs that people are familiar with. *Potential disadvantages* are that the additional display of content also enforces additional interaction (up/down scrolling) and the need to switch between two separate screen layouts for playback and browsing. For example, in our concrete implementation, switching from player to storyboard view is done either by clicking on a related icon at the top of the screen (in pause mode) or by a down gesture starting at the top of the screen (in pause and playback mode). It is also not clear how accurately and easily people can spot relevant parts despite the intuitive layout and how well they are able to handle it in a real-world situation. For example, direct interaction is usually considered more intuitive, but might result in people covering relevant parts of the content with their fingers during operation.

Slider interfaces are usually placed at the horizontal side of the screen (top or bottom, cf. Fig. 1), thus minimizing the risk of covering relevant content during browsing. Yet, when comfortably holding the device with two hands, it still requires people to release one hand from the device during operation. Partly inspired by [15] we therefore decided to place the slider in our enhanced *slider interface design* vertically on the left (cf. Fig. 3). While this makes the interface easier to reach, it even increases the already existing granularity problem due to the screen's aspect ratio (height < width). We aim at coping with this issue by introducing a second slider bar on the right side of the screen for more fine granular navigation. It only represents a fraction of the whole video and displays additional content information in the form of a vertical filmstrip. A small indicator is added to the left slider to illustrate which part of the video is represented by the currently visible filmstrip. Initially, the visible part on the right matches the current video position, but it can be modified by related up and down scrolling gestures. *Potential advantages*, again identified with informal studies, include better visualization of more content during browsing compared to the standard interface (Fig. 1) plus support of higher and lower granularity levels. The left side can be used to get a quick overview of the file or find a larger region of interest (e.g., a particular scene), whereas the right side can be used for more detailed skimming (e.g., to find a specific event in that scene). In addition, the interface can be operated easily while naturally holding the device in two hands. *Potential disadvantages* include a less familiar design with timelines that are placed vertically instead of the usual horizontal arrangement. For example, in the informal studies, we asked what mapping people considered more intuitive – having the video start at the top or at the bottom of the vertical slider. While the majority agreed on starting at the top (mostly due to being used to read from top to bottom), a noticeable number seemed to prefer a reversed design. Common reasons included that "it starts at the same position then as the familiar horizontal design" and that "it resembles filling a glass" when playback is progressing. In the tested version, we decided to place the start at the top of the screen. Similarly to the standard interface, the control elements are hidden during regular playback but appear as soon as a user touches the screen.

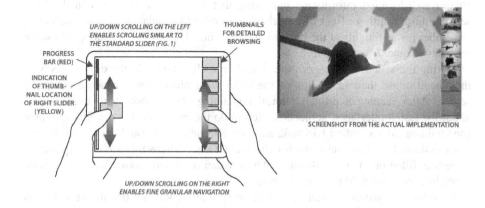

Fig. 3. Enhanced slider interface implementation

4   Comparative Study

In order to verify the presumed advantages and disadvantages described above and to gain further insight with respect to usability, performance, and general interface design for mobile video browsing we compared the storyboard and slider interfaces in a controlled experiment. 24 students volunteered as test subjects (17 male / 7 female, ages 18-27 years with an average of 21.3). 19 were right handed, three left handed, and two had no preference. None of them were involved in any of the informal pre-studies or had any other knowledge of our project. Not all owned a tablet, but all were familiar with them or related touch interaction, for example, via mobile phones or devices owned by friends and family members. No noteworthy difference could be observed in the tests considering age, handedness, gender, experience, or pre-knowledge of the movies used.

4.1   Study Design

Two movies with a length of about 15 minutes were used for the experiment[1]. In a within-subject design, each participant tested both interfaces. Interface order as well as association of the two movies to the two designs was counterbalanced, leading to four setups, each used by six of the 24 subjects. Inspired by the Video Browser Showdown competition [16], users had to perform tasks resembling a Known Item Search (KIS) where they are presented a clip of five seconds from the movie and had to find a frame from that clip using one of the two interfaces. Clips were extracted from the video roughly every two minutes and presented to the subjects in randomized order with the restriction that a following clip cannot be within the same 1/5th portion of the file to avoid targets being too close to each other. For each interface, there were nine tasks of which the first was a training assignment and not considered in the statistical evaluation. The last four of the remaining eight tasks also contained some rough location information indicating that the clip can be found in the first, second, third, fourth, or final fifth of the file in order to simulate situations such as "I remember it was roughly at the beginning of the file." Users were encouraged to solve the tasks as quickly as possible.

During the actual tests, users were seated on a chair and held the device freely in their hand or on their lap. Clips of the tasks were played on a separate computer screen. After reading a printed tutorial containing screen shots with explanations about the tested interfaces, subjects started to do the nine tasks for the first interface (one training, four standard KIS task, and four KIS tasks with rough location information), followed by the nine tasks for the second one. Afterward, a related questionnaire was filled out and an informal interview and discussion took place. The whole procedure took about 20 minutes per person.

In addition to the questionnaire and informal interview, other qualitative data was gathered via observation of the otherwise neutral executor of the experiment who also

---

[1] *Tears of Steel* and *Sintel* from the Blender Foundation, cf.
http://archive.blender.org/features-gallery/movies/index.html

noted comments made during the tests. Quantitative data, such as interactions and time to solve a task, was logged directly on the device. For the tests, we used an Asus Transformer Pad (TF300T) with 10.1-inch screen size (1280x800 pixels resolution) running Android version 4.1. The implementation was optimized for speed using the OpenGL ES 2.0 library. Thumbnails were extracted from the video and generated on the fly, enabling real-time usage of random video files, but resulting in a slight delay during very fast scrolling operations in the storyboard layout. We did not get the impression that this delay had any negative effect on the results, nor could we observe any related issues in the logged data. Yet, it might have had a small impact on the subjective user ratings (cf. below).

4.2   Results

Holding the Device. One of the major advantages we expected for the enhanced slider interface compared to the storyboard design was that people can operate it with their thumbs while comfortably holding the device in both hands. However, analyzing heat maps created from the logged data (indicating which areas people touched during the tests) and observational notes from the neutral observer showed that only one user constantly held the device like this, but all others had it in their non-dominant hand most of the time and used the dominant one for interaction. Only nine subjects operated the slider on the right, which provided a more elaborate and detailed view of the content, with their thumb. One switched midway from thumb to finger operation. This observation is independent of the order in which the interfaces were tested (five of those nine started with the storyboard design). Despite this unexpected behavior, several users explicitly commented positively about the design and placement of the sliders, characterizing them as "easy to reach", "good positioned", and "more handy" than the storyboard. Only one subject made explicit negative comments on the rather uncommon vertical placement.

General Usage and Operation. The storyboard design was mostly used as intended. After switching to storyboard view, subjects used up and down scrolling gestures (mostly with their dominant hand) and visually skimmed the thumbnails in search for content from the played clip. Yet, in 35% of the tasks, users solely relied on the original player's slider bar (Fig. 1) and did not use the storyboard at all. For 17% of the tasks, people exclusively relied on the storyboard, whereas for the rest, subjects used a combination of both (mostly slider first, then switch to storyboard). Surprisingly few participants made comments about how their interactions occluded content and thus interrupted their search process. For the slider interface, in about 65% of the cases subjects did indeed use the interface as intended, i.e., in a mixed approach where the left slider was used to roughly find an area of interest and the right one to further explore this area and find the concrete target. Yet, like with the storyboard design, in almost 35% of the cases people solely used the left slider. These observations for both interface designs indicate the initially identified advantages of the standard slider interface (cf. comments on the related informal studies for the initial player interface at the beginning of Section 3). They further show that people appreciate and need advanced browsing functionality but if and only if it is needed (it should be noted that not all tasks were solvable with the left/bottom slider due to the aforementioned granularity problem).

Search Time. For the slider interface, obviously, the cases where people were able to solely rely on the coarse slider on the left screen side to solve the tasks resulted in a much lower search time than the ones where it was necessary to use the more detailed view provided by the right slider. Average times for a single task were 22.73 seconds for the first case (left slider) versus 54.87 seconds for the second (both sliders used). Observations for the storyboard design are comparable (23.33 sec average search time for exclusive slider usage versus 63.73 seconds for a mixed approach). Average search time for the 17% of the cases where people solely relied on the storyboard was in between (39.72 sec).

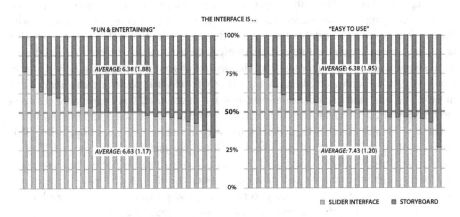

Fig. 4. User ratings for the two tested designs sorted by user with decreasing preference for slider interface design. Values are normalized per user (i.e., 50% corresponds to an identical rating for both designs). Averages (and standard deviation in brackets) are absolute values between 0 (= totally disagree) and 10 (= totally agree).

User Preference and Ratings. When analyzing our observation notes and the logged interaction data we were not able to identify any particular preference or context that might speak in favor of one of the two tested designs over the other. This is generally confirmed by the user ratings. Although people expressed some preferences, in most cases opinions about both designs did not differ by much. In particular, when asked to rate experience ("fun and entertaining") on a scale from 0 (= "totally disagree") to 10 (= "totally agree"), users gave an average rating of 6.63 for the storyboard design versus 6.38 for the slider interface. Figure 4, left, shows the normalized preference on a per user base, illustrating that nine subject had a slight preference for the storyboard, nine for the sliders, and six did not see a difference. Considering usefulness ("easy to use") we observe a slight preference towards the slider design (14 users, cf. Fig. 4, right, versus three neutral ones and seven with a preference for storyboards). This is reflected in the average ratings (7.43 in preference of sliders versus 6.38 for storyboards) but in most cases the differences are only minor (e.g., lower than 10% in 18 of 24 cases). Yet, they are also evident when subjects were asked to directly compare both designs: 14 preferred the slider interface, two were neutral, and only eight preferred the storyboard. When prompted to justify their preference, many subjects indicated the necessary switch between two layouts as a major negative issue of the storyboards. In contrast to this, a fewer, but non-negligible amount of subjects made opposed comments, i.e., expressed appreciation for the storyboard noting that the thumbnails from the slider interfaces were occluding the player window. Although they generally did not consider the slight delay in thumbnail generation for very fast scrolling operations to be a major issue, some users explicitly commented negatively on it, so it might have had a slight influence on the ratings as well. Many people explicitly praised the clearness and clarity of the storyboard design, mentioning its detailed overview. Positive statements often mentioned in relation to the slider interface characterized it as "fast", "easy to use", and "intuitive".

Perceived Search Time. As said above, comparing the average search times per task for the two interfaces, we can observe similar trends and comparable performance for both designs. Likewise, if we compare them on a per user base, 13 were faster with the storyboards, whereas 11 were slightly faster with the slider interface. Yet, when being asked about which interface they assumed to have performed better with, two subjects said that they cannot tell, four assumed they did equally well with both, five suggested a better performance with the storyboard, and 13 thought they were faster with the slider interface. However, in both of the latter cases, users' opinions were frequently wrong compared to their actual performance (3 of 5 for the storyboard, 6 of 13 for the slider interface). Apparently, the slight preference for the slider design expressed in the qualitative ratings and statements discussed above also affected the perceived performance (and vice versa).

4.3   Discussion

Because all but one single task (which contained a scene similar to the target clip elsewhere in the file) have been solved correctly, we can conclude that both interface designs are useful, intuitive, and easy to operate even without any significant training time. No differences in both performance and usage could be identified between the first four and last four tasks (which included additional location information).

Users generally liked the designs, frequently characterizing both options, for example, as "cool", "nice", and visually appealing. Yet, preferences for either of the designs were mixed with different users making contradictory statements to each other. These comments are confirmed by the user ratings and other observations discussed above. No direct relation between performance and user characteristic (e.g., gender, experience) could be identified, thus suggesting that differences are mostly due to personal preference, individual taste, and habits.

Maybe the biggest surprise for us was that few subjects used the slider design in the way we intended, i.e., holding it comfortably with two hands using their thumbs for operation, although they were purposely seated on a chair during the test where they could not put down the device on, for example, a table resembling a more "relaxed" situation (e.g., when sitting in an armchair or on a couch at home). It is unclear if this is due to a still existing laboratory atmosphere during the experiment and its limited duration or if we just overestimated the wish of users to hold devices in a more comfortable way during browsing.

5   Conclusions and Future Work

In contrast to most related work, which either aims at exclusively optimizing and evaluating either storyboard designs or slider interfaces, our research targeted a direct comparison between these two concepts. Yet, the comparative study could neither reveal a clear advantage of one approach over the other, nor identify concrete situations in which either of them should be preferred. It appears that both have their merits thus suggesting a seamless integration of both for optimum interface design. In particular, we draw the following conclusions for the general design of better mobile video browsing interfaces from our observations:

First, we saw that people still rely on the rather simple, but intuitive and easy to operate standard slider interface if search tasks allow for it, suggesting that advanced techniques should only complement such existing designs and need to be seamlessly integrated into the browsing process. While this seems obvious and in line with general interface guidelines, looking at many of the proposed designs introducing new, complex (and useful!) features but neglecting existing yet powerful and established approaches, it also seems noteworthy and important to highlight.

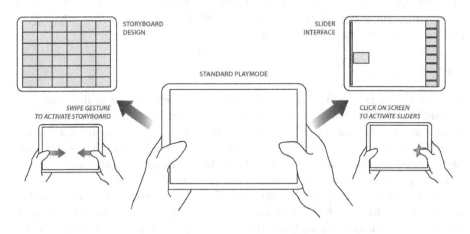

Fig. 5. Proposed design, seamlessly integrating both interaction concepts

Second, this observation also emphasizes the importance of a "smooth", seamless integration of more complex and advanced techniques with the simpler but established and powerful ones. Considering the concretely tested two designs, a combination of both approaches could of course easily be done. In fact, some users even suggested this or asked why we are not just integrating both designs into one. And our observations suggest that such an integration not only could but also actually should be done. Figure 5 illustrates this idea, providing an interface where users could start using simple browsing actions (standard timeline sliders) first, then switch to more advanced ones depending on context, situation, and personal preference (again following commonly accepted design guidelines from other domains, such as Sheiderman's Visual Information-Seeking Mantra [17]).

The proposed design (Fig. 5) also fulfills our design goal of creating an interface that can be operated easily and comfortable when holding the device in two hands, which we assumed to be the most common and preferred way. Yet, as a third and final general conclusion our experiments showed that such intuitive and seemingly reasonable assumptions do not necessarily hold in practice. Although some users commented positively on this, the actual usage data does not clearly verify it and might even suggest different, less expected behavior.

One major aspect for future work is therefore a more detailed investigation of the latter issue, for example, via a long-term study under more realistic conditions (e.g., using the system at home versus in temporary restricted and artificial lab settings). In addition, our experiment was purposely designed to focus on user experience and testing if and how well users can operate the proposed designs, even at the risk of not being able to gain conclusive results considering actual search performance. A more detailed study focusing explicitly on that issue and, for example, aiming at identifying situations and contexts in which one browsing approach might be superior to another would be important. This includes the evaluation of other search tasks than KIS. Finally, further developments of both the actual data visualization (e.g., content-dependent storyboards) and interaction (e.g., different gestures for scrolling) seem like important and interesting options with potential for better mobile video browsing interfaces as well.

Acknowledgement. This work was partially supported by a Yahoo! Faculty Research Grant.

References

1. O'Hara, K., Mitchell, A.S., Vorbau, A.: Consuming video on mobile devices. In: Proceedings of the SIGCHI Conference on Human Factors in Computing Systems (CHI 2007), pp. 857–866. ACM, New York (2007)
2. Schoeffmann, K., Hopfgartner, F., Marques, O., Boeszoermenyi, L., Jose, J.M.: Video browsing interfaces and applications: A review. SPIE Reviews 1(1), 1–35 (2010)
3. Borgo, R., Chen, M., Daubney, B., Grundy, E., Heidemann, G., Höferlin, B., Höferlin, M., Leitte, H., Weiskopf, D., Xie, X.: State of the Art Report on Video-Based Graphics and Video Visualization. Comp. Graph. Forum 31(8), 2450–2477 (2012)
4. Christel, M.G., Hauptmann, A.G., Warmack, A.S., Crosby, S.A.: Adjustable filmstrips and skims as abstractions for a digital video library. In: Proceedings IEEE Forum on Research and Technology Advances in Digital Libraries, pp. 98–104 (1999)

5. Komlodi, A., Marchionini, G.: Key frame preview techniques for video browsing. In: Proceedings of the Third ACM Conference on Digital Libraries (DL 1998), pp. 118–125. ACM, New York (1998)

6. Boreczky, J., Girgensohn, A., Golovchinsky, G., Uchihashi, S.: An Interactive Comic Book Presentation for Exploring Video. In: Proceedings of the SIGCHI Conference on Human Factors in Computing Systems, pp. 185–192. ACM, New York (1998)

7. Schoeffmann, K., Ahlström, D., Hudelist, M.A.: 3D Interfaces to Improve the Performance of Visual Known-Item Search. To Appear in IEEE Transactions on Multimedia (preprint), http://vidosearch.com/?p=1000

8. Wactlar, H.D., Kanade, T., Smith, M.A., Stevens, S.M.: Intelligent access to digital video: Informedia project. Computer 29(5), 46–52 (1996)

9. Hürst, W., Snoek, C.G.M., Spoel, W.-J., Tomin, M.: Size matters! How thumbnail number, size, and motion influence mobile video retrieval. In: Lee, K.-T., Tsai, W.-H., Liao, H.-Y.M., Chen, T., Hsieh, J.-W., Tseng, C.-C. (eds.) MMM 2011 Part II. LNCS, vol. 6524, pp. 230–240. Springer, Heidelberg (2011)

10. Hürst, W., Snoek, C.G.M., Spoel, W.-J., Tomin, M.: Keep moving! Revisiting thumbnails for mobile video retrieval. In: Proceedings of the International Conference on Multimedia (MM 2010), pp. 963–966. ACM, New York (2010)

11. Hürst, W., Darzentas, D.: Quantity versus quality: the role of layout and interaction complexity in thumbnail-based video retrieval interfaces. In: Proceedings of the 2nd ACM International Conference on Multimedia Retrieval (ICMR 2012), article 45, 8p. ACM, New York (2012)

12. Hürst, W.: Video Browsing on Handheld Devices – Interface Designs for the Next Generation of Mobile Video Players. IEEE MultiMedia 15(3), 76–83 (2008)

13. Hürst, W., Götz, G., Welte, M.: Interactive video browsing on mobile devices. In: Proceedings of the 15th International Conference on Multimedia (MULTIMEDIA 2007), pp. 247–256. ACM, New York (2007)

14. Hürst, W., Merkle, P.: One-handed mobile video browsing. In: Proceedings of the 1st International Conference on Designing Interactive user Experiences for TV and Video (UXTV 2008), pp. 169–178. ACM, New York (2008)

15. Hudelist, M.A., Schoeffmann, K., Boeszoermenyi, L.: Mobile video browsing with the ThumbBrowser. In: Proceedings of the 21st ACM International Conference on Multimedia (MM 2013), pp. 405–406. ACM, New York (2013)

16. Schoeffmann, K., Ahlström, D., Bailer, W., Cobarzan, C., Hopfgartner, F., McGuinness, K., Gurrin, C., Frisson, C., Le, D.-D., del Fabro, M., Bai, H., Weiss, W.: The Video Browser Showdown: A Live Evaluation of Interactive Video Search Tools. International Journal of Multimedia Information Retrieval (MMIR) 3(2), 113–127 (2014)

17. Shneiderman, B.: The Eyes Have It: A Task by Data Type Taxonomy for Information Visualizations. In: Proceedings of the IEEE Symposium on Visual Languages, pp. 336–343. IEEE Computer Society Press, Washington (1996)

Performance Evaluation of Students Using Multimodal Learning Systems

Subhasree Basu[1], Roger Zimmermann[1], Kay L. O'Halloran[2], Sabine Tan[2], and Marissa K.L. E.[1]

[1] National University of Singapore, Singapore 117417
{sbasu,rogerz}@comp.nus.edu.sg,elcmekl@nus.edu.sg
[2] Curtin University, Perth, Western Australia 6102
{kay.ohalloran,sabine.tan}@curtin.edu.au

Abstract. Multimodal learning, as an effective method for helping students to understand complex concepts, has attracted much research interest recently. Using more than one media in the learning process typically makes the study material easier to grasp. In the current study, students annotate linguistic and visual elements in multimodal texts by using geometric shapes and assigning attributes. However, how to effectively evaluate student performance is a challenge. This work proposes to make use of a vector space model to process student-generated multimodal data, with a view to evaluating student performance based on the annotation data. The vector model consists of fuzzy membership functions to model the performance in the various annotation criteria. These vectors are then used as the input to a multi-criteria ranking framework to rank the students.

Keywords: Computer Aided Education, Multimodal Learning Software, Fuzzy Vector Representation, E-Assessment, Multicriteria Analysis.

1 Introduction

Computer Aided Education (CAE) usually involves software applications that are aimed at making the concepts easier for students and reducing the workload of instructors. With the evolution of technology, information representation in different media modalities such as text, images and sound have become possible. Such diverse representations help to vividly explain complex concepts in more accessible ways, compared to traditional textbooks. Along with this trend, a variety of multimedia software systems have been developed with the aim of facilitating student learning [5]. These systems have the following advantages over conventional media:

- Various graphical images or video clips are able to hold students' attention.
- Rich multimedia-based materials are able to interest students to actively learn and perform better in class [8].

Along with providing tools to automate the teaching and learning process, it is desirable to automate the evaluation process as well. Not only is the manual evaluation of answer-scripts and assignments time consuming, there also exists the inevitable chance of human errors and subjectivity, which might be avoided with automated systems. It is also easier to handle, assess, store, and reuse assignments as well as tests when the

X. He et al. (Eds.): MMM 2015, Part II, LNCS 8936, pp. 135–147, 2015.

process is automated. Automatic assessment often allows for timely, almost immediate feedback for students, which is an additional motivating factor [1].

In this paper we provide a method to assess and rank students when the teaching material is presented through a multimodal learning software—Multimodal Analysis Image (henceforth MMA) [7]. The software permits students to analyze linguistic and visual elements in texts by creating overlays and attaching attributes (in the form of predefined labels) to these selections. The aim of the current study is to evaluate the students' performance when they undertake such analyses using the software. Two schools (School A - a primary school & School B - a secondary school) were invited to participate in our preliminary study. Students were organized into 10 study groups of 3 to 4 students each and asked to collaborate and interactively annotate portions of texts and images. These annotations were then exported as a set, termed an *analysis*, and they represented the students' understanding of the content learnt during the lesson.

Multimodal systems contain data from different media separately as well as from the interaction of two or more media. Hence vector space models are often used to represent the data exported by these systems. In our model, based on the analysis of the unique data-sets exported by the MMA software, we apply a vector space model to represent each multimodal analysis generated by the collaborative effort of each study group. In this model the teacher's analysis is used as the ground truth. Each study group's analysis is compared with this reference to reveal the overall performance of the student group. The main contributions of our proposed approach lie in

1. determining the factors to consider when comparing the students' analyses,
2. designing the vector using fuzzy membership functions as representative of the students' performance, and
3. determining the best multicriteria analysis framework which will rank the students with the fuzzy vector as the input.

The remainder of the paper is structured as follows. We provide a brief description of the related work in Section 2. In Section 3 we elaborate on our system model. Section 4 provides the experimental setup for extracting the annotation information and Section 5 describes the definition of the vector model along with the fuzzy membership functions used in the model. Then in Section 6 we detail the algorithms used for ranking the students and in Section 7 we report the results. Finally, we conclude with a description of our future work in Section 8.

2   Related Work

Accessing multimodal content and evaluating different educational data-sets are becoming increasingly important [11] as studies show that students enjoy working with multimodal learning materials and are comfortable with an online testing paradigm [8]. However, most of the multimodal learning systems are deficient in the assessment part of their design. This is in part because the available study material is no longer standard text, but information represented as a combination of media. The most popular form of e-assessment is onscreen testing, most of which use Multiple Choice Questions and automated markings [9]. Hence such assessments are often restricted to fields like mathematics or science which have definite answers. Various attempts made at e-evaluation are decision tree method [3] and crowd-sourced grading [8]. Attempts in finding the overlap between students' answers in multimodal sytems have been made using a 3-dimensional indicator matrix [2]. However, none of these applications provide a direct score for the performance of a student using the software. In comparison, we propose to evaluate student performance and provide a rank for each group. Different annotation information (i.e. spatial overlays with linguistic tags) is represented in a unified way as a vector of fuzzy membership values. This facilitates the use of standard analysis techniques in performance evaluation.

3   System Model

We use the MMA software [7] as an example of multimodal learning software for students. The MMA software enables annotation-based study of educational information presented as multimodal files. The multimodal files contain learning material in the form of multimodal texts with linguistic and visual elements. The files also contain a set of possible annotation choices (called *catalog* of *system choices*) to assist in the annotation of the material provided. A teacher prepares both learning material and the catalog, called system-choice space, in advance and presents them to students via the MMA software. After the students (as groups) perform an annotation-based study in the interactive learning situation, their study results are exported to files in a comma-separated format (CSV files). In the two schools participating in our experiments, the material as well as the catalogs used were different. School A used a catalog for language and images to identify simple features of the text (who, what, why, where and how), and School B used a catalog of linguistic devices to identify the different persuasive techniques in the text. We provide a snapshot of our system in Fig. 1. The *Multimodal Learning Engine* represents the MMA software. The MMA GUI is the interface presented to the students as the learning material. They complete the annotation based study and this forms the input to the *Extract Information* procedure of the algorithm (Algorithm 1). The output of this procedure is used as the input to the *Vector Space Representation*. The vector representations are then used to rank the students in *Performance Evaluation*.

Fig. 2 shows two snapshots of the annotations by two different groups placed side by side. In this software, all available system-choices are shown on the right side of screen. Meanwhile, the multimodal text which is to be annotated is presented to the students on the left side. Students are required to annotate parts of the text as well as the image based on their understanding. Each annotation information exported from the MMA software consists of two elements:

- a shape (rectangle) marking an task-relevant object or text, and
- a system choice selected by a student annotating the marked object.

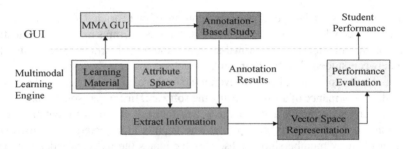

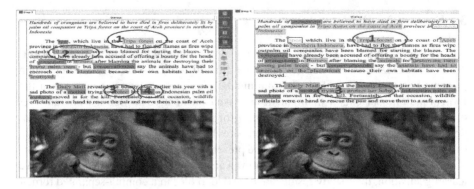

Fig. 1. Flow diagram for student performance evaluation

Fig. 2. Annotation-based multimodal learning

3.1 Algorithm for Processing Student Data

The system stores both the object position (by using a shape covering the object) and its system-choice (selecting from the catalog). The teacher's analysis is taken as the ground truth and the students' analyses are compared with it. The main algorithm for evaluating the the students' analyses is given in Algorithm 1. We propose to evaluate the students on the following three criteria:

1. the difference between the position of a student's annotation and that of the teacher,
2. the difference between the text covered by a student's annotation and the text covered by the teacher's annotation, and
3. the difference between the system choice selected from the catalog by a student and that selected by the teacher.

Each of the three procedures *ExtractAnnotationInformation*(), *DesignFuzzyVectorRepresentation*() and *RankStudents*() are described in details in the Sections 4, 5 and 6, respectively. We formulate the differences mentioned above with the help of fuzzy membership functions and design a vector model for each of the student's annotations. These vectors are then used as inputs to standard multi-criteria decision making techniques like TOPSIS [12] and another technique based on cosine similarity [4]. These two techniques both rank the students based on their fuzzy vector models. The ranks are then reported to the students as their performance.

Algorithm 1. Student Assessment based on Annotation

---

procedure ASSESSSTUDENT(M)
   Γ: a teacher analysis
   Δ_i: Analysis for Student Group i
   while { **do**#(Student Group) $\neq 0$}   ▷ Iterate each group
      while { **do**#(Annotation in Γ) $\neq 0$ }   ▷ Iterate each annotation
         $\langle p_t, c_t \rangle \leftarrow t^{th}$ annotation tuple in Γ.
         p_t is the position, c_t is the attribute.
         Generate a RTRee with Δ_i.
         Extract Annotation Information()
         Design Fuzzy Vector Representation()
      end while
   end while
   Rank Students()
   Return Rankings of the groups as output data.
end procedure

procedure EXTRACTANNOTATIONINFORMATION
   if {there is a $\langle p_k, c_k \rangle$ either contained in or overlaps with $\langle p_t, c_t \rangle$ } **then**
      { $\langle p_k, c_k \rangle$ } is the set of annotations contained in or overlaps with $\langle p_t, c_t \rangle$
      get the text covered by each of { $\langle p_k, c_k \rangle$ }
   else
      get the nearest neighbour $\langle p_l, c_l \rangle$ for $\langle p_t, c_t \rangle$
      get the text covered by $\langle p_l, c_l \rangle$
   end if
end procedure

procedure DESIGNFUZZYVECTORREPRESENTATION(The annotation obtained from ExtractAnnotationInformation)
   find the fuzzy membership value for position of the annotation
   find the fuzzy membership value for the text covered by the annotation
   find the fuzzy membership value for the system choice of the annotation
end procedure

procedure RANKSTUDENTS(Fuzzy Vector Representation of the Groups)
   Rank according to a Multi-Criteria Analysis Framework
end procedure

---

4 Extraction of Annotation Information

In this section, we provide a description of the first procedure *Extract Annotation Information*. It involves extracting the student's annotation that is closest to a particular teacher's annotation. Each of the following subsections highlights the processing of one of the three criteria for the student annotations.

4.1 Annotation Position

The CSV files with the annotation data contain the location of an annotation (the co-ordinates of the vertices). For each of the annotations in the teacher's analysis, we will have to find one or more annotation from the students' analysis that is either contained within, or overlapping with or is a nearest neighbour to the teacher's annotation. Fig. 2 pictorially denotes what we mean by containment, overlap and the neighboring annotations, e.g., the annotations marked *1* in both Group 3 and Group 4 overlap with each other. The annotations marked *3 & 4* in Group 3 are contained in the annotation marked *3* in Group 4. We use the RTree [6] technique to get the containment, overlap and nearest neighbour rectangle from the student's analysis. If there are annotations from the student's analysis that are either contained in or overlap with teacher's annotation, we do not look for the nearest neighbour for this annotation. However, if we do not get containment and/or overlaps, we find the nearest neighbour for this annotation. As often happens, there are multiple annotations from the student's analysis that are contained in/ overlap with the teacher's annotation being scrutinized.

4.2 Annotation Text

The student's annotation may be contained inside the teacher's annotation or it might overlap with the teacher's annotation. However, the text covered by the student's annotation might be much smaller or much greater than the teacher's annotation. Thus we need to compare the text covered by the student's annotation with that covered by the teacher's annotation to arrive at the accuracy of the annotation. We use the open-source OCR software Tesseract [10] for extracting the text. After that we use the Levenshtein distance between the two texts extracted for comparison.

4.3 Annotation System Choice

The system choice the students as well as the teacher assign to the annotations from the catalog are reported in the CSV files exported from the MMA software. For each of the annotations extracted as the ones contained in/ overlapping with/ nearest neighbours for a particular teacher's annotations, we proceed to check whether the system choices assigned to these annotations match those assigned to the corresponding teacher's annotations.

5   Fuzzy Membership Functions

It is very difficult to pinpoint with certainty the exact match of the annotations in terms of position as well as extent. Some annotations lie close to a corresponding teacher's annotation but they cover as much as only 20% of the teacher's annotation. In other cases, there might be annotations overlapping with about 60% of the teacher's text even though the position of the annotation is further than the previous case, when we are analyzing whether the system choices are matching or not, we need to take into consideration that the catalog of system choices contains elements which are synonymous or close in their meaning. Hence we find that there are various uncertainties in the data that we have at hand. To model the uncertainty, we use fuzzy membership functions.

Let the number of annotations in the teacher's analysis be n, *i.e.*, $\#(a^t) = n$ and the number of annotations in the student's analysis be m, *i.e.*, $\#(a^s) = m$. Let us suppose that we are processing the jth student's annotation for the ith teacher's annotation. Hence our vector representation for this annotation is

$$v_i^j = (\mu_{d_i^j}, \mu_{str_i^j}, \mu_{sc_i^j}) \text{ for } i = 1, \ldots, n \text{ , } j = 1, \ldots, m,$$

where $\mu_{d_i^j}$ is the representation of the distance of the students' annotations from the teacher's annotations, $\mu_{str_i^j}$ is the representation of distance between the text covered by the students' annotations from the text covered by the teacher's annotations and $\mu_{sc_i^j}$ is the representation of difference in the system choices selected by the student and by the teacher.

5.1 Membership Function for Position

Here we provide a definition for the $\mu_{d_i^k}$ element of the v_i^k, when a_k^s (annotation k for a student s) is compared to a_i^t (annotation i for teacher t). It has already been reported that the annotation k for student s is either contained in/ overlaps with/ the nearest neighbour for the annotation i of the teacher. We consider the centroid of the rectangles (annotations) as a representative of the position of the rectangles. Let the distance between the centroid of a_i^t and centroid of a_k^s be d. Thus we define,

$$\mu_{d_i^k} = \begin{cases} 1, & \text{if } d = 0. \\ \frac{1}{d}, & \text{otherwise.} \end{cases} \tag{1}$$

5.2 Membership Function for Text

We provide a definition for the $\mu_{str_i^k}$ element of the v_i^k. It represents the overlap of a_k^s with annotation a_i^t. Let the Levenshtein distance between the text covered by a_i^t and that by a_k^s be $lvdist$. Thus we define,

$$\mu_{str_i^k} = \begin{cases} 1, & \text{if } lvdist = 0. \\ \frac{1}{lvdist}, & \text{otherwise.} \end{cases} \tag{2}$$

5.3 Membership Function for System Choice

Let $S = \{sc_z\}$ be the set of system choices in the catalog. Let sc_x^t denote that sc_x is the system choice selected by the teacher and sc_y^s denote that sc_y is the system choice selected by the student. Let $|sc_x^t|$ = the total number of times sc_x is selected by the teacher in the analysis (the same system choice can be selected to annotate more than one annotation). Let $sc_y^s(sc_x) = sc_y$ be the system choice for the annotation a_k^s when it is compared to the a_i^t with the system choice as sc_x. The membership function chosen in this case is,

$$\mu_{sc_i^j} = \begin{cases} 1, & \text{if it is } sc_x^s \text{ for } sc_x^t. \\ |sc_y^s(sc_x)|/|sc_x^t|, & \text{otherwise.} \end{cases} \tag{3}$$

Here $|sc_y^s(sc_x)|$ equates the total number of cases reported when sc_y is chosen by a student for sc_x selected by the teacher in the entire analysis. If there is more than one a_k^s for one a_i^t, the average of the $\mu_{d_i^k}$s for each v_i^ks is the representative $\mu_{d_i^k}$. After computing the μ_{d_i} for each of the teacher's annotations separately for a particular group, the average of all the μ_{d_i}s is the μ_d for the group. This will be the representative fuzzy membership function for distance for that particular group. We follow the same procedure for arriving at the μ_{str} and the μ_{sc} for the group. Thus we get the vector representation for that particular group as $v(G_g) = (\mu_d, \mu_{str}, \mu_{sc})$.

6 Multicriteria Decision Making Framework

We compare the fuzzy vectors as a whole with each other to arrive at a result. The two most common methods to compare vectors are by using cosine similarity [4] and L2

Table 1. RSCS and TOPSIS Algorithms

RSCS	TOPSIS
1. Normalize the decision matrix by the following formula: $R = (r_{ij})_{m \times n}$ where $\dfrac{x_{ij}}{\sqrt{(\sum_{k=1}^{n} x_{ik}^2)}}$	1. Normalize the decision matrix by the following formula: $R = (r_{ij})_{m \times n}$ where $(r_{ij}) = x_{ij} / \text{pmax}(v_j)$ $i = 1, \dots n$, $j = 1, \dots m$,
2. Calculate the weighted normalized decision matrix $T = (t_{ij})_{m \times n} = (w_j r_{ij})_{m \times n}, i = 1, \dots, m,$	2. Calculate the weighted normalized decision matrix $T = (t_{ij})_{m \times n} = (w_j r_{ij})_{m \times n}), i = 1, \dots, m,$ where $w_j = W_j / \sum_{j=1}^{n} W_j, j = 1, \dots, n$ so that $\sum_{j=1}^{n} w_j = 1$, and W_j is the original weight given to the indicator v_j, $j = 1, \dots, n$.
3. Determine the positive ideal(A_+) and the negative ideal(A_-) solution: $A_+ = (t_1^+, t_2^+, \dots, t_m^+),$ $A_- = (t_1^-, t_2^-, \dots, y_m^-)$ where, $t_j^+ = \max\limits_{i=1,2,\dots,n} t_{ij}',$ $t_j^- = \min\limits_{i=1,2,\dots,n} t_{ij}'.$	3. Determine the best alternative (A_+) and the worst alternative (A_-): $A_- = \{ \langle max(t_{ij} | i = 1, 2, \dots, m) | j \in J_- \rangle,$ $\langle min(t_{ij} | i = 1, 2, \dots, m) | j \in J_+ \rangle \} \equiv \{t_{wj} | j = 1, 2, \dots, n\},$ $A_+ = \{ \langle min(t_{ij} | i = 1, 2, \dots, m) | j \in J_- \rangle,$ $\langle max(t_{ij} | i = 1, 2, \dots, m) | j \in J_+ \rangle \} \equiv \{t_{bj} | j = 1, 2, \dots, n\},$ where, $J_+ = \{ j = 1, \dots, n -$ j associated with the criteria having a positive impact$\}$, $J_- = \{ j = 1, \dots, n -$ j associated with the criteria having a negative impact$\}$.
4.Calculate the degree of conflict between an alternative and A_+ as $Cos\theta_{i+} = \dfrac{\sum_{j=1}^{m} t_{ij}' t_j^+}{\sqrt{(\sum_{j=1}^{m} t_{ij}')^2 \sum_{j=1}^{m} (t_j^+)^2)}}.$ The degree of conflict with A_- is given by $Cos\theta_{i-} = \dfrac{\sum_{j=1}^{m} t_{ij}' t_j^-}{\sqrt{(\sum_{j=1}^{m} t_{ij}')^2 \sum_{j=1}^{m} (t_j^-)^2)}}.$	4. Calculate the distance between the alternative i and A_+ $d_{i+} = \sqrt{\sum_{j=1}^{n} (t_{ij} - t_{bj})^2}, i = 1, \dots, m$ the L2-distance between the target alternative i and A_- $d_{i-} = \sqrt{\sum_{j=1}^{n} (t_{ij} - t_{wj})^2}, i = 1, \dots, m,$ where d_{i-} and d_{i+} are L2-norm distances from the target alternative i to the worst and best conditions, respectively.
Determine the degree of similarity between an alternative and A_- as $S_{i-} = \dfrac{\sqrt{(\sum_{k=1}^{m} (t_{ik}')^2)} cos\theta_{i-}}{\sqrt{(\sum_{k=1}^{m} (t_j^-)^2)}}.$	Calculate the similarity as: $s_{i-} = d_{i+}/(d_{i-} + d_{i+}), 0 \leq s_{i-} \leq 1, i = 1, \dots, m.$
Determine the degree of similarity between an alternative and A_+ as $S_{i+} = \dfrac{\sqrt{(\sum_{k=1}^{m} (t_{ik}')^2)} cos\theta_{i+}}{\sqrt{(\sum_{k=1}^{m} (t_j^+)^2)}}.$	$s_{i-} = 1$ if and only if the alternative solution has the best condition $s_{i-} = 0$ if and only if the alternative solution has the worst condition.
5.The Performance Index is: $P_i = \dfrac{S_{i+}}{S_{i+} + S_{i-}}.$	5.The Performance Index is: $s_{i-}, i = 1, \dots, m.$

distance [12]. Both of these methods have been used to design a multi-criteria decision making framework. We call the first method RSCS (Ranking System based on Cosine Similarity) while the second one is known as TOPSIS. We use both techniques to rank the students and compare their performance.

The input to both frameworks is a matrix $A = (x_{ij})$ for $i = 1, \dots, 10$ and $j = 1, \dots, 3$. Here, each row A_i is the vector representation of one of the groups. Hence the number of rows is 10 as there are 10 groups of students in each school. There are three columns in the matrix as we are using the 3 criteria listed in Section 3.1 for the evaluation.

The steps of two algorithms are presented in Table 1 to facilitate a comparative study between the two. As is evident from Table 1, the main *differences* between the two algorithms lie in

- Step 1: The formula for normalizing the elements of the input matrix.
- Step 3: The formula for determining the positive ideal (A_+) and the negative ideal (A_-).
- Step 4: The formula for determining the difference between an alternative and the two ideal solutions (A_+ & A_-).
- Step 5: The definition of the Performance Index (P_i) used to rank the students.

In Step 2, we see that the normalized matrix is multiplied by a weighing vector W = (w_1, w_2, ..., w_m), with one weight for each of the criterion. The weight represents the importance of a particular criterion in the analysis. We can increase or decrease the importance of a criterion by increasing or decreasing the corresponding weight for it. However, in this case we keep the weights as 1 since we consider all the three criteria equally important. We calculate the Performance Index for each alternative and rank the students accordingly, as displayed in Tables 4a and 4b.

7   Experimental Results

The experiments were carried out in two schools - School A and School B. The class selected from School A was a primary 4 class of forty-three students aged 9 or 10. The class chosen from School B was a secondary 3 class, with thirty-two students of ages 15 to 16. The teachers were given the flexibility of choosing the task that they felt was most suitable for the students and yet in line with the research objectives. The teacher in School A assigned students the task of identifying the 5Ws1H (Who, What, Where, When, Why and How) in a simple expository text.

Students were divided into groups of 3 and 4 to complete the task within a 30 minute period during curriculum time. Students in School B were assigned the task of identifying the language and persuasive features (e.g., rhetorical questions, use of emotive language, inclusivity, metaphor, etc.) of a persuasive text. Students were given slightly longer than an hour to complete the task.

We calculated the precision of our method for the extraction of the annotations, which is given in Table 2 to get an idea of the accuracy of the position of the students' annotations when compared with the position of the teacher's annotations. Earlier we defined the vector representation of our analyses in Section 5. Table 3 enumerates them. It is evident from the table that none of the groups scored best across all the three criteria of assessment. These discrepancies are further elaborated in Figs. 3a and 3b.

Table 2. Precision of annotation extraction

	School A	School B
Group 1	0.552	0.674
Group 2	0.588	0.719
Group 3	0.816	0.562
Group 4	0.7	0.178
Group 5	0.846	0.84
Group 6	0.6	0.379
Group 7	0.0345	0.667
Group 8	0.722	1
Group 9	0.485	0.606
Group 10	0.622	0.556

The vectors enumerated in Table 3 are used as the inputs to the two multicriteria analysis algorithms – RSCS and TOPSIS defined in Section 6. Both RSCS and TOPSIS return the Performance Index P_i as their output. We report them in Tables 4a and 4b. Based on the P_i, we rank the group, as reported in the column Rank for the two methods. Thus the tables give a comparative ranking for RSCS and TOPSIS. We report them separately for the two schools as the groups are not comparable across the schools.

Table 3. Vector representations for Schools A & B

	School A	School B
Group 1	(0.056, 0.031, 0.019)	(0.038,0.025,0.081)
Group 2	(0.049, 0.039, 0.034)	(0.012,0.019,0.037)
Group 3	(0.035, 0.045, 0.041)	(0.065,0.073,0.033)
Group 4	(0.093, 0.056, 0.036)	(0.029,0.018,0.011)
Group 5	(0.096, 0.049, 0.045)	(0.026,0.011,0.058)
Group 6	(0.100, 0.027, 0.019)	(0.038,0.035,0.008)
Group 7	(0.003, 0.026, 0.021)	(0.029,0.029,0.041)
Group 8	(0.102, 0.054, 0.028)	(0.095,0.013.060)
Group 9	(0.072, 0.026, 0.046)	(0.046,0.0185,0.053)
Group 10	(0.053, 0.048, 0.026)	(0.034,0.018,0.047)

When a teacher assesses the correctness of a student's annotation a_k^s, he/she does not mark it wrong by just checking whether it coincides with annotation a_i^t or not. The position of a_k^s might shift from a_i^t minutely. In such cases, the students are usually given some credit for being close to the ideal position. Our system takes such cases into account by using the inverse of the distance between the centroids as a metric. In this way, we mark the annotation positions in decreasing order of their distance from the teacher's annotations. The smaller the distance, the higher the score. The same applies for the overlap of the annotations. The lower the Levenshtein distance, the higher they score. Thus we claim that our system very closely models the marking process of a human teacher as most of the time students are given credit for trying to reach the ideal annotation.

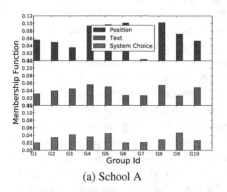

(a) School A

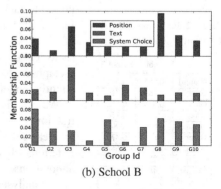

(b) School B

Fig. 3. Comparison of memberships

We also had the teacher's feedback from the exercise to guide us. Tables 5a and 5b give us a comparative study among the three ranks. As none of the ranking systems are a complete match for the ranks given by the teacher, we deduced two similarity criteria for the ranking systems and report them in Table 6. It provides the Kendell's Tau and the Spearman's Footrule for the ranks generated by RSCS and TOPSIS separately. The Spearman's Footrule is the same for both the systems in the case of School A. However,

Table 4. Rankings for schools

(a) School A

	RSCS		TOPSIS	
	P_i	Rank	P_i	Rank
Group 1	0.1749	4	0.3835	9
Group 2	0.0952	7	0.4792	8
Group 3	0.0513	9	0.4843	7
Group 4	0.1592	6	0.8160	2
Group 5	0.1739	5	0.9052	1
Group 6	0.4216	1	0.5602	5
Group 7	0.0236	10	0.0204	10
Group 8	0.1969	2	0.7422	3
Group 9	0.1822	3	0.5998	4
Group 10	0.0911	8	0.4925	6

(b) School B

	RSCS		TOPSIS	
	P_i	Rank	P_i	Rank
Group 1	0.0209	2	0.4603	3
Group 2	0.0196	3	0.1999	9
Group 3	0.0140	10	0.6654	1
Group 4	0.0155	8	0.1405	10
Group 5	0.0218	1	0.2982	6
Group 6	0.0141	9	0.2909	8
Group 7	0.0174	7	0.3064	5
Group 8	0.0176	6	0.5149	2
Group 9	0.0189	5	0.3659	4
Group 10	0.0193	4	0.2958	7

Table 5. Comparison of rankings for schools

(a) Schools A

	Teacher	RSCS	TOPSIS
Group 1	6	4	9
Group 2	7	7	8
Group 3	8	9	7
Group 4	1	6	2
Group 5	5	5	1
Group 6	2	1	5
Group 7	10	10	10
Group 8	4	2	3
Group 9	9	3	4
Group 10	3	8	6

(b) Schools B

	Teacher	RSCS	TOPSIS
Group 1	6	2	3
Group 2	5	3	9
Group 3	7	10	1
Group 4	10	8	10
Group 5	1	1	6
Group 6	8	9	8
Group 7	9	7	5
Group 8	3	6	2
Group 9	2	5	4
Group 10	4	4	7

Table 6. Similarity between ranks

	Kendell's Tau		Spearman's FootRule	
	RSCS	TOPSIS	RSCS	TOPSIS
School A	0.33	0.37	22	22
School B	0.37	0.2	20	28

it is smaller for RSCS in the case of School B. This means the difference between the ranks provided by the teacher and RSCS is less than that between the teacher and TOPSIS. Also, according to the Kendell's Tau, the ranking reported by RSCS is closer to that reported by the teacher for School B (as the Kendell's Tau is closer to 1 for RSCS than TOPSIS). The value of Kendell's Tau is less for RSCS than TOPSIS in the case of School A and the Spearman's Footrule is actually same. So we see that in the cases where the ranks vary, RSCS gives a ranking closer to that provided by the teacher compared to TOPSIS.

8   Conclusions

In this work, we proposed to leverage a vector space model to represent each student and teacher analysis as a 3-dimensional indicator matrix, which facilitates the evaluation of student performance in multimodal learning tasks using standard metrics. Experiments in two different case-studies conducted in two schools confirm that the proposed method is applicable to different educational settings, in cases when only the attribute space is pre-defined. The main contribution here is to accurately represent the multimodal data as a vector of fuzzy membership values. We have also effectively applied the multicriteria decision framework to find a ranking among vectors when the individual elements of the vector were not comparable. We have found a measure to assess the diversity in the positions of the annotations and also the syntactic similarity in the words annotated. Hence such a model can very well be used to assess any system that has to evaluate the differences between positions and/or overlap of annotations. We have also provided a criteria of evaluating the correctness of assigning predefined labels(system choices) to these annotations. It takes care of the closeness of two system choices by the frequency of using one of the choices in the place of the other. Thus we see that our method handles the uncertainties of the students' exercise at each and every level and can effectively be used to model any dataset with uncertainties which require more than one mutually independent criteria to be taken into account.

Acknowledgments.   This research was undertaken for the Multimodal Analysis ONLINE: A Web-Based Software Application for Collaborative Project Work (NRF2012IDM-IDM002-009) project, funded by the Ministry of Education Interactive Digital Media Programme Office under the National Research Foundation in Singapore. Partial support was also provided by the Singapore National Research Foundation under its International Research Centre @ Singapore Funding Initiative and administered by the IDM Programme Office through the Centre of Social Media Innovations for Communities (COSMIC).

References

1. Amelung, M., Krieger, K., Rosner, D.: E-assessment as a service. IEEE Transactions on Learning Technologies 4(2), 162–174 (2011)
2. Basu, S., Yu, Y., Zimmermann, R.: Student performance evaluation of multimodal learning via a vector space model. In: Proceedings of the 1st ACM International Workshop on Internet-Scale Multimedia Management, ISMM 2014. ACM (2014)
3. Bhardwaj, B., Paul, S.: Mining Educational Data to Analyze Students' Performance. International Journal of Advanced Computer Science and Applications 2(6) (2011)
4. Deng, H.: A similarity-based approach to ranking multicriteria alternatives. In: Huang, D.-S., Heutte, L., Loog, M. (eds.) ICIC 2007. LNCS (LNAI), vol. 4682, pp. 253–262. Springer, Heidelberg (2007)
5. Friedland, G., Hürst, W., Knipping, L.: Educational multimedia systems: The past, the present, and a glimpse into the future. In: Proceedings of the International Workshop on Educational Multimedia and Multimedia Education, Emme 2007, pp. 1–4. ACM, New York (2007)

6. Guttman, A.: R-trees: A dynamic index structure for spatial searching. In: International Conference on Management of Data, pp. 47–57. ACM (1984)
7. O'Halloran, K.L., Podlasov, A., Chua, A., Marissa, K.L.E.: Interactive Software for Multimodal Analysis. Visual Communication 11, 363–381 (2012)
8. Phua, Y.C.J., Chew, L.C.: What Do Secondary School Students Think About Multimedia Science Computer Assisted Assessment (CAA). In: Computer Assisted Assessment (2012)
9. Timmis, P.B.S., Oldfield, A., Sutherland, R.: Where is the cutting edge of research in e-Assessment? Exploring the landscape and potential for wider transformation. In: Computer Assisted Assessment (2012)
10. Smith, R.: An overview of the tesseract ocr engine. In: Proceedings of the Ninth International Conference on Document Analysis and Recognition, ICDAR 2007, vol. 02, pp. 629–633. IEEE Computer Society, Washington, DC (2007)
11. Worsley, M.: Multimodal Learning Analytics - Enabling the Future of Learning through Multimodal Data Analysis and Interfaces. In: ICMI 2012, pp. 353–356 (2012)
12. Yoon, K., Hwang, C.: Multiple attribute decision making: an introduction. In: Quantitative Applications in the Social Sciences, pp. 102–104. Sage Publications (1995)

Is Your First Impression Reliable? Trustworthy Analysis Using Facial Traits in Portraits

Yan Yan[1], Jie Nie[2], Lei Huang[1], Zhen Li[1], Qinglei Cao[1], and Zhiqiang Wei[1]

[1] College of Information Science and Engineering, Ocean University of China, Qingdao, China
{yanyan.azj,ithuanglei,lizhen0130,cql.levi}@gmail.com,
weizhiqiang@ouc.edu.cn
[2] Department of Computer Science and Technology, Tsinghua University, Beijing, China
niejie@tsinghua.edu.cn

Abstract. As a basic human quality, trustworthiness plays an important role in social communications. In this paper, we proposed a novel concept to predict people's trustworthiness at first sight using facial traits. Firstly, personality-toward traits were designed from psychology, including permanent traits and transient traits. Then, a mixture of feature descriptors consisting of Histogram of Gradients (HOG), Local Binary Patterns (LBP) and geometrical descriptions were adopted to describe personality traits. Finally, we trained the personality traits by LibSVM to determine trustworthiness of a person using portrait. Experiments demonstrated the effectiveness of our method by improving the precision by 33.60%, recall by 20.33% and F1-measure by 25.63% when determining whether a person is trustworthy or not comparing to a baseline method. Feature contribution analysis was applied to deeply unveil the correspondence between features and personality. Demonstration showed visual patterns in portrait collages of trustworthy people that further proved effectiveness of our method.

Keywords: trustworthy impression, facial traits, portraits, first impression, facial feature extraction.

1 Introduction

As a basic human quality in social activities, trustworthiness gains more and more attentions among friends, couples, and partners in various scenarios such as commercial activities, diplomatic activities and even in life trivia. To build a trustworthy figure at the first sight is quite important factor to start a success relationship. Thus, figuring out what factors influence the trustworthy impression becomes an important issue.

Bar et al. [1] have found that first impressions of people's personalities were often formed by using the visual appearance of their faces and consistent first impressions could be formed very quickly within the first 39 ms. Many researches have

X. He et al. (Eds.): MMM 2015, Part II, LNCS 8936, pp. 148–158, 2014.

demonstrated the important influence of facial appearance, such as Sheila et al. [2] have exploited that baby-faced persons and females were more trustworthy than mature-faced persons and males in communications. Christopher et al. [3] have discovered that rapid judgments about the personality traits of political candidates, based solely on their appearance, could predict their electoral success. All these works have made it clear that facial traits are important for judging person's trustworthiness, which have provided psychological evidences to our work in this paper. Thus, our work intends to find out the correspondence between portraits visual contents and trustworthy impression, which has considerable meaning and wide application. Firstly, finding portraits utilizing personality-toward words can implement image retrieval in personality semantic level. Secondly, adding personality into human-computer interaction can improve intelligence of computer and develop the friendliness of interaction.

So far, there are a good number of works related to our work, which can be split into two groups. The first group is on facial traits level, Asteriadis et al. [4] and Jeng et al. [5] have proposed different approaches for detecting facial features. [6, 7, 8, 9] were mainly on facial expression recognition through different ways. Hoque et al. [10] developed a computer system at MIT that could tell you which kind of smile was showing happiness or frustration. However the first group just focuses on face feature detection staying at physical level, while this paper pays attention to trustworthy analysis using facial traits that have achieved emotion level. The second group is on personality impression agreement analysis. Fitzgerald et al. [11] explored characteristics of the profile photographs and their association with impression agreement. Cristani et al. [12] has shown that visual patterns correlated with the personality traits and the personality traits could be inferred from the images latter posts as "favorite". However, compared to this paper, there are two different aspects: 1) images in the second group are all flexible, including animals, plants, and landscape, while images in this paper are single portrait photos without much background, makeup, dresses and so on. 2) The second group focuses on the agreement of personality impression and actual personality, while this paper just pays attention to the trustworthy impression neglecting actual personality.

Therefore, in this paper we aim to find the relationship between facial traits and trustworthy impression. Firstly, personality-toward traits are designed from psychology, including eleven permanents traits and five transient traits from five main facial features, consisting of eyebrow, eye, nose, mouth and face shape. Then, we extract the face area using Active Shape Model (ASM) and adopt a mixture of feature descriptors consisting of Histogram of Gradients (HOG), Local Binary Patterns (LBP) and geometrical descriptions to describe personality traits. Finally, we train the personality traits by LibSVM to determine trustworthiness of a person using his/her portrait. Experiments compare the effectiveness of our method with a baseline method when determining whether a person is trustworthy or not. Feature contribution analysis is applied to deeply unveil the correspondence between features and personality. Each facial trait, combination of all Permanent traits and combination of all Transient traits are evaluated by Precision, Recall and F1-measure. Demonstrations are used to show visual patterns in portrait collages of trustworthy people that further proved the effectiveness of our method.

2   Methods

2.1   Facial traits

From a psychological perspective, we proposed a novel personality-toward feature combining 11 permanent facial traits and 5 transient facial traits (See Fig.1). All 16 traits are extracted from 5 main facial features, consisting of eyebrow, eye, nose, mouth and face shape (numbered from I to V). Permanent traits are on the left side, which are hereditary, inborn and immutable traits. While transient traits are on the right side, which are temporary and changeable with facial expressions, such as happiness, sadness, fear, disgust, surprise and anger [13]. We described transient traits utilizing the status of Facial Action Units (FAU) referred to [14].

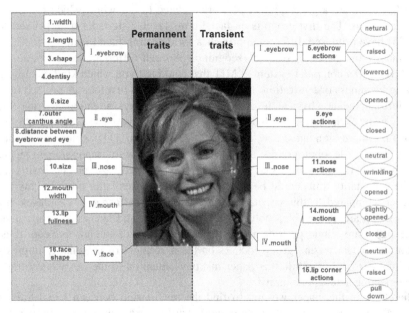

Fig. 1. Facial features are in red and numbered from I to V; feature properties are in blue and numbered from 1 to 16; feature actions are in green

2.2   Features

Feature extraction and description is a pre-process before finding out the correspondence between facial traits and personality. Following Fig.2 shows the workflow of feature extraction and description.

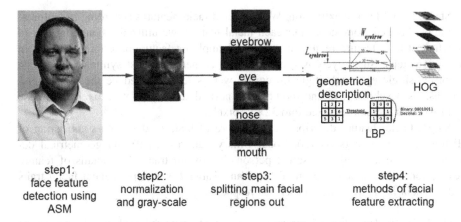

Fig. 2. Workflow of features extraction and description

Step1: Facial feature detection based on Active Shape Model (ASM). We applied method in [15] to detect facial features by utilizing 68 ordered points as shown in Fig.3. By connecting these points in certain order, we obtained the contours of main facial features.

Step2: Face extraction and normalization. We extracted the face area using a bounding box (containing all 68 points) out from a normal portrait. The face area was normalized into 128×128 pixels, and then converted into gray image.

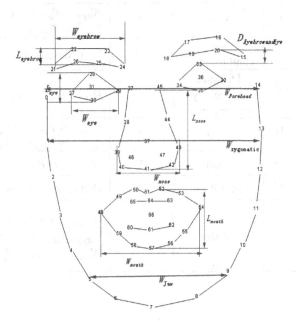

Fig. 3. Facial feature detection using ASMLibrary

Step3: Facial feature extraction. We extracted facial features according to features' contours obtained from step1. For each facial feature, we utilized a bounding box to conclude it as a feature region, in which we applying feature-describing method in following steps. Since eyebrows, eyes and lip corners are most symmetrical, single eyebrow, single eye, and single lip corner were selected to reduce the descriptors' dimensions. Eyebrow, eye and mouth were normalized into 64×32 pixels, and nose and lip corner were normalized into 32×32 pixels.

Step4: Facial feature description. A mixture of features descriptors consisting of Histogram of Gradients (HOG), Local Binary Patterns (LBP) and geometrical descriptions were adopted to describe personality-toward traits. The details of feature description were characterized in Table 1 and Table 2, where permanent facial traits and transient facial traits were described respectively.

Table 1. Detail description of permanent facial traits extracted methods, the column of "Length" represents feature dimensions; $P_i.x$ means the x-coordinate of point i, $P_i.y$ means the y-coordinate of point i, other equations are described in Fig. 3

Regions	No.	Name	Length	Short Description
Eyebrow (I)	1	Width	1	$W_{eyebrow} = P_{25}.x - P_{22}.x$
	2	Length	1	$L_{eyebrow} = P_{29}.y - P_{23}.y$
	3	Shape	19	HOG and LBP descriptors
	4	Density	19	HOG and LBP descriptors
Eye (II)	6	Size	2	$W_{eye} = P_{30}.x - P_{28}.x$, $L_{eye} = P_{31}.y - P_{29}.y$
	7	Outer canthus angle	19	HOG descriptors
	8	Distance between eye and eyebrow	1	$D_{eb} = P_{29}.y - P_{26}.y$
Nose(III)	10	Size	2	$W_{nose} = P_{44}.x - P_{40}.x$, $L_{nose} = P_{42}.y - P_{38}.y$
Mouth(IV)	12	Width	1	$W_{mouth} = P_{55}.x - P_{49}.x$
	13	Lip fullness	24	HOG and LBP descriptors
Face(V)	16	Face shape	3	$W_{forehead} = P_{15}.x - P_1.x$, $W_{zygomatic} = P_{14}.x - P_2.x$, $W_{jaw} = P_{10}.x - P_6.x$

Table 2. Detail description of transient facial traits extracted methods, the column of "Length" represents feature dimensions

Face regions	No.	Name	Length	Short Description
Eyebrow(I)	5	Actions	19	
Eye(II)	9	Actions	19	
Nose(III)	11	Actions	12	HOG and LBP descriptors
Mouth(IV)	14	Mouth actions	24	
	15	Lip corners actions	12	

3   Experiments

In this section, we conduct comprehensive evaluations of our method. The dataset is described first. Then, experiments are performed to evaluate the proposed approach. We divide our experiments into three parts. First, we compare our method with a baseline method. Second, we analyze contributions of each facial trait to trustworthy impression. Finally, a selection of trustworthy impression portraits is provided to demonstrate the results qualitatively.

3.1   Dataset

To evaluate the algorithm objectively, we built a portrait dataset containing 2010 portraits randomly downloaded from "www.google.com" and "www.flickr.com". Each portrait included one person and none of them was wearing sunglasses.

We labeled the ground-truth by about 250 volunteers. For each portrait, the volunteers were asked to label a tag about trustworthiness or untrustworthiness on it at their first impression. If the consistence of trustworthy by different raters achieved 70%, the image was tagged with "trustworthy".

For this dataset, 1404 images were randomly chosen as the training set and the other 606 images were used as the testing set. The details of the training set and testing set are shown in Table 3.

Table 3. Sample number of training set and testing set

Groups	Training set		Testing set		Total
	Trustworthy	Untrustworthy	Trustworthy	Untrustworthy	
All portraits	522	882	273	333	2010

3.2   Experiments and Discussions

Comparison with Method in Reference [16]. Support Vector Machine (SVM) [17], using RBF as kernel, was used as classifier in our experiments. As the issue in this paper is binary classification, one class is trustworthiness and the other is untrustworthiness.

Table 4 shows the performance of reference [16] as baseline and our method. From the results we can see that our method has significantly outperformed method in [16]. This is because that we have combined the permanent traits i.e. (including eyebrow width, eyebrow length, eyebrow shape, eyebrow density, eye size, outer canthus angle, distance between eyebrow and eye, nose size, mouth width, lip fullness and face shape), and transient traits i.e. (including eyebrow actions, eye actions, nose actions, mouth actions and lip corner actions) together, while method in reference [16] only uses transient traits. Our method gains an improvement of 33.60% in precision, a 20.33% improvement in recall and a 25.63% improvement in F1-measure.

Table 4. Performance comparision using Precision, Recall, F1-measure

Methods	Precision	Recall	F1-measure
Reference [16] method	37.50%	16.67%	23.07%
Our method	71.10%	37.00%	48.70%

Through training different feature combinations, we found that using the combination of features (5, 6, 15 and 16) for female and using the combination of features (8, 9, 11, 14 and 16) for male can achieve high performance. So we grouped the dataset into male and female as shown in Table 5. Table 6 lists the relusts for female and male group respectively. Interestingly, we can see that feature 16(face shape) is both useful for female and male group to build trustworthy impression. For male, transient traits (9, 11 and 14) are more useful to build trustworthy impression.

Table 5. Sample number of training set and testing set

Groups	Training set		Testing set		Total
	Trustworthy	Untrustworthy	Trustworthy	Untrustworthy	
Female	219	552	90	150	1011
Male	303	330	183	183	999

Table 6. Performance of this method on female portraits and male portraits

Groups	Precision	Recall	F1-measure
Female portraits	77.78%	72.41%	75.00%
Male portraits	73.34%	54.10%	62.26%

The ROC curves of female, male performance in Table 6 as well as all portraits performance are showed in the following Fig. 4. From the figure we can see that the performances of female and male group are both higher than all portraits without grouping, and the group of female performs best in this experiment, whose TPR achieves 0.82 while FPR is only 0.28. This result unveils that gender is significantly important in judging trustworthy impression, and because of physical difference, female can achieve higher agreement of being judged trustworthiness than male using facial traits.

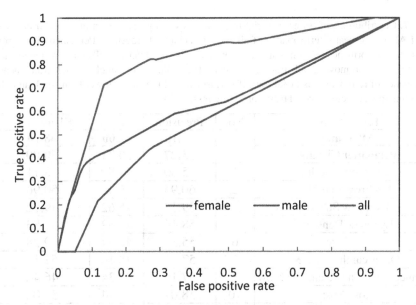

Fig. 4. Performance of trustworthiness prediction on female portraits, male portraits and all portraits

Analysis of Facial Traits Contributions to Trustworthy Impression. To further evaluate the contribution of each facial trait, comparative experiment is performed. See the results demonstrated in Table 7. Transient traits are more useful than permanent traits with the F1-measure decreased about 2%. Interestingly, while the combination of all permanent traits performs worse, single permanent trait (eyebrow width, distance between eyebrow and eye, mouth width and face shape) is useful to form trustworthy impression. This result unveils that specific combination of permanent traits is more useful than all permanent traits together. While the combination of all transient traits performs better, the single transient trait is useless to form trustworthy impression. We can indicate that the transient traits always appear together in facial expression.

Qualitative Evaluations. Furthermore, to demonstrate the results qualitatively, we highlight the performance in Fig. 5. In Fig. 5, it's interesting to see that almost all trustworthy people are smiling with lip corner raised, mouth slightly open. To indicate the qualitative results, we also use quantitative method for further analysis. For female, people with short eyebrows (55.26%), small eye (60%) and big nose (52.63%) are always tagged as trustworthiness. Male with trustworthy impression are always with narrow eyebrows (52.90%), long distance between eye and eyebrow (54.72%) and long (52.12%) face shape. For both trustworthy female and male, they are both with flat eyebrows (52.13%, 52.02% respectively), sparse eyebrows (61.33%, 52.17% respectively), thin lips (57.27%, 52.54% respectively), slightly open mouth (52.70%,

57.81% respectively), raised lip corners (54.46%, 51.12% respectively) and square face shape (52.17%, 52.50% respectively). Hence without cosmetic operation, just changing your transient traits, such as keeping a slightly smile also can improve your trustworthy impression.

Table 7. Contributions of different facial traits to trustworthy impression on all portriats; The row of All Traits means performance of all sixteen facial traits together; Permanent Traits row means the performance of removing all pernanent traits; Transient Traits row means the performance of removing all transient traits; The column of single facial trait means performance of removing this facial trait. If performance of the row is lower than All Traits, the trait in this row is useful to build trustworthy impression (%).

Facial traits	No.	Precision	Recall	F1-measure
All Traits		**71.10**	**37.00**	**48.70**
Permanent Traits		**53.27**	**50.71**	**50.09**
Eyebrow Width	1	55.00	44.80	48.11
Eyebrow Length	2	60.94	46.44	50.96
Eyebrow Shape	3	59.73	45.62	49.72
Eyebrow Density	4	58.16	45.62	49.83
Eye Size	6	55.32	46.44	49.28
Outer canthus angle	7	57.71	48.16	51.13
Distance between eye and eyebrow	8	57.15	42.34	46.46
Nose Size	10	58.07	46.44	50.02
Mouth Width	12	55.08	43.08	47.11
Lip Fullness	13	56.39	46.44	49.36
Face shape	16	53.27	42.34	45.37
Transient Traits		**56.35**	**40.19**	**46.73**
Eyebrow Actions	5	60.61	47.26	51.34
Eye Actions	9	57.36	45.62	49.39
Nose Actions	11	57.86	46.44	50.19
Mouth actions	14	52.55	46.27	49.11
Lip corners actions	15	60.00	43.07	48.50

trustworthy impression →

Fig. 5. Demonstrate trustworthy impression qualitatively

4   Conclusions

In this paper we propose a novel trustworthy impression analyzing method using facial traits in portraits. First, sixteen facial traits including permanent traits and transient traits from face region are introduced. Then, the trait features are described by combing HOG, LBP, etc. Experimental results demonstrate that our method can get satisfactory results. Future work would be interested in features refining and design, and more personality-oriented features will be proposed for further improving the prediction accuracy.

Acknowledgments. This work is supported by the National Nature Science Foundation of China (No.61402428, No.61202208); the Fundamental Research Funds for the Central Universities (No.201413021, No.201513016).

References

1. Bar, M., Neta, M., Linz, H.: Very first impressions. Emotion 6(2), 269 (2006)
2. Brownlow, S., Zebrowitz, L.A.: Facial appearance, gender, and credibility in television commercials. Journal of Nonverbal Behavior 14(1), 51–60 (1990)
3. Olivola, C.Y., Todorov, A.: Elected in 100 milliseconds: Appearance-based trait inferences and voting. Journal of Nonverbal Behavior 34(2), 83–110 (2010)
4. Asteriadis, S., Nikolaidis, N., Pitas, I.: Facial feature detection using distance vector fields. Pattern Recognition 42(7), 1388–1398 (2009)
5. Jeng, S.H., Liao, H.Y.M., Han, C.C., Chern, M.Y., Liu, Y.T.: Facial feature detection using geometrical face model: an efficient approach. Pattern Recognition 31(3), 273–282 (1998)
6. Shan, C., Gong, S., McOwan, P.W.: Facial expression recognition based on local binary patterns: A comprehensive study. Image and Vision Computing 27(6), 803–816 (2009)
7. Albiol, A., Monzo, D., Martin, A., Sastre, J., Albiol, A.: Face recognition using HOG–EBGM. Pattern Recognition Letters 29(10), 1537–1543 (2008)
8. Gritti, T., Shan, C., Jeanne, V., Braspenning, R.: Local features based facial expression recognition with face registration errors. In: 8th IEEE International Conference on Automatic Face & Gesture Recognition, FG 2008, pp. 1–8. IEEE (September 2008)
9. Tian, Y.L., Kanade, T., Cohn, J.F.: Recognizing action units for facial expression analysis. IEEE Transactions on Pattern Analysis and Machine Intelligence 23(2), 97–115 (2001)
10. Hoque, M.E., McDuff, D.J., Picard, R.W.: Exploring temporal patterns in classifying frustrated and delighted smiles. IEEE Transactions on Affective Computing 3(3), 323–334 (2012)
11. Fitzgerald Steele, J., Evans, D.C., Green, R.K.: Is Your Profile Picture Worth 1000 Words? Photo Characteristics Associated with Personality Impression Agreement. Landscape 2, 139 (2009)
12. Cristani, M., Vinciarelli, A., Segalin, C., Perina, A.: Unveiling the multimedia unconscious: Implicit cognitive processes and multimedia content analysis. In: Proceedings of the 21st ACM International Conference on Multimedia, pp. 213–222. ACM (October 2013)
13. Ekman, P., Friesen, W.V.: Constants across cultures in the face and emotion. Journal of Personality and Social Psychology 17(2), 124 (1971)

14. Ekman, P., Friesen, W.V.: Facial action coding system (1977)
15. Wei, Y.: Research on facial expression recognition and synthesis. Master Thesis, Department of Computer Science and Technology, Nanjing (2009)
16. Nie, J., Cui, P., Yan, Y., Huang, L., Li, Z., Wei, Z.: How your portrait impresses people? Inferring personality impressions from portrait contents. In: Proceedings of the ACM international conference on Multimedia (to be published, 2014)
17. Chang, C.C., Lin, C.J.: LIBSVM: a library for support vector machines. ACM Transactions on Intelligent Systems and Technology (TIST) 2(3), 27 (2011)

Wifbs: A Web-Based Image Feature Benchmark System

Marcel Spehr, Sebastian Grottel, and Stefan Gumhold

Chair of Computer Graphics and Visualisation,
Institute of Software and Multimedia Technology, TU Dresden,
Nöthnitzer Strasse 46, 01087 Dresden, Germany

Abstract. Automatic analysis of image data is of high importance for many applications. Given an image classification problem one needs three things: (i) *Training data* and tools to extract (ii) *relevant visual information*—usually image features—that can be used by (iii) *classification algorithms*. For given (i), a multitude of candidates present themselves for (ii) and (iii). Model selection becomes the main issue. We present a web-based feature benchmark system enabling system designers to streamline tool-chains to specific needs using available implementations of candidate tools. Our system features a modular architecture, remote and parallel computing, extensibility and—from a user's standpoint—platform independence due to its web-based nature. Using *Wifbs*[1], image features can be subjected to a sophisticated and unbiased model selection procedure to compose optimized pipelines for given image classification problems.

1 Introduction

Image data accumulate in many different application areas, like biological applications [18], remote sensing [17], or professional image collections [9]. In almost any of those cases semantic content recognition is a key element for a successful exploitation of the corresponding data. If training data in the form of semantic labels is available, recognition systems usually rely on a tool chain of parametrizable feature extractors, data processors and learning algorithms.

Each component of this tool chain can be realized by a number of different techniques. We will refer to this realization as *instantiation* of a component. Each instance may be further configured with numerical parameters (*parametrization*). Under the assumption that implementations of these components suitable for the targeted application and corresponding test data are available, a retrieval system designer must find the optimal instantiation and parametrization within a given design space. Note that, even though the best instantiation is problem specific and not necessarily transferable, the candidate tool set and the procedure of identifying the best parametrization by benchmarking are generic. They are mostly independent of the application and type of image data.

---

[1] http://wifbs.zih.tu-dresden.de

X. He et al. (Eds.): MMM 2015, Part II, LNCS 8936, pp. 159–170, 2015.
© Springer International Publishing Switzerland 2015

Each retrieval system designer usually implements her own pipeline optimization code for the benchmarking procedure anew. We aim at removing the task of reimplementing similar codes. Our system *Wifbs* provides an image feature benchmark application that is ready to use, easy to handle and extend, as well as covers the necessary functionality. Our two main contributions are: first, we model the design process for a discrete multi-class recognition system in an application independent way. Second, we implement the necessary infrastructure which offers benefits lined out in section 2 and 3.

The paper is organized along three points of view. Section 4 explains the administrator's view who is in charge of setting up the computing environment for Wifbs. Section 5 discusses how a developer can extend the benchmarkable candidate tool set. Section 6 elaborates on how to actually employ the system for a benchmarking task by presenting an exemplifying use case.

2   Related Work

Giving machines the ability to recognize visual content is usually a feature driven task. The recognition performance primarily depends on the selected features and available training data. Comparative evaluation of feature algorithms on diverse datasets is hence common to optimize recognition performance and is a relevant and often conducted endeavor. There are four possible scenarios:

1) Evaluating novel features with existing datasets: Novel features are continuously developed, driven by the demand for better recognition performance on problems at hand. Advantages of the novel feature algorithms must be justified by comparing them to precursors [3], [19].

2) Testing existing features with new datasets: High quality and quantity of training data is an important prerequisite for recognition tasks. Hence, the community regularly releases large, novel, labeled datasets. These are usually accompanied by extracted standard features and precomputed classification performance results [14], [22], [4].

3) Contests: Competitions to find the best features for semantic retrieval are a popular way to foster research on specific problems and hence classification performance is boosted [6], [16], [7].

4) Summary benchmarking: Review articles that analyze a selection of existing features on a set of publicly available datasets are an important source of information for retrieval system developers. There exist numerous works that compare features [12], [5] and dissimilarity measures [15] on available datasets.

All these scenarios are related to our system, as they all follow a common work flow. Our system optimizes this work flow and, to our knowledge, is the only free software that offers these capabilities. The above mentioned contests are usually released with a framework to automatically evaluate each single solution. However, there is no support to administrate test runs and parallel execution ability. [10] only offers local benchmarking on a single machine. In spirit, *CloudCV* [1] comes closest to our work. It provides a consistent functional interface to computation resources in a cloud. Details of the execution are hidden as much as possible. However, it includes the concrete implementation of common vision tasks and goes one step further in that the computing infrastructure for running the API code is also supplied. In contrast, we supply a meta-system for the optimization of an image retrieval system. Our system offers a plug-in architecture for new tools based on a standardized interface. This is a major advantage when considering scenario 1) from above.

Our system conveniently offers the ability to compare features across multiple datasets. This is a vital task since many features suffer from over-adaptation to datasets as [20] described as the *domain adaption* problem. Integrating a feature preprocessing before training a classifier adds additional degrees of freedom. This step is skipped in each work referenced above but is included in Wifbs.

3 System Benefits

Administration of Datasets, Users and Results. The process of creating the optimal classification pipeline for a specific task has many degrees of freedom. Wifbs allows administration of multiple image datasets, classification results, data and user management with a comprehensive rights management. Results of hundreds of thousands of test runs can be analyzed using multiple views on classification results with class specific performance listings.

Platform Independence and Remote Job Execution. Feature optimization usually takes place in a heterogeneous workplace environment. To offer platform independent usage and remote access we decided for the web browser for convenient interaction with remotely and asynchronously execute jobs. In addition to the web interface, all functionality can be accessed by command line, allowing sophisticated batch processing. Computational expensive tasks (i.e. feature extraction, nearest neighbor computation, classification) are outsourced to a grid engine for parallel job execution and utilization of distributed computing and storage resources. Wifbs' performance thus scales from single machines to large clusters.

Extensibility. Our system is designed to be easily extensible with respect to all its components (cf. section 5.1). Thus, novel features can be subjected to sophisticated model selection to get a fair comparison to established methods. We supplement our default installation with a set of available feature extractors.

License. Wifbs will be made available under a liberal license which allows it to be freely used and adapted in academic institutions.[2]

4 System Overview

We model the system functionality as a four stage process (see figure 1). I. offers functionality to administrate multiple datasets. II. Afterwards features

---

[2] The system will be made available online at the time of publication.

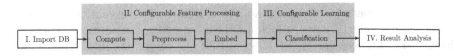

Fig. 1. Four stages of Wifbs which define work flow

are extracted and processed from each image in the selected image databases. The subsequent feature processing serves data normalization (e.g. whitening) and unsupervised pattern recognition purposes (e.g. redundancy reduction). III. Subsequently pattern recognition algorithms are employed to map image features to classes. IV. Finally results are visualized in the web front end to allow insights into the vices and virtues of the tested pipelines.

4.1   Administrators View

We designed the system with modularization and distributed execution in mind. We assume availability of server access to (A) a web server (e.g. *Apache*), (B) a grid engine scheduler (e.g. *Sun Grid Engine, SGE*), (C) a database management system (e.g. *MySQL*), and (D) a file server (e.g. *NFS*). (A) serves the interface to our system. (B) acts as grid engine scheduler that manages job load and work distribution. (C) and (D) store the required data. Given this infrastructure, our four-tiers-based implementation can be employed (cf. table 1):

- The *master* is in charge of running the web portal on (A). This tier also executes the master process of the grid engine (B). In order to complete those tasks access to (A) and (C) is necessary.
- *Compute nodes* are in charge of running jobs assigned by (B) and store results in (C) and (D). No communication between compute nodes occurs.
- The *database server* (C) is responsible for storing the performance results and for access to the system's meta-information. This tier must be accessible from the master and all compute nodes.
- The *file server* (D) provides access to executables and raw data (e.g. images and computed image features). These must be accessible to all compute nodes and (A).

5   Implementation

The primary functions of our system are constructed as pipelines of instantiated and parametrized configurable tools, e.g. feature extraction and classification algorithms. In this section we will detail some aspects from a developer's point of view. For example, we show that novel algorithms can be easily introduced. The implementation of Wifbs is mainly based on *Python* code and the libraries *Shogun* [8], for its considerable machine learning functionality, and *Django*[3] for the web front end and data object serialization.

---

[3] Django: The Web framework for perfectionists with deadlines,
https://www.djangoproject.com/

Table 1. 4-tiers of our system with respective responsibilities and services. Wifbs can be set up in a local environment as well. In this case, all tiers must be provided by a single machine. Even in this set up the SGE holds the advantage of automatic utilization of multiple cores.

Master	(A) Web portal	presents	Information to web browser
		provides	Interface for job execution on compute resources
	(B) SGE Submission Host	creates and sends	Jobs
	(B) SGE Master Host	distributes	Jobs
	(C) DBMS Client	provides access to	Global meta-information
	(D) Network File System Client	provides access to	Image data
			Feature data
			Website code
Compute Nodes	(D) Network File System Client	provides access to	Feature computation tools
			Classification tools
			Image data
			Feature data
	(B) SGE Execution Host	receives and executes	Jobs
	(C) DBMS Client	provides	Upload capabilities for job results
Database	(C) DBMS Server	stores and serves	Global meta-information
File Server	(D) Network File System Server	stores and serves	Classification tools
			Image data
			Feature data
			Website code
			Feature computation tools

5.1   Configurable Tools

There are five different types of configurable tools forming the functional pipeline: *feature-, preprocessing-, embedding-, classification-algorithms* and *kernels*. New feature algorithms are added to the system as stand-alone executables with standardized command line interface and output format. All remaining tools can either be chosen from the 142 algorithms offered by the Shogun library for preprocessing (qty. 12), embedding (15), multi-class classification (23), kernel (59), distance function (17) and normalizer (16) or implemented by inheriting the appropriate Shogun base class. Additionally, the tool meta-data must be added to the system's data base (C). We provide a simple script which imports this meta-data from configuration files in *XML*.

5.2   Parametrization and Data Handling

Each configurable tool relies on parameters for its execution. Our system currently supports 11 types of parameters, ranging from (tuples) of floats, integers, strings to tool references like distances, normalizers and kernels. Finding the optimal parametrization usually involves many feature and classification test runs from which knowledge must be inferred. It is non-trivial to manage tested parameters for selection and parametrization of the configurable tools.

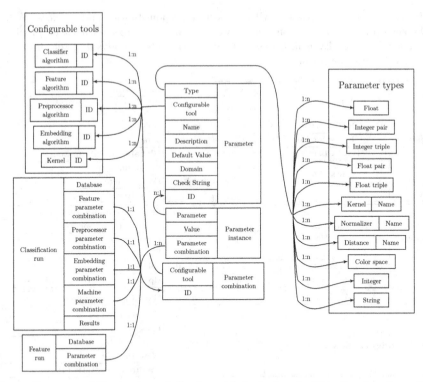

Fig. 2. Structure of *Parameters* of *Configurable tools* stored within the data base. Tool runs necessitate instantiation of their parameters with values from their *Parameter type* domain. These tuples of *Parameter instances* are identified as *Parameter combinations*. Conducted *Feature runs* are identified by a Parameter combination and the dataset ID they were executed on. To keep track of *Classification runs* it is necessary to store the Parameter Combinations of the whole tool chain in addition to the actual results.

Parameter meta-information, test run parameter values and test run results are stored in the data base (C). Technically speaking, this data is persisted using de-/serialization of Django models objects. A parameter is specified by an association of a specific tool, a name, a description, a default value, a valid value domain, and a type. The system keeps track of runs by their specific feature combination, i.e. a tuple of parameter instance objects that reflect the actual values the parameters were set to (cf. figure 2).

Further data is not stored in the data base but in a documented file system structure (D) in simple binary files. This includes raw binary data like images, features and kNN graphs. Not storing this information in the data base simplifies the data access for third party tools.

5.3   System Execution

The system's three main services are: running feature extractors on image datasets, learning in feature space and creating neighbor graphs. These services are supplied as command-line scripts which can be invoked through the web front end. Input forms to start the services are automatically generated using the meta-information stored in the data base. The front end, thus, adapts without further changes, e.g., to newly added algorithms.

The computations of these services can be quite memory and processor intensive. Hence, the master tier (cf. section 4) submits computation tasks to a SGE. The distribution granularity depends on the service itself. While the computation of features is embarrassingly parallel, since each image can be processed independently, classification runs and nearest neighbor computations require access to all data and are thus submitted as single jobs.

6   Usage

This section focuses on the user's point of view. We present an exemplary Wifbs set-up and detail its functionality to optimize the classification pipeline for the *Gist* dataset ([13], 2688 images, 8 intersection free scene classes). Even though datasets are much larger nowadays, it should serve as a good example for the capabilities of *Wifbs*. Our test set-up uses 3 distinct machines as Master, Database and File Server, each additionally serving as computation node. Our machines, due to hyper-thread multi-core processors, provide the SGE with 29 cores for job execution. 161,280 feature runs took 215 minutes. 15,631 classification runs (including preprocessing and embedding) took 4,093 minutes.

We wrapped the implementations of 7 feature algorithms to match the interface of Wifbs.[4] From the Shogun toolbox we included 19 kernel[5], 7 normalizers[6], 11 distances[7], 4 preprocessors[8], 1 embedding algorithm[9] and 2 multi-class classification algorithms[10]. Figure 3 shows the parameter tree that is spanned by the stages of the pipeline. An exhaustive search of the parameter spaces spanned by the numeric parameters of a typical tool chain is not possible, as the cardinality of this parameter space becomes approximately 10^{16}. Thus, our systems offers an adjustable sweep method sampling user-defined parts of the parameter spaces (cf. *II.2+III. Classification*).

---

[4] These are mostly freely available standard implementations of: Gist [13], JCD, CEDD, FCTH [2], Tiny Image [21], Gabor [11], and simple 3D color histograms.

[5] Anova Kernel, Gaussian Kernel, Linear Kernel, Power Kernel, Spline, Chi2, Histogram Intersection, Log, Rational Quadratic, TStudent, Pyramid Chi2, Inverse Multiquadric, Multiquadric, Sigmoid, Wave, Exponential, Jenson Shannon, Poly, Spherical, Wavelet.

[6] Average Diagonal, Tanimoto, Zero Mean Center, Dice, Square Root Diagonalization, Variance.

[7] BrayCurtis, Chebyshew Metric, Cosine, Geodesic, Mahalonobis, Tanimoto, Canberra, ChiSquare, Euclidean, Jensen Metric, Manhatten Metric.

[8] KernelPCA, LogPlusOne, NormOne, PruneVarSubMean.

[9] Diffusion Maps.

[10] KNN, MultiClassLibSVM.

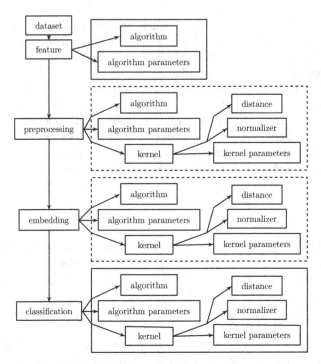

Fig. 3. Parameter tree. *Algorithms, kernel, distances,* and *normalizers* may be instantiated from a set of nominal choices. Types of *algorithm* and *kernel parameters* are shown in figure 2. Dashed areas - - - signify non-permanently generated data, solid bordered areas — represents results that are permanently stored.

I. Dataset Administration: The first step is the import of one or multiple datasets conjointly with available labeling information. To do this for the Gist dataset we copy the images onto the file server (D) and create comma separated value files containing the meta-information. Those include file names (with subdirectories if applicable), class names (scenes that are visually exhibited by the images), classes per image and a dataset description. Through the web portal one can specify the dataset name. The meta-information is imported automatically into the database. The dataset can be deleted via the web site by users possessing adequate rights.

Henceforth, it is possible to browse the Gist dataset on dataset/class/single image level and—after features are computed—visually inspect the computed features.

II.1 Feature Computation: The second essential step of the benchmark procedure is the computation of image features. The web portal offers listings and descriptions of available feature algorithms and already computed descriptors. New feature runs can be parametrized and transparently dispatched to the job scheduler. The user can then observe the progress of job execution and may, if necessary, intervene.

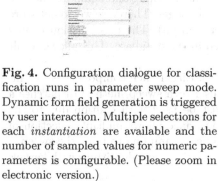

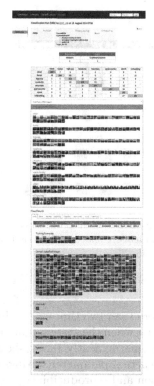

Fig. 4. Configuration dialogue for classification runs in parameter sweep mode. Dynamic form field generation is triggered by user interaction. Multiple selections for each *instantiation* are available and the number of sampled values for numeric parameters is configurable. (Please zoom in electronic version.)

Fig. 5. Detailed view of single classification run with *instantiation* and *parametrization* depicted prominently. Confusion matrix, listings of training sets (due to permutation different for each run), and class-wise performance indicators conjointly with image-wise testing results facilitate thorough result analysis.

For our test scenario we used 60 feature runs with default value parametrization, which is conservatively low number of runs, but which is sufficient for our demonstration. This serves as basis for subsequent fine-adjustment. After initial features are created the user can skip classification and continue with the first step of interpreting the results (see below IV.).

II.2+III. Classification: The configuration of feature and classification runs is conducted similarly (cf. figure 4). Additionally to the selection and parametrization of the classification algorithm one can modify whether and how the data is preprocessed and embedded prior to learning. The latter offers an enormous number of degrees of freedom. Storing the processed data would largely increase the storage requirements. However, contrary to the image features that are usually much more costly to compute, it is seldom necessary to access the processed

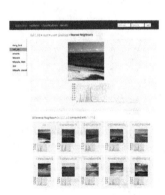

Fig. 6. Nearest neighbor browsing component in feature space with basic feature visualization

Fig. 7. Result summary dataset-wise. There is a distinct sweet spot in parameter space as the graph of sorted *F1*-values of all test runs shows.

data more than two times: once during the benchmarking and again for the real world case when the system is applied to novel images. Therefore we do not store this data permanently (cf. dashed areas in figure 3). The classification results and the actual parameter settings are stored in the database. This allows result inspection and reproducibility.

There are two modes to run classifications. Either each component is instantiated and parametrized explicitly by the user or a sweep method dispatches a bulk of jobs to the SGE. Such sweeps must be used with caution due to the enormous parameter space size. Our web interface allows multiple selection of components and adjustable numerical parameter sampling to limit the sweep.

Our case study includes a sweep over the classification algorithms with kernels and distances and all computed features.

IV. Result Interpretation: The system offers three ways to evaluate the effectiveness of the available features. Usually the first step is to visualize what a feature regards as visually similar. The system allows the user to navigate the nearest neighbors in feature space along with feature visualization and distances (cf. figure 6). This serves to get a first impression of feature usefulness and, e.g., whether images cluster densely in feature space. It is also possible to view the neighbors of an alternative dataset.

The second step is running a classification with default parameters and analyze summary statistics. Here our detailed result page allows insights into common performance indicators like the confusion matrix, precision, recall and *F1* score (cf. figure 5). This enables the user to see strengths and weaknesses of the chosen pipeline broken down to single image instances.

The final step is to find the best parametrization for a given task. After restricting the search space using the previous techniques, one normally wants to fine-tune the pipeline. This can be done via the mentioned parameter sweep method. At this point, there are usually many of classification runs to consider. For this the website offers a tabular view with prepared filter mechanisms for common tasks (e.g. list best runs for a specific dataset).

Figure 7 shows the best results of our use case. It becomes clear that the Gist feature, which accompanied the publication of the dataset, performs very well. However, CEDD, a simple parameter free feature that simply reflects the global color distribution, performs even better.

7   Conclusion

We presented a software system that allows system engineers that face image classification problems to focus on the essential: benchmarking and selecting the best image feature algorithms. Easy extensibility ensures maximal flexibility for a wide range of applications. The implementation will be made available and can be used and adapted without restrictions. We believe our system can serve well as testing base line for future development of algorithms as well as for the presented application scenarios. We plan on further improving Wifbs, especially by adding more sophisticated tools for the results analyses.

The research was financed by the DFG SPP 1335 project "The Zoomable Cell" and by the European Social Fund (ESF) Project #100098171.

References

1. Batra, D., Agrawal, H., Banik, P., Chavali, N., Alfadda, A.: CloudCV: Large-Scale Distributed Computer Vision as a Cloud Service (2013), http://www.cloudcv.org
2. Chatzichristofis, S.A., Boutalis, Y.S.: Compact Composite Descriptors for Content Based Image Retrieval: Basics, Concepts, Tools. VDM Verlag (August 2011)
3. Chatzichristofis, S., Boutalis, Y.: Cedd: Color and edge directivity descriptor: A compact descriptor for image indexing and retrieval. In: Gasteratos, A., Vincze, M., Tsotsos, J.K. (eds.) ICVS 2008. LNCS, vol. 5008, pp. 312–322. Springer, Heidelberg (2008)
4. Deng, J., Dong, W., Socher, R., Li, L.-J., Li, K., Fei-Fei, L.: ImageNet: A large-scale hierarchical image database. In: IEEE Conference on Computer Vision and Pattern Recognition, pp. 248–255 (2009)
5. Deselaers, T., Keysers, D., Ney, H.: Features for image retrieval: An experimental comparison. Information Retrieval 11(2), 77–107 (2008)
6. Everingham, M., Gool, L., Williams, C.K.I., Winn, J., Zisserman, A.: The Pascal Visual Object Classes (VOC) Challenge. International Journal of Computer Vision 88(2), 303–338 (2010)
7. Goëau, H., Bonnet, P., Joly, A., Yahiaoui, I., Barthelemy, D., Boujemaa, N., Molino, J.F.: The ImageCLEF 2012 plant identification task. CLEF 2012 Working Notes (September 2012)
8. Gunnar, R.: SHOGUN - A Large Scale Machine Learning Toolbox. Journal of Machine Learning Research 22, 2006–2006 (2010)
9. International, G.I.: Stock photography, royalty-free photos, video footage & music. (July 2014), http://www.gettyimages.com/

10. Lenc, K., Gulshan, V., Vedaldi, A.: Vlbenchmarks (2011),
    http://www.vlfeat.org/benchmarks/xsxs
11. Manjunath, B., Ohm, J.R., Vasudevan, V., Yamada, A.: Color and texture descriptors. IEEE Transactions on Circuits and Systems for Video Technology 11(6), 703–715 (2001)
12. Mikolajczyk, K., Schmid, C.: A performance evaluation of local descriptors. IEEE Transactions on Pattern Analysis and Machine Intelligence 27, 1615–1630 (2005)
13. Oliva, A., Torralba, A.B.: Modeling the Shape of the Scene: A Holistic Representation of the Spatial Envelope. International Journal of Computer Vision 42(3), 145–175 (2001)
14. Patterson, G., Hays, J.: SUN attribute database: Discovering, annotating, and recognizing scene attributes. In: Proceedings of the IEEE Conference on Computer Vision and Pattern Recognition (CVPR), pp. 2751–2758 (2012)
15. Rubner, Y., Puzicha, J., Tomasi, C., Buhmann, J.M.: Empirical evaluation of dissimilarity measures for color and texture. Computer Vision and Image Understanding 84(1), 25–43 (2001)
16. Russakovsky, O., Deng, J., Huang, Z., Berg, A.C., Fei-Fei, L.: Detecting avocados to zucchinis: what have we done, and where are we going? In: International Conference on Computer Vision (ICCV) (2013)
17. Schroder, M., Rehrauer, H., Seidel, K., Datcu, M.: Interactive learning and probabilistic retrieval in remote sensing image archives. IEEE Transactions on Geoscience and Remote Sensing 38(5), 2288–2298 (2000)
18. Shamir, L., Orlov, N., Mark Eckley, D., Macura, T.J., Goldberg, I.G.: IICBU 2008: a proposed benchmark suite for biological image analysis. Medical and Biological Engineering and Computing 46, 943–947 (2008)
19. Shechtman, E., Irani, M.: Matching Local Self-Similarities across Images and Videos. In: 2007 IEEE Conference on Computer Vision and Pattern Recognition (2007)
20. Torralba, A., Efros, A.A.: Unbiased look at dataset bias. In: 2011 IEEE Conference on Computer Vision and Pattern Recognition (CVPR), pp. 1521–1528. IEEE (2011)
21. Torralba, A., Fergus, R., Freeman, W.T.: 80 million tiny images: a large data set for nonparametric object and scene recognition. IEEE Transactions on Pattern Analysis and Machine Intelligence 30, 1958–1970 (2008)
22. Xiao, J.X.J., Hays, J., Ehinger, K.A., Oliva, A., Torralba, A.: SUN database: Large-scale scene recognition from abbey to zoo. In: 2010 IEEE Conference on Computer Vision and Pattern Recognition (CVPR) (2010)

Personality Modeling Based Image Recommendation

Sharath Chandra Guntuku[1], Sujoy Roy[2], and Lin Weisi[1]

[1] School of Computer Engineering, Nanyang Technological University, Singapore
sharathc001@e.ntu.edu.sg, wslin@ntu.edu.sg
[2] Institute for Infocomm Research, Singapore
sujoy@i2r.a-star.edu.sg

Abstract. With the increasing proliferation of data production technologies (like cameras) and consumption avenues (like social media) multimedia has become an interaction channel among users today. Images and videos are being used by the users to convey innate preferences and tastes. This has led to the possibility of using multimedia as a source for user-modeling, thereby contributing to the field of personalization, recommender systems, content generation systems and so on. This work investigates approaches for modeling personality traits (based on the Five Factor Modeling approach) of users based on a collection of images they tag as 'favorite' on Flickr. It presents several insights for improving the personality estimation performance by proposing better features and modeling approaches. The efficacy of the improved personality modeling approach is demonstrated by its use in an image recommendation system with promising results.

Keywords: semantic features, personality modeling, big five factor.

1   Introduction

Modeling users' personality based on content (images and videos) they like has widespread applications in recommender systems, personalization, novel content generation systems, and target advertising systems and so on. The key factors to address this problem effectively are the source for acquiring users' personality and the model used to facilitate the right mapping between them. In this work we look at positive implicit feedback (likes) on images as the source for modelling users personality, because images are universal in expression even when users speak different languages.

Assessing the personality of users by looking at images they liked has been studied in the literature [1]. To do this, images tagged as 'favorite' by a group of Flickr users were collected. Next the users were asked to answer the BFI-10 questions [2] to get their personality profile based on the Big Five Factor personality modeling approach [3]. 'Psychology experts' were asked to look at the images liked by the users and answer the same BFI-10 questions. The idea was to get the experts' opinion regarding the users' personality profile (a different perspective from self-assessment). The process of automatically assessing the personality profile of users involved learning a regression model (LASSO [4]) mapping low-level image features to their personality profile. A summary of the features used in [1] is depicted Table 1. Each users personality is modeled based on a training set of images liked by the user. The ability of the model to

X. He et al. (Eds.): MMM 2015, Part II, LNCS 8936, pp. 171–182, 2015.

predict personality was evaluated based on a test set of images liked by the user. Difference in self-assessment of personality and the assessment given by experts was studied and the following observations were made.

- With a reasonable level of accuracy, personality profiles assessed by experts can be modelled from low-level features extracted from images users liked.
- However users' self-assessed personality profiles are difficult to model and the learned model does not generalize well.

Based on the above observations it was concluded that because self-assessed personality profiles tend to be more noisy, they are difficult to generalize. The noise is ascribed to the fact that users may not be able to assess their personality properly. However it must be noted that the above conclusions are based on the following two implicit assumptions.

- Users' self assessment of personality profile was based only on the images they have liked.
- Experts assessment of users personality goes beyond the clues presented by the images, based on which they made their assessment.

We note that the above assumptions may not be true because the experts rating of the users personality is based on the limited set of images shown to them. The expert can only work within the bounds of information provided to him. Conversely, when users are assessing their personality, it's based on factors which certainly go beyond the images they like. So what is indicated as noise in the self-assessed personality profiles can very well be additional information that the images do not capture. Note that the users are answering the BFI-10 questions which probably capture more information than that contained in the images. Also it is not really clear that the low-level features used to map from features to personality profiles are indeed representative in predicting highly semantic concepts as the Five Factor (Openness, Conscientiousness, Extraversion, Agreeableness and Neuroticism) personality profiles.

In view of the above observations we first look at answering a few questions that would help us identify a better approach to modelling users personality from a set of images the user has liked and also see how we can generalize better. The questions we attempt to answer are as follows:

- Is the set of features used in [1] sufficient to describe user's interests. How about looking at more semantic features?
- Is there an approach which leverages upon users' likes on images to model their personality, which goes beyond their likes on images?
- If experts can better assess personality by looking at images liked by the user, how can we leverage on this fact to improve personality profile prediction on self assessed data? Does the expert knowledge help generalize well?

After conducting a detailed investigation of the above questions we identify a useful set of features and approach that helps better model the user personality profile. This forms the basis for addressing the recommendation problem which is basically, given a user, answering what kind of images would the user like?

Fig. 1. Need for including Semantic Features. Left: High score on Conscientiousness; Center: Low score on Extraversion; Right: High score on Neuroticism.

Contributions. We conduct a detailed investigation of estimating users personality profile from a collection of images that were liked by the user presenting recommendation for features that need to be considered and also the modeling approach that can generalize well. We show the need for using high level user understandable features and also demonstrate the efficacy of a F2A (Features-to-Answers) +A2P (Answers-to-Personality) approach compared to the usual F2P (Features-to-Personality) approach that has been taken by existing works. Once we have a better approach for modeling personality from images we demonstrate their usefulness in an image recommendation system. Here the Big Five Factors for personality modeling form a latent space for mapping images and users.

Table 1. List of Features (with newly added features are bolded)

List of Features used	
Use of Light	**Head and Upperbody recognition**
HSV statisctics	**Face and Pose recognition**
Emotion-based	**Gender identification**
Entropy	**Scene Classification**
Regions using mean shift segmentation	**Computer Graphics vs. Natural image**
Low Depth of Field (DOF)	**Saliency**
GIST Descriptors	**Black & White vs. Color image**
#Edges	**Visual Clutter**
Tamura	
Wavelet Textures	Colorfulness
Rule of thirds	GLCM-features
Objects: Deformable Parts model	Image Parameters

2   Enhancing Personality Modeling

In this section we investigate how the personality modeling approach based on mapping low-level images features to personality profile can be improved. We also investigate the difference between self-assessed and expert assessed personality profiles in helping model the Five Factor personality profiles.

Fig. 2. Sample of the images 'faved' by users with different personality profiles. Left: High score on Openness; Center: High score on Extraversion; Right side: High score on Agreeableness.

2.1   Adding Semantic Features

As aforementioned, Table 1 lists the features used in [1]. We note that most of these features are low-level features. Hence we first investigate adding more semantic features that might provide a better representation of the users' preferences and hence their personality. Table 1 depicts these additional features in (**bold**).

For images to reflect user's personality, they (or the features extracted from them) should be representative of a universal set of possible images (or features), that help model a diverse set of tastes (associated with different personality profiles). As building a universal set of images is infeasible, the features chosen to represent the limited set of images, should be able to convey semantics that is well representative of characteristics that everybody can relate to.

Depending on a person's psychophysical nature, one is drawn to different 'kinds' of images. To make this distinction, some amount of domain knowledge has to be included to capture the differences in what different people look at. For example, images on the left side of collage (Figure 1) show images liked by a user with a score of 4 on conscientiousness (for a range of [-4,4]). This trait reflects in socially prescribed impulse control and a sense of thoughtful behavior. Images of such users consist of carefully planned and timed shots and many black & white images (which is shown to be a sign of focus, subtlety of tones and versatility [5]). While people with high score on extraversion have images with a lot of people, people with a low score, have opposite preference - consisting of scenic backdrops, without faces or objects to focus upon (images in the middle of the same collage). Also, people with a high score on neuroticism are characterised by anxious and tense behavior. They seldom relax. This is conveyed by the images they like (images on the right side of the collage), most of which have a high level of clutter [6].

Personality traits should ideally be influenced by high level semantic concepts, each dealing with multiple facets of human behavior [7,8]. For example, a person with a high score on Openness factor tends to appreciate art, come up with distinctive-looking work and home environments etc. Figure 2 shows some random images liked by 3 users from the PsychoFlickr data-set [1] (to be described in Section 3.1). The left part of the collage show images 'faved' by a user with a score of 4 on Openness factor. These images show a bias towards dance, music and artistic photography. This is just a representative set. The original dataset can be accessed at [1]. Images of a user having a score of 4 on Extraversion are shown in the middle. This trait reflects in energetic approach towards social and material world. This means that the user is expected to have high affinity towards friends and partners - the images liked by the user can be seen to convey, in this case, a liking towards people of opposite gender. The right part shows images liked by a user with a score of 4 on Agreeableness, which reflects in traits such as altruism, tender-mindedness and trust, and in behavior like consoling friends who are upset etc. The images liked by the user convey a similar inclination.

Each of the five personality traits convey specific innate tastes of users which lead to certain behavioral attributes. Capturing the wide spectrum of personality traits with features extracted from data is a challenge. But, the features can be used to detect the above mentioned behavioral attributes (which are usually less subtler than personality traits) which can give us clues about the user's personality.

Based on this analysis, we have found that while aesthetic features capture a lot of information in the images as mentioned in literature [9], features which are more interpretable are needed to capture a person's tastes. Aesthetics, which formed the majority of the features used in [1], fail to convey these high level semantic characteristics. Especially, when we are trying to find the right user-image match, a more concrete relationship between person's profile and content profile is needed to give the users what they like [10].

2.2   Looking Beyond Existing Approaches

Most of the works in automatic personality modeling have taken the approach of extracting features from data (related to users) and using them directly to model personality of users (termed as Features-to-Personality - F2P). But research in psychology [2] shows that personality is ascertained by the asking users a set of well-designed questions. The answers to these questions are converted into a score, for each of the 5 personality traits (termed as Features-to-Answers, Answers-to-Personality - F2A+A2P). Taking this approach with features extracted from data was shown to be more effective than predicting the personality scores directly [11]. The intuition is that non-experts understand and hence answer the BFI-10 questions better than scoring themselves on a scale of -4 to 4 for personality profiling.

F2A+A2P Model.   Personality prediction is divided into two stages, namely transforming features to answers (F2A) and mapping the answers to trait scores (A2P). For the first stage, i.e, transforming the feature space into answers, a Sparse and Low-rank Transformation (SLoT) algorithm was [11] used here. The motivation for using the sparse and low rank transformation is twofold. (a) The answers to the questions may have some overlap with each other (i.e., a social and outgoing person would want to be thorough in work as this might attract others; a low rank constraint is needed in the regression problem to capture these correlations between the answers). (b) Each answer might only be influenced by a few features (leading to a sparsity constraint). And for the second stage, the BFI 10 scoring scheme [2] is used to predict the personality profiles.

SLoT Formulation. The transformation to BFI-10 answers from features contains both low-rank and sparse structure. In addition to minimizing the regression error, specific matrix norms are therefore used to learn the transformation.

The answers are represented by Z - a $10 \times N_{tr}$ matrix and the features of training images by F_{tr}. Regression model's parameters are represented by W - a $10 \times N_f$ matrix.

Estimation of the transformation matrix W, with sparse and low-rank structure from the training images (F_{tr}, Z), can be formulated as the following objective function,

$$\min_{W} \left\{ \frac{1}{2} \| Z - W F_{tr} \|_F^2 + \lambda_1 \| W \|_* + \lambda_2 \| W \|_1 \right\}. \tag{1}$$

where $\| \cdot \|_F$ is matrix Frobenius norm, $\| \cdot \|_*$ - matrix nuclear norm, a convex surrogate for the matrix rank, and $\| \cdot \|_1$ is the matrix ℓ_1 norm , a convex surrogate for the matrix ℓ_0 norm. Predictions on test samples are obtained as $Z_{tes} = W F_{tes}$, where F_{tes} is the feature matrix for the test samples. Details on solution of Eq. 1 are elaborated in [1].

2.3   Expert vs. Self-assessed Profile

In [1] it was claimed that modeling experts-assessed personality profiles of users performs better than self-assessed profiles. We note that this is subject to what kind of features are chosen as relevant and what learning approach is used to build the personality prediction model. Since the personality profiles are histograms over quantized bins it needs to be understood that the actual value of prediction error (deviation) can have a significant bearing on the the above claim. More specifically, if the error is less than the quantization interval we can expect no error. Hence we believe that with a better choice of features and model learning approach self-assessed personality modeling can be closer to what experts give. This can be used to predict the user's self-assessed scores, even without explicitly asking the users to answer the BFI-10 questionnaire which can be aid in non-intrusive personality acquisition techniques [12].

Moreover it would be interesting to note how well the expert knowledge based model generalizes in being able to predict self-assessed personality models. After all we cannot expect every user to fill up a questionnaire to indicate his profile. Rather we would want to use the knowledge of a few experts to build a model that generalizes well for unknown users for whom only the images they have liked are known. A better approach and better features should help to generalize better.

3   Experiments

In this section we investigate the efficacy of using more semantic features, evaluate a F2A+A2P personality modeling approach unlike the standard F2P approach, study the relationship between self assessed and expert based personality modeling, and finally assess the effectiveness of our investigations on the performance of a personality modeling based image recommender system.

3.1   The PsychoFlickr Dataset

The PsychoFlickr data set consists of Flickr images. The data set consists of 60,000 images where 200 images are tagged as favorite by 300 'pro' users. For every user, personality assessment was done as described in Section 1.

Experimental Set Up. We divide the dataset into training and test sets to validate the results of the different modeling approaches. For all the experiments, unless specified otherwise, the data split is 75% training with 5 fold cross validation and 25% for testing. Results are reported on testing data. In the experiments, 'old features' refers to the features used in [1] and 'new features' refers to the combined set of semantic features and old features (Table 1). The semantic features were extracted using default parameters of softwares provided by the cited works. The traits, for which performance of models is being reported, are: O: Openness, C:Conscientiousness, E:Extraversion, A:Agreeableness and N:Neuroticism. Results obtained are statistically significant ($p < 5\%$).

3.2 Comparing Modeling Approaches

LASSO regression was taken as the baseline approach for modeling users personality based on a their 'faved' images. Note that this was also the model chosen in [1].

Sparse Support Vector Regression. To account for sparsity in the data a sparse SVM modeling approach was chosen to evaluate its efficacy. We also hope that the model captures non-linearities in the data better.

F2A+A2P. The previous two approaches modeled personality scores directly based on the features (F2P). Recent results [11] have shown that the alternative of F2A+A2P gives a better prediction model for modeling profiles of individuals inside video segments. Note that the set of features used in [11] apply to inferring personality profiles of people in the videos whereas in this work we use features to model personality profiles of users who like a set of images. Hence although the modeling approach is the same, the two scenarios are different - a) [11] deals with perception of personality profiles of characters in videos whereas our work deals with both recognition and perception [13] of users personality profile based on the images they like.

Table 2. Results of predicting personality-profiles of users for both F2A+A2P and F2P approaches, represented in terms of RMSE

With Old Features				With New Features			
Trait	LASSO	sparse SVR	F2A+A2P	Trait	LASSO	sparse SVR	F2A+A2P
O	1.698	1.774	**0.982**	O	1.796	1.718	**0.913**
C	1.789	1.684	**0.898**	C	1.704	1.652	**0.830**
E	2.077	1.836	**0.971**	E	1.859	1.820	**0.905**
A	1.669	1.399	**0.831**	A	1.441	1.379	**0.719**
N	2.208	2.356	**1.197**	N	2.299	2.317	**1.114**

However, we must note that the PsychoFlickr data set does not contain the ground truth for users and experts answers to the BFI-10 questions. In the BFI-10 scoring scheme[2], out of the 10 questions, 2 distinct questions contribute to the scoring of every personality trait in a linear relationship as defined below.

$$P_i = (A_i + (6 - A_{i+5}))/2 \ \forall i \in [1,5], \tag{2}$$

Table 3. Measuring actual deviations in prediction of personality scores (which are converted into High, Medium, Low categories based on threshold)

	Classification Accuracy (in %)	
Trait	F2P	F2A+A2P
O	62.40	66.10
C	58.10	**70.50**
E	53.70	**69.70**
A	64.30	**72.30**
N	50.70	**61.50**

where P_i (dependent variable) is the personality score for trait i and A_i (independent variable) is the i-th BFI-10 answer. Note that we know P_i and our goal is to get an estimate of the A_i's. Hence we rewrite the equation with P_i as the independent variable and A_i's as the dependent variables. This gives us a new set of linear equations parameterized by the A_i's which can be represented in their polar form as

$$r_i = |2 * P_i - 6|/\sqrt{2}, \tag{3}$$
$$\theta_i = 135° \text{ if } r_i > 0, \tag{4}$$
$$= 315° \text{ if } r_i < 0, \tag{5}$$
$$= 0 \text{ if } r_i = 0, \tag{6}$$

where $r_i = \sqrt{A_i^2 + A_{i+5}^2}$ is the distance of the line (with P_i as intercept) from the origin and $\theta_i = \pi/2 + \tan^{-1}(A_{i+5}/A_i)$ is the angle the line subtends with the positive x-axis. So for every set answers, we have a corresponding mapping in the polar space. A model is then trained to predict the values of r_i and θ_i which are then converted back into personality scores based on the BFI scoring scheme (from which these values were derived). We trained LASSO, sparse SVR and F2A+A2P models using the old features as the input and self-assessed personality scores as the output. Table 2 shows that F2A+A2P approach is better than F2P approach with significant performance gains. Even RMSE of traits using sparse SVR is lesser than that using LASSO. It also shows that F2A+A2P can bring huge gains in modelling personality scores, reducing the error by 40-50%. The same experiment was repeated using new features. Adding new semantic features led to a 5-15 % increase in performance (Table 2). We also measure the actual deviation w.r.t to the exact personality score values. The scores were divided into low, neutral and high levels. This is to highlight, as discussed previously, how bining influences the actual performance. Table 3 shows that F2A+A2P is better by about 7-15% at predicting all the traits except for Openness. It should be noted that the images were tagged as favorite by Flickr 'pro' users, who pay a fee for using privileged features on Flickr. This means that the users' interests would be highly inclined towards art, photography and the like as mentioned in Section 2.1, thereby skewing the traits' distribution. This bias in the Openness trait is a possible reason for the inferior performance of F2A+A2P.

3.3 Expert vs. Self-assessed

The analysis made in [1] show that a model on personality scores given by experts performed better than model on self assessed personality scores. We tried to see if the increased performance in modeling self-assessed scores using F2A+A2P can decrease the difference between the error in modeling expert-assessed and self-assessed scores. The performance in terms of errors made by the models' is shown in Table 4. Note that the RMSE on both expert and self-assessed scores falls below 1. As in the previous section, we evaluated the accuracy on actual deviation from the three bins as well. On experts scores (training+testing), F2A+A2P could classify almost all the samples in the right range (90% accuracy). And on self-assessed data (training+testing), the performance of F2A+A2P is as shown in Table 3. The accuracy can go as high as 75%. As mentioned in Section 2.3, this deviation is more meaningful to observe and reduce, and we see that F2A+A2P is able to do so for all the traits of both experts and self-assessed scores. F2P approaches used in previous personality modeling literature show good performance for Extraversion and Conscientiousness traits as they can be easily perceived [14]. But when the approach of mapping features to answers is used and personality scores are predicted using the BFI scoring scheme, it results in significant gains in performance for all traits as seen in this section and also previous section. This can be attributed to the fact that the relationship between features and answers are less non-linear and also that we are using the BFI scoring scheme, which is the standard method used in psychology, for personality assessment.

Table 4. Results of F2A+A2P method on modeling experts' scores and self-assessed scores, shown in terms of RMSE

Trait	Experts	Self-assessed
O	0.406	0.913
C	0.062	0.830
E	0.895	0.905
A	0.679	0.719
N	0.194	1.114

Generalization. Often it is found that we have a lot of information about the data related to users, but no information about their personality profiles. It is also infeasible, in many cases, to ask the users to answer the BFI 10 to be able to build their profile. But it is less troublesome to get psychology experts to build the personality profile of the users by asking them to examine the data.

To verify if we can bridge the gap between expert scores and the self-assessed scores, we checked the capability of the F2A+A2P model trained on experts scores as output to generalize on self assessed scores. Comparing Table 4 with Table 5 shows the ability to generalize for all the traits except for Openness and Agreeableness (in line with previous experiments in our work and also in literature). Very low error in conscientiousness can be explained by the observation that the cues associated with this trait, mentioned in Section 2.1, are comparatively easy to notice.

Table 5. Evaluating generalization of training on experts' scores to test on self-assessed scores using F2A+A2P, terms of RMSE. Column under On Experts represents train/test on experts scores and under On Self-assessed represents training on experts' scores and testing on self-assessed scores.

Trait	On Experts	On Self-assessed
O	0.406	1.163
C	0.062	0.146
E	0.895	0.743
A	0.679	0.891
N	0.194	0.482

We see that F2A+A2P can be used to predict the user's self-assessed scores, even without explicitly asking the users to answer the BFI-10 questionnaire. This can be a huge step forward in non-intrusive personality acquisition techniques, which was shown to be preferred the most by users in [12].

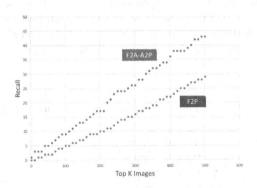

Fig. 3. Mean Recall (for all users) using top-k images retrieved by F2P and F2A+A2P

3.4   Image Recommendation System

For the recommendation problem given a user, we tried to predict which image will the user like. For this, our goal is, given a user profile, evaluate recall at top-K for all test images. Note that in the given data-set, we only have information about the images that were tagged as favorite by every user. We do not know if a user has seen the images tagged as favorite by remaining users. Therefore, although we cannot reject the possibility that the images tagged as favorite by other users could be his 'favorite' as well, we prefer to be more strict in our evaluation and the performance evaluation forms a lower bound on accuracy.

For these experiments we used the models for both F2P (sparse SVM) and F2A+A2P as described in previous sections. We took a total of 2000 images (for which user assignments are known and different from training set) as the test set and ranked the

Fig. 4. Top 10 Images retrieved by F2A+A2P for a sample of users based on their Personality Profile. The images with green border are 'faved' by the user. Retrieved images not belonging to the user's faved set can be seen to have lot of similarity w.r.t cues associated with the corresponding personality traits. Explained in Section 3.4.

images based on RMSE between the predicted profile and users profile. The performance evaluation is based on recall for top-K (where $K \in \{1, 2, \ldots\}$). Figure 3 shows that F2A+A2P is able to retrieve a higher number of images (in top-K) when compared with F2P model. It is definitely a possibility that some images that were highly ranked by the model but were not counted towards recall might have been 'faved' by the user. Even without taking this into account, we see a steep increase in the recall using F2A+A2P when compared to F2P.To verify this we visually inspected the top 10 images retrieved by the model for a few users, out of which 3 are shown along with the corresponding user profile in Figure 4. The images bounded by green are 'faved' by the users. The first user has a high score on Openness and Neuroticism traits and the images are very colorful (which was found to be a common characteristic of images depicting custom work/home environments (image 2 from the left) and artistic photography). Some of the images retrieved have high clutter (images 1 and 10 from the left), which is a characteristic of images 'faved' by neurotic users. Second user has a very low score on extraversion and corresponding images retrieved have scenic backdrop without any explicit objects.

This confirms that a) representing images through semantic features helps in mapping images to users and b) F2A+A2P approach is highly effective in modeling personality profiles from images with high generalization.

4   Conclusions

In this paper, the problem of image recommendation has been studied from a personality modeling perspective. It was seen that F2A+A2P approach outperforms F2P. Also, it was verified that high level semantic features outperform low-level features (used in [1]) in capturing the tastes of users.

References

1. Cristani, M., Vinciarelli, A., Segalin, C., Perina, A.: Unveiling the multimedia unconscious: Implicit cognitive processes and multimedia content analysis. In: Proceedings of the 21st ACM International Conference on Multimedia. MM 2013, pp. 213–222. ACM, New York (2013)
2. Rammstedt, B., John, O.P.: Measuring personality in one minute or less: A 10-item short version of the big five inventory in english and german. Journal of Research in Personality 41, 203–212 (2007)
3. Goldberg, L.R.: An alternative "description of personality": the big-five factor structure. Journal of Personality and Social Psychology 59, 1216 (1990)
4. Tibshirani, R.: Regression shrinkage and selection via the lasso. Journal of the Royal Statistical Society. Series B (Methodological), 267–288 (1996)
5. Rowse, D.: Why Black and White Photography? (April 2014),
   http://digital-photography-school.com/why-black-and-white-photography/
6. Rosenholtz, R., Li, Y., Mansfield, J., Jin, Z.: Feature congestion: a measure of display clutter. In: Proceedings of the SIGCHI Conference on Human Factors in Computing Systems, pp. 761–770. ACM (2005)
7. John, O.P., Srivastava, S.: The big five trait taxonomy: History, measurement, and theoretical perspectives. Handbook of Personality: Theory and Research 2, 102–138 (1999)
8. Ma, K.-T., Sim, T., Kankanhalli, M.: VIP: A unifying framework for computational eye-gaze research. In: Salah, A.A., Hung, H., Aran, O., Gunes, H. (eds.) HBU 2013. LNCS, vol. 8212, pp. 209–222. Springer, Heidelberg (2013)
9. Machajdik, J., Hanbury, A.: Affective image classification using features inspired by psychology and art theory. In: Proceedings of the International Conference on Multimedia, MM 2010, pp. 83–92. ACM, New York (2010)
10. Guntuku, S.C., Zhou, J.T., Roy, S., Lin, W., Tsang, I.W.: Deep representations to model user 'likes'. In: Asian Conference on Computer Vision (November 2014)
11. Srivastava, R., Feng, J., Roy, S., Yan, S., Sim, T.: Don't ask me what i'm like, just watch and listen. In: Proceedings of the 20th ACM International Conference on Multimedia, MM 2012, pp. 329–338. ACM, New York (2012)
12. Hu, R., Pu, P.: Acceptance issues of personality-based recommender systems. In: Proceedings of the Third ACM Conference on Recommender Systems, RecSys 2009, pp. 221–224. ACM, New York (2009)
13. Vinciarelli, A., Mohammadi, G.: A survey of personality computing. IEEE Transactions on Affective Computing 99, 1 (2014)
14. Judd, C.M., James-Hawkins, L., Yzerbyt, V., Kashima, Y.: Fundamental dimensions of social judgment: understanding the relations between judgments of competence and warmth. Journal of Personality and Social Psychology 89, 899 (2005)

Aesthetic QR Codes Based on Two-Stage Image Blending

Yongtai Zhang, Shihong Deng, Zhihong Liu, and Yongtao Wang

Institute of Computer Science and Technology, Peking University,
Beijing, P.R. China 100871
{yt.zhang,dengsh,lzh,wyt}@pku.edu.cn

Abstract. With the popularization of smart phones, Quick Response (QR) code becomes one of the most widely used two-dimensional barcodes. Standard QR code consists of black and white squares (called *modules*) and its noise-like appearance would seriously disrupt the aesthetic appeal of its carrier, e.g., a poster. To make QR code more aesthetic, this paper proposes an automatic approach for blending a visual-unattractive QR code with a background image. This approach consists of two stages: module-based blending and pixel-based blending. At the first stage, a binary aesthetic QR code is generated module by module. At the second stage, a color aesthetic QR code is further generated pixel by pixel. The advantages of our approach are: 1) greatly enhancing the aesthetic appearance of the original QR code, 2) maintaining the error correction capacity of the standard QR code, 3) allowing full-area blending of various photographs, drawings and graphics. Experimental results demonstrate that our approach produces high quality QR codes in terms of both visual appearance and readability.

Keywords: QR code, aesthetic, image blending.

1 Introduction

Quick Response (QR) code is one of the most widely used two-dimensional barcodes which is originally designed for tracking vehicle parts for the automotive industry in Japan. Due to its fast readability and relatively large storage capacity, QR code has quickly been adapted as a convenient method of information acquisition outside the automotive industry. Recently, with the popularization of smart phones, more and more users scan QR codes with their camera-equipped phones to get a URL, email address, phone number or other text information. Despite these advantages, standard QR code's noise-like appearance with randomly distributed black and white modules leads to some problems. For example, when printed on a non-negligible display area, its noise-like appearance can easily disrupt the aesthetic appeal of its carrier. Moreover, users cannot get any intuitive information of the code just by looking at it. Therefore, how to generate more visual-pleasant codes without compromising the machine readability becomes a big challenge.

X. He et al. (Eds.): MMM 2015, Part II, LNCS 8936, pp. 183–194, 2015.

Many manually methods are presented to solve this problem. However, aesthetic QR codes are too expensive and inefficient to be manually designed. For this reason, an automatic method supporting fast generation is preferred. There are several studies addressing the research topic of automatic or semi-automatic QR code beautification, which can be classified into three categories: direct embedding, embedded image modification, and codewords adjustment.

Direct Embedding. A common strategy to generate aesthetic QR codes is to embed a small image directly. Samretwit *et al.* [1] suggested the best location for the superimposed image in a QR code was the center, which is shown in Fig. 2(b). Ono *et al.* [2] used optimization algorithms to find appropriate positions to put illustrations in a QR code. However, their method is still direct embedding method in essence. Obviously, direct embedding introduces invalid codewords and reduces the error correction capacity, thus the size of the embedding area is limited.

(a)   (b)   (c)

Fig. 1. Three aesthetic QR codes generated by our method. By blending an image with a QR code using a two-stage approach, we are able to get a visually pleasant QR code resembling the given image without reducing the error correction capacity.

Embedded Image Modification. Since direct embedding introduces invalid codewords, many researchers try to modify the appearance of modules to beautify the QR code while keeping the error correction capacity. Skawattananon *et al.* [8] used adaption of brightness in image to get lower bit error rate. In [5], Baharav *et al.* changed the value of pixels based on the average luminance of each module to alter the appearance of the code as shown in Fig. 2(e). An error-aware warping technique for deforming the embedded images was proposed in [9] to minimize the bit errors. The method presented in [4] generated a binary embedding by subdividing each QR module into 3 × 3 submodules and binds module's color to the center submodule, while leaving the remaining eight submodules assigned to adapt to target halftone images (see Fig. 2(d)). A method which also utilized halftone masks is from [10]. In this case, the algorithm selected a set of modified pixels and used nonlinear programming techniques to optimize

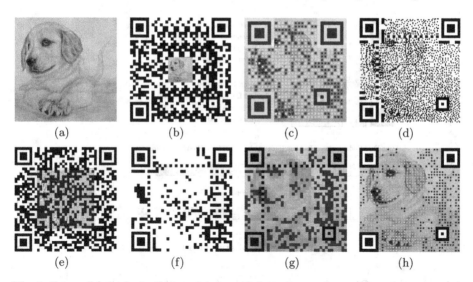

Fig. 2. Image (a) is the background image. Others are aesthetic QR codes generated by: (b) direct embedding. (c) [3], (d) [4], (e) [5], (f) [6], (g) [7], (h) our method.

luminance level. Fang *et al.* [11] proposed an optimization model for embedding images into two-dimensional barcodes. To minimize the visual distance between the background image and the barcode image, it introduced invalid codewords which caused additional bit errors compared with the default barcode. The main idea of the aforementioned approaches is to keep the luminance value of concentric region of each module untouched and blend the neighboring regions with a modified image. However, they just modify the embedded image and do not take advantage of the modification of QR codewords, thus their ability to handle the visual distortion is limited.

Codewords Adjustment. Instead of scribbling on redundant pieces and relying on error correction to preserve the message, several studies take advantage of the positional correspondence between codewords and modules. Wakahara *et al.* [12] modified the values in padding regions to multiplex a dotted picture on the QR code. In this approach, the changeable area is bounded by the area size of padding regions. Cox [6] proposed a complicated algorithm to change the values in padding and error correction regions by appending redundant strings to the original data as shown in Fig. 2(f). Fujita *et al.* [13] proposed a more flexible approach to modify the codes value by introducing the non-systematic encoding of RS codewords. Yu-Hsun *et al.* [7] improved the method in [13] by integrating the visual saliency perception of the embedding image with simulated annealing optimization. However, the background image embedded in a QR code should be resized to fit the resolution of the corresponding QR code in [7] as shown in Fig. 2(g). Thus, when the version number is small (*e.g.*, less than 10), the result will not be satisfactory. It is worth noticing that, a software called *Visualead* [3] was developed to generate aesthetic QR codes, which also used codewords adjustment technology (see Fig. 2(c)).

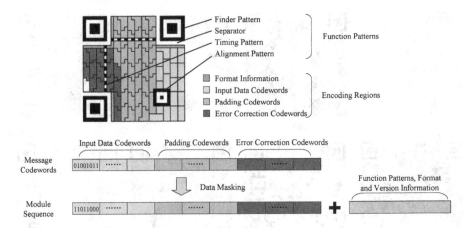

Fig. 3. Structure of standard QR code

For aesthetic QR code generation, the first challenge is that the resultant codes need to be decodable by standard applications. The second one is using maximum possible area to embed images. Regarding the challenges, we propose an automatic method for aesthetic QR codes generation based on two-stage image blending (module-based blending and pixel-based blending). Compared with the existing methods, it allows full-area blending of a variety of photographs, drawings, and graphics to improve the appearance of the codes without sacrificing error correction, and can keep more details of the background image in the resultant QR codes.

The rest of this paper is organized as follows. At first, the basic characteristics of QR code are introduced in Section 2. Next, Section 3 provides a detailed description of the two-stage image blending algorithm. Experimental results are reported in Section 4. Finally, Section 5 concludes this work.

2   Basic Characteristics of QR Code

A QR code symbol can be roughly divided into two parts, function patterns and encoding regions (see Fig. 3). Function patterns include finder patterns, separator, timing and alignment patterns which are required to locate and identify the QR code symbol. Encoding regions are the regions not occupied by function patterns and available for codewords encoding, version and format information. Format information indicates error correction level and mask pattern. There are four error correction levels (L, M, Q and H) which can recover 7%, 15%, 25%, 30% error codewords of the whole QR code.

The initial input data is converted to a bit stream in accordance with the encoding rules and 8 bits make up a codeword. The sequence of message codewords consists of input data codewords, padding codewords and error correction codewords as showed in Fig. 3. QR code utilizes Reed-Solomon (RS) error correction to allow correct reading against burst errors. Generally, a RS code is

Table 1. Summary of Notations

Name	Description
$QR(V,C)$	A QR code whose version number is V and error correction level is C.
c	The length of message codewords (in byte).
k	The length of data codewords (in byte).
d	The length of input data codewords (in byte).
n	The side length of QR codes (in pixel).
m	The side length of each module (in pixel).
I	The background image used for blending.
G	The grayscale image of I.
E	The edge map of G.
S	The saliency map of G.
B	The bit stream carrying the input data after RS encoding.
Q^s	The standard QR code.
Q^i	The ideal QR code.
Q^b	The binary aesthetic QR code.
Q^g	The grayscale aesthetic QR code.
Q^c	The color aesthetic QR code.

denoted as (c, k) where c is the length of the whole message codewords and k is the length of data codewords including input data codewords and padding codewords. The values of c and k in a QR code are already defined for the given version number and error correction level. The left $(c - k)$ codewords are error correction codewords. Standard QR code uses systematic RS encoding where the *systematic* means that the sequence of message codewords should begin with data codewords and end with the calculated error-correction codewords. At the same time, QR codes encoded by non-systematic methods can also be decoded correctly by standard decoding algorithm as long as the input data codewords remain unchanged. With the help of non-systematic encoding, we can freely select locations of the error correction codewords.

The decoding procedure includes four basic steps: grayscale conversion, binarization, detection, and decoding of the bit stream. An acquired QR image is usually in RGB color space and the first step for decoding is to convert it into a grayscale format. During the binarization step, pixels of the grayscale image are segmented into two classes (black modules and white modules) by the predefined threshold. After binarization, a bit stream is constructed by sampling on a grid then mapping the black modules as binary 1 and white modules as binary 0. Notice that only the pixels in the central area of the module are meaningful during sampling. Finally, the original embedded message is extracted from the bit stream following the decoding rules.

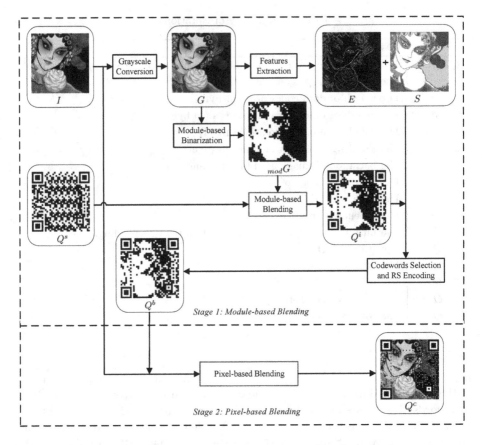

Fig. 4. Overview of two-stage image blending

3   Method for Two-Stage Image Blending

3.1   Overview

For clarity, some important notations are explained at first. We treat an $n \times n$ image I as a sequence of subimages denoted as $_{sub}I$. Each subimage is an $m \times m$ image. So there are $(\frac{n}{m})^2$ subimages in sequence $_{sub}I$ in total. We also define a binary module sequence corresponding to the subimage sequence by using the subscript "$_{mod}$". $_{mod}I_r$ represents the binarization result of the rth subimage $_{sub}I_r$ where 1 means black while 0 means white. Table 1 gives a summary of our notations.

Fig. 4 provides an overview of our two-stage blending method. At the first stage, we blend the background image I with the standard QR code Q^s module by module considering the visual perception of I. Afterwards, we get a binary aesthetic QR code denoted as Q^b. At the second stage, Q^b and I are further blended pixel by pixel. The details of each stage will be introduced in the following two subsections respectively.

3.2   Stage 1: Module-Based Blending

Module-Based Binarization. For the given background image I, we first get its grayscale copy G and then G is divided into $(\frac{n}{m})^2$ subimages of size $m \times m$ pixels. Each subimage is denoted as $_{sub}G_r$, where r is the index of the subimage ranged from 1 to $(\frac{n}{m})^2$. Let $_{mod}G_r$ denote the binarization result of the rth subimage, which is calculated by

$$_{mod}G_r = ROUND\{\frac{\sum_{i,j} {}_{sub}G_r(i,j) \cdot G_W(i,j)}{255}\}, \tag{1}$$

where $G_W(i,j)$ is the weight for pixel at (i,j) in each subimage, and $\sum_{i,j} G_W(i,j) = 1$. In this paper, we choose Gaussian function as G_W given by

$$G_W(i,j) = \frac{1}{2\pi\sigma^2} e^{-\frac{i^2+j^2}{2\sigma^2}}, \tag{2}$$

where $\sigma = \frac{m-1}{6}$.

Module-Based Blending. We blend the binarization result and the input standard QR code Q^s to generate an ideal QR code Q^i. The value of each module in Q^i is assigned by

$$_{mod}Q_r^i = \begin{cases} _{mod}Q_r^s & \text{if } (\ _{mod}Q_r^s \in M\ ) \\ _{mod}G_r & \text{otherwise} \end{cases}, \tag{3}$$

where M denotes the set of modules corresponding to function patterns, format information, version information, and input data codewords. $_{mod}Q_r^s$ and $_{mod}Q_r^i$ are the rth module of Q^s and Q^i respectively.

From eq.(3) we can see that the values of $_{mod}Q^i$ may be different from those of $_{mod}Q^s$. So we have to reset the bit values in B to preserve the correspondence between B and $_{mod}Q^i$. Since the change of a module is equivalent to changing one bit in B, for all the bits in padding and error correction codewords, we modify the values of B as: (1) if $_{mod}Q_r^i$ equals to $_{mod}Q_r^s$, let $B_{L(r)}$ remain unchanged; (2) otherwise, flip the value of $B_{L(r)}$, where $L(r)$ denotes the index of the bit in B corresponding to the rth module. More formally, $B_{L(r)}$ is calculated by

$$B_{L(r)} = B_{L(r)} \oplus {}_{mod}Q_r^i \oplus {}_{mod}Q_r^s. \tag{4}$$

Codewords Selection and RS Encoding. After module-based blending, the obtained ideal QR code Q^i is visually similar to the binarization result of G and contains the encoded bit stream of input data. However, Q^i cannot be decoded by the standard algorithm due to the corresponding bit stream B is not consistent with the encoding rules of RS code. This problem is solved with the help of non-systematic RS encoding, $i.e.$, we select k RS codewords out of c and calculate the values of the remaining $(c - k)$ codewords according to the non-systematic RS encoding rules.

Let Q^b denote the binary aesthetic QR code we get at last which can be decoded successfully, $\eta(r)$ be the visual importance value of the rth module and S_c denote the set of selected k codewords. We obtain S_c by minimizing the visual distortion between Q^b and Q^i defined as

$$\underset{S_c}{argmin} \sum_r \eta(r) \cdot f(_{mod}Q_r^b, _{mod}Q_r^i), \tag{5}$$

where $f(a,b)$ is a function measuring the consistency between a and b given by

$$f(a,b) = 2(a \odot b) - 1. \tag{6}$$

The visual importance of a module is taken into consideration for the selection of S_c. In this paper, we define visual importance by introducing edge and saliency features, $i.e.$, $\eta(r)$ is defined as

$$\eta(r) = \sum_{i,j} \lambda_1 \cdot _{sub}E_r(i,j) + \lambda_2 \cdot _{sub}S_r(i,j), \tag{7}$$

where λ_1 and λ_2 are weighting coefficients ($\lambda_1 = 0.65$, $\lambda_2 = 0.35$ in our tests), E and S are the edge map and saliency map respectively. In our experiments, E is generated by applying the widely used $Sobel$ edge detector, while S is obtained using the method proposed in [14]. In addition, other edge and saliency detectors can be used in our method too.

Since the input d data codewords must be selected into S_c, actually we only need to select $(k-d)$ codewords from the rest $(n-d)$ codewords. Obviously, the search space of S_c is in the order of $\binom{n-d}{k-d}$ which is far too large for exhaustive enumeration. For example, given a QR code $QR(9, L)$, if $d = 20$, the search space is in the order of $\binom{146-20}{116-20} = \binom{126}{96} > 9e + 29$. Existing optimization algorithms such as simulated annealing can achieve an optimal solution. However, it takes a long time to get a good result. In this paper, we adopt a greedy approach to reduce the time complexity. Recall that every codeword contains 8 bits, and each bit corresponds to a module. We calculate the visual importance value of a codeword by summing up the visual importance value of all bits. Then we sort the codewords according to the visual importance value. The top k codewords with the highest visual importance value will be selected into S_c. Experiments show that we can generate an aesthetic QR code in less than one second.

3.3   Stage 2: Pixel-Based Blending

We imitate the sampled result during decoding as a weighted average of the pixel values in a module. The sampling weights are represented by an $m \times m$ matrix denoted as D_W where $\sum_{i,j} D_W = 1, D_W(i,j) \geq 0$. The sampled result of the rth subimage of G, denoted as $\xi(_{sub}G_r)$, is given by

$$\xi(_{sub}G_r) = \sum_{i,j} _{sub}G_r(i,j) \cdot D_W(i,j). \tag{8}$$

Next, we want to generate the color aesthetic QR code Q^c by blending I with Q^b pixel by pixel. Let Q^g denote the grayscale copy of Q^c and $\varphi(_{sub}Q^g_r)$ denote the target sampling result of the rth subimage in Q^g. $\varphi(_{sub}Q^g_r)$ is calculated by

$$\varphi(_{sub}Q^g_r) = \begin{cases} \min\{ \ \tau_b, \ \xi(_{sub}G_r) \ \} & \text{if}(_{mod}Q^b_r = 1) \\ \max\{ \ \tau_w, \ \xi(_{sub}G_r) \ \} & \text{if}(_{mod}Q^b_r = 0) \end{cases}, \tag{9}$$

where τ_b is the upper threshold for the black modules and τ_w is the lower threshold for the white modules. The pixel-based blending is to get Q^c with constraint

$$\xi(_{sub}Q^g_r) \equiv \varphi(_{sub}Q^g_r). \tag{10}$$

To satisfy this constraint, we directly add $\varphi(subQ^g_r) - \xi(_{sub}G_r)$ to each color channel in I to generate Q^c. Let Φ denote one of the RGB channels of Q^c and Ψ denote the corresponding channel of I. The pixel values of Q^c are calculated according to

$$_{sub}\Phi_r(i,j) = _{sub}\Psi_r(i,j) + \frac{[\varphi(subQ^g_r) - \xi(_{sub}G_r)] \cdot D_W(i,j)}{\sum_{i,j} D_W(i,j) \cdot D_W(i,j)}. \tag{11}$$

Because finder patterns and alignment patterns are the most important components of a QR code, our algorithm leaves modules of these patterns untouched and only manipulates the other modules. Moreover, since the readability of a QR code is very sensitive to the pixels in the central region of each module, we adjust the value of central pixels as

$$_{sub}\Phi_r(i,j) = \begin{cases} \min\{ \ \tau_b, \ _{sub}\Phi_r(i,j) \ \} & \text{if}(_{mod}Q^b_r = 1) \\ \max\{ \ \tau_w, \ _{sub}\Phi_r(i,j) \ \} & \text{if}(_{mod}Q^b_r = 0) \end{cases} \tag{12}$$

to enhance readability.

4   Experimental Results

We collected a dataset containing 300 images of different styles (*e.g.*, photographs, drawings, and graphics) for experiments. All images in the dataset are used as background images to generate aesthetic QR codes and nine of the results generated by our method are shown in Fig. 1 and Fig. 5[1]. The message embedded into each QR code is the URL of MMM 2015, i.e., "http://www.mmm2015.org/". For all the results, the version number is 5 and error correction level is L. Each QR code is a 555×555 image. Our algorithm is computationally efficient and takes less than a second to generate an aesthetic QR code on a personal computer with 2.93GHz CPU and 4GB memory.

---

[1] More results are available online at
https://github.com/ZYT2014/CAQR

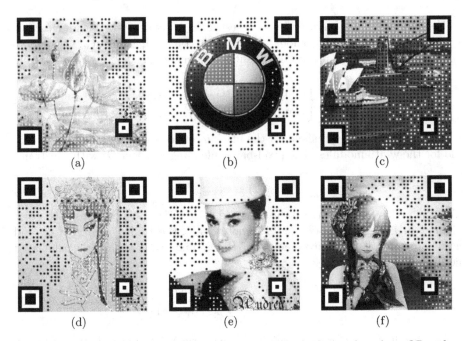

(a)   (b)   (c)

(d)   (e)   (f)

Fig. 5. Six aesthetic QR codes generated by our method. Notice that these QR codes might be decoded a little bit slower because they are zoomed out during the generation of this PDF file. We recommend readers to enlarge the image (*e.g.*, 1.5×) in the digital display to achieve better readability.

4.1 Experiments on Readability

In order to verify the readability of the generated QR codes, we conducted an experiment on several different mobiles phones and QR decoders. During this test, each QR code is displayed at 25%, 50%, 75%, 100% of its original size respectively. The results are reported in Table 2, and we can see that the successful decoding rates are always greater than 96%, which can satisfy most daily applications. Note that our method takes advantage of the unused modules to blend images with QR codes without reducing the correction capacity. Therefore, the successful decoding rate should be 100% if all the codes are displayed at their full size, as verified by our experiment (please see the last column of Table 2). However, the performance becomes a little bit worse when the displayed size reduces. The reason is we assume that the sampling points are located around the central region of each module. When displayed at a small size, the central region is not large enough to cover all the sampling points, which might result in failure of decoding.

4.2 Subjective Evaluation

We designed a scoring system and got assessment results from 25 participants (15 males and 10 females) with ages ranging from 18 to 51 who are not engaged in

Table 2. Results of Successful Decoding Rates.

Mobile Phone	App	Success Rates of Different Displayed Size	
		25%	≥ 50%
iPhone 4S	WeChat	98.6%	100%
	RedLaser	99.6%	100%
iPhone 5S	WeChat	100%	100%
	RedLaser	100%	100%
Google Nexus 5	WeChat	100%	100%
	Barcode Scanner	100%	100%
Samsung Galaxy S3 I9308	WeChat	97.3%	100%
	Barcode Scanner	96.3%	100%
Samsung Galaxy Note 3 N9006	WeChat	100%	100%
	Barcode Scanner	99.6%	100%

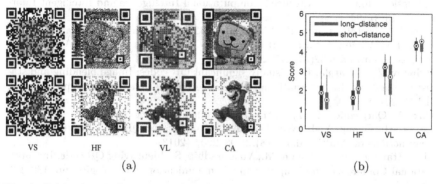

(a)   (b)

Fig. 6. Subjective evaluation results. "VS" is the abbreviation for "Visually Significant QR Codes" generated by [5]. "HF" is the abbreviation for "Halftone QR Codes" generated by [4]. "VL" is the abbreviation for "Visualead QR Codes" generated by [3]. "CA" is the abbreviation for "Color Aesthetic QR Codes" generated by our method. (a) Two sets of QR codes generated by 4 different approaches. (b) Statistical box plots for 4 data sets. Each set of box bars indicates the 100%, 75%, 50%, 25% and minimum scores for each data set with *long-distance* in red and *short-distance* in blue.

this work. There are five levels in the scoring system for QR codes including ugly, relatively ugly, acceptable, relatively beautiful, and beautiful corresponding to the scores ranged from 1 to 5 points respectively. 30 images are randomly selected as background images to generate aesthetic QR codes by different approaches and each resultant QR code is rated by participants from two distances. The subjective evaluation results, presented in Fig. 6, show that the proposed method performs statistically better than other three state-of-the-art methods.

5   Conclusion

This paper presents an efficient method for aesthetic QR codes generation. Given a background image, the proposed method is able to generate a visual-pleasant QR code resembling the given image based on the two-stage image blending approach. Compared with the existing automatic methods, it allows full-area blending of images to improve the appearance of the codes without reducing the error correction capacity. Subjective evaluation results demonstrate that the resultant QR codes are more aesthetic than those of other three state-of-the-art methods. Moreover, experiments on readability indicate that the resultant QR codes can be decoded by standard decoders with high success rates.

References

1. Samretwit, D., Wakahara, T.: Measurement of reading characteristics of multiplexed image in QR code. In: International Conference on Intelligent Networking and Collaborative Systems, pp. 552–557. IEEE (2011)
2. Ono, S., Morinaga, K., Nakayama, S.: Animated two-dimensional barcode generation using optimization algorithms-redesign of formulation, operator, and quality evaluation. Journal of Advanced Computational Intelligence and Intelligent Informatics 13(3), 245–254 (2009)
3. Alva, N., Uriel, P., Itama, F.: Visualead (2012), http://www.visualead.com
4. Chu, H.K., Chang, C.S., Lee, R.R., Mitra, N.J.: Halftone QR codes. ACM Transactions on Graphics 32(6), 217–224 (2013)
5. Baharav, Z., Kakarala, R.: Visually significant QR codes: Image blending and statistical analysis. In: International Conference on Multimedia and Expo, pp. 1–6. IEEE (2013)
6. Cox, R.: Qart codes (2012), http://research.swtch.com/qart
7. Lin, Y.-H., Chang, Y.-P., Wu, J.-L.: Appearance-based QR code beautifier. IEEE Transactions on Multimedia 15(8), 2198–2207 (2013)
8. Skawattananon, C., Ketcham, M., Vongpradhip, S.: Identifying QR code. In: International Conference on Computer and Communication Technologies, pp. 132–135 (2012)
9. Lin, Y.S., Luo, S.J., Chen, B.Y.: Artistic QR code embellishment. In: Computer Graphics Forum, pp. 137–146 (2013)
10. Garateguy, G.J., Arce, G.R., Lau, D.L., Villarreal, O.P.: QR images: optimized image embedding in QR codes. IEEE Transactions on Image Processing 23(7), 2842–2853 (2014)
11. Fang, C., Zhang, C., Chang, E.-C.: An optimization model for aesthetic two-dimensional barcodes. In: Gurrin, C., Hopfgartner, F., Hurst, W., Johansen, H., Lee, H., O'Connor, N. (eds.) MMM 2014, Part I. LNCS, vol. 8325, pp. 278–290. Springer, Heidelberg (2014)
12. Wakahara, T., Samretwit, D., Maki, T., Yamamoto, N.: Design and characteristics of two-dimensional color code editor with dotted images. In: Information Technology Convergence, pp. 323–332. Springer (2013)
13. Fujita, K., Kuribayashi, M., Morii, M.: Expansion of image displayable area in design QR code and its applications. In: Forum on Information Technology, pp. 517–520 (2011)
14. Cheng, M.-M., Warrell, J., Lin, W.-Y., Zheng, S., Vineet, V.: Efficient salient region detection with soft image abstraction. In: International Conference on Computer Vision, pp. 1529–1536. IEEE (2013)

Person Re-identification Using Data-Driven Metric Adaptation

Zheng Wang[1], Ruimin Hu[1,2], Chao Liang[1,2], Junjun Jiang[1], Kaimin Sun[3], Qingming Leng[4], and Bingyue Huang[1]

[1]National Engineering Research Center for Multimedia Software,
School of Computer, Wuhan University, Wuhan, China
[2]Research Institute of Wuhan University in Shenzhen, China
[3]State Key Laboratory of Information Engineering in Surveying, Mapping, and Remote Sensing, Wuhan University, Wuhan, China
[4]School of Information Science & Technology, Jiujiang University, Jiujiang, China
wangzwhu@whu.edu.cn

Abstract. Person re-identification, aiming to identify images of the same person from various cameras configured in difference places, has attracted plenty of attention in the multimedia community. In person re-identification procedure, choosing a proper distance metric is a crucial aspect [2]. Traditional methods always utilize a uniform learned metric, which ignored specific constraints given by this re-identification task that the learned metric is highly prone to over-fitting [21], and each person holding their unique characteristic brings inconsistency. Therefore, it is obviously inappropriate to merely employ a uniform metric. In this paper, we propose a data-driven metric adaptation method to improve the uniform metric. The key novelty of the approach is that we re-exploits the training data with cross-view consistency to adaptively adjust the metric. Experiments conducted on two standard data sets have validated the effectiveness of the proposed method with a significant improvement over baseline methods.

Keywords: Person re-identification, Data-driven metric adaptation, Cross-view consistency.

1 Introduction

Person re-identification (a.k.a. person re-id), namely matching the same person across disjoint camera views, is a valuable task in video surveillance scenarios [1]. Since classical biometric cues, such as face and gait, may be unreliable or even infeasible in uncontrolled surveillance environment [2], the appearance of the individual is mainly exploited for person re-id. However, person re-id remains an unsolved problem due to the challenges caused by low resolution, partial occlusion, motion blur, view change, and illumination variation, making different persons appear more alike than the same person in various cameras [3]. Generally speaking, person re-id can be regarded as an pedestrian-oriented image retrieval problem [4]. Given a probe person image taken from one camera, the algorithm

X. He et al. (Eds.): MMM 2015, Part II, LNCS 8936, pp. 195–207, 2015.

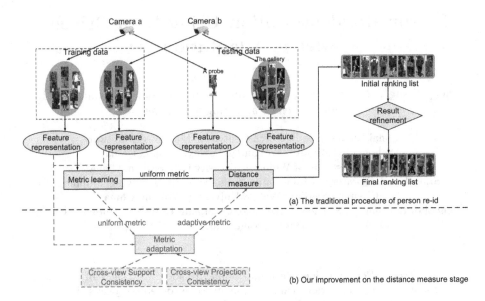

Fig. 1. An overview of the procedure of person re-id problem. The figure (a) concludes the traditional division that a uniform metric is utilized in distance measure stage. The figure (b) indicates our improvement on the distance measure stage that an adaptive metric is used and the training data is re-exploited. As red boxes shown, in this paper, we improve the uniform metric to an adaptive metric.

is expected to search images of the same person from the gallery captured by the other single camera or multiple cameras. It aims at generating a ranking list, in which the top results are more likely of the same person to the probe. As in existing literature [1–8, 10, 12, 16, 17], we consider the problem between two cameras that the probe image comes from one camera, and the retrieved gallery images come from the other.

According to the procedure of person re-id (Fig.1(a)), related work can be usually divided into three categories: feature representation based approaches [1, 6–9], distance measure based approaches and result refinement based approaches [3, 10–12]. This paper focuses on the distance measure stage (rectangle modules as shown in Fig.1) and pays attention to seeking out a proper distance measure [2, 4, 14–17], which is a crucial aspect [2] in person re-identification procedure.

Among various distance measure based methods, supervised metric learning algorithms demonstrate an obvious superiority by learning a uniform metric based on the given labeled training data. Hizer et al. [14] and Dikmen et al. [15] utilized or improved LMNN [13] to learning the optimal metric for person re-id. Zheng et al. [2] learned a Mahalanobis distance metric with a probabilistic relative distance comparison (PRDC) method. Kostinger et al. [16] used Gauss distribution to fit pair-wise samples and got a simpler metric function (KISSME), and then Tao et al. [4] presented a regularized smoothing KISS

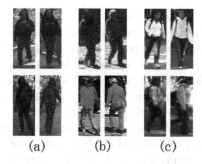

(a)   (b)   (c)

Fig. 2. Examples of image pairs from the VIPeR dataset [18]. Each column has two images of the same person taken from two different camera views. Different columns indicate different persons. All of (a), (b) and (c) reveal the phenomenon that images of the same person may look different but that of different persons may look similar.

metric learning (RS-KISS). Mignon et al. [17] introduced pairwise constrained component analysis (PCCA) to learn distance metric from sparse pairwise similarity/dissimilarity constraints in high dimensional input space. Most of the above methods can be summarized as learning a uniform Mahalanobis-like distance matrix M^*, and exploiting the uniform learned matrix M^* to calculate the distances among testing data and then generate the ranking result [21]. Specifically, given two testing person image feature x_a and x_b, their distance can be defined as $d_U(x_a, x_b) = (x_a - x_b)^\top M^*(x_a - x_b)$.

The performance in above mentioned methods clearly indicate that the uniform learned matrix can be effective and boost the results. However, there still exist specific constraints given by the person re-id task: (1) images showing the same person might not have a similar visual description, whereas images of different persons could be very close in the original feature space (Fig.2). Thus, as suggested in [21], this phenomenon leads to somehow ill-posed problems and highly prone to over-fitting; (2) the essence of the metric learning method is to seek a projection matrix that transforms pairwise feature spaces into a new one [5], and it constructs a uniform relationship between two cameras. The approaches always exploit the established relationship of cameras to evaluate on pairs of specific individuals. As we know, each person holds their unique characteristics. With different persons for learning and evaluating process, it will easily bring inconsistency for the re-identification task. Therefore, it is obviously inappropriate to merely employ a uniform metric.

To conquer the above problem, we introduce an adaptive metric M_A^*, which is particularly and adaptively depending on different data. Then, the distance between x_a and x_b changes to $d_A(x_a, x_b) = (x_a - x_b)^\top M_A^*(x_a - x_b)$. To obtain the dynamic metric M_A^*, a data-driven adaptive factor $f_A(x_a, x_b)$ is used to adjust the uniform metric M^*. Considering that x_a and x_b itself is inadequate for calculating factor $f_A(x_a, x_b)$, the training data (Fig.1(b)) is re-exploited.

In that the training data is derived from two cameras, we exploit properties cross the two different views. The basic idea of our approach is that two images of the same person should not only have similar support set cross views (the support set is characterized by the sparsely selected images that can represent the origin image), but also have similar context set cross views (the context set is characterized by the K-nearest neighbors of the origin image). On the one hand, the intersection of the support sets of the same person cross views should be in high level. We name this cross-view support consistency. On the other hand, two images of the same person should hold similar context not only in camera a, but also in camera b. In this approach, we construct a virtual projection camera to obtain the context similarity of image pair from two different cameras. We call this cross-view projection consistency.

The contribution of this paper can be summarized that we propose a data-driven metric adaptation method (DDMA), which is data-specific and improves the learned uniform metric. The key novelty of our method is that the proposed method re-exploits the training data with the cross-view consistency to obtain the adaptive metric measurement. Experiments conducted on two standard data sets have validated the effectiveness of the proposed method with a significant improvement over baseline methods.

2   The Approach

For the convenience of following discussion, we consider a pair of cameras C_a and C_b with non-overlapping field of views. A set of labeled persons $O = \{o_1, o_2, ..., o_m\}$ crossing the two cameras. We denote the representing image of person o_i captured by C_a (or C_b) as x_a^i (or x_b^i), $x_a^i, x_b^i \in R^d$. Let $X_{a,L}$ and $X_{b,L}$ respectively represent the two labeled training sets captured by C_a and C_b as follow, where m is the number of each training set, and $i = j$ means the same person o_i.

$$X_{a,L} = \{x_a^1, ..., x_a^i, ..., x_a^m\}, 1 \leq i \leq m \tag{1}$$

$$X_{b,L} = \{x_b^1, ..., x_b^j, ..., x_b^m\}, 1 \leq j \leq m \tag{2}$$

Then, let x_a^p stands for a testing probe data from C_a, and $X_{b,U}$ represents the retrieved gallery data from C_b. n is the number of testing data in C_b. Therefore, the purpose of the person re-id task is that for each testing data x_a^p, the algorithm ranks the data in $X_{b,U}$.

$$X_{b,U} = \{x_b^{m+1}, ..., x_b^q, ..., x_b^{m+n}\}, m+1 \leq q \leq m+n \tag{3}$$

The traditional supervised metric learning algorithm learns a discriminative distance function based on $X_{a,L}$ and $X_{b,L}$. Specifically, given two training data x_a^i and x_b^j, their distance can be defined as a Mahalanobis-like distance $d(x_a^i, x_b^j) = (x_a^i - x_b^j)^\top M(x_a^i - x_b^j)$, where M is a positive semi-definite matrix for the validity

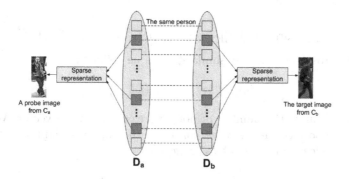

Fig. 3. Cross-view support consistency. Each block represents a feature representation of an image, which means samples from camera C_a and C_b. Two blocks in the same row stand for the same person. The sparse support set of the probe and the target should be similar.

of metric. After learning the uniform metric matrix M^*, the distance between testing data x_a^p and any unlabeled testing data x_b^q in $X_{b,U}$ will be calculated as

$$d_U(x_a^p, x_b^q) = (x_a^p - x_b^q)^\top M^* (x_a^p - x_b^q) \qquad (4)$$

Traditional person re-id methods compute the distances $d_U(x_a^p, x_b^q)$ for every testing data in $X_{b,U}$, and then obtain the ranking list. While, our approach exploits an adaptive metric M_A^* to obtain a data-specific distance $d_A(x_a^p, x_b^q)$ by Equation.5.

$$d_A(x_a^p, x_b^q) = (x_a^p - x_b^q)^\top M_A^* (x_a^p - x_b^q) \qquad (5)$$

2.1 Cross-View Support Consistency

First, we exploit the cross-view support consistency. For a probe image x_a^p from C_a, the approach selects some images sparsely in $X_{a,L}$ to encode the image, meanwhile another gallery image x_b^q from C_b is represented sparsely by images in $X_{b,L}$. The assumption is that two images of the same person will have similar support set cross the two camera views.

We generate the dictionary by $X_{a,L}$, denoted as $D_a = [x_a^1...x_a^i...x_a^m]$, $D_a \in R^{d \times m}$, and then select sparsely from the dictionary to represent x_a^p. Meanwhile, the dictionary $D_b = [x_b^1...x_b^j...x_b^m]$, $D_b \in R^{d \times m}$ is generated by $X_{b,L}$, and x_b^q can be represented sparsely by D_b. It assumes that $x_a^p = D_a w_a^{p*}$, $w_a^{p*} \in R^{m \times 1}$, where w_a^{p*} indicates the selected images and their weights [22]. If the weight is higher than zero, it means that the image is selected as the support data for x_a^p. Here, we use the Equation.6 to compute the sparse representation w_a^{p*}. In the same way, we get the sparse representation w_b^{q*} for x_b^q by Equation.7.

$$w_a^{p*} = \arg\min \|w_a^p\|_0, \quad s.t. \|D_a w_a^p - x_a^p\| < \varepsilon, w_a^p \geq 0 \qquad (6)$$

$$w_b^{q*} = \arg\min \|w_b^q\|_0, \quad s.t. \|D_b w_b^q - x_b^q\| < \varepsilon, w_b^q \geq 0 \tag{7}$$

Then the cross-view support similarity, can be defined as

$$Sim_s(x_a^p, x_b^q) = size(w_a^{p*} w_b^{q*}) \tag{8}$$

where $size(\cdot)$ counts the number of non-zero elements in the vector. With the cross-view support consistency, we construct the support adaptive factor[1] by Equation.9, which is one part of $f_A(x_a^p, x_b^q)$.

$$f_s(x_a^p, x_b^q) = [\frac{1}{1 + Sim_s(x_a^p, x_b^q)}]^\alpha = [\frac{1}{1 + size(w_a^{p*} w_b^{q*})}]^\alpha, \alpha > 0 \tag{9}$$

2.2   Cross-View Projection Consistency

Second, we exploit the cross-view projection consistency. It is valid to assume that images coming from the same person should be alike individually as well as socially [12]. In this work, we exploit the property that a probe image x_a^p from C_a and the target image from C_b should hold a similar context with $X_{a,L}$ and $X_{b,L}$ respectively.

By performing eigenvalue decomposition on M^* with $M^* = L^{*\top}L^*$, the uniform metric can be rewritten as $d_U(x_a, x_b) = (x_a - x_b)^\top L^{*\top} L^*(x_a - x_b) = \|L^* \cdot x_a - L^* \cdot x_b\|^2$, where $(\cdot)^\top$ represents the transpose of a vector or matrix. With this definition, it is easy to see that the essence of the metric based method is to seek a projection matrix that transforms original image features into a new feature space. So we can construct a virtual projection camera C_v to obtain the distance between the data from C_a and the data from C_b. And the distance between the data from the same camera could be calculated by Euclidean distance.

Fig.4(a) shows that when we want to obtain the neighbors of x_a^p in $X_{b,L}$, we should project both of the cameras C_a and C_b into the learned projection camera C_v. So the distance between x_a^p and each training data x_b^j is calculated by $d_U(x_a^p, x_b^j) = (x_a^p - x_b^j)^\top M^*(x_a^p - x_b^j)$. Meanwhile, every testing data x_b^q and the training set $X_{b,L}$ is in the same camera C_b, and Euclidean distance can be utilized. The distance is calculated by $d_U(x_b^q, x_b^j) = (x_b^q - x_b^j)^\top (x_b^q - x_b^j)$.

Through ranking the distances with the training set $X_{b,L}$, the k-nearest neighbors of x_a^p and x_b^q can be acquired respectively. Counting the number of common k-nearest neighbors of x_a^p and x_b^q in $X_{b,L}$, the context similarity is computed. In particular, it assumes that k-nearest neighbors of x_a^p in $X_{b,L}$ is $knn(x_a^p|X_{b,L})$, and that of x_b^q in its ranking list is $knn(x_b^q|X_{b,L})$. Then the context similarity in $X_{b,L}$ can be defined as

---

[1] The parameter α indicates the contribution of the cross-view support consistency. The greater the value is, the greater the impact of metric adaption comes from the similarity of support set. If α is too large, the origin uniform metric will lose efficacy. While, if $\alpha \to 0$, the support adaptive factor will have not effect. We set $\alpha = 0.5$ in this paper unless otherwise specified.

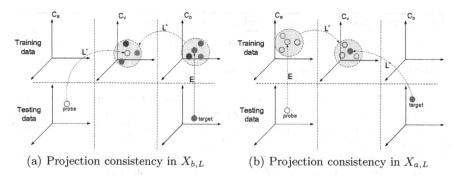

(a) Projection consistency in $X_{b,L}$ (b) Projection consistency in $X_{a,L}$

Fig. 4. Cross-view Projection Consistency. Each point represents a feature vector of image from camera C_a, camera C_b or the virtual projection camera C_v, and different colors represent different persons. In the figure, L^* stands for the projection matrix. After projecting data to a uniform virtual view, the context set cross views can be obtained. And E stands for the Euclidean distance to get the context set in the same view. (a) The context set (k-nearest neighbors) of the probe and the target should be similar in $X_{b,L}$. (b) The context set of the probe and the target should be similar in $X_{a,L}$.

$$Sim_b(x_a^p, x_b^q) = |knn(x_a^p|X_{b,L}) \cap knn(x_b^q|X_{b,L})| \qquad (10)$$

where $|knn(x_a^p|X_{b,L}) \cap knn(x_b^q|X_{b,L})|$ represents the number of common k-nearest neighbors of x_a^p and x_b^q in training data $X_{b,L}$. K is the number of the nearest neighbors.

In this way, as shown in Fig.4(b), we should project both of the cameras C_a and C_b into the learned camera C_v, and then the distance between x_b^q and every training image x_a^i is computed by $d_U(x_b^q, x_a^i) = (x_b^q - x_a^i)^\top M^* (x_b^q - x_a^i)$. So the k-nearest neighbors of x_b^q in $X_{a,L}$, $knn(x_b^q|X_{a,L})$ generates. And in the same feature space C_a, the distance between x_a^p and x_a^i is calculated by $d_U(x_a^p, x_a^i) = (x_a^p - x_a^i)^\top (x_a^p - x_a^i)$, and $knn(x_a^p|X_{a,L})$ generates. Then the context similarity in $X_{b,L}$ can be defined as

$$Sim_a(x_a^p, x_b^q) = |knn(x_a^p|X_{a,L}) \cap knn(x_b^q|X_{a,L})| \qquad (11)$$

With the above two context similarity, we construct the projection adaptive factor [2] by Equation.12, which is the other part of $f_A(x_a^p, x_b^q)$.

$$f_p(x_a^p, x_b^q) = [\frac{1}{1 + Sim_b(x_a^p, x_b^q) + Sim_a(x_a^p, x_b^q)}]^\beta$$
$$= [\frac{1}{1 + |knn(x_a^p|X_{b,L}) \cap knn(x_b^q|X_{b,L})| + |knn(x_a^p|X_{a,L}) \cap knn(x_b^q|X_{a,L})|}]^\beta,$$
$$\beta > 0 \qquad (12)$$

---

[2] The parameter β indicates the contribution of the cross-view projection consistency. Like the parameter α, $\beta = 0.5$ in this paper unless otherwise specified.

Algorithm 1. Data-driven metric adaptation algorithm

Input: probe image x_a^p and a gallery set $X_{b,U} = \{x_b^{m+1}, ..., x_b^q, ..., x_b^{m+n}\}, m+1 \leq q \leq m+n$.

Output: A ranking list for the probe image.

1. **for** each $q \in [m+1, m+n]$ **do**
2.   exploit $X_{a,L}$ to construct the dictionary D_a, and learn the sparse representation weight w_a^{p*} for x_a^p by Equation.6; meanwhile, exploit $X_{b,L}$ to construct the dictionary D_b, and learn the sparse representation weight w_b^{q*} for x_b^q by Equation.7;
3.   generate the support adaptive factor $f_s(x_a^p, x_b^q)$ by Equation.9;
4.   obtain the context similarity $Sim_b(x_a^p, x_b^q)$ by Equation.10 in $X_{b,L}$; meanwhile, obtain the context similarity $Sim_a(x_a^p, x_b^q)$ by Equation.11 in $X_{a,L}$;
5.   generate the projection adaptive factor $f_p(x_a^p, x_b^q)$ by Equation.12;
6.   adjust the uniform metric M* to the adaptive metric M$_A^*$ by Equation.13;
7.   compute the distance between x_a^p and x_b^q with the adaptive metric above by Equation.5;
8. **end for**
9. use the distances to generate the ranking list.

Ultimately, we can obtain the adaptive metric specific for each x_a^p and x_b^q by Equation.13. The whole procedure of DDMA is described in Algorithm.1.

$$\mathrm{M}_A^* = f_A(x_a^p, x_b^q)\mathrm{M}^* = f_s(x_a^p, x_b^q)f_p(x_a^p, x_b^q)\mathrm{M}^* \tag{13}$$

3   Experiments

In this section, the proposed approach is validated on two publicly available datasets, the VIPeR dataset [18], the CUHK Person Re-identification Dataset [19]. These datasets cover a wide range of problems faced in the real world person re-identification applications, e.g. viewpoint, pose, and lighting changes. They provide two labeled image sets of persons captured by two cameras with non-overlapping fields of views, in which the images of the same person have the same label, while the images of the different persons have different labels.

3.1   VIPeR Data Set

The widely used VIPeR dataset [18] contains 1,264 outdoor images obtained from two views of 632 persons. Some example images are shown in Fig.5(a). For evaluation on this data set, we followed the procedure described in [8]. The set of 632 image pairs were randomly split into two sets of 316 image pairs each, one for training and another for testing. The entire evaluation procedure was looped 10 times, Cumulative Matching Characteristic (CMC) curves [9] was used to calculate the average performance, it describes the expectation of finding the true match within the first r ranks. We gave a comparison to three metric

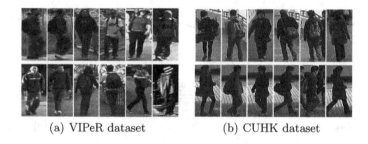

(a) VIPeR dataset   (b) CUHK dataset

Fig. 5. Some typical samples of three public datasets. Each column shows two images of the same person from two different cameras.

learning methods: standard Mahalanobis distance, complicated learning using LMNN [13] that is a classic metric learning method and KISSME [16] which reported a good performance on the VIPeR.

A combination feature descriptor consisting of color and texture features is used to represent images of individuals. Specifically, for each image, the RGB, HSV color histograms and LBP descriptor are extracted from overlapping blocks of size 16×16 and stride of 8×8 [5]. RGB and HSV histograms encode the different color distribution information in the RGB and HSV color space, respectively. The uniform rotation-invariant LBP descriptors [20], encoding the texture feature, are extracted in gray-scale images. The bin numbers of RGB and HSV histograms are 24, and the bin number for LBP descriptor is 59. All of the features are then put together to concatenated to a vector. To accelerate the learning process and reduce noise, we conducted principle component analysis (PCA) to obtain a low-dimension representation as [16], i.e. 100 in this paper.

After generated the descriptor, Mahalanobis metric, LMNN and KISSME were used to measure the distances between pair of images, respectively. Then, the reformed metrics respectively with cross-view support consistency and cross-view projection consistency were also utilized to measure the distances. The obtained results are shown in Fig.6. Moreover, in Table.1 we compare the performance of our approach in the range of the first 50 ranks to state-of-the-art methods. As can be seen, our approach has 1-14% improvement compared with Mahalanobis, LMNN and KISSME. And Table.1 shows that our method outperforms all existing methods over the whole range of ranks.

3.2 CUHK Data Set

CUHK Person Re-identification Dataset is a larger dataset recently proposed by Wang et al. [19] and contains 971 identities from two disjoint camera views. Some example images are shown in Fig.5(b). Each identity has two samples per camera view. Therefore, there are 3,884 images in all. All images are normalized to 160×60. As a single representative image per camera view for each person is considered in this paper, we randomly selected one image from two samples per camera views for each people as the really used dataset. The set of 971 image

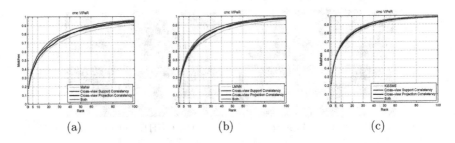

(a)   (b)   (c)

Fig. 6. Comparative results between existing methods and our approach on the VIPeR data set. (a) comparison with Mahalanobis; (b) comparison with LMNN [13]; (c) comparison with KISSME [16].

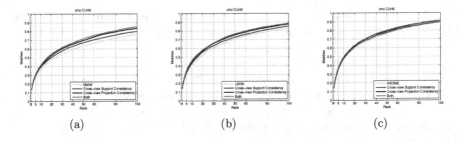

(a)   (b)   (c)

Fig. 7. Comparative results between existing methods and our approach on the CUHK data set. (a) comparison with Mahalanobis; (b) comparison with LMNN [13]; (c) comparison with KISSME [16].

pairs were randomly split into two sets of 485 image pairs each, one for training and another for testing. The re-identification process was repeated 10 times same as the VIPeR data set, and the average result was reported in form of the CMC curve. [3]

The result is presented in Fig.7, our method has evident improvements compared with Mahalanobis, LMNN and KISSME. And in Table.1 we compare the performance of our approach in the range of the first 100 ranks to state-of-the-art methods. Our method has 2-13% ascension than above methods mainly, and exceeds them over the range of ranks entirely.

3.3   Comparing to The-State-of-the-Art Person Re-id Methods

Table.2 summaries the comparing results with the-state-of-the-art person re-id methods on widely used VIPeR dataset with 316 as training set size. For a fair comparison, the results for most of these methods are directly taken from the original public papers. In these methods, SDALF and ELF belong to the feature representation based approaches, PRDC, KISSME and PCCA are the

---

[3] For calculating the context similarity, we set k-nearest neighbors value $K = 30$ when we evaluate on VIPeR data set, and $K = 40$ on CUHK data set.

Table 1. Person re-id matching rates(%) at different ranks on the VIPeR and CUHK data set

Method	rank@1	VIPeR			CUHK			
		10	25	50	10	20	50	100
Mahalanobis	17.91	49.02	66.9	79.81	36.25	47.29	63.44	76.6
DDMA+Mahalanobis	**18.67**	**56.49**	**75.22**	**87.47**	**41.73**	**54.25**	**71.64**	**85.42**
LMNN [13]	18.54	54.59	72.09	85.89	40.74	52.48	70.57	82.97
DDMA+LMNN	**20.51**	**60.16**	**80.28**	**91.99**	**45.97**	**58.37**	**76.62**	**88.71**
KISSME [16]	20.79	63.51	80.51	91.96	46.59	59.44	76.68	88.96
DDMA+KISSME	**22.66**	**66.71**	**85.32**	**95.16**	**49.7**	**62.97**	**80.06**	**91.47**

Table 2. Comparing results with the-state-of-the-art person re-id methods on top ranked matching rate (%)

Method	rank@1	10	25	50
SDALF [1]	19.9	49.4	70.5	84.8
ELF [8]	8.2	36.6	58.2	90.9
PRDC [2]	15.66	53.86	76	87
PCCA [17]	19.27	64.91	83	**96**
KISSME [16]	19.6	62.2	80.7	91.8
Descriptive+Discriminative Model [11]	19	52	69	80
DDMA+LMNN	20.51	60.16	80.28	91.99
DDMA+KISSME	**22.66**	**66.71**	**85.32**	95.16

distance measure based approaches, and the Descriptive+Discriminative Model is a result refinement method. The results clearly show that our approach gives the best performance in most cases.

4   Conclusion

In this work, we present a novel and efficient data-driven metric adaptation method for person re-identification task. The core idea is: we reform the procedure of person re-id framework that optimizing the learned metric by re-exploiting the training data in distance measure stage. Applying with the cross-view support and projection consistency, the approach can obtain a data-specific adaptive metric. Our method can improve the matching result of metric learning method obviously. Extensive experiments compared with three typical methods on two publicly data sets have validated the effectiveness of our proposed method.

Acknowledgement. The research was supported by the National Nature Science Foundation of China (61303114, 61231015, 61170023, 61172173), the Nationl Key Technologies R&D Program (2013AA014602), the Specialized Research Fund for the Doctoral Program of Higher Education (20130141120024), the Technology

Research Project of Ministry of Public Security (2014JSYJA016), the Fundamental Research Funds for the Central Universities (2042014kf0250), the China Postdoctoral Science Foundation funded project (2013M530350), the major Science and Technology Innovation Plan of Hubei Province (2013AAA020), the Key Technology R&D Program of Wuhan (2013030409020109), the Guangdong-Hongkong Key Domain Break-through Project of China (2012A090200007), and the Special Project on the Integration of Industry, Education and Research of Guangdong Province (2011B090400601).

References

1. Farenzena, M., Bazzani, L., Perina, A., Murino, V., Cristani, M.: Person re-identification by symmetry-driven accumulation of local features. In: IEEE Conference on Computer Vision and Pattern Recognition (CVPR) (2010)
2. Zheng, W.-S., Gong, S., Xiang, T.: Person re-identification by probabilistic relative distance comparison. In: IEEE Conference on Computer Vision and Pattern Recognition (CVPR) (2011)
3. Liu, C., Loy, C.C., Gong, S.: POP: Person re-identification post-rank optimisation. In: IEEE International Conference on Computer Vision, ICCV (2013)
4. Li, X., Tao, D., Jin, L., Wang, Y., Yuan, Y.: Person re-identification by regularized smoothing kiss metric learning. IEEE Transactions on Circuits and Systems for Video Technology (TCSVT) (2013)
5. Wang, Y., Hu, R., Liang, C., Zhang, C., Leng, Q.: Camera compensation using feature projection matrix for person re-identification. IEEE Transactions on Circuits and Systems for Video Technology (TCSVT) (2014)
6. Kviatkovsky, I., Adam, A., Rivlin, E.: Color invariants for person reidentification. IEEE Transactions on Pattern Analysis and Machine Intelligence (TPAMI) (2013)
7. Rui, Z., Wanli, O., Xiaogang, W.: Unsupervised salience learning for person re-identification. In: IEEE Conference on Computer Vision and Pattern Recognition (CVPR) (2013)
8. Gray, D., Tao, H.: Viewpoint invariant pedestrian recognition with an ensemble of localized features. In: Forsyth, D., Torr, P., Zisserman, A. (eds.) ECCV 2008, Part I. LNCS, vol. 5302, pp. 262–275. Springer, Heidelberg (2008)
9. Wang, X., Doretto, G., Sebastian, T., Rittscher, J., Tu, P.: Shape and appearance context modeling. In: IEEE International Conference on Computer Vision (ICCV) (2007)
10. Ali, S., Javed, O., Haering, N., Kanade, T.: Interactive retrieval of targets for wide area surveillance. In ACM International Conference on Multimedia (MM) (2010)
11. Hirzer, M., Beleznai, C., Roth, P.M., Bischof, H.: Person re-identification by descriptive and discriminative classification. In: Heyden, A., Kahl, F. (eds.) SCIA 2011. LNCS, vol. 6688, pp. 91–102. Springer, Heidelberg (2011)
12. Leng, Q., Hu, R., Liang, C.: Bidirectional ranking for person re-identification. In: IEEE International Conference on Multimedia and Expo (ICME) (2013)
13. Weinberger, K.Q., Blitzer, J., Saul, L.K.: Distance metric learning for large margin nearest neighbor classification. Journal of Machine Learning Research (2009)
14. Hirzer, M., Beleznai, C., Kstinger, M., Roth, P.M., Bischof, H.: Dense appearance modeling and efficient learning of camera transitions for person re-identification. In: IEEE International Conference on Image Processing (ICIP) (2012)

15. Dikmen, M., Akbas, E., Huang, T.S., Ahuja, N.: Pedestrian recognition with a learned metric. In: Kimmel, R., Klette, R., Sugimoto, A. (eds.) ACCV 2010, Part IV. LNCS, vol. 6495, pp. 501–512. Springer, Heidelberg (2011)
16. Kostinger, M., Hirzer, M., Wohlhart, P., Roth, P., Bischof, H.: Large scale metric learning from equivalence constraints. In: IEEE Conference on Computer Vision and Pattern Recognition (CVPR) (2012)
17. Mignon, A., Jurie, F.: Pcca: A new approach for distance learning from sparse pairwise constraints. In: IEEE Conference on Computer Vision and Pattern Recognition (CVPR) (2012)
18. Gray, S.B.D., Tao, H.: Evaluating appearance models for recognition, reacquisition, and tracking. In: IEEE International Workshop on Performance Evaluation of Tracking and Surveillance (PETS) (2007)
19. Li, W., Zhao, R., Wang, X.: Human reidentification with transferred metric learning. In: Lee, K.M., Matsushita, Y., Rehg, J.M., Hu, Z. (eds.) ACCV 2012, Part I. LNCS, vol. 7724, pp. 31–44. Springer, Heidelberg (2013)
20. Ojala, T., Pietikainen, M., Maenpaa, T.: Multiresolution gray-scale and rotation invariant texture classification with local binary patterns. IEEE Transactions on Pattern Analysis and Machine Intelligence (TPAMI) (2002)
21. Gong, S., Xiang, T.: Person Re-identification. Springer, London (2011)
22. Liu, J., Ji, S., Ye, J.: SLEP: Sparse learning with efficient projections. Arizona State University (2009)

A Novel Optimized Watermark Embedding Scheme for Digital Images

Feng Sha[1], Felix Lo[1], Yuk Ying Chung[1], Xiaoming Chen[2], and Wei-Chang Yeh[3]

[1] School of Information Technologies, University of Sydney, Australia, NSW2006
[2] Institute of Advanced Technology, University of Science and Technology of China
[3] Department of Industrial Engineering and Engineering Management,
National Tsing Hua University, Hsinchu 30013, Taiwan, ROC
vchung@it.usyd.edu.au

Abstract. Many Scientists in Image Processing try to find an efficient way for digital multimedia protection. Although standards and criteria are still in developing, the watermarking which performs mark picture embedding and extraction with original image has been identified as major technology to achieve ownership and copyright protection. This paper is aim to find a more efficient way to embedding watermark into a gray-scale original image using a new algorithm – Artificial Bee Colony to optimize pixel by pixel embedding at different frequency levels (sub-band) with Discrete Wavelet Transform (DWT) Technology in order to enhance the security, invisibility to human visual and robustness of image watermarking. The proposed scheme will take efforts in higher level of DWT decomposition which provide better robustness but low quality of watermarked image and perform better quality of watermarked image and visible watermark compare to random embedding. The proposed new embedding method has been tested against most types of image modifications and in different frequency domain and levels of DWT to provide both high quality watermarked images and superior robustness.

Keywords: Watermarking, ABC, Artificial Bee Colony, DWT, Discrete Wavelet Transform.

1 Introduction

The increases of digital images and multimedia visualization products demand high quality of ownership and copyright protection. Digital Watermarking [1] which provide multiple advantages and effective solution to authorized use of digital media is one of the best way to prevent illegal modification and spread issues.

When embedding a digital watermark into the original image, text, video or audio materials, the watermark image is hiding itself in a proposed pattern in a bottom level of digital sources and it won't be visualized by human eyes or even machine detection. The watermark can still be retrieved using same pattern when watermarked sources have been widely modified or seriously corrupted.

X. He et al. (Eds.): MMM 2015, Part II, LNCS 8936, pp. 208–219, 2015.

There are many techniques for watermarking already existing such as Discrete Cosine Transform (DCT) with Particle Swarm Optimization (PSO) [4]. In this paper, a new method has been proposed to meet the requirements of watermarking as a novel watermarking technique based on Artificial Bee Colony Algorithm (ABC) [5, 6] and Discrete Wavelet Transform (DWT). It decompose frequency levels in DWT which provides high quality of robustness, then using ABC algorithm to optimize and find a seed number to generate a permutation between watermark pixel size and position to be embedded in the frequency domain of original image.

This paper is organized in the following 6 sessions. Firstly, it gives the introduction of digital watermark, then explains the DWT wavelet technology and where to place watermark. Session 3 describes the basic concepts of ABC optimization. Session 4 will explain how the proposed new approach can achieve the good quality of watermarking. Finally, the experimental results and conclusion based on the new method will be shown in session 5 and 6 respectively.

2   Introduction to DWT

Discrete Wavelet Transform is a technique used to divide signals into 4 separate sub-bands discretely (i.e. no overlapping of details between different sub-bands), namely LL (containing approximation details), HL (horizontal details), LH (vertical details) and HH (diagonal details) sub-bands as illustrated in Figure 1, and therefore the size of each sub-band will be ¼ of the original signal [8]. Each sub-band can be subdivided even further using the same DWT technique, and original image can be retrieved using inverse discrete wavelet transform.

This property allows us to embed watermark into one of the sub-band without totally destroying the cover image, rather a distortion will become visible to human eye as the original image may still be retrieved using other sub-banded information.

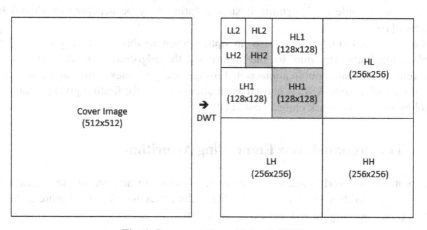

Fig. 1. Representation of 3-level DWT

In this paper, level 3 sub-band HH2 (64x64) which provide higher quality of robustness of watermark and level 2 sub-band HH1 (128x128) which provide lower robustness but higher imperceptibility value will be chosen to demonstrate the ABC algorithm embedding and extraction.

Advantages of using DWT to embed watermark includes:

1. Similar to JPEG compression technology and hence reduce the need for implementing other algorithms [9].
2. By separating in LL to HH band, we are able to exploit the characteristic that human visual system is less sensitive to HH sub-band (where edges and brightness are mainly stored in HH)[7].

3   Introduction to Artificial Bee Colony Algorithm

The artificial bee colony algorithm (ABC) is defined by Karaboga in 2005 to mimic the food hunting behaviour of honey bees [10]. It is an optimization algorithm used for finding global minima and maxima within a set number of iteration cycles. Possible solutions are named to be "Food sources" at the start of algorithm, and by defining the bee colony size we then use bees to find the best value food source within the ranges.

Bees are categorized as follows: Employed, Onlooking and Scout; at the start of the algorithm, we define the proportion of employed bees and onlooking bees to be 1:1. Employed bees are the ones that are currently settled in one particular food source and will not be leaving until it had exhausted the food source (as specified by the generic variable "trial"), when exhausted it will become a scout bee. Onlooking bees are the ones that search around the employed bees for further food sources, and scout will be ones that are will search at no particular space for new solutions. ABC works with the idea of feedback (fitness), and by using such fitness we are able to determine if such solution may be accepted or should be rejected [10].

Compare to PSO, ABC is also an optimization method using exploration and exploitation, however due to the nature of the algorithm yields performance differences. Multiple publications stated using ABC generates results at a faster rate, based on such results ABC is chosen as an alternate way for finding pixel position on pixel-by-pixel watermark embedment [11][12].

4   The Proposed New Embedding Algorithm

The proposed method contains two parts of activities to achieve the watermarking, one is the embedding process and the other is the extraction process. Figure 2 shows the proposed new digital watermarking algorithm.

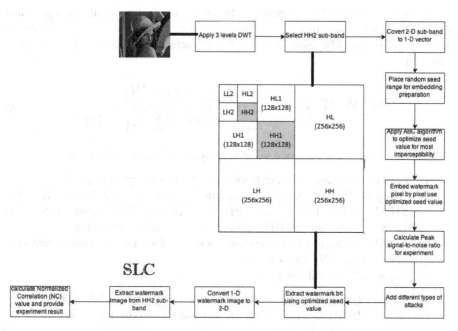

Fig. 2. The proposed new watermarking algorithm

We calculate the fitness of food source h by using the following equation:

$$fit_h = \begin{cases} 0, & PSNR(h) \leq 32 \text{ if HH3}, \ \leq 38 \text{ if HH2} \\ 1, & PSNR(h) > 32 \text{ if HH3}, \ > 38 \text{ if HH2} \end{cases} \tag{1}$$

By using such fitness we are able to determine the probability of selection of the food source as:

$$\text{Prob} = 0.9 \times \frac{\text{Fitness}}{max(\text{Fitness})} + 0.1 \tag{2}$$

Suppose we have cover image c and watermarked cover image c' both have image size $M*N$, at each position (i,j) we are able to calculate the mean square error (MSE) and hence PSNR of the overall image.

$$\text{PSNR} = 10 \times \log_{10} \frac{255^2}{\text{MSE}} \text{ (dB)} \tag{3}$$

$$\text{MSE} = \frac{\sum_{ij} |c'_{ij} - c_{ij}|^2}{N \times N} \tag{4}$$

Finally we quantify the differences between the original watermark image w and the extracted watermark image w' as normalized correlation (NC) value. If NC=1 then it implies no difference between the extracted watermark and the original watermark.

$$NC(W, W') = \frac{\sum_{i=1}^{n} w(i)w'(i)}{\sqrt{\sum_{i=1}^{n} w(i)^2}\sqrt{\sum_{i=1}^{n} w'(i)^2}} \tag{5}$$

The procedures of the proposed watermarking algorithm are as follows:

1. Lord cover image and watermark image to start watermarking
2. Using DWT Technology to decompose cover image in 3 levels (LL->LL1->HH2) and 2 levels (LL->HH1) and choose HH2 sub-band for first watermarking case and HH1 sub-band for second watermarking case.
3. In HH2 sub-band, perform conversion to transform two dimensions matrix to one dimension Vector (1-64x64)
4. Begin ABC embedding [13]

    a. Generate initial random seeds randomly Sh, h=1,2,3…10000
    b. Evaluate fitness of such random seeds according to equation (1)
    c. Cycle = 1
    d. Repeat iterations
    e. FOR (employed phase) {

        i. Produce a new set of random seeds
        ii. Evaluate fitness using equation (1)
        iii. Apply selection process by choosing maximum value of PSNR in corresponding seeds}

    f. Calculate the probability equation (2)
    g. FOR (onlooker phase){

        i. Select another set of Neighbor solutions from Sh
        ii. Produce a new solution Si using seed contained within Neighbor set
        iii. Calculate fitness by equation (1) for Si
        iv. Apply selection process by choosing maximum value of PSNR in corresponding seeds}

    h. IF (scout phase)

        There is an abandoned solution for the scout depending on "limit"
        THEN
        Repeat it with a new one which will by randomly produced

        i. Memorize the best solution so far
    j. Cycle=Cycle+1

> Stop if Cycle=MGN
> Calculation END.

5. Locate most efficient seed (ABC output optimal solution) for pixel by pixel embedding.
6. Calculate the PSNR base on Equation (3) for watermarked image
7. Add 11 types of Attacks on watermarked image and calculate their PSNR values.
8. Select the same level of DWT decomposition to extract watermark base on actual seed value to extract watermark and calculate the NC values using Equation (5) for all attacks.
9. Reconstruct and convert the watermark pixel from one dimension to two dimensions and Provide test outputs.
10. END

5 Experimental Results

In the experiment tests, Figure 3(a) Lena.jpg 512x512 8 bit gray-scale image and 3(b) dmg1.tif 64x64 1 bit image which represent the original cover image and the watermark respectively. Various arracks are used to test the imperceptibility and robustness of the watermark for ABC algorithm involved DWT level2 and level3 watermark embedding and extraction. All tests are performed at Matlab2014a environment.

(a)   (b)

Fig. 3. (a) Lena.jpg 512x512 8 bit gray-scale image and (b) dmg1.jpg 64x64

5.1 The Proposed ABC Embedding and Extraction Algorithms

Base on the pixel by pixel embedding from source watermark image, tests on to level 3 DWT decomposition (LL->LL1->HH2) high frequency HH2 shows the efficient of the random seed embedding is hard to control and the quality of watermarked image performs lower value than other technologies like DWT-SVD [5] and PSO [4] by using PSNR value. And the random seed value is required to locate watermark for extraction.

Table 1 illustrates the embedding and extraction result for a random seed number (100) under different attack types and from this table, the quality of watermarked images are significant difference among each other under different attack types. And the attack affects a lot with the extraction watermark by Normalized Correlation (NC) evaluation.

Table 1. DWT level 3 (LL->LL1->HH2) Random seed (100) embedding

No Attack		Wiener_filtering_3x3		Gaussian_Noise_mean_0_variance_0.0005	
PSNR=34.34	NC=1	PSNR=32.58	NC=0.72	PSNR=30.66	NC=0.76
Intensity_adjustment_from_0.2_0.8_to_0_1_		Crop_image_100_400_150_450_		Crop_1_4___th___image	
PSNR=18.90	NC=0.85	PSNR=11.46	NC=0.80	PSNR=13.43	NC=0.86
Resize_512x512_256x256_512x512_		Row_column_blanking_10_70_100_160_190_230_400_450_500		Row_column_copying___10_30__40_70__100_120__50_160__250_200__330_300__470_400	
PSNR=30.71	NC=0.80	PSNR=22.06	NC=0.93	PSNR=27.03	NC=0.94
Gaussian_low_pass_filter_9x9_sigma_0.5		Salt___pepper_noise_1%_		Salt___pepper_noise_5%___median_filter_3x3	
PSNR=33.98	NC=0.88	PSNR=24.83	NC=0.87	PSNR=30.94	NC=0.74

Table 2. DWT level 3 (LL->LL1->HH2) ABC algorithm embedding

No Attack		Wiener_filtering_3x3		Gaussian_Noise_mean_0_varianc e_0.0005	
PSNR=34.36	NC=1	PSNR=32.58	NC=0.72	PSNR=30.63	NC=0.76
Intensity_adjustment_from_0.2_ 0.8 to 0_1		Crop_image_100_400_150_450		Crop_1_4___th__image	
PSNR=18.90	NC=0.86	PSNR=11.46	NC=0.81	PSNR=13.43	NC=0.86
Resize_512x512_256x256_512x 512_		Row_column_blanking_10_70_ 100_160_190_230_400_450_50 0		Row_column_copying_10_30_ 40_70__100_120__50_160__250 200__330_300__470_400	
PSNR=30.72	NC=0.80	PSNR=22.06	NC=0.93	PSNR=27.03	NC=0.95
Gaussian_low_pass_filter_9x9_ sigma_0.5		Salt___pepper_noise_1%_		Salt___pepper_noise_5%___medi an_filter_3x3	
PSNR=33.99	NC=0.88	PSNR=24.66	NC=0.87	PSNR=30.85	NC=0.74

By implement of ABC algorithm, the test result demonstrate a better embedding quality of watermarked image. The colony size is defined as 20 which represents the employee bee and onlooker bee, the food number is half of the colony size, the limit trials is 100, and runtime is 1. More numbers of iteration cycle performed, better result presented. Table 2 illustrates the ABC algorithm embedding PSNR result after 500 cycles (Compare to 10) for foraging and their NC values by locating the embedding position seed number. Figure 4 displays an increasing trend for PSNR value though different ABC iterations.

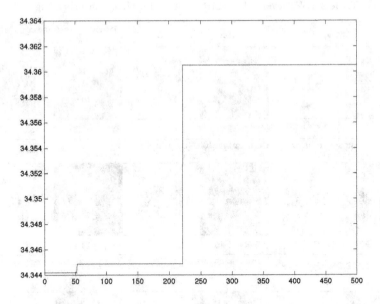

Fig. 4. Line chart for PSNR in Level 3 DWT with ABC 500 iterations

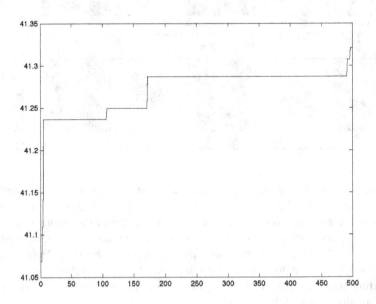

Fig. 5. Line chart for PSNR in Level 2 DWT with ABC 500 iterations

5.2   ABC Embedding and Extraction on Level 2 DWT

Tests result on to level 2 DWT decomposition (LL->HH1) high frequency HH1 shows better embedding quality of watermarked image with the same parameters setting in ABC Algorithm. Table 3 illustrates the level 2 DWT decomposition ABC algorithm embedding PSNR result after 500 cycles for foraging and their Normalized Correlation (NC) values by locating the embedding position seed number. Figure 5 also displays an increasing trend for PSNR value though different ABC iterations in level 2 DWT.

Table 3. DWT level 2 (LL->HH1) ABC algorithm embedding

No Attack		Wiener_filtering_3x3		Gaussian_Noise_mean_0_variance 0.0005	
PSNR=41.32	NC=1	PSNR=35.18	NC=0.65	PSNR=32.43	NC=0.76
Intensity_adjustment_from_0.2_0.8 to 0_1_		Crop_image_100_400_150_450		Crop_1_4___th__image	
PSNR=18.97	NC=0.85	PSNR=11.46	NC=0.81	PSNR=13.45	NC=0.86
Resize_512x512_256x256_512x512_		Row_column_blanking_10_70_100_160_190_230_400_450_500		Row_column_copying_10_30_40_70_100_120_50_160_250_200_330_300_470_400	
PSNR=32.20	NC=0.78	PSNR=22.26	NC=0.95	PSNR=27.68	NC=0.96
Gaussian_low_pass_filter_9x9_sigma_0.5		Salt___pepper_noise_1%_		Salt___pepper_noise_5%___median_filter_3x3	
PSNR=38.60	NC=0.91	PSNR=24.82	NC=0.96	PSNR=32.52	NC=0.72

6   Conclusion

The reliable outcomes of the watermarked material and clear extracted watermark are the essential requirements for implementation of technology and algorithm. These requirements refer to two measurement criteria, the imperceptibility and the robustness [1] to verify possible watermarking algorithms and technologies. When the imperceptibility or perceptual transparency value is high, the visual quality of watermarked material is high to provide a good similarity to the original material. When robustness value is good, that means the watermarked image can perform good capacity to hold against unauthorized modifications and the watermark within is difficult to remove by illegal changes which does not make significant degradation in whole quality of the material. Therefore the higher imperceptibility value will affect the robustness quality [2, 3].

In this paper, a new image watermarking system using Artificial Bee Colony Algorithm has been proposed. The proposed scheme has demonstrated to have a better imperceptibility in level 3 DWT decomposition, and provide good quality watermarked image that not visible to human visual system. The ABC algorithm performs well to resist several attack types, such as Row column blanking and Row column copying in level 3's decomposition and provide an average 0.85 NC value. In level 2 DWT decomposition, the proposed method has proved to have a better PSNR value to the similarity between watermarked image and original image, and the NC value increased significantly in watermark extraction.

When we compared with level 2 and level 3 DWT using ABC algorithm, our results showed that the level-3 ABC embedding provides more robustness to various attacks. By increasing the number of iteration cycle of the ABC algorithm, the results provide a possible trend that the more iteration the ABC algorithm cover, the better quality of watermarked image could be generated in DWT decomposition watermarking and better extraction robustness. In the future research work, there is still high potential for Artificial Bee Colony Algorithm to participate in digital watermarking technology.

References

1. Yusof, Y., Khalifa, O.O.: Imperceptibility and robustness analysis of DWT-based digital image watermarking. In: International Conference on Computer and Communication Engineering, ICCCE 2008, pp. 1325–1330. IEEE (2008)
2. Hsieh, M.-S.: Perceptual Copyright Protection Using Multiresolution Wavelet-Based Watermarking And Fuzzy Logic. arXiv preprint arXiv:1007.5136 (2010)
3. Wang, F.-H., Pan, J.-S., Jain, L.C.: Innovations in Digital Watermarking Techniques. Springer (2009)
4. Rohani, M., Nasiri Avanaki, A.: A watermarking method based on optimizing SSIM index by using PSO in DCT domain. In: 14th International CSI on Computer Conference, CSICC 2009, pp. 418–422 (2009)

5. Karaboga, D., Basturk, B.: Artificial bee colony (ABC) optimization algorithm for solving constrained optimization problems. In: Melin, P., Castillo, O., Aguilar, L.T., Kacprzyk, J., Pedrycz, W. (eds.) IFSA 2007. LNCS (LNAI), vol. 4529, pp. 789–798. Springer, Heidelberg (2007)
6. Karaboga, D.: http://mf.erciyes.edu.tr/abc/
7. Basheer, N.M., Abdulsalam, S.S.: Digital Image Watermarking Algorithm in Discrete Wavelet Transform Domain Using HVS Characteristics. In: Proceedings of the IEEE International Conference on Information Technology: Coding and Computing, pp. 122–127 (2011)
8. Mistry, D., Banerjee, A.: Discrete Wavelet Transform using MATLAB. International Journal (2013)
9. Chowdhury, M.M.H., Khatun, A.: Image Compression Using Discrete Wavelet Transform. IJCSI International Journal of Computer Science Issues 9 (2012)
10. Karaboga, D.: An idea based on honey bee swarm for numerical optimization. Technical report-tr06, Erciyes university, engineering faculty, computer engineering department (2005)
11. Akay, B.: A study on particle swarm optimization and artificial bee colony algorithms for multilevel thresholding. Applied Soft Computing 13, 3066–3091 (2013)
12. Turanoğlu, E., Özceylan, E., Kiran, M.S.: Particle Swarm Optimization and Artificial Bee Colony Approaches to Optimize of Single Input-Output Fuzzy Membership Functions. In: The 41st International Conference on Computers & Industrial Engineering (2011)
13. Bae, C., Yeh, W.-C., Shukran, M., Chung, Y.Y., Hsieh, T.-J.: A novel anomaly-network intrusion detection system using ABC algorithm. Int. J. Innov. Comput. Inf. Control 8, 8231–8248 (2012)

User-Centred Evaluation to Interface Design of E-Books

Yang-Cheng Lin

Department of Arts and Design, NATIONAL Dong Hwa University, Hualien, 974, Taiwan
lyc0914@mail.ndhu.edu.tw

Abstract. A good interface design of e-books is convenient for users, whereas a poor design can disorientate them. In this study, we conduct an experiment on four test scenarios, including the Zinio iPad version, Zinio PC version, MagV iPad version, and MagV PC version. This study includes 48 subjects (including 12 men and 36 women), with the majority being long-term Internet users. We use the performance measurement, retrospective testing, and a semistructured questionnaire to analyse the operation time of five experimental tasks, error frequency, and subjective satisfaction. The result shows the Zinio interface design provides a user-friendly device, and the iPad version (using a touchscreen) facilitates engaging experimental tasks compared with the PC version (with a mouse and/or a keyboard). In addition, most participants prefer to combine two or three options to complete the experimental tasks. The result can provide designers with a useful insight into designing a proper user model that best meets user requirements.

1 Introduction

Because of the development of Wi-Fi technology, mobile devices (or handy devices) have become some of the most desired and popular commercial products, including smart phones, e-book readers, ultrabooks (laptops), and tablets. According to a marketing report [2], the number of mobile device users is increasing by 10%-20% each year. It is a fantastic outcome for manufacturers and product designers. In addition, a trend has been noted in the development of mobile devices. The e-book is gradually changing the reading habits of mobile users and the manner in which they retrieve information. For example, 70% of users have read an e-book in the past three months, and 50% have downloaded it from websites [3]. In addition, certain universities in the United States have adopted e-books as their teaching texts, whereas some elementary schools in Japan teach students to complete their homework by using e-books [2].

However, can e-books meet users' demands for reading? Is reading an e-book the same as reading a real book (the convention)? Can users adapt to the differences in the interaction interface between e-books and real books? Are there any differences in learning performance among different user models [1, 5]? In other words, do mobile devices have a good user interaction interface for enabling the reading of e-books compared with real books? These are critical research topics for researchers to discuss. Specifically, the competitive markets are fill with multifunctional mobile devices

X. He et al. (Eds.): MMM 2015, Part II, LNCS 8936, pp. 220–226, 2015.

(handy devices). For example, tablet mobile phones (or called phablets) with a large touchscreen (and some with a small keyboard) are used as tablets as well as mobile phones. For another example, a 'Transformer Pad' or 'Flip PC' is not only a laptop but also a tablet with a touchscreen and a keyboard. Users can choose a specific model based on their requirements. In other words, users can use their fingers on the touchscreen to increase the size of an image on an e-book, and can also use the keyboard to type a title or a keyword, to search for a specific topic of interest. Because multifunctional mobile devices are available, users have more options to choose from. However, are these devices user friendly? What types of input devices or instruments (e.g. a touchscreen, mouse, and keyboard) do users prefer? Which is a suitable user model [1]? Which model is better used for reading e-books, with higher subjective satisfaction [4]?

We conduct a user-centred experiment on e-books to answer these questions. In this study, we choose two input devices (instruments): the iPad version with a touchscreen, and the PC version with a mouse and/or a keyboard. In order to collect additional data for further analysis, we choose two popular e-book systems: Zinio and MagV, due to their wider user population and cross-platform attributes [2]. Thus, there are four test scenarios in this study, including Zinio iPad version (Z-Pad), Zinio PC version (Z-PC), MagV iPad version (M-Pad), and MagV PC version (M-PC), particularly. The subsequent sections present our experimental study involving the use of several analytical techniques, followed by the study results, a discussion, and a conclusion.

2   Method

A good interface design of e-books is convenient for users, whereas a poor design can disorient them. This indicates that users lose track of the location and directions when browsing or navigating e-books [6]. Fig. 1 shows the interface design of the four test scenarios.

This study includes 48 participants who are long-term Internet users (more than 20 hours per week). These 48 participants are divided into two groups. The first group (including seven men and 17 women, with an average age of 22.8 years) is asked to test two iPad versions (Z-Pad, and M-Pad) by using a touchscreen, and the second group (including five men and 19 females, with an average age of 21.5 years) is asked to test two PC versions (Z-PC, and M-PC) by using a mouse and/or a keyboard. These 48 participants are asked to perform five experimental tasks, as shown in Table 1. Table 1 lists the corresponding functions of operating e-books when the participants are engaging in the experimental tasks. We video-record the sessions simultaneously. After the experiment, a semistructured questionnaire is used to collect information pertaining to the 'subjective satisfaction' [4] of reading e-books. The procedure of the experiment comprises the following steps:

Step 1: Divide 48 participants into two groups.
Step 2: Conduct five experimental tasks by five design experts with more than 10 years of interface design experience.

(a) Z-Pad   (b) Z-PC

(c) M-Pad   (d) M-PC

Fig. 1. The interface design of four test scenarios

Step 3: Perform the five experimental tasks by two groups. The first group tests two e-book systems by using the touchscreen (Z-Pad, and M-Pad). The participants are randomly assigned to test the Zinio or MagV e-book system. After the test, they take a short break (about 10 minutes), and then test the other system (MagV or Zinio, depending on which is taken first) until the experimental tasks are completed. The second group is asked to test the PC version by using a mouse and/or a keyboard (Z-PC, and M-PC).

Step 4: Record the entire experimental process, including the operation time and errors that occurred during the five tasks.

Step 5: Distribute the semistructured questionnaire to collect quantitative (i.e. the subjective satisfaction of participants) and qualitative data (i.e. the participants' suggestions regarding the e-book system).

Step 6: Analyze the numerical data, including the operation time, error frequency, and subjective satisfaction.

Step 7: Discuss and conclude.

Table 1. Description of the five experimental tasks and the corresponding operating functions

Task No.	Task Description	Corresponding Operating Functions
T1	Please find the AA article on the Content Page of the BB book, and point out the exact position.	● Scale ● Drag Page
T2	Please find the first page of the AA article, and point out the exact position.	● Content Page ● Drag and Scroll ● Up/Down Page
T3	Please find the CC title in the AA article, and point out the exact position.	● Drag and Scroll ● Up/Down Page
T4	Please read the DD point of suggestions aloud in the CC title, and answer the EE question.	● Scale ● Drag Page
T5	Please return to the initial model, and choose another FF book.	● Content Page ● Choose Book

3   Results and Discussion

Table 2 shows the operation time, error frequency, and subjective satisfaction of the 48 participants as they perform the five experimental tasks. T1-T5 in the second row of Table 2 represent Tasks 1-5, respectively.

Table 2. Result of the operation time, error frequency, and subjective satisfaction

	Time						Error						Satisfaction	
	T1	T2	T3	T4	T5	Total	T1	T2	T3	T4	T5	Total	S1	S2
Z-Pad	12.4	21.2	16.2	8.2	22.6	80.6	0.2	0.1	0.1	0.0	0.3	0.7	3.9	3.9
Z-PC	11.8	20.6	25.9	27.3	41.8	127.4	0.1	0.3	0.4	0.2	1.1	2.1	3.9	3.7
M-Pad	24.6	28.4	5.9	3.3	22.8	85.0	0.3	0.3	0.0	0.0	0.3	0.9	3.8	3.3
M-PC	19.1	73.7	28.0	31.0	86.1	237.9	0.2	1.2	0.3	0.2	0.9	2.8	3.6	3.1

3.1   Quantitative Analysis (Operation Time and Error Frequency)

Table 2 shows that the operation time of Z-Pad (80.6 seconds) is the lowest, followed by those of M-Pad (85.0 s), Z-PC (127.4 s), and M-PC (237.9 s). The result means the iPad version (using the touchscreen) is easier to operate compared with the PC version (with a mouse and/or a keyboard). In addition, with the same input devices (instruments), the Zinio e-book system requires less operation time than does MagV (Z-Pad (80.6 s) < M-Pad (85.0 s), and Z-PC (127.4 s) < M-PC (237.9s)). The result indicates that the interface design of Zinio facilitates usability. Hence, a good interface design of e-books makes it user-friendly.

The error frequency of Z-Pad (0.7 times) is the lowest, followed by those of M-Pad (0.9 times), Z-PC (2.1 times), and M-PC (2.8 times), indicating that the interface design of the iPad version results in fewer errors. This also shows that the manner in which the touchscreen is used is more intuitive for users, and thus, helps in preventing errors. As compared to the operation time, the Zinio e-book system also results in a lower error frequency than MagV (Z-Pad (0.7 times) < M-Pad (0.9 times), and Z-PC (2.1 times) < M-PC (2.8 times)).

For further analysis, we conduct the retrospective testing [4] to identify differences between the operation time and the error frequency while participants are engaging in the experimental tasks. For example, three options (operating functions) are chosen when the users perform Task 2. The users can adopt only the 'Content Page' to find the first page of the AA article; otherwise, they can use 'Drag and Scroll' or 'Up/Down Page' to achieve the objective. The users could also combine these options. Table 3 lists the percentage of options chosen by the participants, and that most of the participants prefer combining two or three options (e.g. 'Up/Down Page' with 'Drag and Scroll') to read an e-book. In addition, the participants (except in their use of Z-PC) also like to navigate the article (i.e. page by page) in a similar manner to when reading a real book. The result provides useful insights for designers in designing a suitable user model that can best meet user requirements and demands.

Table 3. The percentage of options chosen by the participants

	Options / Operating Functions			
	Content Page	Drag and Scroll	Up/Down Page	Combined Options
Z-Pad	25.0%	4.2%	37.5%	33.3%
Z-PC	37.5%	8.3%	4.2%	50.0%
M-Pad	0.0%	0.0%	62.5%	37.5%
M-PC	29.2%	8.3%	37.5%	25.0%

3.2   Qualitative Analysis (Subjective Satisfaction and Recommendations)

After the experimental tasks, we distribute the semistructured questionnaire to collect quantitative and qualitative data. The participants are asked to assess the interface design of the e-book system. They answer two questions: 'Q1: How do you feel regarding the visual and aesthetical style of the interface design [3]?' and 'Q2: How do you feel regarding the operation (operating efficiency) of the interface design?' A 5-point scale is adopted in the semistructured questionnaire, ranging from 1 (the lowest satisfaction) to 5 (the highest satisfaction). The last two columns (S1 and S2) of Table 2 list the participants' subjective satisfaction. The S1 result reveals a slight difference (only M-PC being slightly lower) among the four different test scenarios. However, the S2 result shows that the Zinio e-book system has a slightly higher subjective satisfaction (Z-Pad of 3.9 and Z-PC of 3.7) compared with MagV (M-Pad of 3.3 and

Table 4. A part of the participants' recommendations on the e-book systems

The Zinio E-Book System	The MagV E-Book System
• It is difficult to navigate smoothly while the page is scaled up.	• The Content Page should mark the page number of the article.
• The framework of the interface design is too complex and confusing.	• The operation is not very smooth.
• The page number should be consistent with the Scroll page.	• Page numbers should be visible as users scroll the page.
• The Scroll function is unclear.	• The interface design of the 'Choose Book' is too dazzling and unclear to follow.
• Personalise the 'Choose Book' function, and add the personal favourite category.	• Books should be categorised into proper sections, so that users can adapt or modify them.
• The indicator of the Drag Page is occasionally slow and unclear.	• The 'Memory' function is good, but if users could choose to adapt it, this would be better.

M-PC of 3.1). The result indicates that the Zinio interface design is more user-friendly and efficient, although the visual and aesthetical style of Zinio is identical to that of MagV. This is also reflected by the operation time and the error frequency described in Section 3.1. For a further discussion, the participants are asked to write down their opinions and suggestions regarding the two e-book systems, as given in Table 4.

4   Conclusion

In this paper, we have presented an experimental study on two e-book systems with two input devices to address whether different test scenarios result in the same learning performance in reading e-books. The experimental result has demonstrated that the iPad version with a use of a touchscreen is easier to operate compared with the PC version with the use of a mouse and/or a keyboard. This means that users find operating a touchscreen more intuitive, which helps in avoiding errors. In addition, the Zinio e-book system is found to be easier to operate and has a lower error frequency compared with the MagV e-book system. The result of this paper provides useful insights for designers considering an improved interface design for e-books.

Acknowledgements. This research is supported in part by the Ministry of Science and Technology, Taiwan under Grant Nos. NSC101-2410-H-259-043, NSC102-2410-H-259-069, and MOST103-2420-H-259-015.

References

1. Fischer, G.: User Modeling in Human–Computer Interaction. User Modeling and User-Adapted Interaction 11, 65–86 (2001)
2. Lin, Y.-C., Huang, S.-W., Yang, M.-C.: Applying Usability Approach to Interface Design of e-books. In: Proceeding of 2012 Conference of Kansei Design, pp. 559–563 (2012)
3. Lin, Y.-C., Yeh, C.-H., Wei, C.-C.: How Will the Use of Graphics Affect Visual Aesthetics? A User-Centered Approach for Web Page Design. International Journal of Human-Computer Studies 71, 217–227 (2013)
4. Nielsen, J.: Usability Engineering. Academic Press, United Kingdom (1993)
5. Pu, P., Chen, L., Hu, R.: Evaluating Recommender Systems from the User's Perspective: Survey of the State of the Art. User Modeling and User-Adapted Interaction 22, 317–355 (2012)
6. Ruddle, R.A., Howes, A., Payne, S.J., Jones, D.M.: The Effects of Hyperlinks on Navigation in Virtual Environments. International Journal of Human-Computer Studies 53, 551–581 (2000)

A New Image Decomposition and Reconstruction Approach -- Adaptive Fourier Decomposition

Can He[1], Liming Zhang[1], Xiangjian He[2], and Wenjing Jia[2]

[1] Faculty of Science and Technology, University of Macau,
Avenida da Universidade, Taipa, Macau, China
[2] Faculty of Science & Information Technology
University of Technology, Sydney, Australia
rahxphoon@gmail.com, lmzhang@umac.mo
{xiangjian.he,wenjing.jia}@uts.edu.au

Abstract. Fourier has been a powerful mathematical tool for representing a signal into an expression consist of sin and cos. Recently a new developed signal decomposition theory is proposed by Pro. Tao Qian named Adaptive Fourier Decomposition, which has the advantage in time frequency over Fourier decomposition and without the need for a fixed window size problem such as short-time frequency transform. Studies show that AFD can fast decompose signals into positive-frequency functions with good analytical properties. In this paper we apply AFD into image decomposition and reconstruction area first time in the literature, which shows a promising result and gives the fundamental prospect for image compression.

Keywords: Adaptive Fourier decomposition, signal processing, image compression, image decomposition, mono-components.

1 Introduction

Fourier transform expands a signal into infinite series, after it stops through some given threshold, we can get the reconstructed signal by the inverse transform [1,2,3]. Adaptive Fourier Decomposition (AFD) is a newly proposed signal processing theory that has significant influence in signal denoising and control theory [4]. 'Mono-components' is brought up here to represent signals decomposed at each level. There are two ways that can decompose signal into mono-components with positive frequencies [5], which makes it possible to have some applications that are related to some mathematical analysis of signals. Maximal Selection Principle (MSP) is used here to get the mono-components [6,7]. AFD first generates a large pool of mono-components, then it decomposes the signal by using these components in the pool. Fourier decomposition is a special case of AFD, when all the corresponding parameters in AFD are all chosen to be zero; it will be in the form of Fourier transform. By doing the inverse transform, AFD can also get the approximate signal [8].

X. He et al. (Eds.): MMM 2015, Part II, LNCS 8936, pp. 227–236, 2015.
© Springer International Publishing Switzerland 2015

AFD keeps most good characteristics of Fourier transform. In general, AFD can be applied in all applications in which Fourier transform can be used. The advantage of AFD is that the decomposed components have distinct division for different frequencies, which can be further used for distinguishing different frequency components of signals. Signals decomposed by Fourier Transform (FT) in the sum of trigonometric functions do not have such good time-varying time-frequency instantaneous frequency (IF).

AFD can provide many potential applications in image processing. With the better and faster converging properties, AFD can serve for some applications such as speech recognition, image denoising [9,10,11], edge detection, image compression and so on. For a given signal, by summing up its Fourier series components, we can get the approximate signal. The more components participate in accumulation, the more likely they are. Noises are usually in the form of high frequencies. They come later no matter in Fourier expansion or AFD. Adding up some components decomposed in the preceding levels will take away the high frequency part. In another word, noise elimination can be done. Edge detection continues in this way, certain components which have some particular frequency from the AFD are selected, by adding them together, we can get the desired signal -- edges.

Transformations are usually used in image compression. Discrete Cosine Transform (DCT) is used in the famous JPEG compression before quantization. Same as Fast Fourier Transformation (FFT), DCT is the transform from time domain to frequency domain. The difference is that there exists no complex number in the result. DCT first partitions the original image into 8×8 blocks. Every 8×8 blocks turns to be another 8×8 group through basis function. DCT concentrates most of the energy (low frequency) in the upper left corner of an image and lower right corner with less energy (high frequency). JPEG is a lossy compression that it deducts the high frequencies that are insensitive to our human eyes. Removing 50% of the high frequencies may only loss 5% of the encoding information. So if there are other transformations converging faster than the one currently used in JPEG, there would be less redundancy generated. AFD is one of the new transformations, which converges faster than Fourier transform does.

This paper is organized as follow. The principle of the AFD based image decomposition and reconstruction is introduced in Section 2. The experiment results are shown in Section 3. Conclusions are drawn in Section 4.

2   Principles of AFD Based Image Decomposition and Reconstruction Approach

2.1   Brief Overview of AFD

AFD is based on the rational orthogonal system, or the Takenaka-Malmquist system [4]. For real valued signal G, first project it into Hardy Space $H^2(D)$ to get the projection signal G^+. Then decompose the signal into corresponding portions (1).

$$G^+(z) = \sum_{k=1}^{\infty} C_k B_k(z) \tag{1}$$

C_k is the coefficient and B_k is the basis determined by a_k of each decomposition, where k stands for the decomposition level (2).

$$B_k(z) = B_{\{a_1,\dots,a_k\}}(z) := \frac{1}{\sqrt{2\pi}} \frac{\sqrt{1-|a_k|^2}}{1-\overline{a_k}z} \prod_{l=1}^{k-1} \frac{z-a_l}{1-\overline{a_l}z}, k=1,2,\dots, \tag{2}$$

Here B_k is what we called Takenaka-Malmquist (TM) system, a_k is in the unit disk $a_k \in D, D = \{z \in C : |z| < 1\}$, C is the complex plane and z is the boundary of the plane.

a_k is chosen by the decomposed signal [5,6]. Maximal Projection Principle (MPP) is used here to select a_1 first, during each iteration, AFD needs an evaluator to maximize the energy restored. $e_{\{a\}}$ consists of elementary functions is used here to calculate energy gain during each iteration (3)

$$e_{\{a\}}(z) := B_{\{a\}}(z) = \frac{1}{\sqrt{2\pi}} \frac{\sqrt{1-|a|^2}}{1-\overline{a}z}, a \in D \tag{3}$$

For the first decomposition, set $G = G^+ = G_1$, a_1 is obtained by calculating the inner product of G_1 and $e_{\{a_1\}}$ (4) (5).

$$< G, e_{\{a\}} > = \sqrt{2\pi}\sqrt{1-|a|^2} G(a), a \in D \tag{4}$$

$$\begin{aligned} a_1 &= \arg\max\{|< G_1, e_{\{a\}} >|^2, a \in D\} \\ &= \arg\max\{2\pi(1-|a|^2)|G_1(a)|^2, a \in D\} \end{aligned} \tag{5}$$

After the first step, G_1 is represented as (6):

$$G_1 = < G_1, e_{\{a_1\}} > e_{\{a_1\}} + R_1 \tag{6}$$

R_1 is the reminder between original signal and signal recovered from first decomposition (7). AFD trys to minimize every R in order to maximize the energy restored.

$$R_1 = G_2 \frac{z - a_1}{1 - \overline{a_1} z} \tag{7}$$

Combine formula (7) with (6) we get G_2 (8):

$$G_2 = (G_1 - <G_1, e_{\{a_1\}}> e_{\{a_1\}}) \frac{1 - \overline{a_1} z}{z - a_1} \tag{8}$$

By taking (6) and (7) into generalization, after N times decomposition, the original signal can be defined as (9):

$$G_1 = \sum_{k=1}^{N} <G_1, e_{\{a_1\}}> e_{\{a_1\}} + R_k$$

$$= \sum_{k=1}^{N} <G_1, e_{\{a_1\}}> B_{\{k\}} + G_{k+1} \frac{z - a_k}{1 - \overline{a_k} z} \tag{9}$$

Where every a_k is selected under MPP principle:

$$a_k = \arg \max \{ |<G_k, e_{\{a\}}>|^2, a \in D \}$$

$$= \arg\max \{ 2\pi (1 - |a|^2) |G_k(a)|^2, a \in D \}$$

By discarding the last remainder R_k, AFD can get the reconstructed signal G' (10):

$$G' = 2 \operatorname{Re}(G') - C_0 \tag{10}$$

C_0 here is the mean value of original signal. In other words, C_0 is the first Fourier coefficient of G.

AFD will stop either the algorithm has reached to the desired level or the energy difference has come to the accuracy ε (12).

$$\left\| G^+ \right\| - \sum_{k=1}^{N} C_k B_k < \varepsilon \tag{11}$$

In our case, we set a_1 to be zero and the nth parameter is calculated through the previous arguments.

2.2 Decomposition Comparison between AFD and FT

In Fourier transform which decomposes any signals into the same basic trigonometric functions, the entrance e^{ikt} for the Fourier expansion of a given signal may arrive late, which is an important part of the total energy. So the convergence is not so ideal [4].

In AFD, greedy algorithm has been applied. AFD decomposes the given signal into different mono-components [5,6]. The mono-components decomposed are selected based on the given signal by using Maximal Selection Principle (MSP). MSP means that for a given signal, the AFD algorithm starts with selecting a mono-component that is most close to the original signal in energy sense, which starts from the low frequency to high frequency. Then at each continuous selection, it applies the same energy principle to find each mono-component that draws near the remainder. It is the reason that the decomposition is said to be adaptive. The decomposition usually leads to fast convergence than what Fourier decomposition does [7].

2.3 The Algorithm of the Proposed Approach

The flowchart of the proposed algorithm is illustrated in Fig. 1.

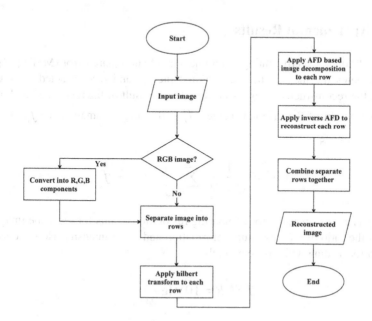

Fig. 1. The flowchart of the proposed method

An image can be regarded as different portions of signals according to its rows. So we disassemble the original image into different row and then apply AFD row by row. First we apply Hilbert transform to the rows to make the primitive function in a complex plane, and then handle it with AFD. The ultimate step is the combination of different rows into a whole image. The detailed algorithm is shown below. AFD code is released in Pro. Qian's Homepage:

```
http://www.fst.umac.mo/en/staff/documents/fsttq/afd_for
m/Index.html
```

Algorithm 1. Image decomposition with AFD
Input: Original image *img, m×n*, 8bits/pixel
Output: Approximated image output *New_image*
  1: Segment *img* into different rows.
  2: If the number of decomposition level is *N*
  3: for *i* = *1:m* **do**
  4: Apply hilbert transform to *i*-th row of *img*;
  5: for *k* = 1:*N* **do**
  6: decompose using AFD and get the *k*-th mono-component B_K
  7: **end for**
  8: Using reverse transform to get the reconstructed signal *img'*
  9: **end for**
  10: Combine *img'* for different rows together to get the result *New_image*

---

3   Experimental Results

I. The traditional objective rating of an image is Mean Square Error (MSE) [12]. MSE first calculates the mean square of original picture and reconstructed picture, then evaluate the reconstructed image according to the result of the results. M and N is the width and height of the original image, f_{ij} is the original image and f_{ij}' is the reconstructed image.

$$MSE = \frac{1}{M \times N} \sum_{0 \le i < N} \sum_{0 \le j < M} (f_{ij} - f_{ij}')^2$$

II. PSNR is a good measure for comparing restoration results for the same image [12]. PSNR is the ratio of the maximum information and noise intensity. Here we assign L with 255 for dealing with discrete pixels.

$$PSNR = 10 \times \lg \frac{L \times L}{MSE}$$

III. Contour Volume (CV) can tell the whether the image is clear or fuzzy, the bigger number indicates the good quality. First we use a 3×3 Laplacian window to extract the edges. Then count the pixels of edges, summing up the absolute value of the pixels and see the results.

IV. Other evaluation criteria are also included here to enhance our results: structural similarity (SSIM) [13], Shannon Entropy and Figure Definition.

3.1 Experiments with Grayscale Images

Fig. 2. 'Lena' reconstructed by Fourier and AFD. (a) Original image. (b) 50th decomposition by Fourier. (c) 50th decomposition by AFD.

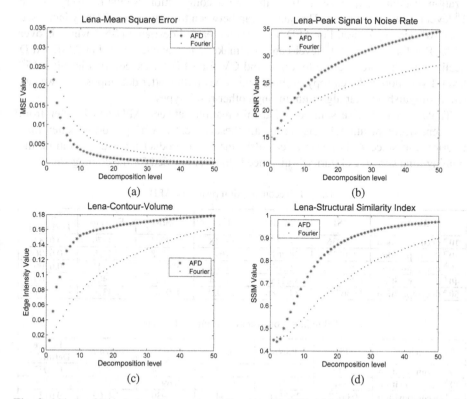

Fig. 3. 6 kinds of Discriminant parameter of Fourier and AFD. (a) Mean square error. (b) Peak signal to noise rate. (c) Contour volume. (d) Structural similarity index. (e) Shannon entropy. (f) Figure definition.

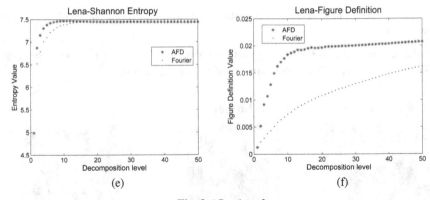

Fig. 3. (*Continued*)

6 kinds of experiments are conducted in this paper. The results are illustrated in Fig. 3. The classical 'Lena' (Fig. 3) with size 512×512, 8 bits/pixel is used here. Fig. 3 shows image quality assessment in terms of PSNR, MSE and CV from level 1 to level 50. For the first decomposition, they all start at the same point. AFD is a generalization of Fourier, a_1 equals to 0 for the first decomposition. So the recovery for the 1st level is always the same. The quality gap between these two significantly increased after 10th decomposition. For AFD, images can reach a desirable PSNR which is over 33db. But for Fourier, 50th level can only make it just above 25db. For MSE, AFD declines quickly pretty close to x axis and CV for AFD far exceed Fourier after 15th level. Even though the entropy of these two varies in the latter decomposition, we can still distinguish a better algorithm through other result types.

Table 1. and 2. give some detailed information between AFD and Fourier from different decomposition levels. From all these aspects, AFD performs better than Fourier. But since AFD needs to calculate the inner product for every iteration, the running time surpasses Fourier, which needs to be simplified for the coming study.

Table 1. Reconstruction results of AFD

AFD	MSE	PSNR	CV	SSIM	Entropy	Figure Definition
10th decomposition	0.0036	24.4793	0.1518	0.7044	7.4670	0.0183
20th decomposition	0.0013	28.7611	0.1641	0.8718	7.4479	0.0196
30th decomposition	7.2924e-04	31.3634	0.1707	0.9331	7.4441	0.0200
40th decomposition	4.7559e-04	33.2158	0.1754	0.9607	7.4407	0.0203
50th decomposition	3.4367e-04	34.6386	0.17925	0.9748	7.4379	0.0207

Table 2. Reconstruction results of Fourier

Fourier	MSE	PSNR	CV	SSIM	Entropy	Figure Definition
10th decomposition	0.0087	20.5921	0.0773	0.5474	7.3845	0.0073
20th decomposition	0.0042	23.7185	0.1133	0.6906	7.4528	0.0108
30th decomposition	0.0027	25.7526	0.1345	0.7930	7.4709	0.0130
40th decomposition	0.0019	27.2116	0.1514	0.8574	7.4735	0.0148
50th decomposition	0.0014	28.5042	0.1630	0.9049	7.4711	0.0162

3.2 Experiments with Color Images

(a)   (b)   (c)

Fig. 4. Color image comparison of AFD and Fourier. (a) Original image (NO. 020). (b) 50th decomposition by Fourier. (c) 50th decomposition by AFD.

McGill Calibrated Colour Image Database is used here to test the algorithm. We apply 50th decomposition from five kinds of animal picture and see the results below.

Table 3. 50th decomposition of AFD

	MSE	PSNR	CV	SSIM	Entropy	Figure Definition
NO.0125	0.0011	29.5678	0.2390	0.9362	7.1684	0.0869
NO.0281	7.3252e-04	31.3518	0.1486	0.9641	7.6111	0.0497
NO.020	2.8835e-04	35.4008	0.0805	0.9141	7.2959	0.0266
NO.022	0.0040	24.0298	0.2201	0.8519	7.0567	0.0820
NO.037	0.0021	26.7717	0.1655	0.9067	6.9220	0.0601

Table 4. 50th decomposition of Fourier

	MSE	PSNR	CV	SSIM	Entropy	Figure Definition
NO.0125	0.0025	26.0155	0.2072	0.9029	7.1276	0.0683
NO.0281	0.0023	26.3689	0.1300	0.9442	7.5532	0.0383
NO.020	0.0030	25.2369	0.0758	0.8527	7.3815	0.0231
NO.022	0.0062	22.0857	0.1732	0.8017	7.0514	0.0559
NO.037	0.0034	24.6269	0.1325	0.8693	6.9458	0.0413

The table above also shows outstanding performance of AFD applied to color images. The PSNR value is not so ideal, but the real picture is fair enough for our human perception.

4   Conclusions

This paper presents the principle of the AFD based image decomposition and reconstruction approach. Our experiment results show that AFD converges faster than Fourier transform does. Discrete cosine transform (DCT) is the real part of the Fourier transform, which is the foundation of JPEG compression. Since DCT has proved its advantages in JPEG, AFD can also find its corresponding transform to replace DCT, which can be another promising method for lossy image compression.

Acknowledgement. This study is supported by multi-year research fund of UM: MYRG144(Y1-L2)-FST11-ZLM.

References

1. Fienup, J.R.: Reconstruction of an object from the modulus of its Fourier transform. Optics Letters 3(1), 27–29 (1978)
2. Matej, S., Fessler, J.A., Kazantsev, I.G.: Iterative tomographic image reconstruction using Fourier-based forward and back-projectors. IEEE Transactions on Medical Imaging 23(4), 401–412 (2004)
3. Tsai, D.M., Chiu, W.Y.: Motion detection using Fourier image reconstruction. Pattern Recognition Letters 29(16), 2145–2155 (2008)
4. Qian, T., Zhang, L.M., Li, Z.X.: Algorithm of adaptive Fourier decomposition. IEEE Transactions on Signal Processing 59(12), 5899–5906 (2011)
5. Qian, T., Wang, Y.B., Dang, P.: Adaptive decomposition into mono-components. Advances in Adaptive Data Analysis 1(04), 703–709 (2009)
6. Qian, T.: Mono-components for decomposition of signals. Mathematical Methods in the Applied Sciences 29(10), 1187–1198 (2006)
7. Qian, T., Li, H., Stessin, M.: Comparison of adaptive mono-component decompositions. Nonlinear Analysis: Real World Applications 14(2), 1055–1074 (2013)
8. Qian, T., Wang, Y.B.: Adaptive Fourier series—a variation of greedy algorithm. Advances in Computational Mathematics 34(3), 279–293 (2011)
9. Zhang, L.M.: Adaptive Fourier Decomposition Based Signal Instantaneous Frequency Computation Approach. World Academy of Science, Engineering and Technology, International Science Index 68 6(8), 1946–1951 (2012)
10. Zhang, L.M.: A New Time-Frequency Speech Analysis Approach Based On Adaptive Fourier Decomposition. World Academy of Science, Engineering and Technology, International Science Index 79 7(7), 1782–1786 (2013)
11. Zhang, L.M.: Adaptive Fourier Decomposition Based Time-Frequency Analysis. Journal of Electronic Science and Technology of China (2), 201–205 (2014), doi:10.3969/j.issn.1674-862X.2014.02.012
12. Wang, Z., Bovik, A.C.: A universal image quality index. IEEE Signal Processing Letters 9(3), 81–84 (2002)
13. Channappayya, S.S., Bovik, A.C., Heath, R.W.: Rate bounds on SSIM index of quantized images. IEEE Transactions on Image Processing 17(9), 1624–1639 (2008)

Graph-Based Browsing for Large Video Collections

Kai Uwe Barthel, Nico Hezel, and Radek Mackowiak

HTW Berlin, University of Applied Sciences – Visual Computing Group
Wilhelminenhofstraße 75a, 12459 Berlin, Germany
barthel@htw-berlin.de

Abstract. We present a graph-based browsing system for visually searching video clips in large collections. It is an extension of a previously proposed system *ImageMap* which allows visual browsing in millions of images using a hierarchical pyramid structure of images sorted by their similarities. Image subsets can be explored through a viewport at different pyramid levels, however, due to the underlying 2D-organization the high dimensional relationships between all images could not be represented. In order to preserve the complex inter-image relationships we propose to use a hierarchical graph where edges connect related images. By traversing this graph the users may navigate to other similar images. Different visualization and navigation modes are available. Various filters and search tools such as search by example, color, or sketch may be applied. These tools help to narrow down the amount of video frames to be inspected or to direct the view to regions of the graph where matching frames are located.

Keywords: Content-based Video Retrieval, Exploration, Image Browsing, Visualization, Navigation.

1 Introduction

The task of the Video Search Showcase 2015 (VSS2015) is to quickly find short video clips. The data set will consist of about 100 hours of video content of various BBC programming. The search has to be performed visually without text-queries, initial query images are not available, OCR is not allowed. VSS2015 consists of two search categories, visual KIS (known item search): the moderator presents a short video clip, and descriptive KIS: the moderator presents a textual description about a specific segment. The video segment randomly selected from the entire video collection has to be found within 3 minutes.

The past Video Browser Showdown has shown the importance of good visual search capabilities in combination with an easy to use interface allowing for fast inspection of the retrieved video clips [1]. Especially for descriptive KIS it is important to use an approach allowing to view many video frames at the same time without losing overview and to support the user in fast browsing through large collections.

As human perception is limited to 20 to 50 unsorted images at a time, overview is quickly lost if too many images are shown. If however the images are sorted by their visual similarities, then several hundreds of images can be viewed simultaneously [2].

X. He et al. (Eds.): MMM 2015, Part II, LNCS 8936, pp. 237–242, 2015.

In [3] we have proposed *ImageMap,* a system for fast visual browsing through huge image sets using a hierarchical pyramid where the images are positioned according to their visual and semantic similarities. Although this approach often works very well, one of the major drawbacks is the problem that the complex relationships between all images cannot be preserved with a 2D-projection. This loss of information about the relationship can result in similar images being separated and close borders of unrelated images. In this paper we propose an extension of our previous approach maintaining the easy visual browsing capabilities without losing the relationships between the images. The image relationships are preserved by representing them with a hierarchical set of high dimensional graphs where edges connect related images. The lowest level graph connects very similar images whereas connected images on higher levels are still related but less similar. Visual browsing consists in traversing this graph or in changing the level of the graph.

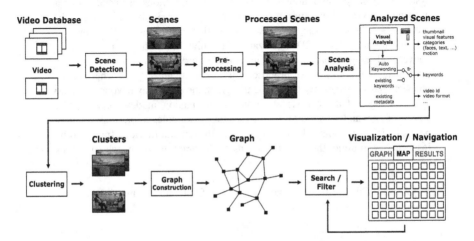

Fig. 1. Architecture of the video search system

2   Building a Graph for Large Video Collections

While there has been a lot of effort to improve CBIR, there is not much support for exploratory image or video search. Our graph-based video browser unifies search and exploration in one combined approach.

Figure 1 shows the architecture of our proposed scheme. First we detect the different scenes of a video clip. In the next step these scenes are preprocessed to remove mattes and to perform a basic color correction for frames with very low quality. The scene analysis extracts all available information from that scene. Since using still images instead of videos is more effective for video exploration [4], we determine the most representative key frame for each scene. The visual appearance is described with a low-level visual feature vector including color histogram and distribution, texture and edge information. For filtering purposes different analysis steps are performed to determine high-level categories such as number of faces, is text displayed, camera motion, audio type, etc.

We use automatic keywording to enrich the frames with high-level semantics. By means of a huge collection of annotated images we try to propagate the keywords from known images to the unknown frames. Although these keywords are not always correct they still help to better model the image similarities.

After the frames have been analyzed we cluster very similar frames as they might occur in interview situations or news programs. The clusters are represented by one single image, which is visually marked to indicate that it represents several frames.

The next step consists of building a hierarchical image graph by connecting similar images, allowing other related images to be reached by navigating the edges of the graph. We examined different graph construction techniques. An easy approach would connect all images with similarity above a threshold. However this leads to disconnected, very unbalanced and irregular graphs. In order to allow an easy visualization and navigation it is desirable to have a regular graph with a constant number of connections (edges) per image. We build the graph in two steps. First we perform a two dimensional sorting of all images. This 2D-sorting serves as initial graph which is optimized in the next step by swapping edges.

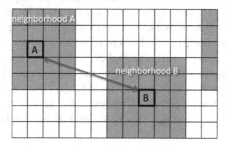

Fig. 2. Swapping of images depending on the neighborhood similarities

The 2D-sorting can be achieved using a self-organizing map, however we obtained better results with a discrete optimization process that involves swapping of images depending on their similarities, which are computed by fusing visual and keyword similarities. Initially all images are randomly positioned on a rectangular grid. Then for randomly chosen image pairs the similarities between these images and their neighboring images are computed. Two images are swapped if the total similarity of the swapped positions is greater than the initial similarity: $Sim_{AB} + Sim_{BA} > Sim_{AA} + Sim_{BB}$ (Figure 2). We start with a large neighborhood radius which is gradually reduced. The sorting process is stopped when the similarity improvements get small.

After the 2D-sorting is finished, the positions of the images serve as initial graph such that every image is connected with its four adjacent neighbors (figure 3 left). Then again we perform a random swapping optimization. However this time we swap edges instead of images. We swap two non-touching edges if the sum of the similarities increases with a swap, i.e. if $sim_{AX} + sim_{EY} > sim_{AE} + sim_{XY}$ (figure 3 center). With each swap the total similarity of the graph (the sum of the similarities of all edges) increases. To speed up the swapping procedure we focus on edges with low similarity values. Again, swapping is stopped when the improvements get too small.

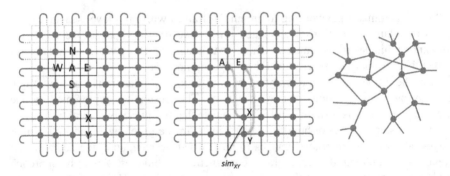

Fig. 3. Left: the initial graph is based on the image position of the 2D-sorting, center: edges are swapped if this increases the total similarity, right: detail of an example of a final graph

The last step of the graph building process consists of generating a hierarchical version of this graph. For very similar images it is not necessary to display them all, one image of a higher level can represent the lower level images. Reducing the graph is achieved by successively removing vertices (images) from the graph that are very similar to their connected neighbors. After removing an image (and the four corresponding edges) there are three possibilities to reconnect the disconnected four images. We choose those connections with the highest similarities (see figure 4). For each hierarchy level we remove ¾ of the images as this corresponds to a resolution reduction by a factor of two in each direction in the 2D case. The total number of hierarchy levels depends on the total number of images. We aim at having a few hundred images at the highest level, because this is the maximum number which still can be perceived by the user.

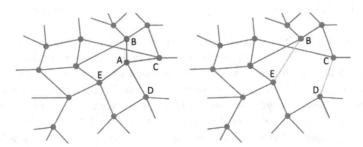

Fig. 4. A reduced version of the graph is generated by successively removing images with high edge similarities. If image A is removed we reconnect the images B, C, D, and E with two new edges depending on the maximum of $sim_{BC} + sim_{DE}$, $sim_{BD} + sim_{CE}$ and $sim_{BE} + sim_{CD}$.

3   Graph Visualization and Navigation

In order to visually browse and search video clips different use cases have to be distinguished. There are situations where sequential lists make most sense. This is true for viewing all key frames of one clip in temporal order. Also the results of a visual query often should be displayed as a list, with the most similar results at the top. In some situations if there are too many result images or too many key frames in a single video then visual sorting helps to get a better overview. Last year winner [5] showed the key frames of five scenes per page and query. If there are many similar scenes the user might has to scroll several pages further.

Visual exploration however works best by navigating the hierarchical graph. Compared to a pure 2D-sorting the images are much better connected in the graph, however there is no obvious way to display the graph in an intuitive way that can easily be navigated. In an effort to achieve this the graph-neighborhood of the actual image is transformed to a minimal spanning tree which then is projected onto a 2D plane for display. We have developed two different display modes, between which the user might switch according to his needs:

Fractal tree map: When traversing the graph, each image will always have four connected images: the previous image and three successors with its corresponding subtrees. Using the arrow keys the user may quickly choose which subtree to pursue (center, left, right or back to the previous view) (figure 5). In cases where images have edges to an image that is already displayed, we choose other images by going further in the graph. This avoids having repeated subtrees in the display. The fractal tree map allows a fast navigation within the graph. However one problem is the different sizes of the images. 120 images can be displayed with four rows of images.

Sorted graph projection: To be able to display more images with constant size, we have developed another display mode. Again we take the neighborhood of the graph and project it on a 2D-display. The considered image is centered and the images of the four subtrees are visually sorted around it. By clicking an image the user may select a new center image. This will lead to the fact that some images remain visible, new related images will be loaded while others disappear. We animate the transition from the old view to the new view such that the user does not get confused.

Fig. 5. Fractal tree map display. Traversing the graph, each image has three successors. The user may quickly navigate (choose the subtree) with the arrow keys. In this example the left direction was chosen.

4   Using the Graph-Based Video Browser

The current user interface shown in Figure 6 allows the user to search by color, sketch or example images and category filters. Different searches can be combined in a search sequence, to find scenes where visual appearance of multiple frames and there order are known. The previous and upcoming frames for a selected image are shown at the bottom, next to a video player. To identify possible scene boundaries two smaller frames are displayed above the player showing time shifted (earlier and later) versions of the same video.

Fig. 6. User interface of the proposed prototype. Left: Browsing mode. Right: Search mode.

In case the content of a scene is described verbally, less or no visual information may be available, making a search nearly unfeasible. Usually the user still has an idea how the scene might look like. With the help of ImageMap and a fractal tree map display in browsing mode, it is possible to quickly navigate and check potential key frames. Both displays are connected to each other. While the 2D-based ImageMap gives a better overview of all key frames, the graph-based display has more degrees of freedom to show visually similar images, finding frames that might be placed at different locations on the 2D map.

References

1.   Schoeffmann, K., et al.: The Video Browser Showdown: A Live Evaluation of Interactive Video Search Tools. International Journal of Multimedia Information Retrieval (MMIR) 3(2), 113–127 (2014)
2.   Barthel, K.U.: Automatic Image Sorting using MPEG-7 Descriptors. In: ICOB 2005, Workshop on Immersive Communication and Broadcast Systems (2005)
3.   Barthel, K.U., Hezel, N., Mackowiak, R.: ImageMap - Visually Browsing Millions of Images. Submitted to 21st International Conference on Multimedia Modelling (2015)
4.   Haubold, A.: Indexing and browsing unstructured videos using visual, audio, textual, and facial cues. Doctoral Dissertation, Columbia University New York, NY (2008)
5.   Lokoč, J., Blažek, A., Skopal, T.: Signature-based video browser. In: Gurrin, C., Hopfgartner, F., Hurst, W., Johansen, H., Lee, H., O'Connor, N. (eds.) MMM 2014, Part II. LNCS, vol. 8326, pp. 415–418. Springer, Heidelberg (2014)

Enhanced Signature-Based Video Browser

Adam Blažek, Jakub Lokoč, Filip Matzner, and Tomáš Skopal

SIRET research group, Department of Software Engineering,
Faculty of Mathematics and Physics, Charles University in Prague
blazekada@gmail.com, {lokoc,skopal}@ksi.mff.cuni.cz, floop@floop.cz

Abstract. The success of our Signature-Based Video Browser presented last year at Video Browser Showdown 2014 (now renamed to Video Search Showcase) was mainly based on effective filtering using position-color feature signatures, while browsing in the results comprising matched keyframes was based just on a simple sequential search approach. Since the results can consist of highly similar keyframes (e.g., news studio scenes) making the browsing more difficult, we have enhanced our tool with more advanced browsing techniques considering also homogeneous result sets obtained after filtering phase. Furthermore, we have utilized improved search models based on feature signatures to make the filtering phase more effective.

1 Introduction

In order to find the most promising approaches for specific video retrieval tasks, events like Video Search Showcase (VSS) are organized to directly compare approaches implemented by various video retrieval tools. As demonstrated at VSS 2014, the diversity of the approaches used by the tools can be considerably high, without a clear winner outperforming all the other approaches in all the retrieval tasks. The retrieval tasks at VSS 2014 consisted of Known-Item Search tasks in a single video and video archive, where especially the video archive scope has remained challenging for all the competing approaches. Let us note, textual queries are not allowed at VSS and thus the participants are forced to develop innovative browsing interfaces, which makes the event even more interesting.

The teams competing at VSS 2014 employed various approaches that represent a decent overview of current state-of-the-art in the respective field. While some teams utilized automated high-level concept detection [6, 7, 11] which can compensate the not-provided annotation, other teams based their tools on suitable low-level visual descriptors [1, 2, 4]. Since detection techniques for high-level concepts (e.g. faces, objects) are still far from perfection and also the concepts that could be detected reliably simply may not occur at all in the videos of interest, we consider the high-level concept detection only as a supportive feature rather than a key-stone in our video retrieval approach.

During the past decades, a variety of low-level feature descriptors suitable for searching audiovisual content have emerged (e.g., based on dominant colors of video frames, position-color distribution descriptors [5] or detected camera/object motions [10]). In order to enable effective and efficient filtering, in

X. He et al. (Eds.): MMM 2015, Part II, LNCS 8936, pp. 243–248, 2015.
© Springer International Publishing Switzerland 2015

our approach, we utilize position-color feature signatures [8] to represent video frames as sets of positioned colored circles. We assume, the representation is more or less intuitive for users and thus the users can express their search intents (e.g., memorized color stimuli) in a form of simple sketches. An example of the feature signature for a video frame and a simple sketch-drawing tool is depicted in Figure 1. We may observe that a feature signature can be considered as an intuitive rough approximation of the content of the video frame and also the frames represented by feature signatures can be easily queried using simple sketches comprising a few drawn colored circles.

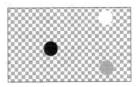

Fig. 1. a) A video frame and the visualization of its feature signature. b) A simple sketch-drawing tool.

Such simple sketch-based filtering approach was already employed in our Signature-Based Video Browser [4] (SBVB) participating at VSS 2014. Along with the demonstrated effectiveness of the approach, the utilized 5-dimensional feature space comprising just position (x, y) and color (L, a, b) information[1] enabled efficient grid-like indexing guaranteeing real-time responses of our system. The tool also enabled specification of a second sketch for filtering results obtained by the first sketch, which we found useful when searching within the archive scope. When retrieving the results, the video key-frames were ranked according to the Euclidean distance from the currently drawn sketch circles (for more details about the retrieval model see [3]). The best matching key-frames were presented together with a limited number of the following and preceding key-frames from the video (Fig. 2 – left).

The results of VSS 2014 confirmed the potential of feature signature-based video retrieval model — SBVB was able to compete with and even outperform other state-of-the-art browsing techniques even though SBVB used just simple browsing capabilities. However, we have also noted many limiting factors in both the filtering and browsing part of the tool, and thus in this paper we provide its enhanced version. At first, we describe the incorporated preprocessing steps in Section 2. The enhancements of the filtering including adjustment of the metric function are discussed in Section 3 and finally, we enumerate the novel features supporting the browsing capabilities in Section 4 and conclude the paper.

---

[1] The weights of the *centroids* were not utilized.

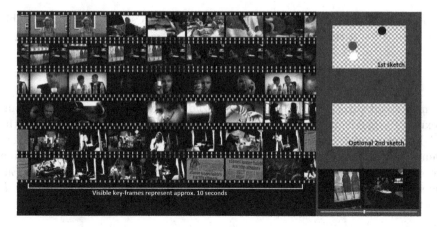

Fig. 2. A screenshot of the original SBVB. Matched key-frames (marked with red dots) are presented together with the preceding and following key-frames.

2 Preprocessing

The preprocessing phase starts with uniform sampling of the video frames[2], where the extraction of feature signatures is the same as in our previous version. More specifically, given a video frame, all pixels are transformed into a 5-dimensional position-color feature space (x, y, L, a, b) and an adaptive variant of k-means clustering is performed resulting in a set of k clusters, where k depends on the complexity of the video frame. The cluster centers – *centroids* (i.e., positioned colored circles) form the feature signature and hopefully match the centers of distinct color regions of the video frame. Each *centroid* also has a weight w that denotes the number of pixels assigned to the *centroid* (see Figure 1). Beside feature signature extraction, we perform additional tasks in the preprocessing phase.

It is common practice to detect the shot boundaries in video and utilize this information in both the retrieval model and browsing framework. Although there are quite accurate techniques for shot boundaries detection (SBD) available, we have employed a simple threshold-based technique using feature signatures similarity model[3] and thus re-using the already extracted descriptors. Having the sampled video key-frames and their feature signatures, the distances between the feature signatures of consecutive key-frames are calculated. The shot boundaries are estimated using these distances and a threshold.

In an extra step, we also estimate the 5 most dominant colors for each key-frame. Moreover, our tool supports plug-in like extensions possibly enriching the

---

[2] In practice, one second of video results in approx. two key-frames.
[3] The performance of this technique might not be optimal in diverse video content and also the threshold must be estimated manually; however, the precision of SBD is not crucial in our approach.

preprocessing with concept detectors leaving users to select the most suitable ones ad-hoc for particular videos.

3   Enhanced Filtering

As mentioned earlier, the number of *centroids* varies with respect to the complexity of a particular key-frame (demonstrated in Fig. 3). Thus, omitting the weights(radii) of the *centroids* introduces a noticeable handicap for the key-frames with fewer *centroids* which are more likely to be spatially farther from the defined sketch circles. For this reason, we have replaced the original Euclidean distance with a distance function utilizing the weights of the *centroids* and thus correcting the discussed bias.

Fig. 3. The number of *centroids* in a particular feature signature reflects the complexity of the respective key-frame

Another problem is introduced along with the dense key-frame sampling. Consecutive key-frames happen to be quite similar and therefore receiving similar rankings and creating a cluster of matches. To avoid the redundancy in the results, our approach selects only the best match within each such cluster. The cluster boundaries are estimated so that the initially displayed key-frames in the results do not overlap and is further extended to the detected shot boundaries.

Furthermore, users are now enabled to specify the time interval between the searched scenes in the case of a two-sketch query. For example, when we search for a scene followed immediately by another one, setting the time interval to 0-1 seconds can improve the filtering effectiveness and thus provide more relevant results and also lower the response time (compared to a larger time interval).

4   Enhanced Browsing

During the VSS 2014, practically all the teams experienced difficulties when dealing with a large amount of very similar scenes (e.g., TV studio scenes). Since these scenes can be discriminated by their context (different following and preceding scenes), it is desirable to include this context in the results so that the user can quickly identify the correct one without the need of any further investigation. To maximize the chance that the context will be visible right away, we utilize the detected shot boundaries and display the key-frames in a more compact way (see Figure 4). Within each shot, only one key-frame is

Fig. 4. A screenshot of the enhanced SBVB. Note that we also removed the video player which seemed to be unnecessary.

fully visible while others are significantly cropped. Moving the cursor over the key-frames moves the "zoomed" area adequately.

Furthermore, a row representing one matched key-frame (or one pair of matched key-frames in the case of a two-sketch query) can be moved freely in the usual drag-drop manner in order to explore the wider neighborhood.

Any result can be further examined in the detail row (Fig. 4 - bottom) enriched with an interactive navigation summary [9] of the respective video. The summary captures the extracted dominant colors of the key-frames, highlights the displayed chunk of the video and marks the positions of the rest of the best matches. This gives users an overview of the searched videos and the ability to quickly investigate the promising video regions.

We have also observed that the sketches may unintentionally match several key-frames, clearly not the searched ones, which are being retrieved repetitively between the best matches. This phenomenon obviously prolongs the search process as users are forced to investigate more results. Therefore, we enable users to exclude any retrieved scene (and any other similar enough) from further displaying. In addition, a whole video can be excluded from the search or a user can restrict herself to the most promising video when searching within a video archive.

5 Conclusion

In this paper, we have presented an enhanced Signature-based Video Browser. The tool allows effective yet efficient filtering of video content based on position-color feature signatures. With improved browsing capabilities, SBVB helps to rapidly identify the searched scene even among highly similar content.

248   A. Blažek et al.

Acknowledgments. This research has been supported in part by Czech Science Foundation projects P202/11/0968, P202/12/P297 and SVV project 2014-260100.

References

[1] Bailer, W., Weiss, W., Schober, C., Thallinger, G.: Browsing linked video collections for media production. In: Gurrin, C., Hopfgartner, F., Hurst, W., Johansen, H., Lee, H., O'Connor, N. (eds.) MMM 2014, Part II. LNCS, vol. 8326, pp. 407–410. Springer, Heidelberg (2014)

[2] Cobârzan, C., Hudelist, M.A., Del Fabro, M.: Content-based video browsing with collaborating mobile clients. In: Gurrin, C., Hopfgartner, F., Hurst, W., Johansen, H., Lee, H., O'Connor, N. (eds.) MMM 2014, Part II. LNCS, vol. 8326, pp. 402–406. Springer, Heidelberg (2014)

[3] Loko, J., Blaek, A., Skopal, T.: On effective known item video search using feature signatures. In: ICMR, p. 524 (2014)

[4] Lokoč, J., Blažek, A., Skopal, T.: Signature-based video browser. In: Gurrin, C., Hopfgartner, F., Hurst, W., Johansen, H., Lee, H., O'Connor, N. (eds.) MMM 2014, Part II. LNCS, vol. 8326, pp. 415–418. Springer, Heidelberg (2014)

[5] Manjunath, B.S., Ohm, J.-R., Vasudevan, V.V., Yamada, A.: Color and texture descriptors. IEEE Transactions on Circuits and Systems for Video Technology 11(6), 703–715 (2001)

[6] Moumtzidou, A., et al.: VERGE: An interactive search engine for browsing video collections. In: Gurrin, C., Hopfgartner, F., Hurst, W., Johansen, H., Lee, H., O'Connor, N. (eds.) MMM 2014, Part II. LNCS, vol. 8326, pp. 411–414. Springer, Heidelberg (2014)

[7] Ngo, T.D., Nguyen, V.H., Lam, V., Phan, S., Le, D.-D., Duong, D.A., Satoh, S.: NII-UIT: A tool for known item search by sequential pattern filtering. In: Gurrin, C., Hopfgartner, F., Hurst, W., Johansen, H., Lee, H., O'Connor, N. (eds.) MMM 2014, Part II. LNCS, vol. 8326, pp. 419–422. Springer, Heidelberg (2014)

[8] Rubner, Y., Tomasi, C.: Perceptual metrics for image database navigation, vol. 1. Springer (2000)

[9] Schoeffmann, K., Boeszoermenyi, L.: Video browsing using interactive navigation summaries. In: Seventh International Workshop on Content-Based Multimedia Indexing, CBMI 2009, pp. 243–248 (June 2009)

[10] Schoeffmann, K., Lux, M., Taschwer, M., Boeszoermenyi, L.: Visualization of video motion in context of video browsing. In: IEEE International Conference on Multimedia and Expo, ICME 2009, pp. 658–661 (June 2009)

[11] Scott, D., et al.: Audio-visual classification video browser. In: Gurrin, C., Hopfgartner, F., Hurst, W., Johansen, H., Lee, H., O'Connor, N. (eds.) MMM 2014, Part II. LNCS, vol. 8326, pp. 398–401. Springer, Heidelberg (2014)

VERGE: A Multimodal Interactive Video Search Engine

Anastasia Moumtzidou[1], Konstantinos Avgerinakis[1], Evlampios Apostolidis[1],
Fotini Markatopoulou[1,2], Konstantinos Apostolidis[1], Theodoros Mironidis[1],
Stefanos Vrochidis[1], Vasileios Mezaris[1], Ioannis Kompatsiaris[1], and Ioannis Patras[2]

[1] Information Technologies Institute/Centre for Research and Technology Hellas,
6th Km. Charilaou - Thermi Road, 57001 Thermi-Thessaloniki, Greece
{moumtzid,koafgeri,apostolid,markatopoulou,kapost,mironidis,
stefanos,bmezaris,ikom}@iti.gr
[2] School of Electronic Engineering and Computer Science, QMUL, UK
i.patras@eecs.qmul.ac.uk

Abstract. This paper presents VERGE interactive video retrieval engine, which is capable of searching into video content. The system integrates several content-based analysis and retrieval modules such as video shot boundary detection, concept detection, clustering and visual similarity search.

1 Introduction

This paper describes VERGE interactive video search engine[1], which is capable of retrieving and browsing video collections by integrating multimodal indexing and retrieval modules. VERGE supports Known Item Search task, which requires the incorporation of browsing, exploration, or navigation capabilities in video collection.

Evaluation of earlier versions of VERGE search engine was performed with participation in video retrieval related conferences and showcases such as TRECVID, VideOlympics and Video Browser Showdown (VBS). Specifically, ITI-CERTH participated in the TRECVID Search tasks in 2006, 2007, 2008, and 2009, in the Known Item Search (KIS) task in 2010, 2011, and 2012, in the Instance Search (INS) task in 2010, 2011, 2013 and 2014 and in the VideOlympics event in 2007, 2008 and 2009. VERGE has also participated in VBS 2014. The proposed version of VERGE aims at participating to the KIS task of the Video Search Showcase (VSS) Competition 2015 which was formerly known as Video Browser Showdown [1].

2 Video Retrieval System

VERGE is an interactive retrieval system that combines advanced retrieval functionalities with a user-friendly interface, and supports the submission of queries and the accumulation of relevant retrieval results. The following indexing and retrieval

---

[1] More information and demos of VERGE are available at: http://mklab.iti.gr/verge/

X. He et al. (Eds.): MMM 2015, Part II, LNCS 8936, pp. 249–254, 2015.

modules are integrated in the developed search application: a) Visual Similarity Search Module; b) High Level Concept Detection; and c) Hierarchical Clustering.

The aforementioned modules allow the user to search through a collection of images and/or video keyframes. However, in the case of a video collection, it is essential that the videos are pre-processed in order to be indexed in smaller segments and semantic information should be extracted. The modules that are applied for segmenting videos are: a) Shot Segmentation; and b) Scene Segmentation;

Thus, the general framework realized by VERGE in case of video collection is depicted in Figure 1. This framework contains all the aforementioned segmenting and indexing modules.

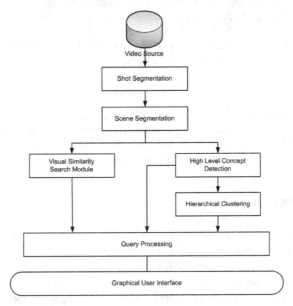

Fig. 1. Framework of VERGE

2.1   Shot Segmentation

The video temporal decomposition module defines the shot segments of the video, i.e., video fragments composed by consecutive frames captured uninterruptedly from a single camera, based on a variation of the algorithm proposed in [2]. The utilized technique represents the visual content of the each frame by extracting an HSV histogram and a set of ORB (Oriented FAST and Rotated BRIEF) descriptors (introduced in [3]), being able to detect the differences between a pair of frames, both in color distribution and at a more fine-grained structure level. Then both abrupt and gradual transitions are detected by quantifying the change in the content of successive or neighboring frames of the video, and comparing it against experimentally specified thresholds that indicate the existence of abrupt and gradual shot transitions. Erroneously detected abrupt transitions are removed by applying a flash detector, while false alarms are filtered out after re-evaluating the defined gradual transitions with the

help of a dissolve and a wipe detector that rely on the algorithms introduced in [4] and
[5] respectively. Finally, a simple fusion approach (i.e. taking the union of the detected abrupt and gradual transitions) is used for forming the output of the algorithm.

2.2   Scene Segmentation

Drawing input from the analysis in section 2.1, the scene segmentation algorithm of
[6] defines the story-telling parts of the video, i.e., temporal segments covering either
a single event or several related events taking place in parallel, by grouping shots into
sets that correspond to individual scenes of the video. For this, content similarity (i.e.,
visual similarity assessed by comparing HSV histograms extracted from the keyframes of each shot) and temporal consistency among shots are jointly considered
during the grouping of the shots into scenes, with the help of two extensions of the
well-known Scene Transition Graph (STG) algorithm [7]. The first one reduces the
computational cost of STG-based shot grouping by considering shot linking transitivi-
ty and the fact that scenes are by definition convex sets of shots, while the second one
builds on the former to construct a probabilistic framework that alleviates the need for
manual STG parameter selection. Based on these extensions, and as reported in [6],
the employed technique can identify the scene-level structure of videos belonging to
different genres, and provide results that match well the human expectations, while
the needed time for processing is a very small fraction (<3%) of the video's duration.

2.3   Visual Similarity Search

The visual similarity search module performs content-based retrieval based on global
and local information. To deal with global information, MPEG-7 descriptors are extracted from each keyframe and they are concatenated into a single feature vector.
More specifically, the colour related descriptors Colour Structure (CS), Colour Layout (CL) and Scalable Colour (SC) are used. Regarding the case of local information,
SURF features are extracted. Then K-Means clustering is applied on the database
vectors in order to acquire the visual vocabulary and VLAD encoding for representing
images is realized [8].

For Nearest Neighbour search we implement three different approaches between the
query and database vectors that is described in [8]. In each case, an index is first constructed for database vectors and K-Nearest Neighbours are then computed from the
query file. An index of lower-dimensional PCA-projected VLAD vectors, an ADC index
and an IVFADC index were constructed from the database vectors as in [8]. Exhaustive
search is deployed in the first two cases using a Symmetric Distance Computation (SDC)
and Asymmetric Distance Computation (ADC) for Nearest Neighbour calculation, while
a faster solution is suggested in the third one, where an inverted file system is combined
with ADC instead. Based on the experiments realized in [9], the approach that performs
the best is IVFADC. Therefore, in our implementation, we will apply this approach but
we are going to investigate the other two methods as well. It should be noted that this
indexing structure is utilized for both descriptors (i.e. global and local). Finally, a web
service is implemented in order to accelerate the querying process.

2.4   High Level Concepts Retrieval Module

This module indexes the video shots based on 346 high level concepts (e.g. water, aircraft). To build concept detectors a two-layer concept detection system is employed. The first layer builds multiple independent concept detectors. The video stream is initially sampled, generating for instance one keyframe per shot by shot segmentation. Subsequently, each sample is represented using one or more types of appropriate local descriptors (e.g. SIFT, RGB-SIFT, SURF, ORB etc.). The descriptors are extracted in more than one square regions at different scale levels. All the local descriptors are compacted using PCA and are subsequently aggregated using the VLAD encoding. These VLAD vectors are compressed by applying a modification of the random projection matrix [10] and served as input to Logistic Regression (LR) classifiers. Following the bagging methodology of [11] five LR classifiers are trained per concept and per local descriptor (SIFT, RGB-SIFT, SURF, ORB etc.), and their output is combined by means of late fusion (averaging). When different descriptors are combined, again late fusion is performed by averaging of the classifier output scores. In the second layer of the stacking architecture, the fused scores from the first layer are aggregated in model vectors and refined by two different approaches. The first approach uses a multi-label learning algorithm that incorporates concept correlations [12]. The second approach is a temporal re-ranking method that re-evaluates the detection scores based on video segments as proposed in [13].

2.5   Hierarchical Clustering

This module incorporates a generalized agglomerative hierarchical clustering process [14], which provides a structured hierarchical view of the video keyframes. In addition to the feature vectors described in section 2.3, we extract vectors consisting of the responses of the concept detectors for each video shot. The hierarchical clustering is applied to these representations to cluster the keyframes into classes, each of which consists of keyframes of similar content, in line with the concepts provided.

3   VERGE Interface and Interaction Modes

The modules described in section 2 are incorporated into a friendly user interface (Figure 2) in order to aid the user to interact with the system, discover and retrieve the desired video clip. The existence of a friendly and smartly designed graphical interface (GUI) plays a vital role in the procedure. Within this context, the GUI of VERGE has been redesigned in order to improve the user experience. The new interface, similarly to older one, comprises of three main components: a) the central component, b) the left side, c) the lower panel. We have incorporated the aforementioned modules inside these components, in order to allow the user interact with the system and retrieve the desired video clip during known item search tasks. In the sequel, we describe briefly the three main components of the VERGE system and then present a simple usage scenario.

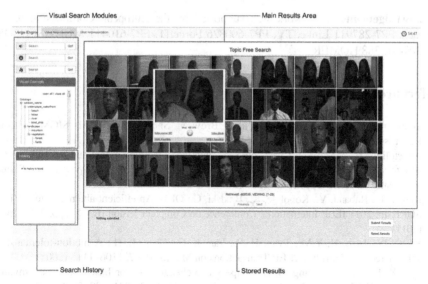

Fig. 2. Screenshot of VERGE video retrieval engine

The central component of the interface includes a shot-based or scene-based representation of the video in a grid-like interface. When the user hovers over a shot keyframe, the shot preview is visible by rolling three to five different keyframes that constitute the shot. Moreover, when the user clicks on a shot a pop-up frame appears that contains a larger preview of the image and several links that support her in viewing adjacent shots or all video shots, the frames constituting the shot and in searching for visually similar images. On the left side of the interface, the search history, as well as additional search and browsing options (that include the high level visual concepts and the hierarchical clustering) are displayed. Finally, the lower panel is a storage structure that holds the shots selected by the user.

Regarding the usage scenario for the known-item task, we suppose that a user is interested in finding a clip where 'a man hugs a woman while both of them have dark-skin' (Figure 2). Given that there is a high level concept called "dark-skinned people", the user can initiate her search from it. Then, she can either use the visual similarity module if a relative image is retrieved during the first step or if she finds an image that possibly matches the query; she can browse the temporally adjacent shots and retrieve the desired clip. Finally the user can store the desirable shots in a basket structure.

4   Future Work

Future work includes fusion of high level visual concepts in order to allow for retrieval of video shots that can be described equally with more than one concept. Another feature that could be implemented is the capability of querying the video collection with one or more colors found in specific place of the shot. However, this requires knowledge of the specific location of the color in the image.

Acknowledgements. This work was supported by the European Commission under contracts FP7-287911 LinkedTV, FP7-600826 ForgetIT, FP7-610411 MULTISENSOR and FP7-312388 HOMER.

References

1. Schoeffmann, K., Bailer, W.: Video Browser Showdown. ACM SIGMultimedia Records 4(2), 1–2 (2012)
2. Apostolidis, E., Mezaris, V.: Fast shot segmentation combining global and local visual descriptors. In: 2014 IEEE International Conference on Acoustics, Speech and Signal Processing (ICASSP), pp. 6583–6587 (2014)
3. Rublee, E., Rabaud, V., Konolige, K., Bradski, G.: ORB: An efficient alternative to SIFT or SURF. In: 2011 IEEE International Conference on Computer Vision (ICCV), pp. 2564–2571 (2011)
4. Su, C.-W., Liao, H.-Y.M., Tyan, H.-R., Fan, K.-C., Chen, L.-H.: A motion-tolerant dissolve detection algorithm. IEEE Transactions on Multimedia 7, 1106–1113 (2005)
5. Seo, K.-D., Park, S., Jung, S.-H.: Wipe scene-change detector based on visual rhythm spectrum. IEEE Transactions on Consumer Electronics 55(2), 831–838 (2009)
6. Sidiropoulos, P., Mezaris, V., Kompatsiaris, I., Meinedo, H., Bugalho, M., Trancoso, I.: Temporal video segmentation to scenes using high-level audiovisual features. IEEE Transactions on Circuits and Systems for Video Technology 21(8), 1163–1177 (2011)
7. Yeung, M., Yeo, B.-L., Liu, B.: Segmentation of video by clustering and graph analysis. Computer Vision and Image Understanding 71(1), 94–109 (1998)
8. Jegou, H., Douze, M., Schmid, C., Perez, P.: Aggregating local descriptors into a compact image representation. In: Proc. CVPR (2010)
9. Jegou, H., Douze, M., Schmid, C.: Product quantization for nearest neighbor search. IEEE Transactions on Pattern Analysis and Machine Intelligence 33, 117–128 (2011)
10. Mandasari, M.I., McLaren, M., van Leeuwen, D.A., Bingham, E., Mannila, H.: Random projection in dimensionality reduction: Applications to image and text data. In: 7th ACM SIGKDD Int. Conf. on Knowledge Discovery and Data Mining, pp. 245–250 (2001)
11. Markatopoulou, F., Moumtzidou, A., Tzelepis, C., Avgerinakis, K., Gkalelis, N., Vrochidis, S., Mezaris, V., Kompatsiaris, I.: ITI-CERTH participation to TRECVID 2013. In: TRECVID 2013 Workshop, Gaithersburg, MD, USA (2013)
12. Markatopoulou, F., Mezaris, V., Kompatsiaris, I.: A comparative study on the use of multi-label classification techniques for concept-based video indexing and annotation. In: Gurrin, C., Hopfgartner, F., Hurst, W., Johansen, H., Lee, H., O'Connor, N. (eds.) MMM 2014, Part I. LNCS, vol. 8325, pp. 1–12. Springer, Heidelberg (2014)
13. Safadi, B., Quénot, G.: Re-ranking by local re-scoring for video indexing and retrieval. In: 20th ACM Int. Conf. on Information and Knowledge Management, pp. 2081–2084 (2011)
14. Johnson, S.C.: Hierarchical Clustering Schemes. Psychometrika 2, 241–254 (1967)

IMOTION — A Content-Based Video Retrieval Engine

Luca Rossetto[1], Ivan Giangreco[1], Heiko Schuldt[1], Stéphane Dupont[2],
Omar Seddati[2], Metin Sezgin[3], and Yusuf Sahillioğlu[3]

[1] Databases and Information Systems Research Group,
Department of Mathematics and Computer Science, University of Basel, Switzerland
{firstname.lastname}@unibas.ch
[2] Research Center in Information Technologies, Université de Mons, Belgium
{firstname.lastname}@umons.ac.be
[3] Intelligent User Interfaces Lab, Koç University, Turkey
{mtsezgin,ysahillioglu}@ku.edu.tr

Abstract. This paper introduces the IMOTION system, a sketch-based
video retrieval engine supporting multiple query paradigms. For vector
space retrieval, the IMOTION system exploits a large variety of low-
level image and video features, as well as high-level spatial and temporal
features that can all be jointly used in any combination. In addition,
it supports dedicated motion features to allow for the specification of
motion within a video sequence. For query specification, the IMOTION
system supports query-by-sketch interactions (users provide sketches of
video frames), motion queries (users specify motion across frames via
partial flow fields), query-by-example (based on images) and any combination of these, and provides support for relevance feedback.

1 Introduction

The IMOTION content-based video search engine is being developed in the con-
text of the Chist-Era project IMOTION [2] (Intelligent Multi-Modal Augmented
Video Motion Retrieval System), a joint effort of the Numediart Institute for
Creative Technologies at the University of Mons (Belgium), the Intelligent User
Interfaces Lab at Koç University (Turkey), and the Databases and Information
Systems Research Group at the University of Basel (Switzerland). The IMO-
TION system is based on Cineast [7], which has originally been designed and
implemented for sketch-based known-item video retrieval applications. It is ca-
pable of retrieving video sequences based on either single frames or rough color-
or edge-sketches. However, it also supports additional retrieval modes like query-
by-example, or motion queries.

The IMOTION system does that by combining multiple low-level features
(e.g., color, edge, and motion features) with high-level spatial and temporal
features (e.g., keyframe content, motion). The various features and the meta
data are all stored in the database and information retrieval system ADAM [1],
a storage engine built upon PostgreSQL, that jointly supports Boolean retrieval
(exact search) and vector space retrieval (similarity search).

X. He et al. (Eds.): MMM 2015, Part II, LNCS 8936, pp. 255–260, 2015.

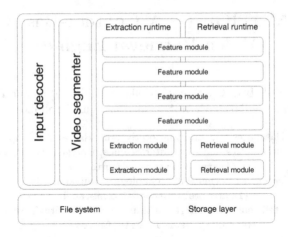

Fig. 1. Architectural overview of IMOTION

The user interface of the IMOTION system is browser-based and provides a sketching canvas for query specification. It has been optimized for use on a tablet computer, but can also be used on any other hardware.

The remainder of this paper is structured as follows: Section 2 introduces the system architecture of IMOTION. In Section 3, we briefly present the features and in Section 4 the retrieval modes supported by the IMOTION system. Section 5 concludes.

2   IMOTION System Architecture

From a conceptional point of view, the IMOTION system can be divided into an off-line part, responsible for feature extraction and management, and an on-line part which handles the actual retrieval. Figure 1 illustrates the architecture of the IMOTION system.

The *on-line* part consists primarily of a module runtime which manages the individual feature modules of the features supported by IMOTION. These modules work in parallel and independently of each other. The module runtime handles the initialisation of the modules, provides them with the input information they need, manages their outputs and shuts them down when they are no longer needed. During retrieval, the module runtime receives a query object which it passes to the retrieval modules. The modules, having been initialised with a connection to the storage layer, query the storage engine for shots matching the query object. Each module independently returns result information which is finally combined by the runtime to determine the overall result set of a query.

The *off-line* part consists of an input decoder, a video segmenter and the individual extraction modules for the IMOTION features. Input decoding logic provides the video segmenter with a continuous stream of data which it segments into shots. These shots are then passed to the feature extraction modules,

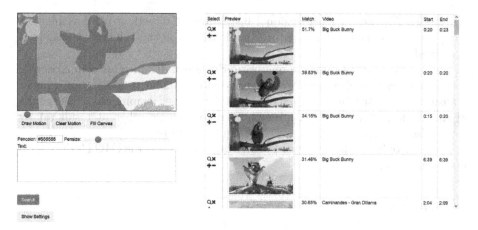

Fig. 2. Web-based user interface of the IMOTION system

which then perform the extraction of the corresponding features which are finally handed off to the storage layer.

Usually, a module performs both the retrieval and extraction task because the logic for transforming video information into feature representation is the same in both phases. An exception to this are those retrieval modules which use data already generated by other extraction modules.

The IMOTION system provides a browser-based web-interface offering a painting canvas where existing images can be pasted and possibly also manipulated and where sketches can be provided (either edge sketches or color sketches). Furthermore, it also allows to specify flow-fields for the specification of motion. The results of a search are presented in a list which contains a representative keyframe of the sequence as well as additional meta information. Figure 2 shows an example of a query and its results in the browser-based web-interface.

3   Image and Motion Features

The IMOTION system makes use of a multitude of features which are described in more detail in the following sections.

3.1   Low-Level Features

The set of low-level features contains features concerned with color and edges (representing spatial information) and motion (representing temporal information). The color features describe global as well as regional color properties of a shot, such as average or median color of a specific region. The edge features consider the regional distribution as well as directionality of edges. Finally, the motion features produce regional histograms of directions of movement. A complete list of the low-level features implemented in the IMOTION system is presented below.

All low-level features use the quadratic Euclidean distance for comparison of feature vectors. The distances are converted into a similarity score using a linear transformation for which the maximum is determined empirically. The results of the single feature modules are combined to a single coherent result set by computing a weighted average over the similarity scores.

A complete list of the low-level features implemented in the IMOTION system is presented below.

Global features

- *Average / Median color*
- *Dominant shot colors*: centre-points of the three largest color clusters
- *Chroma / Saturation*: average of all chroma / saturation values of a shot
- *Color histogram*
- *Shot position*: relative position of a shot with respect to the entire video

Regional color features

- *Color moments*: channel-wise statistical moments over regional partitions (uniform grid, angular radial partitioning) of an aggregation over all frames of a shot
- *Registered color grid*: grid of fuzzy quantized colors registered during retrieval.
- *Color layout descriptor* [3]
- *Color element grids*: grids containing partial color information in various representation (average saturation, variance of hue, etc.)
- *Subdivided color histogram*: fuzzy color histograms of image partitions

Regional edge features

- *Partitioned edge image*: regional ratios of edge- and non-edge pixels
- *Edge histogram descriptor* [6]
- *Dominant edge grid*: regional dominant edge direction quantised into 5 categories.

Motion features

- *Directional motion histograms*: regional normalised histograms of motion quantized into 8 directions.
- *Regional motion sums*: regional sums of the lengths of all motion vectors.

3.2 High-Level Features

The category of high-level features makes use of state-of-the-art machine learning approaches based on deep neural networks to extract relevant descriptors. Here too, two categories of features are available, representing either spatial (keyframe appearance), or temporal (video shot motion) information. Such approaches are able to efficiently encode natural image (and motion) key characteristics and similarities.

For the spatial component, we use a neural network architecture similar to the one proposed by Krizhevsky et al. [4]. The training is conducted on the image dataset ImageNet [8], which contains 1000 categories and about 1.2 million images. The available diversity and natural characteristics in the training data are very important to reach to a system better able to generalize to unseen image content. For the spatial component, dedicated to extracting motion-related features, we also use a similar architecture but rather than image pixels, the input relies on optical flow extracted from the video shots, and which direction components are considered as if they were image channels. Training is performed on the following video data sets: KTH [9], UCF101 [10], HMDB-51 [5].

Multiple techniques are applied to calculate the optical flow locally in time, and a sequence of these micro-movements (representing motion within the video shot) is passed as input to the neural network. The outputs of the last hidden layer of these two neural networks are then used as high-level features amenable to the vector space retrieval approach used in the IMOTION system.

4 Retrieval Modes

The IMOTION system supports known-item search by offering users three different modes of retrieval. The first mode, *query-by-sketch*, is a direct user input mode in which the user provides a rough hand-drawn sketch (either a line drawing or a color sketch) to search within the video collection. The second mode offers a *query-by-example* interaction where a user provides a query object via drag-and-drop.

This approach can be used when a user wants to find sequences similar to a previously retrieved one. It is achieved by using the internal representation of a shot as input and otherwise proceeding as usual. The third mode, *motion queries*, allows a user to specify the motion of objects across consecutive frames via (partial) flow-fields. These three retrieval modes are complemented with *relevance feedback* for refining the query results (i.e., by marking either relevant or irrelevant elements from the result list of a previous search).

The resulting query will then produce results which are similar to the relevant set, but not to the non-relevant set. Note that IMOTION has a stateless behavior and does not employ learning methods; feedback from previous iterations is, thus, discarded and no session is kept for queries.

5   Conclusion

In this paper, we have presented the IMOTION system, a content-based video retrieval engine for known-item searches using exemplary images or sketches. Since the IMOTION system was developped to support a wide variety of different kinds of video and implements many diverse features (both low-level and high level) and query paradigms that can be flexibly combined, it provides effective support for known-item search in large video collections.

Acknowledgements. This work was partly supported by the Chist-Era project IMOTION with contributions from the Belgian Fonds de la Recherche Scientifique (FNRS, contract no. R.50.02.14.F), the Scientific and Technological Research Council of Turkey (Tübitak, grant no. 113E325), and the Swiss National Science Foundation (SNSF, contract no. 20CH21_151571).

References

1. Giangreco, I., Kabary, I.A., Schuldt, H.: ADAM — A Database and Information Retrieval System for Big Multimedia Collections. In: Proc. Int. Congr. on Big Data 2014 (BigData 2014), Anchorage, USA. IEEE (2014)
2. IMOTION project, https://imotion-project.eu/
3. Kasutani, E., Yamada, A.: The MPEG-7 Color Layout Descriptor: A Compact Image Feature Description for High-Speed Image/Video Segment Retrieval. In: Proc. Int. Conf. on Image Processing (ICIP 2001), Thessaloniki, Greece, pp. 674–677. IEEE (2001)
4. Krizhevsky, A., Sutskever, I., Hinton, G.E.: ImageNet Classification with Deep Convolutional Neural Networks. In: Pereira, F., Burges, C.J.C., Bottou, L., Weinberger, K.Q. (eds.) Advances in Neural Information Processing Systems 25: Proc. Conf. on Neural Information Processing Systems (NIPS 2012), Lake Tahoe, USA, pp. 1097–1105 (2012)
5. Kuehne, H., Jhuang, H., Garrote, E., Poggio, T., Serre, T.: HMDB: A Large Video Database for Human Motion Recognition. In: Proc. Int. Conf. on Computer Vision (ICCV 2011), Barcelona, Spain, pp. 2556–2563. IEEE (2011)
6. Park, D.K., Jeon, Y.S., Won, C.S.: Efficient Use of Local Edge Histogram Descriptor. In: Proc. Ws. on Multimedia, Los Angeles, USA, pp. 51–54. ACM (2000)
7. Rossetto, L., Giangreco, I., Schuldt, H.: Cineast: A Multi-Feature Sketch-Based Video Retrieval Engine. In: Proc. Int. Symp. on Multimedia (ISM 2014), Taichung, Taiwan. IEEE (December 2014)
8. Russakovsky, O., Deng, J., Su, H., et al.: ImageNet Large Scale Visual Recognition Challenge. CoRR, abs/1409.0575 (2014)
9. Schüldt, C., Laptev, I., Caputo, B.: Recognizing Human Actions: A Local SVM Approach. In: Proc. Int. Conf. on Pattern Recognition (ICPR 2004), Cambridge, England, pp. 32–36. IEEE (2004)
10. Soomro, K., Zamir, A.R., Shah, M.: UCF101: A Dataset of 101 Human Actions Classes from Videos in the Wild. CoRR, abs/1212.0402 (2012)

A Storyboard-Based Interface for Mobile Video Browsing

Wolfgang Hürst, Rob van de Werken, and Miklas Hoet

Department Information & Computing Sciences, Utrecht University, The Netherlands
huerst@uu.nl

Abstract. We present an interface design for video browsing on mobile devices such as tablets that is based on storyboards and optimized with respect to content visualization and interaction design. In particular, we consider scientific results from our previous studies on mobile visualization (e.g., about optimum image sizes) and interaction (e.g., human perception and classification performance for different scrolling gestures) in order to create an interface for intuitive and efficient video content access. Our work aims at verifying if and to what degree optimized small screen designs utilizing touch screen gestures can compete with browsing methods on desktop PCs featuring significantly larger screen estate as well as more sophisticated input devices and interaction modes.

Keywords: Mobile interfaces. Mobile video browsing. Interactive multimedia.

1   Introduction

The ubiquity of handheld mobile devices such as tablets combined with the increasing popularity of mobile video playback and the possibility to access larger video archives via fast network connections results in an increasing need for better interface designs for mobile video search and browsing. Yet, interaction design for such devices – especially for rather complex tasks such as quick and efficient video browsing – is difficult for several reasons. First, the devices' form factor results in limited screen estate, which in turn limits, for example, the ability to visualize a video's content (e.g., via storyboards) and meta-information about a video (e.g., text annotations). Second, the predominant input modes for such devices, i.e., touch and tilting actions (e.g., via touch screen and accelerometers, respectively) are often lacking the flexibility and accuracy of input devices commonly used in desktop PC environments (such as keyboard and mouse). While we can therefore not expect video browsing systems on mobile devices to achieve a similar performance as interfaces optimized for desktop PCs, scientific studies (e.g., [4, 5, 6]) as well as prototypes and concrete interface designs (e.g., [1, 3]) suggest that high video browsing performance can be achieved if such a mobile system is optimized for the task at hand and considers the presumably limiting factors in the interface design.

For example, in our preceding research, we evaluated how the size of thumbnails used to represent video content influences video search performance [5, 6]. Our results indicate that surprisingly small sizes are actually sufficient in order to achieve a high search performance, thus suggesting that the small screen sizes of mobile

X. He et al. (Eds.): MMM 2015, Part II, LNCS 8936, pp. 261–265, 2015.

devices might be much less limiting for the interface design than commonly assumed. Likewise, touch interaction has obvious disadvantages, for example, when it comes to entering content, such as typing a query on an onscreen keyboard that lacks the tactile feedback of its physical counterpart and utilizes valuable screen estate. They also often lack the accuracy of controllers or mouse interfaces in tasks that require precision and accurate placement. Yet, touch gestures have be proven to be very intuitive, efficient, and considering performance maybe even better than traditional interaction modes in situations where quick navigation of large amounts of content is required – a characteristic which can obviously be very useful for quick video browsing if related interactions and gestures are implemented appropriately. For example, in [4] we compared how a paged versus continuous navigation of storyboards via touch gestures influences video search performance, resulting in related guidelines for mobile video browsing interface design.

Encouraged by such promising results, we proposed two interface designs – one utilizing a filmstrip style visualization integrated in vertically mounted timeline sliders placed on the left and right side of the screen, and one with a storyboard design utilizing our previous related research results [4, 5, 6]. Both designs have been evaluated in a comparative study [3] illustrating their usefulness, but also demonstrating complementary strengths and weaknesses. Consequently, we propose a new interface integrating both concepts into one single design with the ability to easily switch between the two interaction modes. While our studies so far have verified the design's usability for mobile video search, it will be interesting to evaluate it in comparison to more complex desktop PC systems as part of the Video Search Showcase (VSS) 2015 event in order to gain more insight into how well mobile systems can perform compared to such traditional setups, to identify their potential and also possible boundaries. The interface that we present is based on the one for single video browsing introduced and evaluated in [3], and extended in order to also support parallel browsing in video archives of up to ten individual files, as specified in the tasks of this year's VSS competition.

2   Interface Designs

Figure 1 illustrates the storyboard-based interface design used in the comparative study in [3]. Thumbnails extracted from the video are temporally sorted and presented in a 5x5 grid layout that can extend to the top and bottom beyond the screen. Scrolling to parts of the video before or after the currently visible area is done via up and down gestures, respectively. In order to illustrate the location of the currently visible part within the whole video, a scrollbar-style icon is added to the right side of the screen.

Figure 2 shows the aforementioned filmstrip style visualization which appears when the vertically timeline slider on the right side of the screen is used. Compared to the traditionally used horizontal orientation of such a slider, the vertical placements on the left and right side of the screen enables easier access and operation when holding the device with two hands during interaction (cf. illustration on the left side of the figure), a design decision that was also utilized in the interface design presented in

[2]. In our case, the slider on the left side of the screen covers the whole content of the video, enabling quick access to a certain part of it if and only if the related position is mapped on the (rather short) slider timeline representing the whole length of the video. For longer videos, the slider bar on the right can be used, which illustrates only a fraction of the whole file thus enabling browsing at a finer granularity level.

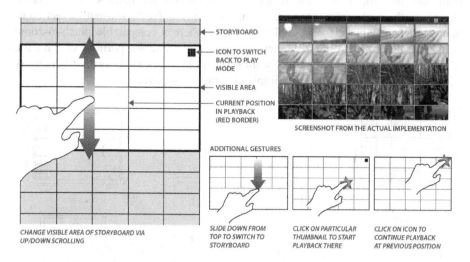

Fig. 1. Storyboard design implementation (from [3])

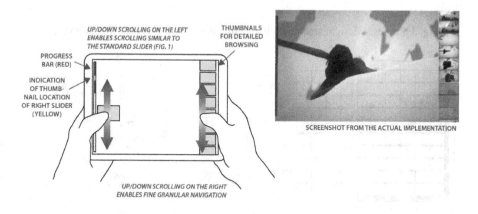

Fig. 2. Enhanced slider interface implementation (from [3])

In a comparative study using tasks slightly modified to the ones utilized in previous Video Browser Showdown competitions [7], both interfaces demonstrated their usability and power for video search (for detailed results we refer to [3]). Yet, both designs also revealed limitations and disadvantages – some of them opposed to each other. For example, the sliders obviously offer a faster access to searched locations if and only if those are directly accessible, whereas the storyboard design often outperforms the slider interface when a more sophisticated inspection of the presented content is needed. Consequently, we propose a design that seamlessly integrates both interfaces, as illustrated in Figure 3, which we will present in the 2015 edition of the Video Search Showcase. Using gestures, users can easily activate either of the two scrolling modes (i.e., storyboard and filmstrip view) and switch between them. In particular, the storyboard view is activated by moving the thumbs of both hands slightly to the center of the screen, resembling an intuitive "zoom out" effect commonly used on tablets, for example, for maps where a comparable pinch-to-zoom gesture is also used to gain a higher-level overview of larger portions of the data. Clicking on a thumbnail in the storyboard activates playback mode again, where users can activate the filmstrip slider by simply clicking on the screen.

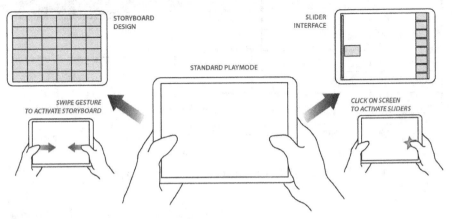

Fig. 3. Proposed design, seamlessly integrating both interaction concepts (from [3])

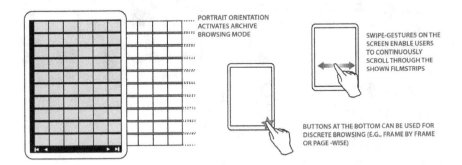

Fig. 4. Browsing video archives (ten videos in parallel) in portrait mode

Our studies confirm that this design enables quick and efficient video browsing within *one* video file and is thus well suited for tasks such as the Known Item Search (KIS) in single video files that was part of the Video Search Showcase in previous years. In order to deal with this year's tasks, which require search in ten videos from a larger archive, we propose the design illustrated in Figure 4. Turning the device from landscape to portrait mode activates video archive browsing, i.e., the simultaneous navigation within ten video files shown as filmstrips. Navigating the content is done either by a simple left-right swiping gesture on the screen or by using buttons on the bottom of the display that result in a discrete motion. Both interactions enable a simultaneous movement of all filmstrips, so users can visually browse and inspect the content of all videos by just using these simple gestures. While our initial tests confirm that people are indeed able to simultaneously browse all videos with this approach, it should be noted that for larger archives than the ten videos used in this year's Video Browser Showdown obviously some sort of pre-filtering (e.g., via querying that creates a ranked list of video search results) is required, and part of our future research.

References

1. Cobârzan, C., Hudelist, M.A., Del Fabro, M.: Content-Based Video Browsing with Collaborating Mobile Clients. In: Gurrin, C., Hopfgartner, F., Hurst, W., Johansen, H., Lee, H., O'Connor, N. (eds.) MMM 2014, Part II. LNCS, vol. 8326, pp. 402–406. Springer, Heidelberg (2014)
2. Hudelist, M.A., Schoeffmann, K., Boeszoermenyi, L.: Mobile video browsing with the ThumbBrowser. In: Proceedings of the 21st ACM International Conference on Multimedia (MM 2013), pp. 405–406. ACM, New York (2013)
3. Hürst, W., Hoet, M.: Sliders versus storyboards – Investigating interaction design for mobile video browsing. In: Proceedings of the 21st International Conference on Advances in Multimedia Modeling, MMM 2015 (2015)
4. Hürst, W., Darzentas, D.: Quantity versus quality: the role of layout and interaction complexity in thumbnail-based video retrieval interfaces. Proceedings of the 2nd ACM International Conference on Multimedia Retrieval(ICMR 2012), article 45, 8p. ACM, New York (2012)
5. Hürst, W., Snoek, C.G.M., Spoel, W.-J., Tomin, M.: Size Matters! How Thumbnail Number, Size, and Motion Influence Mobile Video Retrieval. In: Lee, K.-T., Tsai, W.-H., Liao, H.-Y.M., Chen, T., Hsieh, J.-W., Tseng, C.-C. (eds.) MMM 2011 Part II. LNCS, vol. 6524, pp. 230–240. Springer, Heidelberg (2011)
6. Hürst, W., Snoek, C.G.M., Spoel, W.-J., Tomin, M.: Keep moving! Revisiting thumbnails for mobile video retrieval. In: Proceedings of the International Conference on Multimedia (MM 2010), pp. 963–966. ACM, New York (2010)
7. Schoeffmann, K., Ahlström, D., Bailer, W., Cobarzan, C., Hopfgartner, F., McGuinness, K., Gurrin, C., Frisson, C., Le, D.-D., del Fabro, M., Bai, H., Weiss, W.: The Video Browser Showdown: A Live Evaluation of Interactive Video Search Tools. International Journal of Multimedia Information Retrieval (MMIR) 3(2), 113–127 (2014)

Collaborative Browsing and Search in Video Archives with Mobile Clients

Claudiu Cobârzan, Manfred Del Fabro, and Klaus Schoeffmann

Alpen-Adria-Universität Klagenfurt
9020 Klagenfurt, Austria
{claudiu,manfred,ks}@itec.aau.at

Abstract. A system for collaborative browsing and search within video archives is proposed. It comprises of mobile clients and a back-end server. The server is responsible for inter-client communication as well as for archive partitioning according to the active client's population. The participating clients employ a GUI designed and optimized for Nexus 7 tablets.

1 Introduction

Exploring a video archive with either highly specialized tools or simple video players is a hard task for expert and non-expert users alike. Most tools, either high-end or simplistic, disregard any form of collaborative features or provide only limited ones. Building on our previous experience with collaborative video browsing and search on mobile clients [2], we propose a new system that tries to enhance the collaborative aspects of exploring a video archive, taking advantage of the high availability and popularity of tablet computers. While our previous tool focused on individual video exploration, the current system is aimed at archives and tries to exploit the power of collaboration in solving a specific search task. That translates in a divide and conquer approach when it comes to data exploration: a server system is responsible for the coordination of a collaborative search task and partitions the archive equally among all mobile clients participating in a search task at any moment in time. Primarily, each client only concentrates on assigned data, but can also explore the rest of the archive. Clients can reduce the search space by sending filtering requests based on color, motion, and face detection to the server. Filter settings as well as filtering results can be manually shared among clients (on request). State information, e.g. video segments currently being explored, the degree of exploration, etc., is automatically exchanged between clients in real time.

2 System Architecture

The proposed architecture employs mobile clients, a server as well as a database. The server is responsible for the coordination of search tasks as well as for the communication between the clients. The clients communicate only with the

X. He et al. (Eds.): MMM 2015, Part II, LNCS 8936, pp. 266–271, 2015.
© Springer International Publishing Switzerland 2015

server via Wi-Fi using a proprietary UDP based protocol. Pull and push techniques are combined in order to perform different tasks like selecting a video segment for search, iterating through key frames, setting filters and retrieving the corresponding results, sharing filters as well as key frames sets, etc. Details regarding the client's as well as the server's implementation are presented in the following subsections.

2.1 Server

The main tasks of the server are as follows: (a) assign and update the search space for each active client taking into consideration the client activity (new clients entering the system, clients leaving the system); (b) collect status information from the active clients (selected video, selected segment, selected keyframe) and share it among all participants; (c) retrieve query results according to client's filtering requests; (c) ensure message passing among clients in order to support collaborative operations.

The search archive is held by the server which also holds the results of offline keyframe selection, content analysis and filtering within a proprietary database. The server assigns segments from the video archive to each client entering the system. The client is responsible for searching within those segments but it is able to perform specific operation in all available segments either at will or in response to another client's specific request (share a search filter, respectively share a keyframe set). We consider segments of fixed lengths, namely 15 minutes of playback video.

Our keyframe selection approach operates on a sub-shot level and is able to create several keyframes for shots with varying content over time, such as camera pans, rotates, zooms etc. This will ensure that – in contrary to common shot detection algorithms [6] – we will not miss any important content, even for shots of longer duration.

The keyframe selection method consists of three stages. First, it inspects the optical flow [1] over successive frames. The algorithm starts with a densely sampled grid of keypoints in the first frame and tries to track these keypoints over consecutive frames. Keypoints that can't be tracked anymore are removed and if the ratio of "still trackable" keypoints falls below a certain threshold, the current frame is selected as keyframe candidate and the optical flow estimation starts with a fresh grid of keypoints. In the second stage, the neighborhood of such candidate keyframes is further inspected and only the sharpest frame in this neighborhood is selected as keyframe. We use a Difference-of-Gaussian (DoG) approach to determine the blurriness (i.e., sharpness) of a frame. Finally, in the third stage consecutive keyframes are checked for their visual similarity by a simple histogram comparison. Such consecutive keyframes that are too similar are removed from the keyframe set.

For the content analysis part, we use a tool for determining the dominant color as well as the color layout for each previously detected keyframe. We also employ a face detector in order to detect keyframes containing 1, 2, 3 or more faces. Motion is detected at segment level. For color analysis we use the MPEG-7 color layout descriptor [4]. When filtering for a certain dominant color, all segments containing keyframes that match the dominant color will be returned.

For face detection we use boosted cascades of Haar-like features [7]. We employ the classifiers that ship with the OpenCV library[1]. Faces are detected and counted on all the previously obtained keyframes.

For motion analysis we use the motion histogram proposed in [5], which is applied on segment level. Both color-based and motion-based segment detection have been successfully applied to search tasks of the Video Browser Showdown 2013 [3].

2.2   Mobile Clients

The clients joining the system are automatically assigned a portion of the archive, which is searched in a way that is intended to ensure fairness (e.g. if there are only 2 active clients, each will have assigned half of the video collection - the first user will search within the first half of the collection, while the second will search within the second half). They can browse not only the assigned archive part but the entire archive as well. They can also issue search requests and then browse throughout the results. Each client will begin to search within the default allocated segments. If the allocated segments do not match the search request, a client can also collaborate with another client by searching within the segments assigned to that client.

The interface is divided into multiple task specific areas:

- *target segment selection, segment assignment and current activity overview area* (see Figure 1D): allows the users to have an overview of the whole archive as well as on the parts that are assign to each individual user. The list of active users is presented on the left side. Each active user is assigned a number (U1 for user 1, U2 for user 2, etc.) and a color for identification. The same color is used to mark the segments that have been assigned to him by the server. At the finest granularity, each square represents a segment of 15 minutes of playback video. The arrow buttons on the right side of the area, as well as the plus and minus buttons are used to navigate within the archive and to modify the current viewpoint by increasing/decreasing the virtual segmentation granularity of the video archive. Decreasing the segment granularity level has a zoom-out effect: each two consecutive segments from the previous granularity level are merged into a new "virtual" segment (e.g. two 15 minutes segments will be combined to form a new 30 minutes segment).
- *segment exploration area* (see Figure 1B): when a user selects a segment from the target selection area (Figure 1D), the keyframes comprised within that particular segment start being displayed as thumbnail pictures. All the keyframes from one particular segment form what we call a *keyframe set*. The number of keyframes that are displayed at one particular moment can

---

[1] http://opencv.org/ (last access 2014-09-25)

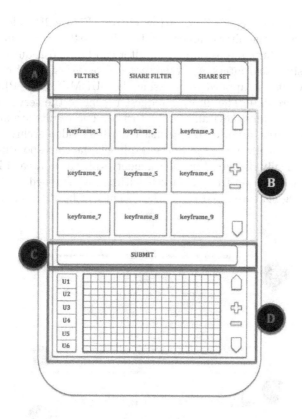

Fig. 1. Sketch of the client interface

be modified by pressing the plus/minus button on the right hand side. The arrow buttons are used for navigation within the keyframe set in a page-by-page manner (e.g. the first 9 images in the set, the next 9 images when the down arrow is pressed and so on). The displayed keyframes act also as entry points within the corresponding video: double tapping a keyframe opens a basic video player with minimum controls (play/stop button and a seeker bar) whose current playback position is set to match the selected keyframe. Pressing the SUBMIT button (Figure 1C) sends the current frame to the server for checking whether it belongs to the target scene or not.

– *query construction area* (see Figure 2): by pressing the FILTERS button (Figure 1A), a user is able to build a filter that can include either single or multiple components that are grouped into 3 categories: (1) faces, (2) color and (3) motion (Figure 2A). A user can filter for keyframes containing 1, 2, 3 or more faces. One can also indicate the desired dominant color as well as the direction and speed of motion. All filters are applied on the content-based analysis of keyframes and videos segments, which is performed on the server in advance. A user can combine filters from the 3 filter groups. Pressing the down arrow button (Figure 2B) adds the current selected filter to the filter composition bar (Figure 2C) and numbers it. Once added to the filter composition bar, a filter can be repositioned with respect to previously added filters or can be removed by pressing the corresponding buttons. A filter is applied by pressing the SUBMIT FOR PROCESSING button (Figure 2D). The filtering request is sent to the server and the corresponding segments/keyframes are returned. The results are highlighted in the overview area (Figure 1D) against the whole video archive. If segments with matching keyframes exist in the result set, the corresponding keyframes are automatically displayed in the segment exploration area (Figure 1B). If a filtering request returns no match within the assigned partition of the dataset, a segment assigned to another client can be chosen manually.

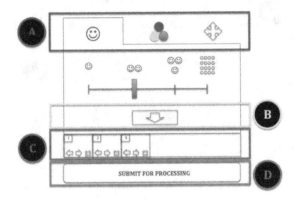

Fig. 2. Sketch of the *query construction area* (face filtering section)

The collaborative aspect of the proposed system includes sharing the most recent built filter with other users, sharing the current keyframes dataset as well as being able to view in real-time the segments that are being explored by the other participants. All clients can ask for help in exploring the results of one particular filter by pressing the SHARE FILTER button (Figure 1A). This translates in a notification being sent to all participants. The ones that accept the notification will have the segments returned as filter results highlighted in the corresponding area (Figure 1D). The server also shares among all participants the current segment being explored by each individual user. Those segments are correspondingly highlighted in the activity overview area (Figure 1D). Segments whose keyframes have been partially or full viewed in the segment exploration area are also highlighted. This allows collaborative clients to join a search task and focus mainly on unexplored or only partially explored segments as well as monitor the current browse activity within the archive. Similarly, help can be requested when exploring a *keyframe set* that comprises all the frames from a selected segment by pressing the SHARE SET button (Figure 1A). Clients that accept the request will have the keyframes of the set displayed in the segment exploration area (Figure 1B).

3   Conclusion

We presented a new collaborative video browsing and search system targeted on archive exploration for the Video Search Showcase 2015 competition. Similar to our previous approach [2], we concentrate on mobile clients and try to explore and take advantage of the interaction capabilities of the most recent generation of tablets. The focus is on client, server and communication features as well as data organization that enhance the collaborative aspects of video exploration within large video archives.

Acknowledgments. This work was funded by the Federal Ministry for Transport, Innovation and Technology (bmvit) and the Austrian Science Fund (FWF): TRP 273-N15 and by Lakeside Labs GmbH, Klagenfurt, Austria, and funding from the European Regional Development Fund (ERDF) and the Carinthian Economic Promotion Fund (KWF) under grant KWF-20214/22573/33955.

References

1. Bouguet, J.-Y.: Pyramidal Implementation of the Affine Lucas Kanade Feature Tracker Description of the Algorithm. Intel Corporation 5, 1–10 (2001)
2. Cobârzan, C., Hudelist, M.A., Del Fabro, M.: Content-Based Video Browsing with Collaborating Mobile Clients. In: Gurrin, C., Hopfgartner, F., Hurst, W., Johansen, H., Lee, H., O'Connor, N. (eds.) MMM 2014, Part II. LNCS, vol. 8326, pp. 402–406. Springer, Heidelberg (2014)
3. Del Fabro, M., Münzer, B., Böszörmenyi, L.: Smart Video Browsing with Augmented Navigation Bars. In: Li, S., et al. (eds.) MMM 2013, Part II. LNCS, vol. 7733, pp. 88–98. Springer, Heidelberg (2013)
4. Kasutani, E., Yamada, A.: The MPEG-7 color layout descriptor: a compact image feature description for high-speed image/video segment retrieval. In: Proceedings of 2001 International Conference on Image Processing. IEEE, Thessaloniki (2001)
5. Schoeffmann, K., Lux, M., Taschwer, M., Böszörmenyi, L.: Visualization of Video Motion in Context of Video Browsing. In: Proceedings of the IEEE International Conference on Multimedia and Expo (ICME). IEEE, New York (2009)
6. Smeaton, A.F., Over, P., Doherty, A.R.: Video shot boundary detection: Seven years of TRECVid activity. In: Computer Vision and Image Understanding, vol. 114 (2010)
7. Viola, P., Jones, M.: Rapid object detection using a boosted cascade of simple features. In: Proceedings of the 2001 IEEE Computer Society Conference on Computer Vision and Pattern Recognition (CVPR). IEEE, Kauai (2001)

The Multi-stripe Video Browser for Tablets

Marco A. Hudelist[1] and Qing Xu[2]

[1]Klagenfurt University, Universitaetsstr, 65-67, 9020 Klagenfurt, Austria
[2]School of Computer Science and Technology, Tianjin University, Tianjin, China
marco@itec.aau.at, qingxu@tju.edu.cn

Abstract. We present a prototype for video search and browsing in large video collections optimized for tablets. The content of the videos is organized into sub-shots, which are visualized by frame stripes of different configurations. Moreover, all videos can be filtered by color layout and motion patterns to reduce search effort. An additional overview mode enables the parallel inspection of multiple filtered or unfiltered videos at once. This mode should be both easy to use and still efficient, and therefore well-suited for novice users.

1 Introduction

The Video Search Showcase (VSS), formerly known as Video Browser Showdown (VBS), is an annual live evaluation of video browsing tools [13,15,14]. It consists of visual and textual Known-Item Search (KIS) tasks, similar to the one performed in TRECVID from 2010-2012 [12], which are tested for expert users (the developers of the tools) as well as novice users (volunteers from the audience). Most participants of the last years (e.g. the winners of 2014 [9] and 2013 [8]) concentrated on video browsing on traditional desktop computers. Although mobile devices like tablets or smartphones are ubiquitous only few mobile-centric submissions took part in the competition so far.

For VSS 2015 we propose a tool developed for tablet computers, which is based on our previous work on the KNT-Browser [5]. The KNT-Browser focused on exploring a single video and used frame stripes with sub-shots for content visualization. Due to space limitations we omit the details about the interface and the sub-shot segmentation of the KNT-Browser, but the interested reader can find them in [5]. The configuration of the frame stripes differed in terms of how much of the frames was visible, ranging from a rather condensed representation to a full-frame visualization. To adjust to the preconditions of large video collections – as use by VSS 2015 – we added content-based filtering (color layout, motion) as well as a new overview mode for multi-video exploration.

This paper is structured as follows: we first discuss related work of mobile video browsing and then continue with detailed information about our new prototype. We discuss the new filtering mechanisms as well as the before mentioned overview mode. Finnally, we end with a short conclusion.

X. He et al. (Eds.): MMM 2015, Part II, LNCS 8936, pp. 272–277, 2015.
© Springer International Publishing Switzerland 2015

2   Related Work

Video browsing/retrieval is a very active research field ([16], [2]). We will focus on works that address video browsing/retrieval optimized for mobile devices like smartphones or tablets. One of the most prominent examples is the *Mobile ZoomSlider* proposed by Hürst et al. [7]. The traditional seeker bar concept is extended by adding the ability to define navigation granularity with the click position. An alternative to the typical seeker bar is shown by the *Wipe'n'Watch* interface proposed by Huber et al. [4], which focuses on videos of e-lectures. Vertical swipes across the screen let users navigate between keyframes of a video. Swiping horizontally switches between videos. Cobarzan et al. present a similar video browsing interface that also includes collaboration means [3]. Users are able to set content-based filters and browse through the results by either using a 3D ring visualization of video segments or a 2D storyboard representation. Moreover, a search team is able to collaborate by sharing search information like *already visited frames, already used searched queries* of each member and sharing of *interesting video segments* across the team. A hierarchical browsing approach based on a storyboard-like visualization is presented by Hürst and Darzentas [6]. A grid of thumbnails representing video segements is displayed. Tapping on one of the thumbnails causes a transition to a lower hierarchy level which can be continued until a frame-by-frame level is reached.

3   The Interface

We present a video browsing tool that uses a compact "stripe image" visualization to quickly communicate the content of several video files to the user. The interface operates in two modes: *detail mode* and *overview mode*. The modes can be switched by changing the orientation of the device. Holding the tablet in landscape orientation activates *detail mode*. Turning the tablet to portrait orientation switches to *overview mode*. The modes are intended for different search scenarios. For content-based filtering we analyze the videos offline (i.e., on a server computer) and create meta-data and thumbnails for each video, which are then used in the prototype.

3.1   Detail Mode

In this mode users have the possibility to either search the video collection sequentially or set several filters to reduce the amount of to be inspected data. The preview window at the top left can be used to get a better idea about the currently selected video segment (see Figure 1). Below the preview window three buttons (play/pause, fast forward, fast reverse) can be used to navigate in the video segment. At the top right the filter area is placed where users can set filters for *color layout* and and *motion layout*.

The *color layout* filter will screen all available videos for keyframes that match the specified color pattern. The control consists of a visual representation of a

Fig. 1. Detail mode with filter and navigation controls visible

frame and a color palette. The frame representation is divided into nine sub-sections, each assignable with a selected color by the user. This is achieved by first marking one or several sub-sections with a single tap and then tapping the the desired color on the right panel.

With the *motion layout* filter users are able to define a proportional distribution pattern of motion directions similar to what was presented by Schoeffmann et al. [17]. For example, a user might want to define that 60% of the motion goes up but another 20% of the motion goes to the right side, and 20% is still/no-motion. The motion control is positioned below the color layout control and consists out of a motion circle as well as an indicator for the no-motion option. The indication cirlce is divided up into four sections: up, down, left and right. In its initial state 100% are assigned to the no-motion option. Users can set motion percentages by tapping and dragging their fingers across the sections of the motion circle. The further their finger is positioned to the outside of the circle the more percentage points are assigned to that direction. The percentage points are first taken from the no-motion option until it is at 0% percent. After that percentage points are subtracted equally from each not currently altered section that has more than zero percentage points. The same is true if users want to reassign percentage points to the no-motion option. This can be done by touching and dragging the percentage bar on the left side of the motion circle up or down.

The bottom area of the interface is user for *video content navigation* by compact keyframe visualization. Navigation can be performed either sequentially through all videos or based on the results of filtering. With the displayed seeker bar users can navigate through all of the available data, starting with the first frame of the first video and ending with the last frame of the last video. The top frame stripe shows slices of video frames in a very condensed fashion. Each of a detected sub-shot [10] is represented by its middle frame. We only show a slim slice of each frame to present as much information as possible in the available screen real estate. Further-more, we select the most salient part of the frame as vertical slice, by utilizing the method proposed by Achanta et al. [1]. The second stripe shows the same frames but in full size for more detailed inspection. The stripes are scrolled by drag ges-tures. Selection of a video segment is done by a single tap on its slice. Both stripes always synchronize their position in relation to each other, so that always the same frame is position below the playing head (indicated by a red line). The same is true for the seeker bar thumb and the preview window.

3.2 Overview Mode

The overview mode should give users great insight into the content of multiple videos at once. It is activated when users turn their device to portrait orientation.

Fig. 2. Overview mode – displaying the frame stripes of the first ten videos

For each individual video now a separate frame stripe is displayed (as can be seen in Figure 2). Ten of the stripes are always grouped together to a page. When users scroll the interface up or down they always are positioned to the previous or next page of stripes. Moreover, the stripes can be scrolled left or right by drag gestures. If any of the strips are scrolled like that, the same amount of scrolling is also applied to all other visible strips. In this way, users are able to sequentially inspect ten videos at once from their beginning to the end. When users scroll to another page the stripes are repositioned to the beginning. This is done in order to enable a workflow, where users start inspecting the first ten videos, reach their end, scroll downwards and are immediately able to continue to search from the beginning of the next ten videos. The overview mode also adjusts to filters set in the detail mode. Each stripe continues to represent one video, but sequences that don't match the filter criteria are hidden making each stripe shorter and therefore faster to inspect. When users tap on one of the slices they are automatically transferred to the detail mode for further inspection of the selected video sequence. We expect that the overview mode is especially suitable for non-expert users, i.e., novices that are part of the test sessions in the Video Search Showcase, and allow both easy and efficient navigation in a video collection.

4   Conclusions

We introduced a new interactive video search tool for tablets, intended for participation in the Video Search Showcase (VSS) at the International Conference on MultiMedia Modelling (MMM 2015) [11]. Users are able to explore video collections based on sub-shots by compact stripe-based frame visualizations and content-based filtering. Moreover, users are able to explore the contents of multiple videos at once by utilizing an overview mode that visualizes multiple video stripes at once.

Acknowledgements. This work was funded by the Federal Ministry for Transport, Innovation and Technology (bmvit) and Austrian Science Fund (FWF): TRP 273-N15 and the European Regional Development Fund and the Carinthian Economic Promotion Fund (KWF), supported by Lakeside Labs GmbH, Klagenfurt, Austria. The work by Qing Xu has been funded by Natural Science Foundation of China (61471261, 61179067, U1333110).

References

1. Achanta, R., Hemami, S., Estrada, F., Susstrunk, S.: Frequency-tuned salient region detection. In: IEEE Conf. on Computer Vision and Pattern Recognition (CVPR), pp. 1597–1604 (2009)
2. Borgo, R., Chen, M., Daubney, B., Grundy, E., Heidemann, G., Höferlin, B., Höferlin, M., Leitte, H., Weiskopf, D., Xie, X.: State of the art report on video-based graphics and video visualization. Comp. Graph. Forum 31(8), 2450–2477 (2012)

3. Cobârzan, C., Hudelist, M.A., Del Fabro, M.: Content-based video browsing with collaborating mobile clients. In: Gurrin, C., Hopfgartner, F., Hurst, W., Johansen, H., Lee, H., O'Connor, N. (eds.) MMM 2014, Part II. LNCS, vol. 8326, pp. 402–406. Springer, Heidelberg (2014)

4. Huber, J., Steimle, J., Lissermann, R., Olberding, S., Mühlhäuser, M.: Wipe'n'watch: spatial interaction techniques for interrelated video collections on mobile devices. In: Proc. of the 24th BCS Interaction Specialist Group Conf., BCS 2010, pp. 423–427. British Computer Society, Swinton (2010)

5. Hudelist, M.A., Schoeffmann, K., Xu, Q.: Improving interactive known-item search in video with the keyframe navigation tree. In: MultiMedia Modeling, Lecture Notes in Computer Science. Springer (to appear)

6. Hürst, W., Darzentas, D.: History: A hierarchical storyboard interface design for video browsing on mobile devices. In: Proc. of the 11th Int. Conf. on Mobile and Ubiquitous Multimedia (MUM), pp. 17:1–17:4. ACM, NY (2012)

7. Hürst, W., Götz, G., Welte, M.: Interactive video browsing on mobile devices. In: Proc. 15th Int. Conf. on Multimedia, MULTIMEDIA 2007, pp. 247–256. ACM, NY (2007)

8. Le, D.-D., Lam, V., Ngo, T.D., Tran, V.Q., Nguyen, V.H., Duong, D.A., Satoh, S.: NII-UIT-VBS: A video browsing tool for known item search. In: Li, S., et al. (eds.) MMM 2013, Part II. LNCS, vol. 7733, pp. 547–549. Springer, Heidelberg (2013)

9. Lokoč, J., Blažek, A., Skopal, T.: Signature-based video browser. In: Gurrin, C., Hopfgartner, F., Hurst, W., Johansen, H., Lee, H., O'Connor, N. (eds.) MMM 2014, Part II. LNCS, vol. 8326, pp. 415–418. Springer, Heidelberg (2014)

10. Luo, X., Xu, Q., Sbert, M., Schoeffmann, K.: F-divergences driven video key frame extraction. In: IEEE Int. Conf. on Multimedia & Expo (ICME). IEEE (2014)

11. MMM2015. Video Search Showcase (2014), http://www.videobrowsershowdown.org (accessed September 25, 2014)

12. Over, P., Awad, G., Michel, M., Fiscus, J., Sanders, G., Shaw, B., Kraaij, W., Smeaton, A.F., Quénot, G.: Trecvid 2012 - an overview of the goals, tasks, data, evaluation mechanisms and metrics. In: Proc. of TRECVID (2012)

13. Schoeffmann, K.: A user-centric media retrieval competition: The video browser showdown 2012-2014. IEEE Multimedia Magazine, 1–5 (to appear, 2014)

14. Schoeffmann, K., Ahlström, D., Bailer, W., Cobârzan, C., Hopfgartner, F., McGuinness, K., Gurrin, C., Frisson, C., Le, D.-D., Del Fabro, M., Bai, H., Weiss, W.: The video browser showdown: a live evaluation of interactive video search tools. Int. Journal of Multimedia Information Retrieval 3(2), 113–127 (2014)

15. Schoeffmann, K., Bailer, W.: Video browser showdown. ACM SIGMultimedia Records 4(2), 1–2 (2012)

16. Schoeffmann, K., Hopfgartner, F., Marques, O., Boeszoermenyi, L., Jose, J.M.: Video browsing interfaces and applications: a review. SPIE Reviews 1(1), 018004 (2010)

17. Schoeffmann, K., Lux, M., Taschwer, M., Boeszoermenyi, L.: Visualization of video motion in context of video browsing. In: IEEE Int. Conf. on Multimedia and Expo (ICME), pp. 658–661 (June 2009)

NII-UIT Browser:
A Multimodal Video Search System

Thanh Duc Ngo[1], Vinh-Tiep Nguyen[2], Vu Hoang Nguyen[1]
Duy-Dinh Le[3], Duc Anh Duong[1], and Shin'ichi Satoh[3]

[1] University of Information Technology - VNUHCM, Vietnam
{thanhnd,vunh,ducda}@uit.edu.vn
[2] University of Science - VNUHCM, Vietnam
nvtiep@fit.hcmus.edu.vn
[3] National Institute of Informatics, Japan
{ledduy,satoh}@nii.ac.jp

Abstract. We introduce an interactive system for searching a known scene in a video database. The key idea is to enable multimodal search. As the retrieved database is getting larger, using individual modals may not be powerful enough to discriminate a scene with other near duplicates. In our system, a known scene can be described and searched by its visual cues or audio genres. Templates are given for users to rapidly and exactly describe the scene. Moreover, search results are updated instantly as users change the description. As a result, users can generate a large number of possible queries to find the matched scene in a short time.

Keywords: Multimodal Approach, Know Item Search, Interactive Tool.

1 Introduction

Searching a certain scene in videos is a great demand from users, especially when video datasets are getting larger and larger. Typical search systems usually require example images from users as queries to search. Without such examples, searching known items or scenes is an extremely challenging task. The main reason is about query formation - the process of converting user intentions to queries in specific forms. An ideal search system is therefore expected to provide interactive tools for users to exactly describe what they want to search for.

A video scene (or a video shot) basically can be described by its visual information such as dominant colors, color distribution, color signatures and textures. Existing systems have shown the advantages of using color information [1,2]. However, the larger the video dataset is, the more scenes have similar visual clues. This make finding only one true matched scene among sets of thousands similar scenes very difficult. Hence, more information about the scene to differentiate it with the others is needed. In this paper, we introduce a search system with interactive tools which enable users to describe and search a scene by both visual and audio information. In general, following functions are available:

X. He et al. (Eds.): MMM 2015, Part II, LNCS 8936, pp. 278–281, 2015.
© Springer International Publishing Switzerland 2015

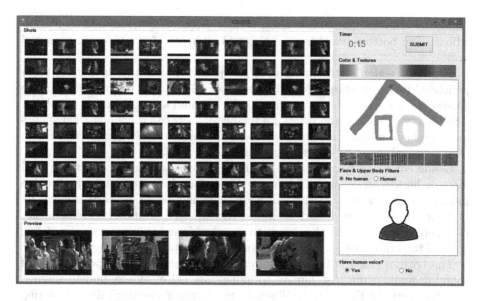

Fig. 1. NII-UIT Browser

- **Visual Search.** Users can prepare a sketch by drawing non-rigid shapes or circular signatures with different colors and textures. Or, they can focus on videos having human appearances by using another sketch for human description (i.e. positions of faces and upper-body parts, colors of the upper-body parts).
- **Audio Search.** Users can search videos with specific types of sound or audio genres.

2   System Overview

2.1   User Interface

The user interface of our system is illustrated in Figure 1. There are two main regions in the interface. The left one includes top and bottom panels. The top panel with the largest area is for displaying search results. The results will be ranked by their similarities with the given query. The most similar scene with highest confidence score will be shown at the top-left corner. Other scenes from left to right and from top to bottom are displayed with descendant order of confidence scores. Each thumbnail image represents the middle frame of a 10-second shot of a video.

As users move the mouse over a thumbnail image, 2 other thumbnail images of its 2 neighboring 10-second shots in the same video will be shown in a pop-up window. This is to provide additional information for users to differentiate near-duplicate shots. When returned shots are ranked by their similarities with the query, shots in top positions of the list are likely identical. Hence, the most important clues to differentiate them is their neighboring scenes and their transitions. The bottom panel is used to show all shots of the video in temporal order.

The right region of the interface is for query formation. It includes 3 main panels: *Colors & Textures*, *Face & Upper-body Filters*, and *Sound Filter*. In the *Colors & Textures* panel, we provide a color bar and a set of texture templates. After selecting a color and a texture template, users can draw non-rigid shapes and circular signatures on the *white-board*. The drawn *white-board* will be then used as visual query for further searching.

Beside color and texture, we also allow users to search videos with special focus on human occurrence. The *Face & Upper-body Filter* is for users to describe human-related queries. Given a half-body shape in a board, users can change its position and size in the board to express the expected position and relative size of a person in the scene they want to search. Moreover, users can also define color of the body part (i.e. T-shirt color) as additional cue for filtering.

The bottom panel is for sound filtering. Users can select a specific type of sound or audio genre (e.g. human speaking, city noise, silence, human singing, etc.) to search. In Figure 1, we illustrate the "human speaking" filter. Filtering and ranking results of different modals can be merged (i.e. OR) or intersected (i.e. AND) to provide final search results. Search results are instantly updated as users change their description e.g. change positions of the shapes. Hence, users can try many queries in a short time to find the matched scene.

2.2   Off-line Processing

Given a video dataset, we pre-process all videos. For each video, we generate a set of 10-second shots by equally dividing the video. The middle frame of a shot represent for the whole shot. It is used for feature extraction, content analysis and thumbnail display. By dividing a video into shots every 10 seconds, we expect to achieve a balance between recall and efficiency. Finer division results in more shots and information for accurate searching. However, its trade-off is a higher computational cost.

Given the shots and their representative frames, we first extract texture and color information from frames. We divide each frame into a 30×40 grid. For each cell of the grid, we compute the most dominant color. An index of frames based on colors at its cells will be created for fast access at the online stage. Instead of using all possible colors, we only select 40 most popular colors. Similarly, we only consider 20 types of textures. Textures features (SIFT descriptor [6][7]) extracted from cells will be assigned to the most similar texture template. Then, an yet another index of frames based on textures in cells is generated. Hashing and inverted indexing techniques are applied.

In the next pre-processing step, we detect and index positions of human faces and upper-body parts in frames. A robust face detector is applied [4]. For detecting upper-body, we develop a detector based on Deformable Part Model (DPM)[3] and HOG-LBP features. In the final step, we classify shots by its sound and audio genres. We build classifiers based on Support Vector Machine

(SVM) with Mel-frequency Ceptral Coefficient (MFCC) features [5]. All training samples are crawled from the Internet with manual selection. With each audio genre or sound type, 100-120 positive samples are selected. One-versus-All training and testing methodology is applied. Given classification results, an inverted index of sound types and shots are prepared for online stage.

2.3 Online Process

Given the sketch, we have a list of cells and colors in cells. Using the list and the index generated in the off-line process, we obtain a list of frames having colors and textures similar to the sketch. In a very similar way, we can get lists of candidate frames with the human-related sketch and the sound filter. These lists then can be merged or intersected to provide the search result to users.

3 Conclusion

We introduce an interactive system for searching a known scene in a video database. Compared to previous interactive search systems, we incorporate following key features to help users in query formation and searching. First, besides color, we enable texture description by providing texture templates. Second, we support face and upper-body search with color information. Finally, our system utilizes audio information in videos for filtering and searching.

References

1. Lokoč, J., Blažek, A., Skopal, T.: Signatured-based Video Browser. In: Gurrin, C., Hopfgartner, F., Hurst, W., Johansen, H., Lee, H., O'Connor, N. (eds.) MMM 2014, Part II. LNCS, vol. 8326, pp. 415–418. Springer, Heidelberg (2014)
2. Ngo, T.D., Nguyen, V.H., Lam, V., Phan, S., Le, D.-D., Duong, D.A., Satoh, S.: NII-UIT: A Tool for Known Item Search by Sequential Pattern Filtering. In: Gurrin, C., Hopfgartner, F., Hurst, W., Johansen, H., Lee, H., O'Connor, N. (eds.) MMM 2014, Part II. LNCS, vol. 8326, pp. 419–422. Springer, Heidelberg (2014)
3. Felzenszwalb, P., Girshick, R., McAllester, D., Ramanan, D.: Object Detection with Discriminatively Trained Part Based Models. IEEE Transactions on Pattern Analysis and Machine Intelligence (TPAMI), 1627–1645 (2010)
4. Zhu, X., Ramanan, D.: Face detection, pose estimation and landmark localization in the wild. In: Computer Vision and Pattern Recognition (CVPR), pp. 2879–2886 (2012)
5. Lee, C.H., Soong, F., Juang, B.H.: A segment model based approach to speech recognition. In: International Conference on Acoustics, Speech and Signal Processing (ICASSP), pp. 501–541 (1988)
6. Lowe, D.G.: Distinctive image features from scale invariant keypoints. International Journal Computer Vision (IJCV), 91–110 (2004)
7. Thomas, D., Daniel, K., Herman, N.: Features for Image Retrieval: An Experimental Comparison. Journal Information Retrieval, 77–107 (2008)

Interactive Known-Item Search Using Semantic Textual and Colour Modalities

Zhenxing Zhang, Rami Albatal, Cathal Gurrin, and Alan F. Smeaton

Insight Centre for Data Analytics
School of Computing, Dublin City University
Glasnevin, Co. Dublin, Ireland
{zzhang,ralbatal,cgurrin,asmeaton}@computing.dcu.ie
{rami.albatal}@dcu.ie
https://www.insight-centre.org

Abstract. In this paper, we present an interactive video browser tool for our participation in the fourth video search showcase event. Learning from previous experience, this year we focused on building an advanced interactive interface which allows users to quickly generate and combine different styles of query to find relevant video segments. The system offers the user a comprehensive search interface which has as key features: keyword search, color-region search and human face filtering.

Keywords: Multimedia Indexing, Deep Learning, Human Face Detection, Interactive Interface.

1 Introduction

The Video Search Showcase is an annual live video search evaluation, where teams evaluate and demonstrate the efficiency of their interactive video search tools live, in front of an audience. Two different search categories have been proposed in this year's Video Browser Showcase which are variations of Known-Item Search (KIS) namely Visual KIS and Descriptive KIS. In order to address these advanced evaluation procedures, the classic video retrieval approaches based on query images (or sketch images) may not necessarily be sufficient. Hence, we decided to improve our video retrieval system from last year's participation [1]; in this work we propose a more comprehensive, effective and efficient video browser tool that contains three main retrieval modalities: *a)* Keyword-based Retrieval, *b)* Color-based Retrieval, *c)* Human Face Filtering. By combining multiple retrieval modalities with a flexible and interactive user interface, the system enhances the user's ability to quickly locate the required video segments.

The rest of the paper is organized as follows. Section 2 discusses the different proposed retrieval modalities; then in Section 3 we describe our retrieval interface under development; finally section 4 concludes the work.

X. He et al. (Eds.): MMM 2015, Part II, LNCS 8936, pp. 282–286, 2015.
© Springer International Publishing Switzerland 2015

2 Retrieval Modalities

In order to address the two search categories of this year's challenge, two dimensions are considered when analysing the video collection: a visual and a semantic dimension, hence we provide multiple retrieval modalities that explore the content of the collection from both visual and/or from semantic point-of-views. By adopting state-of-the-art content-based image retrieval technologies, users can employ one or more of the following available retrieval modalities.

2.1 Keyword-Based Retrieval

A keyword-to-image approach is developed to help users in formulating textual queries by proposing keywords from a pre-defined vocabulary. The suggested keywords correspond to pre-trained classifiers used to index the collection. We employ machine learning technologies to analyse the visual content of each video segment, and to index it using a vocabulary that contains two major parts: 1,000 semantic concepts (e.g. screen, sky, indoor, pizza... etc.) and around 100 highly discriminative visual objects (brand logos and alpha-numerical characters). This keyword search modality can be used for both Visual KIS and Descriptive KIS search categories.

Semantic Concept Indexing. In order to allow users to formulate semantic queries, we employed pre-trained deep convolutional neural networks (CNN) (using the Caffe software [8]) to identify concepts in video sample frames. We use the output judgements of the pre-trained model "CaffeNet" to calculate ranked list results. This model is trained on images from ImageNet [9] (an image database organized according to the nouns of the WordNet hierarchy where each node is depicted by an average of +500 images), and it is able to classify 1,000 concepts.

Visual Object Indexing. Unlike the previous approach, which is based on pre-trained models for 1,000 concepts, in this approach we are adding to this list of 1,000 concepts other objects and words extracted from the collection using visual recognition techniques. Here we focus on the occurrence of known visual elements that share the same appearance over different video segments; these elements can be categorised into:

- logos for known pre-selected brands and products; or
- alphanumerical characters.

The identification of these elements is based on extracting HOG features [10] from query objects (that are mined from external visual data sources) then training a linear SVM classifier which is able to identify the occurrence of the query object in the data set sampled key-frames.

For the identification of logos, we are using an approach inspired from the work in [2]. Figure 1 shows that the occurrence of *Starbucks Logo* in two images

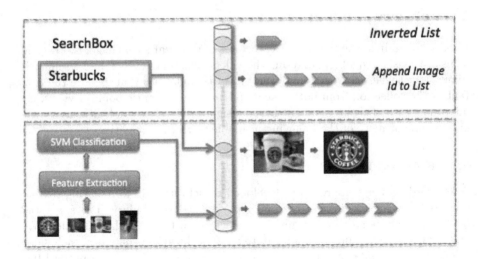

Fig. 1. Inverted Index Structure Based on Semantic Concepts and Visual Object Vocabulary

has led into adding them to the same entry (Starbucks entry) of an inverted index for online search, so both of them can be found now by type "Starbucks" as a query (knowing that Starbucks is not among the 1,000 pre-trained concept in 2.1).

The identification of alphanumeric characters is used in order to estimate what are the possible words that exist in a keyframe. For that we are using the approach of [3] that uses an external lexical base and a specific algorithm for candidate word estimation from identified letters and numbers. In Figure 2, we present examples of estimated words from keyframes taken from last year's query videos. While searching, the user just types an alphanumerical strings that s/he saw in the query video and the system will search for videos containing these strings.

All the video segments are indexed using the Semantic Concept and the Visual Object Indexing approaches; both approaches assign to each segment a score between 0 and 1 indicating the computed degree of presence or absence of a concept or a visual object in the segment. The result is saved in an inverted index file [7] which groups the video segments according to the vocabulary words. During the online retrieval process, a ranking score will be computed based on the decreasing value of presence score.

2.2   Color-Based Retrieval

The signature-based video browser tool from Lokoč et al. [4] performed well in last year's VBS. In their approach, color regions and their position have been extracted from video frames and feature signatures are generated to represent those video frames offline. When interactively searching, users need only to memorise

Fig. 2. An example of identified alphanumeric strings that are used to index keyframes, and thus can be used in the text query to retrieve relevant segments (Images are taken from last year's query videos [1]). Left: *LAAT* and Right: *10, 2010*.

a scene with significant color spot and then draw simple sketches as input to the ranking system. This approach is conceptually simple, yet surprisingly powerful as a method for visual known item search scenario. This approach is successful for two reasons: firstly, it allows users to pick a frame with only few significant color regions by quickly viewing the query video segment; and secondly, it has an interactive interface which help users to produce a simple but flexible sketch query by drawing a few color spots. In this way, we add a color-region based retrieval modality to the system.

2.3   Human Face Filtering

Human face filtering functionality is provided to allow users to filter the videos according to their inclusion of a human faces. This filtering functionality will be be useful for queries involving people and has been considered effective in previous editions of the VBS. To this end, we provide a filter check-box to toggle face requirements. This operates as a filter and if checked, the interface will limit the results to videos that contains human faces, all those that do not contain human faces will be filtered out. We use the Viola-Jones face detector [6] to detect human faces in the videos.

3   System Interactive Interface

The user interface is interactive and flexible, which allows a user to produce an effective search query. The interface is designed to employ multiple retrieval modalities. As a standard framework, a text input box allows users to type keywords and a panel displays the sample frames of the top 100 video segments from the ranking . Users are allowed to scroll left or right to understand the context of each video segment. In addition, the interface includes a canvas panel to help users to quickly draw a sketch or to toggle to the color based retrieval modality. Users are able to choose color, different types of brush and adjust contrast and so on. In addition, the human face filter can be turned on or off for queries which involve people.

4   Conclusion

This paper presents our fourth consecutive participation in the Video Browser Showdown. Learning from the past experience and by taking into account both searches categories of this year (Visual KIS and Descriptive KIS), we propose a video browsing and retrieval system that incorporate different retrieval modalities in order to explore the test video collection from multiple angles.

Acknowledgments. This publication has emanated from research conducted with the financial support of Science Foundation Ireland (SFI) under grant number SFI/12/RC/2289.

References

1. Scott, D., et al.: Audio-Visual Classification Video Browser. In: Gurrin, C., Hopfgartner, F., Hurst, W., Johansen, H., Lee, H., O'Connor, N. (eds.) MMM 2014, Part II. LNCS, vol. 8326, pp. 398–401. Springer, Heidelberg (2014)
2. Malisiewicz, T., Gupta, A., Efros, A.A.: Ensemble of Exemplar-SVMs for Object Detection and Beyond. In: 2011 IEEE International Conference on Computer Vision (ICCV), Barcelona, Spain (2011)
3. Mishra, A., Alahari, K., Jawahar, C.V.: Image Retrieval using Textual Cues. In: 2013 IEEE International Conference on Computer Vision (ICCV), Sydney, Australia (2013)
4. Lokoč, J., Blažek, A., Skopal, T.: Signature-based video browser. In: Gurrin, C., Hopfgartner, F., Hurst, W., Johansen, H., Lee, H., O'Connor, N. (eds.) MMM 2014, Part II. LNCS, vol. 8326, pp. 415–418. Springer, Heidelberg (2014)
5. Mathias, L., Chatzichristofis, S.A.: Lire: Lucene Image Retrieval An Extensible Java CBIR Library. In: Proceedings of the 16th ACM International Conference on Multimedia, Vancouver, Canada, pp. 1085–1088 (2008)
6. Viola, P., Jones, M.: Rapid object detection using a boosted cascade of simple features. In: Proceedings of the IEEE Computer Vision and Pattern Recognition, CVPR, pp. 511–518 (2001)
7. Manning, C.D., Raghavan, P., Schütze, H.: Introduction to Information Retrieval. Cambridge University Press (2008)
8. Jia, Y.: Caffe: An Open Source Convolutional Architecture for Fast Feature Embedding, http://caffe.berkeleyvision.org/ (last accessed September 2013)
9. Deng, J., Dong, W., Socher, R., Li, L.-J., Li, K., Fei-Fei, L.: ImageNet: A Large-Scale Hierarchical Image Database. In: Proceedings of the IEEE Computer Vision and Pattern Recognition, CVPR (2009)
10. Dalal, N., Triggs, B.: Histograms of Oriented Gradients for Human Detection. In: Proceedings of the IEEE Computer Computer Vision and Pattern Recognition (CVPR), pp. 886–893 (2005)

ImageMap - Visually Browsing Millions of Images

Kai Uwe Barthel, Nico Hezel, and Radek Mackowiak

HTW Berlin
University of Applied Sciences – Visual Computing Group
Wilhelminenhofstraße 75a, 12459 Berlin, Germany
barthel@htw-berlin.de

Abstract. In this paper we showcase ImageMap - an image browsing system to visually explore and search millions of images from stock photo agencies and the like. Similar to map services like Google Maps users may navigate through multiple image layers by zooming and dragging. Zooming in (or out) shows more (or less) similar images from lower (or higher) levels. Dragging the view shows related images from the same level. Layers are organized as an image pyramid which is build using image sorting and clustering techniques. Easy image navigation is achieved because the placement of the images in the pyramid is based on an improved fused similarity calculation using visual and semantic image information. Our system also allows to perform searches. After starting an image search the user is automatically directed to a region with suiting results. This paper describes how to efficiently construct an easily navigable image pyramid even if the total number of images is huge.

Keywords: Exploration, Image Browsing, Visualization, Navigation, CBIR.

1   Introduction

The increasing amount of digital images has led to a growing problem of finding particular images. Typically images are searched by keywords. One-keyword searches often lead to too many results, whereas searches with more keywords may find too few or no results at all. If an example image is available, content-based image retrieval (CBIR) can be used to retrieve visually similar images.

Human perception is limited to 20 to 50 unsorted images at a time. Overview is quickly lost if more images are shown. Due to this fact most websites show about 20 images per page. Because most users do only look at the first few pages of an image search result, only a tiny fraction of the entire search result is viewed. For many stock photo agencies the image amount has become so huge that nobody can ever see all images. Most search systems do not offer the possibility to visually browse or explore the entire image collection.

If images are sorted by their visual similarities, then up to several hundreds can be viewed simultaneously [1]. This paper describes a system that allows fast visual navigation through millions of images using a hierarchical pyramid structure of images sorted by visual and semantic similarities. Users can explore subsets of the images through a viewport at different pyramid levels or positions which can be changed by zooming or dragging with the mouse (see Figures 1 and 2).

X. He et al. (Eds.): MMM 2015, Part II, LNCS 8936, pp. 287–290, 2015.
© Springer International Publishing Switzerland 2015

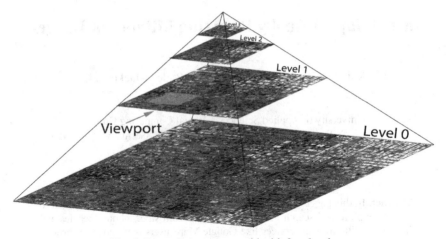

Fig. 1. Example image pyramid with four levels

2   Related Work

While there has been a lot of effort to improve CBIR, there is not much support for exploratory image search. An overview of various visual browsing models for CBIR is given in [3]. Some authors use visual attributes to split images into subsets, others use techniques like storyboards, ISOMAP or SOMs to generate visually sorted arrangements of search results [6, 7, 8]. Approaches like in [6] and [7] help to get a better overview but suffer from unequally positioned and overlapping images. In all cases only a tiny fraction of the entire image set is shown and there is no way to experience relationships with other images. Graph-based techniques like Google Image Swirl are addressing this issue by using image networks [4, 5]. A similar approach to ours has been proposed in [2]. In this article we present a system which combines image search and visual navigation. Our system allows to handle much larger image sets and achieves a far superior visual sorting quality than systems described above.

3   Method

The main idea of our system is to represent millions of images in a way similar to map services like Google Maps where a viewport shows a part of the image map at a certain scale. Zooming in shows more similar images whereas zooming out leads to a better overview. In order to apply this map-approach for searching and browsing images, a hierarchical image organization is required as well as the following two requirements must be satisfied:

1. Images on the lowest level of the image pyramid need to be arranged in such a way that similar images are positioned close to each other.
2. Images of higher levels need to be chosen such that they are good representatives of the corresponding images from the lower levels.

Fig. 2. On the left: a region showing different food items, on the right the zoomed regions of the red marked images (beer, nuts, and oranges) are shown

Obviously it is impossible to perfectly project the complex relationships between all images onto a 2D-plane. However, even if the image sorting in only two dimensions will result in some similarity discontinuities, it still helps the user to visually navigate and find similar images. For sorting the images by their similarities we use a hierarchical torus-shaped SOM. This leads to smoother image transitions and avoids sharp straight visual borders as in [2]. In addition we compute image similarities by fusing visual features and image keywords instead of using visual features only. This helps to better approximate the "real" image similarities. We use a quadtree structure for the pyramid, where each higher level is reducing the number of images by a factor of four. If 16x16 images are displayed on the top, only six more levels are required to represent one million images.

The image pyramid is constructed in four steps. To reduce the set of images to a quantity which the SOM can handle, we first group all images into approximately equally sized clusters using the Generalized Lloyd Algorithm. Then for each cluster the best representative image (the generalized median) is chosen. In the next step only those images are sorted with the SOM. The SOM positions are aligned on a 2D grid where each position may only be occupied once by an image. In the last step the levels below the sorted level are filled with the remaining images from the corresponding clusters. Since these images are very similar it does not matter how they are mapped. For the higher levels the generalized median image of the four images below is used.

Our system also allows for a new way to present the results of image searches. The spatial distribution of the result images is visualized as a heatmap. The position of the maximum of the heatmap selects the initial viewport position. Other positions may be chosen by the user (see Figure 3).

4   Experiments and Conclusion

In our prototype implementation we use 1 million images from Fotolia. By the time of the conference our web-based demo will be using 10 million images. Figure 2 shows views from lower levels when zooming the marked positions of a higher level.

Figure 3 illustrates the result of an image search using the keyword "cow". The heatmap displays regions of images matching the search term. A set of 3x3 images representing those regions are shown below the heatmap. By clicking such an image or the heatmap the user may navigate the viewport to the corresponding region.

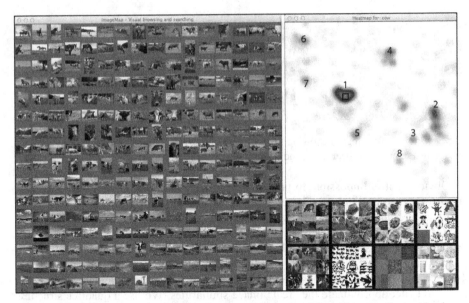

Fig. 3. "Cow"-Image search showing view port, heatmap and typical image clusters (such as 1: cows on meadows, 2: raw and 3: cooked meat, 4: cow comic figures, etc.)

Our presented system is very effective for visually searching and browsing huge numbers of images. It is easy to use and helps to quickly visually search for other images with related concepts. For the reason that the viewport always shows only a limited number of images, this approach can be realized as a web-based application.

References

1. Barthel, K.U.: Automatic Image Sorting using MPEG-7 Descriptors. In: ICOB 2005, Workshop on Immersive Communication and Broadcast Systems (2005)
2. Chen, J., Bouman, C., Dalton, J.: Similarity pyramids for browsing and organization of large image databases. In: SPIE/IST Conf. on Human Vision and Electronic Imaging III (1999)
3. Heesch, D.: A survey of browsing models for content based image retrieval. Multimedia Tools Appl. 40(2) (2008)
4. Jing, Y., Rowley, H., Rosenberg, C., Wang, J., Zhao, M., Covell, M.: Google image swirl, a large-scale content-based image browsing system. In: IEEE ICME (2010)
5. Qiu, S., Wang, X., Tang, X.: Visual Semantic Complex Network for Web Images. In: ICCV 2013 (2013)
6. Strong, G., Hoque, E., Gong, M., Hoeber, O.: Organizing and Browsing Image Search Results Based on Conceptual and Visual Similarities. In: Bebis, G., et al. (eds.) ISVC 2010, Part II. LNCS, vol. 6454, pp. 481–490. Springer, Heidelberg (2010)
7. Wang, J., Jia, L., Hua, X.: Interactive browsing via diversified visual summarization for image search results. Multimedia Systems, 17 (2011)
8. Schoeffmann, K., Ahlstrom, D.: Similarity-Based Visualization for Image Browsing Revisited. In: ISM, pp. 422–427. IEEE Computer Society (2011)

Dynamic Hierarchical Visualization of Keyframes in Endoscopic Video

Jakub Lokoč, Klaus Schoeffmann, and Manfred del Fabro

SIRET Research Group, Department of Software Engineering,
Faculty of Mathematics and Physics, Charles University in Prague, Prague,
Czech Republic
Klagenfurt University, Universitaetsstr. 65-67, 9020 Klagenfurt, Austria
lokoc@ksi.mff.cuni.cz, {ks,manfred}@itec.aau.at

Abstract. The after-inspection of endoscopic surgeries can be a tedious and time consuming task. Physicians have to search for important segments in the video recording of an intervention, which may have a duration of several hours. Automatically selected keyframes can support physicians in this task. The problem is that either too few keyframes are selected, missing some important information, or too many keyframes are selected, which overwhelms the user. Furthermore, keyframes of endoscopic videos typically show highly similar content. It is hence difficult to keep track of the temporal context of selected keyframes if they are presented in a grid view. To overcome these limitations, we present a dynamic hierarchical browsing technique for large sets of keyframes that preserves the temporal context in the visualization of the frames.

1 Introduction

In the field of medical endoscopy over the last years clinicians have adopted to archive recordings from endoscopic interventions. The reasons for this development are manifold. First, the videos show a first-hand perspective of the surgeons' work and are hence a good source of information for future interventions of the same patient. Secondly, it can be used as explanatory material for training of young surgeons. Moreover, the recording and archival of endoscopic interventions are enforced by law in some countries (e.g., The Netherlands)[3].

Since an archive of endoscopic video is typically growing quickly, with lots of new and long videos added on a daily basis in a hospital, it is especially important to provide expressive and distinctive keyframes to facilitate later browsing of the video material [5,1]. However, unlike classical broadcast videos the endoscopic videos are far more complex data for automatic content-based analysis, which makes the selection of a good set of representative keyframes a challenging problem. The content does not contain any shot boundaries, but segments with the same content over a long time period, as well as re-appearing content. Therefore, keyframe extraction with uniform sampling or common keyframe selection strategies [4] typically produce a large result set with hundreds of keyframes showing highly similar content. Browsing such a large result set with a lot of

X. He et al. (Eds.): MMM 2015, Part II, LNCS 8936, pp. 291–294, 2015.

redundancies is an inconvenient process for clinicians who want to get a quick overview of the most important segments in the video. Moreover, as selected keyframes might be non-linearly distributed along the timeline, the physicians can also lose the notion of the corresponding time location, which makes the orientation in the result set even more complicated.

Therefore, we propose a dynamic hierarchical visualization of keyframes for browsing endoscopic video at different levels of detail. More precisely, our approach first samples keyframes based on motion-flow [2] in the endoscopic video content and then performs a hierarchical clustering with a uniformity constraint and a special visualization that shows where the keyframes are located in the video. The visualization uses a compact layout and starts at the root level of the clustering tree but allows the user to zoom in and out at any temporal position in the video. To optimally preserve the browsing context, results of navigation actions (i.e., appearance of new and disappearance of old keyframes) are smoothly animated, which allows for convenient browsing of endoscopic video even for longer recordings with hundreds of keyframes.

2   Keyframe Extraction and Hierarchical Visualization

In order to present potentially highly ranked representatives in a user-friendly way and keeping fluency of the browsing, our summary browser combines several approaches described in the following paragraphs.

In the preprocessing phase, keyframes are detected in the endoscopic video based on tracking of keypoints with the KLT tracker [2]. We start with a dense grid of keypoints [6] and try to track them over time as long as possible. Due to motion in the video, more and more keypoints get lost over time and cannot be tracked further. As soon as the number of initial keypoints falls below a specific threshold (e.g., less than 50%), we extract a keyframe and restart keypoint tracking with a new grid of densely sampled keypoints.

Next, a hierarchical clustering is performed on the set of selected representatives resulting in a dendrogram that is used to assign a priority to each keyframe. More specifically, we employ an agglomerative hierarchical clustering where the cluster similarity is based on the Ward's method

$$L_2^2(M(C_i), M(C_j)) \cdot (|C_i| \cdot |C_j|)/(|C_i| + |C_j|),$$

where C_i, C_j are two clusters, $M(C)$ denotes the mean of all keyframes in the cluster and L_2^2 is the squared euclidean distance defined over two keyframes of the same size, where each pixel is represented by three values from the Lab color space. Furthermore, we allow just merging of two adjacent clusters, where the adjacency is based on the keyframe timestamp. For a new cluster, the timestamp is computed as weighted combination of two merged clusters. As a result of the clustering, a binary dendrogram for a set of representatives is obtained. Then the priority of each representative is assigned in such a way that the left most object (the lowest timestamp) in each cluster is assigned the highest priority.

The browser also employs timestamps for timeline-based visualization of actually visible top k representatives (based on their priority), where the actually visible part/range of the video is controlled by a browsing user. The displayed images are pinned to the timeline based on their real position in the video. Let us note, placement of displayed representatives considers minimal overlap of the already depicted keyframes, because, as the representatives are distributed non-uniformly, a trivial placement technique would result in overlaps.

3 Dynamic Browsing of Clustered Keyframes

In order to browse the keyframes, the user can employ several browsing operations. As an initial view, the user sees the whole time-line period with top k representatives (Figure 1). At any position of this visualization, the user can use the mouse wheel for zooming in or out, which results in a smoothly animated visualization of more or less keyframes for the selected time region. After each zoom in/out operation, the middle timeline panel adjusts its border timestamps while the top and bottom panels just depict the temporal position of the actually visible range using black rectangles. In order to navigate the user to parts with yet invisible images, the timeline panels depict also positions of the images using gray circles.

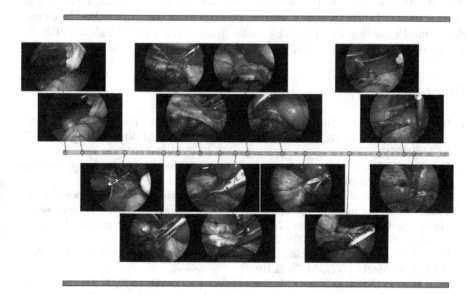

Fig. 1. Browsing endoscopic videos through an interactive hierarchical visualization of keyframes

The zoom out operation is implemented as simple undo operation. The zoom in/out operations try to preserve the context of the browsing using animations that move already displayed objects to their new locations (some may disappear if they reach the border) and after the objects are moved, the new objects with a high ranking in an actually visible range are placed and displayed. The user can also simply shift left/right using simple left/right mouse click, shifting the border timestamps by a given constant.

4   Conclusions

In this demo paper we have presented a video browser for endoscopic videos, which is based on a novel extraction and visualization of keyframes. The keyframe extraction uses significant motion changes in order to detect distinctive frames in the whole endoscopic video containing highly similar content. The visualization uses a compact layout of keyframes that shows the temporal positions the currently displayed keyframes as well as an indication of further available keyframes at higher levels of detail. A user can navigate within the tree of keyframes by using the mouse wheel and thus navigate deeper into the hierarchy for a specific temporal segment. The proposed video browser conveniently allows (1) to see a quick overview of representative frames in endoscopic video and (2) to see the details for specific temporal regions when selected by the user.

Acknowledgement. This work was supported by Lakeside Labs GmbH, Klagenfurt, Austria, by Czech Science Foundation project P202/12/P297 and by funding from the European Regional Development Fund and the Carinthian Economic Promotion Fund (KWF) under grant KWF - 20214 22573 33955. We would also like to thank to AKTION Czech Republic – Austria programme.

References

1. Hürst, W., Meier, K.: Interfaces for timeline-based mobile video browsing. In: Proceedings of the 16th ACM International Conference on Multimedia, pp. 469–478. ACM (2008)
2. Lucas, B.D., Kanade, T., et al.: An iterative image registration technique with an application to stereo vision. IJCAI 81, 674–679 (1981)
3. Münzer, B., Schoeffmann, K., Böszörmenyi, L., Smulders, J.F., Jakimowicz, J.: Investigation of the impact of compression on the perceptional quality of laparoscopic videos. In: Proceedings of the 27th IEEE International Symposium on Computer-Based Medical Systems (CBMS 2014), pp. 1–6 (2014)
4. Peng, J., Xiao-Lin, Q.: Keyframe-based video summary using visual attention clues. IEEE MultiMedia 17(2), 64–73 (2010)
5. Schoeffmann, K., Del Fabro, M., Szkaliczki, T., Böszörmenyi, L., Keckstein, J.: Keyframe extraction in endoscopic video. In: Multimedia Tools and Applications, pp. 1–20 (to appear, 2014)
6. Shi, J., Tomasi, C.: Good features to track. In: Proceedings of 1994 IEEE Computer Society Conference on Computer Vision and Pattern Recognition, CVPR 1994, pp. 593–600 (June 1994)

Facial Aging Simulator by Data-Driven Component-Based Texture Cloning

Daiki Kuwahara[1], Akinobu Maejima[2], and Shigeo Morishima[3]

[1]Waseda University,
[2]Waseda University(currently working at OLM Digital Inc.),
[3]Waseda Research Institute for Science and Engineering
kuwahautbois@fuji.waseda.jp, shigeo@waseda.jp

Abstract. Facial aging and rejuvenation simulation is a challenging topic because keeping personal characteristics in every age is difficult problem. In this demonstration, we simulate a facial aging/rejuvenating only from a single photo. Our system alters an input face image to aged face by reconstructing every facial component with face database for target age. An appropriate facial components image are selected by a special similarity measurement between current age and target age to keep personal characteristics as much as possible. Our system successfully generated aged/ rejuvenated faces with age-related features such as spots, wrinkles, and sagging while keeping personal characteristics throughout all ages.

Keywords: Facial Aging Simulation, Face Image Synthesis, Seamless Texture Synthesis.

1 Introduction

Because a human face shows unique characteristics of a person strongly, many research into expressing and recognizing facial information has been proposed in computer graphics and computer vision. Among them, synthesis of an aged or rejuvenated face image of a person is an important and still challenging problem for criminal investigations, kidnapping victim searching, a long term face authentication, an entertainment, and so on.

One of the most typical approach to generate aged facial texture and shape is based on regression of 3D Morphable Model(3DMM) parameters[5]. This approach, however, cause blurs on the synthesized image because 3DMM generate a face as a linear combination of faces in database. In addition, it can generate only one face for each age. Because of indetermination of aging progress, to propose some possible candidates is more desired. Another approach is to reconstruct an aged facial image with other persons' images directly. Maejima et al. [3] proposed an aging method by facial texture reconstruction with age-specific small square patches each of which is a part of another person in the target age group. Though its result image has age related features such as spots and pigments of facial skin, such small patch-based texture synthesis has a difficulty to reflect aging effect to each component in the face. On the other hand,

X. He et al. (Eds.): MMM 2015, Part II, LNCS 8936, pp. 295–298, 2015.

Suo et al. [6] takes a component-based reconstruction method in which aging process is modeled by a Malkov Chain and the simulation is gradually performed, which possibly cause the great difference between the input image and the result one. Moreover, their method needs not only images but also manually obtained labels of the wrinkles of the images in database.

In this paper, we propose a novel facial aging/rejuvenating simulator. Our technical contribution is that we adopt a component-based reconstruction approach with a constraint to keep the personality of the input face. In addition, our system needs only face image database with age labels. Considering the variation of aging, some aged/rejuvenated faces are automatically generated with the rank of plausibility.

In our system, the aged/rejuvenated faces are simulated as follows. First, facial feature points of the input face are detected [2]. Second, appropriate components images for the reconstruction are searched in database (section 2). Then the input face are warped to reflect aging changes in facial shape, and selected components are cloned into the warped input face (section 3).

2   Personality-Keeping Aging Dynamics

To describe the aging dynamics, it is ideal but difficult to get photos of a person over long time. Therefore, we learned the aging dynamics across many individuals in database based on [6] with some modification to keep the identity of a person more strongly. In Suo's dynamic model for an aging simulation [6], the aging process is represented as Malkov Chain and a component of a face changes to the best matching one in next age group photometrically and geometrically. In this way, however, it is possible that the selected component after an interval of several age groups is far from the beginning because of accumulation of difference.

To overcome this problem, we compare the facial component not only to the last selected one but also the one of input image in two steps. First, our system selects components of several persons as candidates by photometric and geometric similarity. Then, we compare each of them to input image by the same standard, and the most similar one is the chosen for our simulation.

We decompose a face to 3 components, "mouth", "eyebrows", and "whole face". "Whole face" is meant for the expression of age-related whole face features such as spots and wrinkles. The features we used in component selection is weighted summation of follows: thin-plate spline distortion of facial feature points between two part, Euclid distance of Histogram of Oriented Gradients(HOG) and the intersection of color histogram around the points.

3   Natural Image Synthesis by Texture Cloning

In this section, we describe how to synthesis natural aged/rejuvenated face images with the components selected in section 2.

First, to reflect the effect of the whole face shape change by aging, the input face image is warped. The amount of change is the distance between the average shape of the faces in an] group database and the average shape of the ones in a target age group database. Then, the component images selected in section 2 are warped to fit the warped input image.

After the adjustment of shape, the components are cloned into the input image. This cloning process has three steps. In the first step, the "whole face" component is seamlessly cloned into the warped input image as same as Maejima's method [3], which is basically based on a Modified Poisson Blending by Tanaka et al. [7] but modified to keep the color of the target image. Additionally, we turn the source image monochrome beforehand. This cloning method is robust to the difference of lighting or racial differences from the images in database. Secondary, the "mouth" and "eyebrows" are cloned into the input image by Original Poisson Blending [4], which keeps the edges of the source image more clearly than Modified Poisson Blending. Finally, we partially restore the warped input image to keep the personality strongly because, in our experience, eyes and a nose have great importance to recognize the personality.

4   Results and Future Works

We tested our system with database consists of frontal face images of Japanese 184 males and 164 females in the range between 20 and 80. Fig. 1 shows results generated as rank 1 by this system. Yellow outlined images are original images. We can see that our method enabled natural aged/rejuvenated face synthesis, i.e. the age-related features such as wrinkles, spots and sagging gradually appear through the aging progress keeping his/her personal characteristics. Other results are shown in Fig. 2. These are challenging situation because the database only contains Japanese, frontal face images. Even the difficulty, the results look sufficiently plausible. We are sure that the better results are given with more variegated database. In addition, we can get several other probable results using the similarities calculated in section 2 interactively through our GUI in Fig. 3.

In the future work, we plan to take a method to replace hair into our system. We would also like to experiment with public database and evaluate our system.

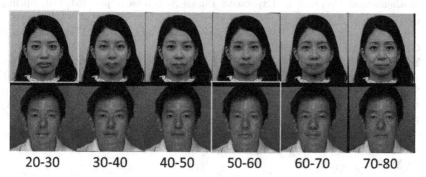

20-30   30-40   40-50   50-60   60-70   70-80

Fig. 1. Simulated Results. Images outlined yellow are original input images.

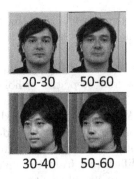

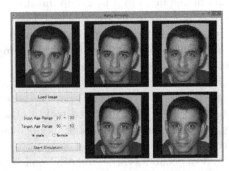

Fig. 2. Results in challenging situation. These are generated without non-frontal images or non-Japanese ones.

Fig. 3. GUI of our system. From an image, current and target age groups, our system generates four probable age changed images. This input image is from FGNET Aging Database [1].

Acknowledgement. This work was supported by R&D Program for Implementation of Anti-Crime and Anti-Terrorism Technologies for a Safe and Secure Society, Funds for integrated promotion of social system reform and research and development of the Ministry of Education, Culture, Sports, Science and Technology, the Japanese Government.

References

1. Fg-net aging database (2002), http://www-prima.inrialpes.fr/FGnet/
2. Irie, A., Takagiwa, M., Moriyama, K., Yamashita, T.: Improvements to facial contour detection by hierarchical fitting and regression. In: 2011 First Asian Conference on Pattern Recognition (ACPR), pp. 273–277. IEEE (2011)
3. Maejima, A., Mizokawa, A., Kuwahara, D., Morishima, S.: Facial aging simulation by patch-based texture synthesis with statistical wrinkle aging pattern model. In: Mathematical Progress in Expressive Image Synthesis I, pp. 161–170. Springer (2014)
4. Pérez, P., Gangnet, M., Blake, A.: Poisson image editing. ACM Transactions on Graphics (TOG) 22, 313–318 (2003)
5. Scherbaum, K., Sunkel, M., Seidel, H.-P., Blanz, V.: Prediction of individual non-linear aging trajectories of faces. Computer Graphics Forum 26(3), 285–294 (2007)
6. Suo, J., Min, F., Zhu, S., Shan, S., Chen, X.: A multi-resolution dynamic model for face aging simulation. In: IEEE Conference on Computer Vision and Pattern Recognition, CVPR 2007, pp. 1–8. IEEE (2007)
7. Tanaka, M., Kamio, R., Okutomi, M.: Seamless image cloning by a closed form solution of a modified poisson problem. In: SIGGRAPH Asia 2012 Posters, p. 15. ACM (2012)

Affective Music Recommendation System Based on the Mood of Input Video

Shoto Sasaki[1], Tatsunori Hirai[1], Hayato Ohya[1], and Shigeo Morishima[2]

[1] Waseda University,
[2] Waseda Research Institute for Science and Engineering,
3-4-1 Okubo Shinjuku-ku Tokyo, Japan, 169-8555

Abstract. We present an affective music recommendation system just fitting to an input video without textual information. Music that matches our current environmental mood can enhance a deep impression. However, we cannot know easily which music best matches our present mood from huge music database. So we often select a well-known popular song repeatedly in spite of the present mood. In this paper, we analyze the video sequence which represent current mood and recommend an appropriate music which affects the current mood. Our system matches an input video with music using valence-arousal plane which is an emotional plane.

Keywords: music recommendation, image processing, valence, arousal.

1 Introduction

Music that matches current environmental mood can enhance a deep impression. However, we do not know which music best matches our present mood. We have to listen to each song, searching for music that matches our mood. As it is difficult to select music manually, we need a recommendation system that can operate affectively. Most recommendation methods, such as collaborative filtering or content similarity, do not target a specific mood. In addition, there may be no word exactly specifying the mood. Therefore, textual retrieval is not effective. In this paper, we assume that there exists a relationship between our mood and videos because visual information affects our mood when we listen to music. We now present an affective music recommendation system using an input video without textual information. Our system provides a new way to enjoy music by listening to music that matches our current mood.

Our system matches an input video with music using an emotional plane. Russell proposed the valence/arousal model of affect[1]. Valence refers to positive-negative associations with affective phenomena, whereas arousal refers to energetic-calm associations. All affective stimuli can thereby be defined as a combination of these two independent dimensions. In addition, as the distance on emotional plane is close, they are similar impressions. Thus, it is reasonable to use this emotional plane (V-A plane) for matching music and a video.

X. He et al. (Eds.): MMM 2015, Part II, LNCS 8936, pp. 299–302, 2015.

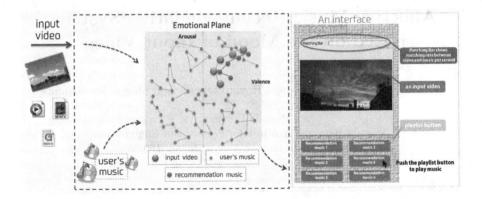

Fig. 1. Outline of our system

2 Proposed System

The system consists of three parts. Figure 1 outlines the system.

2.1 Setting an Input Video

The first part involves setting the image on the V-A plane using color and texture features. Patricia and Albert demonstrated by means of equations the relationship of saturation S, brightness B with valence V_1, arousal A_1[2]. These equations are given below.

$$V_1 = 0.69B + 0.22S \tag{1}$$

$$A_1 = -0.31B + 0.60S \tag{2}$$

We use average of each pixel of saturation and brightness. By this relationship, the system reflects the color impression of the image.

Meanwhile, we can establish associations of coarseness C, direction D with valence V_2, arousal A_2 using canonical correspondence analysis(CCA) as follow.

$$V_2 = 4.57(D - 0.41) + 3.95 \tag{3}$$

$$A_2 = -0.29(C - 44.93) + 4.26 \tag{4}$$

The learning data is the International Affective Picture System(IAPS)[3], which is a standard affective image set used in psychology. By this relationship, the system reflects the texture's impression of the image.

We calculate V-A values every 1 second using these correlations. In addition, the system integrates (V_1, A_1) with (V_2, A_2) on the basis of distances from the origin of the V-A plane. This is based on impressions being stronger toward the edge of the V-A plane than at the origin.

In this way, the system determines the impression of the image by setting on V-A plane.

2.2   Setting the User's Music

The second part involves setting the user's music onto the V-A plane. Eerola et al. have shown relationship of acoustic features of 29 dimensional with valence, arousal [4]. And they demonstrated using principal components analysis(PCA) that acoustic features can be related to valence and arousal. In this paper, we use the RWC Music database, which has equally populated sets of music categories [5]. And acoustic features f are normalized using following equation.

$$f' = \frac{f - f_{ave}}{f_{std}} \tag{5}$$

f_{ave} is average of acoustic features. f_{std} is standard deviation of acoustic features. Coefficients are computed by PCA with the acoustic features of the RWC Music database. We then multiply these coefficients by the acoustic features of the user's music, thus obtaining V-A values every 1 second for that music.

In this way, the system determines the impression of the music by setting on V-A plane.

2.3   Recommendation

The third part is musical recommendation. Figure 1 shows our system. The system calculates the Euclidean distances of transition between an input video and the user's music on the V-A plane. The system then arranges the distances in ascending order, thereby creating a recommended playlist.

3   Evaluation

Our proposed method was assessed using subjective evaluation by the random selection method. Six questions were asked to a group of 11 male and female examinees, all in their twenties. The subjects evaluated at five levels on which music is better matched with the video on display. A score of 5 indicated that recommended music matched the video on display. A score of 1 indicated that randomly selected music matched the video on display. Figure 2 shows the results of this experiment. The average score is 3.95. Therefore, our proposal technique is confirmed to be effective.

The results show that questions 1 have not obtained good scores with the proposed method. The system sets question 1 at the origin of the V-A plane, indicating neutral emotion. Hence, video or music in this area tend to affect individuals differently. Therefore, an interactive system is necessary to deal with individuality. For instance, if a user pushes a skip button, the system can learn the user's preferences and make subsequent recommendations by taking them into account.

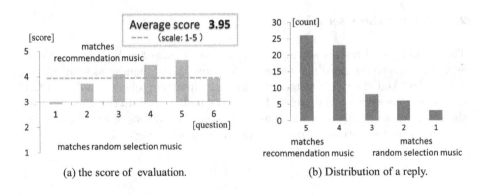

(a) the score of evaluation.   (b) Distribution of a reply.

Fig. 2. Result of evaluation experiment

4   Conclusion

In this paper, we have presented an affective music recommendation method based on an input video using an emotional plane without textual information. It is possible to easily check, impressions of color and texture are what effect on music. Further work on adding an interactive system and improvement in the accuracy by detection of the attention object by image processing, such as segmentation, are being considered.

References

1. Russell, J.: A circumplex model of affect. Journal of Personality and Social Psychology 39(6), 1161–1178 (1980)
2. Patricia, V., Albert, M.: Effects of color on emotions. Journal of Experimental Psychology: General 123(4), 394–409 (1994)
3. Lang, P.J., Bradley, M.M., Cuthbert, B.N.: International aective picture system (IAPS): Aective ratings of pictures and instruction manual. Technical report A-6 University of Florida Gainesville (2008)
4. Eerola, T., Lartillot, O., Toiviainen, P.: Prediction of multidimension alemotional ratings in music from audio using multivariate regression models. In: Proceedings of the 10th International Conference on Music Information Retrieval, pp. 621–626 (2009)
5. Goto, M., Hashiguchi, H., Nishimura, T., Oka, R.: RWC music database: Music Genre Database and Musical Instrument Sound Database. In: Proceedings of the 4th International Conference on Music Information Retrieval, pp. 229–230 (2003)

MemLog, an Enhanced Lifelog Annotation and Search Tool

Lijuan Marissa Zhou, Brian Moynagh, Liting Zhou, TengQi Ye,
and Cathal Gurrin

Insight Centre for Data Analytics,
Dublin City University, Dublin 9, Ireland
mzhou@computing.dcu.ie

Abstract. As of very recently, we have observed a convergence of technologies that have led to the emergence of lifelogging as a potentially
pervasive technology with many real-world use cases. While it is becoming easier to gather massive lifelog data archives with wearable cameras
and sensors, there are still challenges in developing effective retrieval
systems. One such challenge is in gathering annotations to support user
access or machine learning tasks in an effective and efficient manner. In
this work, we demonstrate a web-based annotation system for sensory
and visual lifelog data and show it in operation on a large archive of
nearly 1 million lifelog images and 27 semantic concepts in 4 categories.

1 Introduction

Digital recording of life experience is gaining popularity with the emergence of
a new generation of wearable sensors. Lifelogging refers to the idea that an individual can utilise wearable sensors to generate a rich archive of life experience
[2]. Typically this is done using a number of sensors, including a wearable camera. Developing effective lifelog search engines and management tools is very
challenging for reasons such as the lack of available test collections and the difficulty of searching through multimodal data archives. Central to this challenge
is the idea that sufficiently detailed annotations of lifelog data, both for use in
supporting human analysis of lifelog data, or as a source of training data for
machine-learning-based semantic enrichment tools, are difficult, expensive and
time-consuming to generate.

In this paper, we present a lifelog management framework called MemLog,
which is based on extensive experience of managing and annotating lifelog archives.
Using the MemLog system, lifelog data can be uploaded from a variety of lifelogging devices into a cloud-based service and annotated in an efficient manner, using both shared and personal concepts. Instead of simply displaying a sequential
list of images for annotation, which can number many thousands per day, Mem-
Log can compress consecutive images for efficient annotation and segment their
lifelog recordings into multiple life events. Given a starting set of user annotations,
MemLog suggests multiple most likely labels for each event based on an underlying machine learning technique. Although MemLog is designed for lifelog data

X. He et al. (Eds.): MMM 2015, Part II, LNCS 8936, pp. 303–306, 2015.

management, it but also can be applied to other temporally organised multimedia content.

2   Background

Lifelogging is becoming a normative activity and much of the research into lifelogging has used wearable cameras, along with other sensors, to capture life experience in high-fidelity. Wearable cameras, such as the SenseCam are designed to automatically capture all aspects of wearer's day[3] in rich visual detail. When hung on a lanyard around the neck, the SenseCam could take photos automatically of the activities of the individual, from their viewpoint. Up to 4,000 pictures could be captured daily. There are many commercial offerings also, such as the Narrative Clip or OMG Autographer, along with smartphone-based offerings.

Hence, this has provided us with the ability to gather and store large volumes of personal data in a straightforward manner, using off-the-shelf tools. Once the lifelog data is gathered, to be useful, it must be indexed in some manner and made available through an interface. There have been a number of lifelog data browsing interfaces developed. The first of these was the SenseCam image browser which facilitates annotations of images of interest and also rapid playback of image sequences from Hodges et al. [3]. Other systems include the event-based SenseCam image browser which automatically segments lifelog data into events or episodes and then allows users to manually annotate those events [1]. The authors have previously developed the ShareDay lifelog system that supports reminiscence through incorporating event segmentation and group sharing [5]. From all these tools, the only annotation tool for large lifelog archives is from Doherty et al. [1], which simply presents a sequence of images for multipass annotation. The contribution of the system proposed in this paper is that it is flexible to the individual annotation needs of the user, easy to manage and it actively utilises the user annotations in order to generate additional suggested concept annotations.

3   MemLog Overview

The MemLog system is an end-to-end lifelog management and annotation system and a screen-shot of MemLog is shown in Figure 1, in which an annotator can be seen annotating a piece of content. On the left of the screen is the calendar to access the lifelog data, on the right is the annotation panel and the main part of the screen is the data panel, which displays (in this case) a full-sized playback view of the event being annotated. MemLog has been designed with multiple attributes that meet practical research purposes, including both local and remote data uploading, manual and automatic data annotation, and quality checking etc. These attributes have been gathered after extensive consideration of the needs of large-scale lifelog annotation efforts. Aside from multiple user types, uploader tools for multiple lifelogging devices and annotation quality checks and

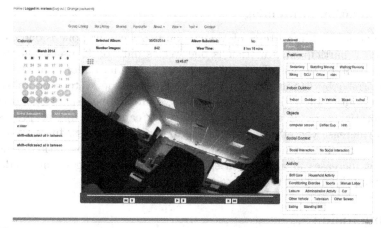

Fig. 1. MemLog in Use, the main interface with data picker and tag picker

validations, the most relevant and important novel attributes of MemLog can be summarised thus:

- *Extensible Categories and Concept Labels.* Experiences of large-scale annotation suggests that a mix of shared and user-specific concepts is needed. Hence, a core set of 27 concepts in 5 categories are available. For personalised concept annotation, users can add their own personal label sets to the predefined concepts. Concepts can be hierarchical or independent.

- *Automatic Label Recommendation.* A lifelog for an individual user can surpass 4,000 images per day and poses major challenges for manual annotation [2], hence Memlog integrates the first lifelogging event-based automatic concept label suggestion tool, which is considered necessary for efficient and effective management of large volumes of content[4]. Events represent repeating life patterns and these events are detected automatically (using supervised machine learning) by MemLog and by computing the visual similarity of lifelog events, as well as the repeating patterns of user activity, labels for current lifelog events are recommended using previously annotated labels. Annotators can check and modify these recommended labels, reviewing them in a dedicated review screen for validation and update.

- *Event Reviewing.* Event reviewing shows storyboards of daily events to the user for review or annotation. All events are presented to users chronologically with key-frames of each event. Each event can be demonstrated either in a conventional thumbnail view or in movie view (sequential playback of lifelog images). Selecting an image allows the user overview all annotations assigned to that event, including more general details of events like start/end time and number of pictures in that event. Users can modify an annotation if they think it is not correct for that event and add new concepts for that event if it is not automatically added.

3.1 Evaluation of Memlog

We have performed inital tests of Memlog on two data-sets, one is one month's data from 7 participants, another is one week's data of 46 users. Our results have shown it to be an effective annotation tool. The annotation evaluation was performed by 6 non-expert interns who tested the system by uploading, annotating and conducting quality checking. Our findings show that dividing the annotation process into different categories can speed up the process by 70% and increase the effectiveness of the annotation process. In addition, the event segmentation technique with annotation label recommendation allows for significant increases in annotation speed as a benefit of tag recommendation. Quantifying this impact is one of our future research tasks.

4 Conclusion

There is a lack of a useful visual data annotation engine in the lifelog research area. In this paper, we introduced MemLog, a lifelog management system for lifelog researchers to annotate large volumes of lifelog data in a quick and efficient manner. We believe that the system is not only applicable for lifelog researchers, but also more generally for image crowdsourcing organizations, for reasons such as data annotation, training data gathering for machine learning and enhanced quality multimedia data retrieval. In the future, we are planning to continue using this system to facilitate additional multi-modal data research projects and are exploring how we can potentially make this software available for others to use.

Acknowledgements. This publication as emanated from research conducted with the financial support of Science Foundation Ireland, under grant nos. 11/RFP.1/CMS/3282 and 13/TIDA/I2875(FT).

References

1. Doherty, A.R., Moulin, C.J.A., Smeaton, A.F.: Automatically assisting human memory: A sensecam browser. Memory 19(7), 785–795 (2011)
2. Gurrin, C., Smeaton, A.F., Doherty, A.R.: Lifelogging: Personal big data. Foundations and Trends® in Information Retrieval 8(1), 1–125 (2014)
3. Hodges, S., Williams, L., Berry, E., Izadi, S., Srinivasan, J., Butler, A., Smyth, G., Kapur, N., Wood, K.: SenseCam: A retrospective memory aid. In: Dourish, P., Friday, A. (eds.) UbiComp 2006. LNCS, vol. 4206, pp. 177–193. Springer, Heidelberg (2006)
4. Sarin, S., Fahrmair, M., Wagner, M., Kameyama, W.: Holistic feature extraction for automatic image annotation. In: 2011 5th FTRA International Conference on Multimedia and Ubiquitous Engineering (MUE), pp. 59–66. IEEE (2011)
5. Zhou, L.M., Caprani, N., Gurrin, C., O'Connor, N.E.: ShareDay: A novel lifelog management system for group sharing. In: Li, S., El Saddik, A., Wang, M., Mei, T., Sebe, N., Yan, S., Hong, R., Gurrin, C. (eds.) MMM 2013, Part II. LNCS, vol. 7733, pp. 490–492. Springer, Heidelberg (2013)

Software Solution for HEVC Encoding and Decoding

Shengbin Meng, Jun Sun, and Zongming Guo

Institute of Computer Science & Technology, Peking University
No.5 Yiheyuan Road, Beijing, China
{shengbin,sunjun,guozongming}@pku.edu.cn

Abstract. In this demonstration, we showcase a complete software encoding and decoding solution for the new High Efficiency Video Coding (HEVC) standard. The encoder is optimized for x86 processors using SSE instruction set extension and multi-thread technology, and achieves high efficiency at a significantly reduced computation load. We have integrated the encoder library into the widely-used media framework FFmpeg and developed transcoding and recording applications for HEVC. The decoder is highly optimized for both x86 and ARM architecture. With novel single-instruction-multiple-data (SIMD) algorithms and a frame-based parallel framework for multi-core CPUs, decoding speed of 46FPS for 1080p videos on ARM Cortex-A9 1.5GHz dual-core processor and 75FPS for 4K (3840x2160) videos on Intel i7-2600 3.4GHz quad-core processor can be achieved. We have also integrated the decoder library into FFmpeg and made an Android video player based on that. The software solution can well meet the demand of producing and watching HEVC videos on existing devices, showing promising future of HEVC applications.

Keywords: HEVC, codec, software implementation, SIMD, optimization.

1   Introduction

The new video coding standard High Efficiency Video Coding (HEVC) [1] introduces some enhanced coding tools and manages to save about 50% bit-rate at the same video quality, comparing with its predecessor H.264/AVC. However, the computational complexity has also increased and become an inevitable obstacle for HEVC's popularity. In order to increase the amount of HEVC video content, efficient transcoding and recording, or basically, encoding, for HEVC has practical demand and becomes the prerequisite of the HEVC industrialization. On the other side, for HEVC experience to reach large scale of users, it's necessary to achieve real-time UHD/HD video decoding under the limited capacity of existing personal computers and mobile devices. Before hardware HEVC encoding/decoding chips are produced and dominate the market, the implementation of fast software HEVC encoder/decoder is essential and challenging, and will be a long-lasting demand. To the best of our knowledge, existing published work related to improving software HEVC encoding or decoding speed is mostly based on the reference software HM [2]. And apart from

X. He et al. (Eds.): MMM 2015, Part II, LNCS 8936, pp. 307–310, 2015.
© Springer International Publishing Switzerland 2015

general performance evaluation, these works provide neither technical details nor any source code for reference, thus being less beneficial for practical HEVC application.

In this demo paper, we present a complete software solution for HEVC, including optimized encoder and decoder libraries, integration into well-known open source media framework and some applications.

2   Implementation, Optimization and Performance

2.1   The Encoder

The reference software HM [2] contains an implementation of encoder. However, this implementation is not suitable for practical application due to its redundant structure and inefficient code. So we start from scratch and implement a totally new HEVC encoder, which is written in C language and then highly optimized for x86 processors. The optimization mainly includes novel data-level and task-level methods.

On data level, optimal SIMD algorithms are designed for the enhanced coding tools. We rewrite the time-consuming modules (e.g., motion compensation, integer transform, deblocking) using Streaming SIMD Extensions (SSE) instructions [3], which make the encoder about 2~3 times faster than the C version. On task level, an Inter-Frame Wavefront (IFW) method based on HEVC's native Wavefront Parallel Processing (WPP) [4] design is introduced to parallelize the encoding process of Coding Tree Blocks (CTBs), and achieves corresponding encoding speedup when running on multi-core CPUs with multiple threads.

The performance evaluation is conducted on Intel Xeon E5620 2.40 GHz processor under Windows Server 2008 operating system. We choose the x265 encoder [5], successor of the best-in-class H.264/AVC encoder x264, to be the baseline of the performance comparison. The best quality preset (with slowest encoding) and the 8-bit HEVC standard test sequences from Class C (832x480 resolution) [6] are used to test the encoding quality and speed. From the data in Table 1 it can be seen that, compared with x265, the proposed encoder (named Lentoid, as shown in the table) not only achieves about -20% BD-rate (coding gain) [7], but also runs nearly 8 times faster.

2.2   The Decoder

In our HEVC decoder solution, to guarantee the effectiveness of optimization, an efficient decoder prototype other than HM is designed. For the most time-consuming decoding modules, novel SIMD algorithms are designed and implemented for both x86 processors, using the SSE instruction set, and ARM processors, using the NEON instruction set [8]. We then use a frame-based parallel framework to achieve several times decoding speedup with multi-thread strategies on multi-core CPUs.

For x86 architecture, the performance evaluation is conducted on Intel i7-2600 3.4GHz quad-core processor with 8GB memory and Microsoft Windows 7 operating system. For ARM, the experiments are conducted on Xiaomi Mi2 [9] smartphone with Qualcomm Snapdragon S4 Pro APQ8064 Quad-core 1.5GHz Cortext-A9 processor, 2GB memory and Android 4.1 operating system. Considering the trending application scenarios, we present the test data for 4K (3840x2160) video decoding on the x86 device and 1080p (1920x1080) video decoding on the ARM device, in Table 2 and Table 3, respectively. The 1080p videos are selected from HEVC standard test sequences; and the 4K videos are downloaded from [10]. It can be concluded that, with the highly optimized decoder, real-time (24 FPS) playback is available for 4K videos on the x86 PC and for 1080p videos on the quad-core smartphone.

Table 1. Performance evaluation of the proposed encoder, comparing with x265 (both running at the best quality preset)

832x480 Sequence	QP	Bitrate (kbps)		Y-PSNR		Encoding FPS		BD-rate	Speedup
		x265	Lentoid	x265	Lentoid	x265	Lentoid		
BasketballDrill (500 frames)	30	1443.73	1428.51	35.20	35.99	0.208	1.548	-18.03%	7.11
	33	932.61	951.60	33.51	34.37	0.259	1.797		
	36	582.48	650.77	31.46	32.84	0.347	2.375		
	39	511.12	452.69	30.41	31.37	0.385	2.770		
PartyScene (500 frames)	30	2860.43	2686.94	32.20	33.45	0.161	1.431	-33.06%	7.85
	33	1586.59	1678.95	29.68	31.49	0.226	1.706		
	36	810.82	1034.41	26.93	29.57	0.305	2.393		
	39	583.29	624.60	25.88	27.69	0.409	2.914		

Table 2. Performance evaluation of the optimized HEVC decoder over x86 processor for 4K videos

3840x2160 Sequences	Bit-rate (kbps)	FPS
Coastguard	10000	44.40
	5000	48.97
	7500	56.94
Foreman	10000	46.96
	5000	54.53
	7500	65.00
Mobile	10000	40.46
	5000	46.75
	7500	56.23

Table 3. Performance evaluation of the optimized HEVC decoder over ARM processor for 1080p videos

1920x1080 Sequences	Bit-rate (kbps)	FPS
Cactus	5661	29.20
	2634	37.59
	1347	45.36
BQTerrace	8083	21.35
	2209	34.33
	856	46.18
BasketBall Drive	6349	25.98
	3028	32.25
	1629	39.83

3   System Integration and Demonstration

For easy application, we have integrated the optimized encoder and decoder into the well-known media framework FFmpeg [11], which is widely used in the open source society and industry. We provide patches to FFmpeg, and enable it to encode and decode HEVC videos using our optimized codec as external libraries.

The demonstration of encoder includes applications such as transcoding and recording. For transcoding, the command line tool *ffmpeg* which comes with FFmpeg can be directly used. For recording, we provide an Android application which achieves 15FPS HEVC recording for CIF (352x288) resolution on a tablet with Intel Atom Quad-core 1.5GHz processor and a digital camera.

To demonstrate the decoder for x86, we use the simple player *ffplay* which come with FFmpeg to playback HEVC videos on an Intel PC. To demonstrate the decoder for ARM, we have made an Android application to play HEVC videos on Android phones and tablets with ARM CPUs.

The proposed implementation of HEVC encoder and decoder is evolving towards a commercial product and can be downloaded at [12]. More test results about the codec performance and the source code of some applications are also available at the website www.xhevc.com.

4   Conclusion

In this paper, we demonstrate a complete solution for software HEVC encoding and decoding. The encoder and decoder are highly optimized and can achieve significant performance promotion compared with existing implementations. The codec libraries are also integrated into widely-used media framework FFmpeg and ready for HEVC application development.

Acknowledgments. This work was supported by National Natural Science Foundation of China under contract No. 61271020, National High-tech Technology R&D Program (863 Program) of China under Grant 2014AA015205 and Beijing Natural Science Foundation under contract No.4142021. Jun Sun is the corresponding author.

References

1. Sullivan, G.J., Ohm, J.-R., Han, W.-J., Wiegand, T.: Overview of the High Efficiency Video Coding (HEVC) standard. IEEE Trans. Circuits Syst. Video Technol. 22(12), 1649–1668 (2012)
2. Joint Collaborative Team on Video Coding (JCT-VC) Reference Software
   svn://hevc.kw.bbc.co.uk/svn/jctvc-hm/
3. Intel Corp., Intel® 64 and IA-32 Architectures Software Developers Manual
4. Henry, F., Pateux, S.: Wavefront Parallel Processing, document JCTVC-E196, JCT-VC, Geneva, Switzerland (March 2011)
5. http://www.videolan.org/developers/x265.html
6. Bossen, F.: Common test conditions and software reference configurations, document JCTVC-L1100, JCTVC, Geneva (January 2013)
7. Pateux, S.: Tools for proposal evaluations. ISO/IEC JTC1/SC29/WG11, JCTVC-A031 (April 2010)
8. The ARM Architecture, ARM Co. Ltd., Cambridge, UK,
   http://www.arm.com/files/pdf/ARM_Arch_A8.pdf
9. http://www.phonearena.com/phones/Xiaomi-Mi-Two_id7427
10. http://www.elementaltechnologies.com/resources/4k-test-sequences
11. FFmpeg, http://ffmpeg.org
12. http://www.xhevc.com/en/downloads/downloadCenter.jsp

A Surveillance Video Index and Browsing System Based on Object Flags and Video Synopsis

Gensheng Ye[1,2], Wenjuan Liao[2], Jichao Dong[2,3], Dingheng Zeng[1], and Huicai Zhong[1]

[1] Institute of Microelectronics of Chinese Academy of Sciences, Beijing, China 100029
[2] Chinese Academy of Sciences R&D Center for Internet of Things, Wuxi, China 214135
[3] Institute of Automation, Chinese Academy of Sciences, Beijing, China 100190
{gensheng.ye,wenjuan.liao}@outlook.com,
{zengdingheng,zhonghuicai}@ime.ac.cn, jichao.dong@nlpr.ia.ac.cn

Abstract. This paper demonstrates a novel retrieval and browsing system based on moving objects for surveillance video. Under the pressure of digital video surveillance generalization, massive data with ever-increasing volume has been involved. How to effectively and efficiently employ the surveillance videos is strategically important in practical applications. In order to improve the availability of videos, intelligent applications contain object extraction, video indexing, video retrieval, and fast browsing. Specifically, This system includes two retrieval browsing sub-systems: (1) as for the retrieval browsing based on moving objects, it can achieve the "browsing with object storage" and "browsing with object classification"; (2) as for the retrieval browsing based on video synopsis, it can achieve the "browsing with playback synopsis" and "browsing with customized synopsis". As shown in demos, video index and synopsis browsing can be flexibly and efficiently realized in this system.

Keywords: Surveillance video, retrieval synopsis, video browsing, moving objects.

1 Introduction

Within the last few years there has been an explosion in the number of digital video surveillance around the world. Techniques like video coding and browsing, which are essential in video surveillance applications, are cause for concern. In fact, most videos are captured by fixed cameras; even for unfixed cameras, large amount of video segments are still captured with fixed angle and location in a certain period. Therefore, by using this feature, the video techniques will provide video surveillance a more efficient solution.

In surveillance video storage, researchers dedicate themselves to improving the storage efficiency and adaptability of mass videos. [1] has improved the video coding efficiency by using the features of fixed visual angle and static background in surveillance video. Scalable coding technology [2] has also been lucubrated for satisfying different channels, terminals, and demands. However, the conventional scalable coding techniques only support the frame-based browsing of temporal scalability, spatial

X. He et al. (Eds.): MMM 2015, Part II, LNCS 8936, pp. 311–314, 2015.

scalability, and quality scalability. The problem how to retrieve and browse moving objects more conveniently still remains to be resolved to some extent.

In the result, video retrieval and browsing technologies have been blended into video surveillance. In this paper, by incorporating background modeling and video synopsis, moving objects can be quickly extracted and original video can be efficiently retrieved in the proposed retrieval synopsis system. Moreover, three retrieval modes: playback retrieval mode, variable-fidelity retrieval mode, and attribute retrieval mode, are customized for surveillance video to meet users' demands. Therefore, this paper presents a comprehensive solution for retrieving and browsing surveillance video.

The rest of the paper is organized as follows: Section 2 discusses the technical details based on object flags and video synopsis. Section 3 introduces the functions of this system.

2   Algorithms for Video Index and Browsing

As a summary of a long video, video abstraction can provide access to large volumes of video content in a relatively short time [3]. Therefore, it is a useful approach for browsing surveillance video. Unlike the conventional method of video abstraction, a dynamic video synopsis for video abstraction is proposed in [4], it can compress original video into a short video with the most informative contents reserved by simultaneously showing several object activities, whose originally occurring time is different.

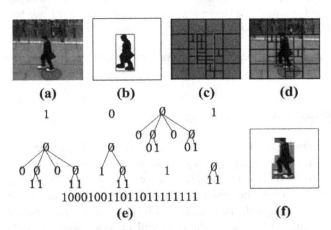

Fig. 1. Flag bits based on masks of object and macroblock partition information: (a) Original video, (b) Mask information, (c) Macroblock partition information with rectangle, (d) ROI to be extracted, (e) Marking based on quadtree, (f) Obtained ROI

The introduced system in this paper is established on the technical framework in [4]. What is different is that [4] focuses on the elaboration and verification of core algorithm while the proposed method incorporates video coding [5], object labeling

[6] and several other technologies based on this core algorithm to build a complete surveillance video index and browsing system by using video synopsis and object flags. The particular scheme is represented as below:

As shown in Fig.1 and Fig.2, the object flags, which acquired by synopsis analysis [4], efficiently represent the moving object regions and their mapping relationship between original video and synopsis video. Object flags consist of object region flags and object mapping flags [6]. They are elaborated as follows:

1) Object region flags: first object mask can be extracted by background modeling and moving object detection [7] to describe object moving region and set a rectangle for moving region. Then the rectangle and mask information are combined with macroblock partition information for object flag coding.

2) Object mapping flags: as described in [4], synopsis video can be obtained by reassigning the start playtime of each object in original video. Therefore, object mapping flags can record mapping relationship between original video and synopsis video and determine the display order of objects in synopsis video.

In the index and browsing system, object region flags are mainly used to support the sub-system browsing based on object flags. This system not only can directly proceed to index and browse for moving objects but also can index and browse by categories after moving objects are classified [12]. Object mapping flags are employed to support the sub-system browsing based on video synopsis. This system not only can realize playback browsing with synopsis video but also can browse customized synopsis video through scalable synopsis [5] and various moving object detecting technologies [8-11].

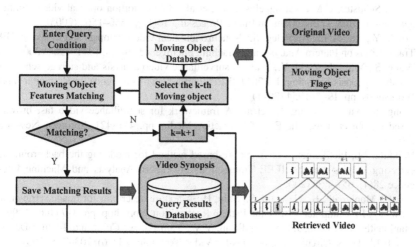

Fig. 2. Schematic diagram of retrieval browsing based on extracted moving objects

3   System Demonstration

By combining the region flags, the mapping flags and object ID together at the encoder, a group of object flags can be obtained, and then writing the sets of flags into the scalability extension parameter sets of H.264/AVC bitstreams in original video with lossless coding. After written into bitstreams, object region flags and object mapping flags can be regained at the decoder. As these flags and original video both can be obtained from decoder, surveillance video bitstream is indexed based on object flags, and then fast retrieval as well as synopsis browsing are achieved in different applications, such as sample retrieval [8], color retrieval [9], shape retrieval [10], trajectory retrieval [11], and fast browsing [5].

When object flags are obtained at the decoder, the moving object region information in the original video can be easily achieved according to object region flags. Simultaneously, the priority of object region can be acquired and initial synopsis video mapping scheme can be generated with those object flags. Then, there can perform retrieval-based video synopsis. Therefore, surveillance video retrieval and browsing could be realized in a convenient and rapid way. The flow diagram of the algorithm in decoder retrieval system is as shown in Fig. 2.

References

1. Zhang, X., Huang, T., Tian, Y., et al.: Hierarchical-and-Adaptive Bit-Allocation with Selective Background Prediction for High Efficiency Video Coding (HEVC). In: Data Compression Conference 2013, pp. 535–535. IEEE (2013)
2. Heiko, S., Marpe, D., Wiegand, T.: Overview of the scalable video coding extension of the H. 264/AVC standard. IEEE Transactions on Circuits and Systems for Video Technology 17(9), 1103–1120 (2007)
3. Li, Z., Schuster, G.M., Katsaggelos, A.K., et al.: Rate-distortion optimal video summary generation. IEEE Transactions on Image Processing 14(10), 1550–1560 (2005)
4. Pritch, Y., Rav-Acha, A., Peleg, S.: Nonchronological video synopsis and indexing. IEEE Transactions on Pattern Analysis and Machine Intelligence 30(11), 1971–1984 (2008)
5. Wang, S., Yang, J., Zhao, Y., et al.: A surveillance video analysis and storage scheme for scalable synopsis browsing. In: 2011 IEEE International Conference on Computer Vision Workshops, pp. 1947–1954 (2011)
6. Wang, S., Xu, W., Wang, C., et al.: A framework for surveillance video fast browsing based on object flags. The Era of Interactive Media, pp. 411–421. Springer, New York (2013)
7. Heikkila, M., Pietikainen, M.: A texture-based method for modeling the background and detecting moving objects. IEEE Transactions on Pattern Analysis and Machine Intelligence 28(4), 657–662 (2006)
8. Wu, P., Manjunath, B., Newsam, S., Shin, H.: A texture descriptor for image retrieval and browsing. In: Computer Vision and Pattern Recognition Workshop, pp. 3–7 (June 1999)
9. Manjunath, B.S., Ohm, J.-R., Vasudevan, V.V., Yamada, A.: Color and Texture Descriptors. IEEE Trans. Circuits and Systems for Video Technology 11(6), 703–715 (2001)
10. Hu, M.: Visual pattern recognition by moment invariants. IRE Trans. Inform. IT-8(2), 179–182 (1962)
11. Hsieh, J., Yu, S., Chen, Y.: Motion-based video retrieval by trajectory matching. IEEE Transactions on Circuits and Systems for Video Technology 16(3), 396–409 (2006)
12. Zhang, T., Lu, H., Li, S.: Learning semantic scene models by object classification and trajectory clustering. In: IEEE Conference on Computer Vision and Pattern Recognition, CVPR 2009, pp.1940–1947. IEEE (2009)

A Web Portal for Effective Multi-model Exploration*

Tomáš Grošup, Přemysl Čech, Jakub Lokoč, and Tomáš Skopal

SIRET Research Group, Department of Software Engineering,
Faculty of Mathematics and Physics, Charles University in Prague
{lokoc,grosup,cech,skopal}@ksi.mff.cuni.cz

Abstract. During last decades, there have emerged various similarity models suitable for specific similarity search tasks. In this paper, we present a web-based portal that combines two popular similarity models (based on feature signatures and SURF descriptors) in order to improve the recall of multimedia exploration. Comparing to single-model approach, we demonstrate in the game-like fashion that a multi-model approach could provide users with more diverse and still relevant results.

1 Introduction

The content-based similarity search has become a standard retrieval technique for users searching multimedia collections. Depending on the data and similarity search tasks, there have emerged many similarity models (modalities) employing various object descriptors. For example, SURF descriptors [2] enabling matching of the same but partially rotated/scaled/occluded objects, or color-position-texture (CPT) feature signatures suitable for matching of objects with similar distribution of texture and color. Although the models can reach state-of-the-art precision on specific similarity search tasks, they hardly reach satisfactory precision in general, especially in cases the models are used for unknown data without thorough and often expensive model training. Hence, instead of imprecise retrieval of top k objects from such data, users could prefer exploration of the collection to get at least an insight of the data. Furthermore, we believe a combination of more models (i.e., multi-model approach) could result in more diverse yet relevant set of images which could further improve the user experience.

In our demo application [1], we implement combinations of two similarity models, more specifically, CPT feature signatures with signature quadratic form distance [3] (denoted as FS+SQFD) and SURF descriptors with L_1 distance (denoted as SURF+L_1). In order to present and test two different combinations of these two models, we have utilized our previously presented web portal [6] where users can search annotated Profimedia dataset [4] in a game-like fashion. The web portal enables users to explore multimedia collections by clicking objects of user interest (starting in a page zero view), where at each iteration

---

* This research has been supported in part by Czech Science Foundation projects P202/11/0968, P202/12/P297 and SVV project 2014-260100.

X. He et al. (Eds.): MMM 2015, Part II, LNCS 8936, pp. 315–318, 2015.

similar objects to the clicked object appear there, considering also the underlying exploration structure. In order to improve the presentation of the results, the objects are placed using the particle physics model [5] where similar objects are attracted while dissimilar objects are repulsed. Furthermore, the web portal already contains two exploration techniques[1] that can be compared with the two new techniques based on combination of the models.

2   Multi-model Approaches

In this section, we present two multi-model approaches we have implemented in our demo application – distance weighting and result mixing. These approaches are utilized after a user selects a new query image to create a new result set.

2.1   Distance Weighting

In order to implement distance weighting, each object stores the original distances to the query object in the basic similarity models we are combining together, where $\delta_{\text{SURF}+L_1}$ ($\delta_{\text{FS}+SQFD}$ respectively) represents the distance from the returned object to the query object in the SURF+L_1 (FS+SQFD, respectively) similarity model. The ratio of the original distances determines the new weighting scheme for the current exploration step. To smooth the differences of almost identical objects, a small constant value is added to both of the distances, so the ratio never reaches extreme values and both underlying similarity models always participate in the combined similarity model at least partially. The formula for the weighting scheme is as follows:

$$\alpha = \frac{0.05 + \delta_{\text{SURF}+L_1}}{0.05 + \delta_{\text{FS}+SQFD}}$$

$$\text{new weighting scheme} = \alpha \times (\text{FS} + \text{SQFD}) + 1 \times (\text{SURF}+L_1)$$

This dynamic approach can be seen as a relevance feedback technique that detects user's search intention by inclining more to one or the other similarity model. For example, when the browsing user focuses on objects with similar color distribution, the FS+SQFD returns smaller values for distances, as the signatures are likely to be similar as well. The SURF+L_1 model, on the other hand, does not focus on color at all and will likely return high distance values if there are not common keypoints between compared images. This will produce a high value of α (for example ten or more) and incline to the FS+SQFD model more, since even small differences in this model will outweigh the differences in the SURF+L_1 model.

---

[1] Exploration using M-Index and PM-Tree using similarity model based on feature signatures and signature quadratic form distance.

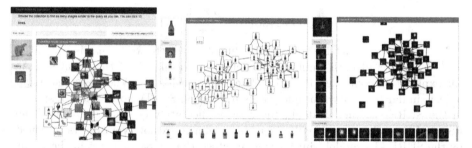

Fig. 1. An example of initial (zero) view of the demo application is depicted on the left. Then follow showcases of the distance weighting approach (in the middle) and the result mixing strategy (on the right).

2.2   Result Mixing

The second approach does not form a new similarity model. Instead, it issues queries to the underlying FS+SQFD and SURF+L$_1$ models and mixes the returned results together. This way it effectively deals with the mismatch of meaning of the values returned by distance functions, and it does not need any weighting scheme to adjust the distances. The only thing that has to be decided is how much objects should be displayed for each model.

Both models together should fill the screen with $k = 50$ objects. By default, the FS+SQFD similarity model returns $k/2$ nearest neighbors to the query object. The SURF+L$_1$ similarity model is used to fill the result set with nearest neighbors which are not already present up to total number of k objects. For each image in the result set, we store metadata M with the positions (denoted as $M_{\mathrm{SURF+L_1}}$ or $M_{\mathrm{FS+SQFD}}$ respectively) within a list of all database objects ordered by the distance to the current query object for both similarity models. Let us note, to do so we have to compute the distances to all database objects using a sequential scan and order the results. When an object with the metadata M is used as a query object, number of nearest neighbors for the FS+SQFD model is computed using this metadata. The better the position of the query object in $FS + SQFD$ model was compared to the position in SURF+L$_1$, the more objects can FS+SQFD model return in this exploration step and vice versa. The SURF+L$_1$ similarity model is always used to fill the rest of the result set. Since the Profimedia collection has a size of 21 993 images, ratios between positions within individual similarity models can be quite huge. To soften this ratio, we used fourth root of the ratio instead. To always mix at least some results from both similarity models, the ratio is kept within the range $< \frac{1}{9}; 9 >$. In the case of $k = 50$ images per screen, this means the minimum of 5 objects per similarity model and the maximum of 45 objects per similarity model. The formula for calculating the ratio ρ from metadata M and to calculate the number of nearest neighbors $k_{\mathrm{FS+SQFD}}$ for the FS+SQFD model is as follows:

$$\rho = \min\left(\max\left(\sqrt[4]{\frac{M_{\mathrm{SURF+L_1}}}{M_{\mathrm{FS+SQFD}}}}, \frac{1}{9}\right), 9\right)$$

$$k_{\text{FS+SQFD}} = \frac{\rho \times 50}{\rho + 1}$$

We can observe that for the initial value of $\rho = 1$, the $k_{\text{FS+SQFD}}$ is computed as $\frac{1 \times 50}{1+1} = 25$ and both similarity models have to fill half of the result set. As the exploration continues, the ratio changes for each step. When user selects an object similar to the current query only in the FS+SQFD model and not similar in the SURF+L$_1$ model at all, the ratio immediately changes in favor of the FS+SQFD model. User is still given the option to explore via objects returned by the SURF+L$_1$ model, since it is guaranteed that it participates in the result set (at least 10% of displayed objects).

Both multi-model approaches are depicted in the Figure 1 together with illustration of demo application results.

3 Conclusions

In this demo paper, we presented two new approaches to multi-model exploration of multimedia collections. We implemented both of them in our demo application which is available online. According to our preliminary experiments, both multi-model approaches achieve better results than our previous exploration techniques utilizing only a single similarity model. In the future, we plan to combine these approaches with data structures like M-Index or PM-Tree to improve the efficiency of the search. We also plan to investigate different combinations of more similarity models and compare them in an user study.

References

1. A Web Portal For Effective Multi-model Exploration, SIRET Research Group (2014), http://herkules.ms.mff.cuni.cz/find-the-image
2. Bay, H., Tuytelaars, T., Van Gool, L.: SURF: Speeded up robust features. In: Leonardis, A., Bischof, H., Pinz, A. (eds.) ECCV 2006, Part I. LNCS, vol. 3951, pp. 404–417. Springer, Heidelberg (2006)
3. Beecks, C., Uysal, M.S., Seidl, T.: Signature quadratic form distance. In: Proceedings of the ACM International Conference on Image and Video Retrieval, CIVR 2010, pp. 438–445. ACM, New York (2010)
4. Budikova, P., Batko, M., Zezula, P.: Evaluation platform for content-based image retrieval systems. In: Gradmann, S., Borri, F., Meghini, C., Schuldt, H. (eds.) TPDL 2011. LNCS, vol. 6966, pp. 130–142. Springer, Heidelberg (2011)
5. Lokoč, J., Grošup, T., Skopal, T.: Image exploration using online feature extraction and reranking. In: Proceedings of the 2nd ACM International Conference on Multimedia Retrieval, ICMR 2012, pp. 66:1–66:2. ACM, New York (2012)
6. Lokoč, J., Grošup, T., Čech, P., Skopal, T.: Towards efficient multimedia exploration using the metric space approach. In: 2014 12th International Workshop on Content-Based Multimedia Indexing (CBMI), pp. 1–4 (June 2014)

Wearable Cameras for Real-Time Activity Annotation

Jiang Zhou[1], Aaron Duane[1], Rami Albatal[1], Cathal Gurrin[1,2],
and Dag Johansen[2]

[1] Insight, Dublin City University, Dublin, Ireland
[2] Computer Science, UiT, The Arctic University of Norway

Abstract. Google Glass has potential to be a real-time data capture
and annotation tool. With professional sports as a use-case, we present a
platform which helps a football coach capture and annotate interesting
events using Google Glass. In our implementation, an interesting event
is indicated by a predefined hand gesture or motion, and our platform
can automatically detect these gestures in a video without training any
classifier. Three event detectors are examined and our experiment shows
that the detector with combined edgeness and color moment features
gives the best detection performance.

1 Introduction

Video annotation has played a very important role in multimedia information
retrieval, medical image processing, sports performance analysis and many other
domains. However, the real-time annotation has received little consideration [7].
In this paper, we present a platform built with Google Glass which enables
us to capture and annotate videos in real-time. There are two major reasons
motivating us applying the real-time annotation. First, post-capture annotation
is laborious and error-prone. Second, it would be an intractable task to build
special purpose detectors for each interesting event a priori due to a wide variety
of potential events [6]. Our platform manages to provide the real-time annotation
by using a predefined hand gesture or motion such that the play itself is not
disturbed on the field and rewinding a video for manually tagging is not required.

The fundamental principle that enables real time video annotation with
Google Glass is the *hindsight recording* [4]. A person observes an entire situation unfolding and determines afterwards whether it was a notable event worth
capturing or not. The annotation works as a stop button in a video, indicating
the end of a sequence worth capturing. The net effect of this hindsight evaluation
process in real-time is that it is less likely important events can be missed, and
there is no need for labor-intense manual tagging of entire videos. Only a detector needs to be built to extract footage of interesting events fully automatically
by exploiting the annotation as input a definition of interesting events.

Event detection has been intensively studied in recent decades and many
methods depend on building classifiers of specific instances to infer the events

X. He et al. (Eds.): MMM 2015, Part II, LNCS 8936, pp. 319–322, 2015.
© Springer International Publishing Switzerland 2015

[3]. For example, Lai [5] learns an instance-level event detection model based on video-level labels, and Aarflot [1] applies face detection to reduce the need of ever deleting digital objects from a digital library. However, building a prior classifier for each interesting event would not be practical or necessary for real-time annotation. Therefore, our platform identifies signals provided from real-time annotations as an indication of interesting events, without building any classifiers. A 10 second video segment before each annotation point is extracted as the footage of the interesting event. The footage will then be viewed through a web interface to recall activity performance, diagnose problems and give feedback.

The major contribution of our work presented in this paper is combining the Google Glass, a state of the art hardware, with event detection techniques for real-time annotation, which can be used for many real-world applications. Taking the specialty of Google Glass into consideration, a region importance mask is created to reduce noisy information in the event detection. An event detector built on the combination of edgeness and color moment features is proposed and has shown very promising detection performance.

Fig. 1. The platform web interface

2   The Platform

Our platform consists of video acquisition using Google Glass, an interesting events detection tool and a web interface. Our platform is evaluated with the coach of a soccer team in Norway. In the training, whenever there was a "good" or "bad" example of play, the coach would wave his hand close to the camera of Google Glass. This would insert an annotation point to indicate an event of interest and generate a keyframe. As shown in the figure 1, the extracted event footage can be reviewed by the coach clicking any keyframe.

2.1   Interesting Event Detection

Our interesting event detection is designed to identify hand-wave gestures. A hand-wave gesture would cause sudden changes in color and significant decrease in edges. Therefore, the event detector is built with a combination of grid edge-ness and color moment of each frame, which can be computed very efficiently. We noticed that the coach's fixated point is almost around a quarter of the height from the top in the picture. Hence, a region importance mask is applied to each frame, with weight 1 at a quarter of the height from the top and gradually de-creases to 0 toward the top and bottom of the image, to relatively enhance the potential important information in each frame. Hence, the edgeness and hue de-viation are calculated in a cell-grid. The edgeness and second order hue moment of each frame are then obtained by summing up the values from all cells and the negative edgeness value represents the final edgeness.

As we can see in figure 2, neither the edgeness nor the color moment can give a good indication of interesting events alone. Therefore, we propose to fit both features into a sigmoid function with equal weight 0.5. Both features are first scaled into a range of $(-6, 6)$ such that the function would have an approximate probability output. As shown in figure 2, the combination of edgeness and color moment is discriminative. The peaks in the plot strongly suggest the occurrence of interesting events. However, non-standard hand-wave movement may intro-duce fluctuation on the top of a peak or even double peaks in a very shot time period due to the hand wave-up and wave-down movement. Thus, we apply a maximum filter on the function output to make sure there is only one rise and fall for an event of interest. The middle frame within a rise and fall window is then regarded as the annotated point.

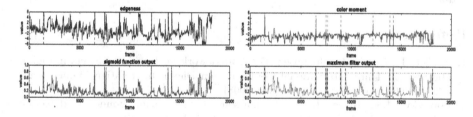

Fig. 2. Event detection with multiple features; red vertical dotted lines indicate ground truth and blue vertical dot lines indicate annotation points (threshold 0.8)

Another impression from the conspicuous changes over frames due to the hand-wave movement is that the changes have effects analogous to the fade-in and fade-out shots in movies. Therefore, we also experiment on re-deploying the camera-shot detection algorithms [2] for our interesting event detection. Table 1 shows the results of the three detection methods we proposed. From the results, it can be seen that the combined-feature detector gives the best performance. The combined-feature detector finds all interesting events and only a small amount of false positive events are introduced. A camera-shot detection algorithm can struggle to make a good balance between precision and recall; our experiment also shows that the camera-shot detection algorithms are vulnerable when the Glass camera is in an unstable state or during white balance adjustment.

Table 1. Experimental results

event detection method	precision	recall
combined-feature	88.9%	100.0%
fade-in shots	80.0%	69.6%
fade-out shots	61.1%	95.7%

3   Conclusions

In this paper we propose an annotation platform for real-time activity logging using Google Glass. Hand-wave signals are detected in a Glass video stream using three detectors: a combined-feature detector, a fade-in shot detector and a fade-out shot detector. Our experimental results show that the combined-feature detector can give very promising detection performance with 100% recall rate and 88.9% precision rate. Our current platform was a proof-of-concept application. We envisage that this can be employed to many real-time annotation tasks, where the annotation requirement is sufficiently straightforward to be represented by one, or a small number of, hand gestures.

Acknowledgement. This work has been performed in the context of the iAD center for Research-based Innovation project number 174867 funded by the Norwegian Research Council. A special thanks to TromsøIL and coach Morten Pedersen.

References

1. Aarflot, T., Gurrin, C., Johansen, D.: A framework for transient objects in digital libraries. In: Third International Conference on Digital Information Management, ICDIM 2008, pp. 138–145 (November 2008)
2. Boreczky, J.S., Rowe, L.A.: Comparison of video shot boundary detection techniques. Journal of Electronic Imaging 5(2), 122–128 (1996)
3. Jiang, Y.G., Bhattacharya, S., Chang, S.F., Shah, M.: High-level event recognition in unconstrained videos 2(2), 73–101 (2013)
4. Johansen, D., Stenhaug, M., Hansen, R., Christensen, A., Hogmo, P.M.: Muithu: Smaller footprint, potentially larger imprint. In: 2012 Seventh International Conference on Digital Information Management (ICDIM), pp. 205–214 (August 2012)
5. Lai, K.-T., Yu, F.X., Chen, M.-S., Chang, S.-F.: Video event detection by inferring temporal instance labels. In: IEEE Computer Society Conference on Computer Vision and Pattern Recognition (CVPR), oral, Columbus, OH (June 2014)
6. Over, P., Awad, G., Michel, M., Fiscus, J., Sanders, G., Kraaij, W., Smeaton, A.F., Quenot, G.: Trecvid 2013 – an overview of the goals, tasks, data, evaluation mechanisms and metrics. In: Proceedings of TRECVID 2013. NIST, USA (2013)
7. Stenhaug, M., Yang, Y., Gurrin, C., Johansen, D.: Muithu: A Touch-Based Annotation Interface for Activity Logging in the Norwegian Premier League. In: Gurrin, C., Hopfgartner, F., Hurst, W., Johansen, H., Lee, H., O'Connor, N. (eds.) MMM 2014, Part II. LNCS, vol. 8326, pp. 365–368. Springer, Heidelberg (2014)

Making Lifelogging Usable: Design Guidelines for Activity Trackers

Jochen Meyer[1], Jutta Fortmann[2], Merlin Wasmann[1], and Wilko Heuten[1]

[1] OFFIS Institute for Information Technology, Germany
[2] University of Oldenburg, Germany

Abstract. Of all lifelogging tools, activity trackers are probably among the most widely used ones receiving most public attention. However, when used on a long-term basis e.g. for prevention and wellbeing, the devices' acceptance by the user and its usability become critical issues. In a user study we explored how activity trackers are used and experienced in daily life. We identified critical issues with regard not just to the HCI topics wearability, appearance of the device, and display and interaction, but also to aspects of modeling and describing the measured and presented data. We suggest four guidelines for the design of future activity trackers. Ideally, activity tracking would be fulfilled by a modular concept of building blocks for sensing, interaction and feedback that the user can freely combine, distribute and wear according to personal preferences and situations.

1 Introduction

The evolution of wearable devices for data capture in the last few years has changed the typical user of such devices from the researchers and early adaptors to normal persons using them in their daily lifes. While multimedia devices with video, audio and image capture such as Google Glass are spectacular and represent the technological state of the art, it is the fairly lo-fi activity trackers that are currently probably most widely being used and receiving the most public attention.

Products such as the Fitbit One[1], Nike Fuelband[2], or Jawbone UP[3] are attractive, easy to use and fairly low-cost consumer products for monitoring daily activity. They are not just a valuable source of data on their own right, e.g. for health monitoring and behavior change [5], but also provide metadata for other applications such as activity recognition in life logging.

Using these trackers as part of lifelogging or for prevention and wellbeing, however, is a long-term effort possibly covering many years. In these cases, user acceptance and usability of monitoring devices in users' daily lives also in an ergonomical and aesthetic sense become important aspects.

---

[1] https://www.fitbit.com/one

[2] http://www.nike.com/us/en_us/lp/nikeplus-fuelband

[3] https://jawbone.com/up

X. He et al. (Eds.): MMM 2015, Part II, LNCS 8936, pp. 323–334, 2015.
© Springer International Publishing Switzerland 2015

We therefore explored how activity trackers are used in daily life. We researched user experiences and examined usability and perceived comfort. From these observations we identified factors that influence the use of activity tracking devices in daily life. Based on these we discuss design implications and a vision for the future design of activity trackers.

2   Related Work

While lifelogging was initially understood as capturing primarily images, nowadays it is more and more seen that lifelogging must in fact be understood broader. [2] presents six classes of data that are relevant for lifelogging, namely passively as well as actively captured media, mobile context and activity data, computer activity, and biometric information, where the latter also includes behaviors such as physical activity. Ryoo et al. differentiate between wearable lifelogging devices with high data rate (e.g. video) and low data rate (e.g. GPS signal) [16]. Lifelogs have been used to capture images [9], track context information [10], or monitor actions taken at a computer [7].

[2] also discusses the main requirements for lifelogging devices. He particularly addresses technical issues, including stability, reliability, battery life, and storage, but also identifies unobtrusive wearability as a necessity and mentions the intrusive nature of multiple wearable devices, taking into account issues such as direct skin contact needed by some devices. We continue this discussion by focussing particularly on state-of-the-art activity trackers as specific lifelogging devices.

Activity trackers are routinely used in interventions, e.g. to encourage physical activity, as well as in public health studies [17] or to monitor elderly persons' behaviors (e.g. [15], [8]). In this context the feasibility and usability of activity trackers are frequently discussed. Many studies have been done about the precision and expressive value of the measured data (e.g. [3]). However, end-users ask more differentiated questions about their data [13], therefore precision alone is not the key. A positive attitude of the user towards monitoring (e.g. [6]) is helpful to ensure compliance to interventions, and it's a prerequisite for unsolicited use of activity trackers as part of lifestyle and wellbeing management. Equally important are the design requirements for mobile interventions, such as [4], and there are a number of evaluations, e.g. of Android based pedometer apps [11], or comparison of two approaches for personal health research [18]. In previous work [14] we gained first insights into the usability of activity trackers in daily life. We continue this work by systematically exploring how users interact and work with recent activity trackers.

3   Activity Trackers

Activity trackers vary e.g. in form factor and size, types of data measured, display type and size, interaction design, connectivity to portals and mobile phones, price, and motivational and persuasive measures. Also numerous activity tracking apps for smartphones are available. In this study we focussed on the investigation of form factor, input methods and the presentation of feedback, taking into account both, dedicated activity tracking devices, and smartphone apps.

Form factor: Clip-shaped activity trackers can be attached to clothes, e.g. a trouser pocket or a blouse, or can be worn invisibly inside a pocket. Bracelet-shaped activity trackers are worn on the wrist like a watch.

Input: A single button is used to navigate through the display of different types of information or to switch between different modes, such as the sleep monitoring mode and the daily activity mode. Various patterns of tapping on the device are used to initiate the display of information or to switch between different modes.

Feedback: A numeric display on the device shows numbers such as the steps done or the calories burnt so far. A graphic display on the device uses abstract visualizations to display results, such as a colored progress bar to show the progress towards a pre-defined goal, or a motivational image. Vibration may be used to represent e.g. the achievement of a goal or to confirm certain inputs.

Apps use a mobile phone's sensors such as the accelerometer or the GPS receiver to measure activity. Smartphones are normally carried in a pocket, attached to the arm or belt, or hold in the hand during activities. The phone's touch screen or physical buttons are used to make input. Feedback is typically presented via the phone's high-resolution display and can be complemented by sound and vibration signals. As apps are cheap or even free, and quick and easy to download and install, the inhibition threshold to use them is low.

4   Study Design

We aimed to understand which factors influence the use or non-use of activity trackers in daily life. Particularly we wanted to know:

- Which form factor do people prefer and why?
- Do the users understand the devices' measurements in the way they are presented?
- Which are motivating and hindering factors for using or not using the devices?

We therefore conducted an exploratory study in which we investigated how popular activity trackers are used and experienced under real-life circumstances.

We acquired 12 participants (7 males) from a participant database, through public announcements, and personal contacts. They were aged between 25 and 70 (M = 43.7, SD = 18.9). None of the participants suffered from serious health problems. All of them stated to be interested in health and in monitoring their behavior. All participants had access to and experience in operating a computer. Nine participants stated that they owned a smartphone.

We chose four activity trackers that prototypically represent today's typical products capturing biometric data including, but not limited to physical activity

Table 1. Activity trackers used in the study

Activity Tracker	Form Factor	Input	Feedback	Synchronization
Fitbit Ultra / One	Clip	One-Button	Numeric and simple graphic display	Wireless through PC docking station to fitbit.com
Fitbit Flex	Wristband (can be worn in pocket)	Tapping	5 point progress bar; vibration	Wireless through PC dongle or Bluetooth to fitbit.com
Nike Fuel-Band	Wristband	One-button	Numeric and simple graphic display	USB connection via PC to nike.com
Runtastic Pedometer	Smartphone app	Touch, button	Hi-res graphic display, sound, vibration	Automatically to runtastic.com (optional)

(see Table 1). They vary in form factor, input method, and feedback display to cover a broad range of the design space of current activity trackers. To keep the technical barriers as low as possible, we avoided trackers that required specific mobile phones for synchronization.

Fig. 1. From left to right: Fitbit One[1], Fitbit Ultra[4], Fitbit Flex[5], Nike FuelBand[2], Runtastic Pedometer App[6]

Participants took part in an initial meeting in which we explained the procedure of the study and introduced all activity trackers that were used in the study. Each participant received a Fitbit Ultra[4] or One, a Fitbit Flex[5] and a Nike FuelBand (see Figure 1). In the first, four-day phase the participants were asked to install and use each of the activity trackers for one day in their daily life. Owners of an iOS or Android phone were also asked to install and use the Runtastic Pedometer app[6] for a day (see Figure 1). This phase aimed at making

---

[4] https://help.fitbit.com/customer/de/portal/articles#product_ultra
[5] http://www.fitbit.com/us/flex
[6] https://www.runtastic.com/en

the participants familiar with all devices. During this phase, participants documented their initial experiences with each activity tracker in a short protocol. Afterwards, in the second, 10-day phase the participants were asked to use at least one of the activity trackers as continuously as possible in their daily life, as long as they felt comfortable. Participants were free to choose the activity tracker(s) they liked most from the three/four we gave them. During this phase, participants filled out a daily diary in which they briefly documented their experiences.

After the second phase, participants took part in individual meetings that began with the completion of the System Usability Scale (SUS) [1] and the Comfort Rating Scales (CRS) [12] for each activity tracker they had used continuously in the second phase. Afterwards, we conducted a semi-structured interview. We asked about the participants' choice of device, about their use of the measured data and whether they found the data sufficient and understandable. We asked about the data presentation on the device, the use of the portal, and whether they believed they would use the activity tracker over a longer period of time. The interview notes and the participants' diaries were coded by three experts and jointly clustered.

5 Study Results

5.1 Quantitative Results

For the 10-day phase, 9 participants chose to use a Fitbit Ultra or One, 6 used a Nike Fuelband, 5 chose a Fitbit Flex, and 1 used the Runtastic app. Several participants chose to use two different devices simultaneously. The usability of the devices was in general assessed very well (Median SUS for Fitbit Flex: 92.5; Fitbit Ultra/One: 92.5; Fuelband: 85; Runtastic: 72.5). The Fuelband and the Runtastic app received considerably lower values than the Fitbit trackers (see

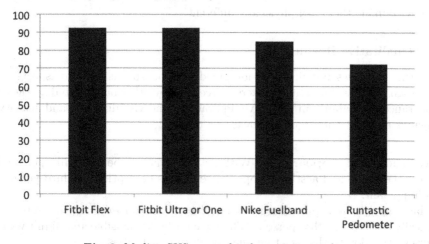

Fig. 2. Median SUS scores for the activity trackers

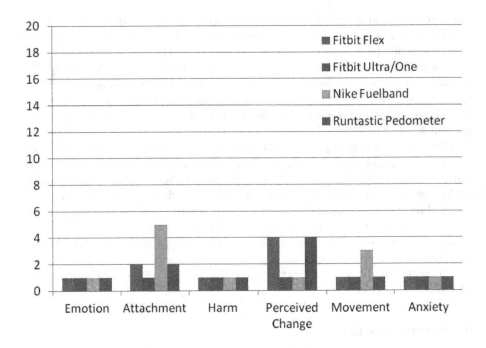

Fig. 3. Median CRS scores for the activity trackers

Figure 2). The Comfort Rating Scales show that in general all activity trackers were rated well with regard to comfort (see Figure 3). The Fuelband performed slightly weaker in attachment (Mdn = 5) and movement (Mdn = 3). These scales reflect how the users felt the device on their body and how it affected the way they moved. Fitbit Flex and the Runtastic app received slightly worse rates in perceived change (Mdn = 4). This scale describes in how far the users felt that the device made them feel physically different.

5.2 Qualitative Results

From the interviews and diaries we identified four concepts as relevant aspects for the design and use of activity trackers: perception by other persons and aesthetic appearance, wearability in daily live, interaction and visualization, and activity measures and the validity of the measured data.

Perception and Appearance. With activity trackers being highly personal devices the visual and aesthetic appearance and its perception by third persons was important.

Many participants appreciated when devices were *inconspicuous* and were not or hardly noticed by other persons. They liked it being invisible and didn't want to draw attention to the fact that they track their activities.

However, some participants liked that wristbands were noticed as a *fashionable accessory*. The Fuelband's design was mentioned as a positive example.

When trackers, both clips and wristbands, would be visible, the *aesthetic appearance* became important. The devices would need to be fashionable and fit to the clothes. This included different and customizable colors for the wristbands as well as the clips.

Wearability. Wearability means the practical consequences of having to wear and use a device 24/7 and in a person's different contexts throughout daily live.

Regarding the *form factor* wrist bands were often felt to interfere with daily activity, whereas clips were in general better accepted, also because they could be worn in a pocket. This, however, was also dependent on the clothes worn: Not all clothes have appropriate pockets. And attaching the clip e.g. to the belt was also a matter of aesthetics. A modular concept allowing use as both a wristband and a "bit", as implemented by the Flex, was appreciated for its flexibility; but having to handle many parts was found to be annoying.

Robustness of the trackers was an important property. Nine participants were concerned about damaging the device. They took it off during gardening work, while swimming or taking a shower, even if it was said to be waterproof. The smaller clip devices that were appraised for inconspicuousness were felt to be easily lost. The stiff design of the Fuelband was positively perceived as being robust and safe. On the other hand it was also reported to be disturbing e.g. when working at a computer: *"It felt like having an elephant on the arm"* (P11).

Four participants considered *smartphone apps* not to be as practical as dedicated activity tracking devices because they would not wear a smartphone continuously and close enough to the body which resulted in incorrect and incomplete data.

Intensity of Interaction. The presentation of tracked data on the device, and how to interact with the device e.g. to change tracking modes was another important topic.

Most participants preferred the *attractive, possibly colorful data visualization* with text, graphics, or animations e.g. on the Fuelband over more minimalistic displays such as the 5-dot-display of Fitbit Flex. They liked getting detailed data, and some liked to use a wristband tracker as a watch replacement.

This demand for visualization might, however, also *change over time*: One participant mentioned that after a longer period of use, a less detailed display would be enough.

Overall, participants found *interacting with the devices* easy. They could easily manage the one-button interface as implemented by the Fitbit One/Ultra and Fuelband. A button-less tap interface like on the Fitbit Flex caused problems. Six participants said it was not intuitively usable and they sometimes made misentries. Sometimes taps were not recognized as such, and sometimes taps were misleadingly recognized from normal arm movements in daily life.

Participants reported they would regularly monitor their progress throughout the day and considered the *continuous feedback at-a-glance* that they received from the devices' displays an important feature.

Direct feedback on the *achievement of a daily goal* was considered very effective: *"Five points on the Flex!"*, *"Vibration of the wristband is motivating"* (P9).

Two participants wanted to be more flexible in defining daily goals or more complex goals such as fitness beyond daily activity.

Activity Measures and Validity of Data. The measured activity data as the primary outcome of using a tracker, raised important questions and issues for the participants.

Seven participants confirmed that monitoring in itself is already motivating and increases the awareness of one's own activity behavior.

An appropriate *level of abstraction* was important to ensure that the data was understandable for the user. Most users instantly understood the concept of step counts that were naturally accepted as the primary measure. However, they were also basically aware that step count as such was already an abstract measure for different types of activities and was not applicable e.g. to cycling. In general, they therefore appreciated alternative activity measures beyond step count. Calories burnt that are calculated by virtually all trackers were found to be interesting, but confusion was caused when it was not clear whether the calories count includes the resting metabolic rate or not. Participants had difficulties in *understanding more abstract measures* for activity such as the Nike Fuelpoints or the flower of the Fitbit One/Ultra.

The *validity of data* was a concern for many participants. They reviewed the results of different activities and compared multiple devices' measurements. They criticized that non-walking activities, such as swimming or biking were not adequately reflected. Incomplete data such as when forgetting to wear or to activate the device was an important aspect that annoyed the participants.

Participants appreciated getting *detailed views* and analyses of their activity data. For most participants the web portals which they would check every couple of days were appropriate. Some preferred an app over a portal, few used the on-device display only.

Beyond the scope of this study were general observations on technical problems with regard to the installation of and daily interaction with the device. In general, users were able to operate the devices, but we didn't research that in detail.

6   Guidelines for the Design of Activity Trackers

Our qualitative results show a broad range of users' preferences. Little surprisingly and like in most HCI designs there is no "one size fits all" device that everybody likes. Rather the appropriate design needs to take into account differing preferences, contexts and sometimes conflicting requirements. For activity trackers we identify four guidelines that should be taken into account:

Make it either invisible, or make it fashionable: Most users liked the trackers to be not noticed by others. The best tracker would therefore be worn invisibly. However, pocket-worn or clip-on devices may not be appropriate to the user's dressing habits. Trackers that are worn visibly are perceived as a fashionable accessory, much like a watch or jewelry. In these cases, designs must match the user's taste and therefore considerably vary between different users and period of use. Adding familiar functions beyond activity tracking such as a watch might mitigate acceptance problems.

Intervention defines interaction, and intervention changes over time: There is a trade-off between unobtrusiveness of wearing and rich interaction. Most users were at first exited about the detailed data they could instantly see on the devices, and they accepted more obtrusive trackers when they provided richer information. But, after some time of use they were satisfied with more limited instant feedback if this attended a device that was easier to use in daily life. Ultimately, the users' decision on how they want to use the device defines what the appropriate interaction method is. A tracker's interaction design may therefore either be intentionally limited to a given intervention and would probably not be used beyond, or it must be adaptable to the changing needs.

Perceived robustness is as important as actual robustness: Trackers should ideally be worn 24/7, and particularly in phases of increased physical activity such as labor, household and gardening, or transportation. There it is wet or dirty, physical force may act on the device, and there is the risk of loss. Although today's devices are in fact quite robust and water-resistant, users were very eager to protect their devices and avoid loss or damage. Consequently they occasionally put off the devices in such contexts, leading to incomplete data. It is therefore not enough to just *build* the device as robust. Rather the users must also *perceive* the device as being reliable and not easily broken or lost. Alternatively, trackers could be cheap single-use devices similar to e.g. contact lenses that are disposed of after use.

Fuzzy but reliable data is better than pseudo-precise but possibly wrong data: Users were highly interested in the data they collected, and they were highly annoyed if the data didn't reflect their behaviors correctly. This is not a simple matter of technically precise measurements: We found too simplistic measures such as step count inherently insufficient, as these measures were *per se* not applicable when accounting for e.g. cycling. Also such measures may make different assumptions about when exactly a certain movement counts as a step and when it's just quivering with a leg. Users notice such deviations between the trackers' measurements and their own perception and will lose confidence in the trackers. The trackers therefore need alternative measures that must be both, easy to understand, and perceived correct for the user. This is primarily a challenge of data modeling. The "'traditional"' step count may be too fine-grained and not appropriate for all activities, but active minutes a day may on the other be too coarse and simplistic. It seems likely that measures must be user-specific to account for a user's individual behaviours and preferences. This requires modeling the user's needs with respect to activity tracking, and matching them with what can be measured.

Taking into account these conflicting requirements it seems unlikely that one single device will be able to fulfill them. Not just do preferences vary between users, but also a single user has different needs over time and even throughout the day. An approach could be to understand an activity tracker not as one single and monolithic system, but a combination of multiple sensing and feedback devices of varying precision, interactivity, and obtrusiveness. For long term monitoring the user might choose a very small sensor with no feedback at all, whereas for a specific health behavior change intervention more precision and a rich interaction would make a more obtrusive system acceptable for a limited period of time. The user would also have the choice to change devices throughout the day, using an invisible one when working, and wearing a sophisticated wristband in the evening. This approach calls for an ecosystem of interconnected low-cost devices. Specific parts of today's devices may in the future be integrated into smart clothes. In the long term, even implanted devices that are injected under the skin or integrated into a tooth would be thinkable.

7   Conclusion

In a qualitative field study we identified four concepts as relevant for the design of activity trackers: perception and appearance, wearability, intensity of interaction, and activity measures and validity of data. Of these we derived four concrete design guidelines, where the first three are closely related to HCI, and the fourth is a data modeling challenge. Our guidelines extend existing work such as [2] that identifies general requirements for lifelogging devices, with usability being just one of multiple aspects. While some points identified there relate well to our results, our focus on usability allows us to be much more concrete and derive concrete recommendations.

Our results are based on feedback from 12 participants. We believe that this sample size is large enough for sound qualitative results. However we obtained the results in an unsupervised diary-based approach; therefore we had to rely on the participants' reports and were unable to double-check the results.

We conclude that "the universal" all-purpose activity tracking device does not exist. Rather there will be a need for different designs, concepts, and data models, depending on the personal preferences of the user and the intended use, taking into account our design implications. While initially our results relate to activity trackers only, most of the rationale behind the guidelines e.g. on appearance, wearability, or robustness apply to general lifelogging devices, too. Therefore an important next step would be to examine the usability of other lifelogging devices such as cameras in order to validate and possibly adapt the results. In future work the guidelines could be applied in the design process of a new lifelogging device to prove their validity.

References

1. Brooke, J.: SUS - A quick and dirty usability scale. In: Jordan, P.W., Thomas, B., McClelland, I.L., Weerdmeester, B. (eds.) Usability Evaluation in Industry. CRC Press (1996)
2. Byrne, D., Kelly, L., Jones, G.J.F.: Multiple multimodal mobile devices: Lessons learned from engineering lifelog solutions. In: Handbook of Research on Mobile Software Engineering: Design, Implementation and Emergent Applications (2012)
3. De Cocker, K.A., De Meyer, J., De Bourdeaudhuij, I.M., Cardon, G.M.: Non-traditional wearing positions of pedometers: Validity and reliability of the omron hj-203-ed pedometer under controlled and free-living conditions. Journal of Science and Medicine in Sport 15(5), 418–424 (2012)
4. Consolvo, S., Everitt, K., Smith, I., Landay, J.A.: Design requirements for technologies that encourage physical activity. In: Proc. of CHI 2006, pp. 457–466 (2006)
5. Consolvo, S., McDonald, D.W., Toscos, T., Chen, M.Y., Froehlich, J., Harrison, B., Klasnja, P., LaMarca, A., LeGrand, L., Libby, R., Smith, I., Landay, J.A.: Activity sensing in the wild: A field trial of ubifit garden. In: Proc. of CHI 2008, pp. 1797–1806. ACM (2008)
6. Fausset, C.B., Mitzner, T.L., Price, C.E., Jones, B.D., Fain, B.W., Rogers, W.A.: Older adults use of and attitudes toward activity monitoring technologies. Proc. of the Human Factors and Ergonomics Society Annual Meeting 57(1), 1683–1687 (2013)
7. Gemmell, J., Bell, G., Lueder, R., Drucker, S., Wong, C.: Mylifebits: Fulfilling the memex vision. In: Proceedings of the Tenth ACM International Conference on Multimedia, MULTIMEDIA 2002, pp. 235–238. ACM, New York (2002)
8. Giehoff, C., Wallbaum, T., Rahn, A., Rolf, C., Boerner, R.J., Neitzel, L., Link, I.: Support of dementia patients by home tele monitoring (Unterstützung von Demenz-Betroffenen durch Home-Tele-Monitoring). In: 58. Annual Symposium of the German Society for Medical Informatics, Biometrics and Epidemiology (58. Jahrestagung der Deutschen Gesellschaft für Medizinische Informatik, Biometrie und Epidemiologie (GMDS) e.V.) (2013)
9. Hodges, S., Williams, L., Berry, E., Izadi, S., Srinivasan, J., Butler, A., Smyth, G., Kapur, N., Wood, K.: Sensecam: A retrospective memory aid. In: Dourish, P., Friday, A. (eds.) UbiComp 2006. LNCS, vol. 4206, pp. 177–193. Springer, Heidelberg (2006)
10. Joki, A., Burke, J., Estrin, D.: Compaignr: A Framework for Participatory Data Collection on Mobile Phones. Technical report, Centre for Embedded Network Sensing, University of California, Los Angeles (2007)
11. Khan, D.U., Ananthanarayan, S., Le, A.T., Siek, K.A.: Evaluation of Physical Activity Monitoring Applications for Android. In: UbiComp 2012 Workshop on Evaluating Off- the-Shelf Technologies for Personal Health Monitoring (2012)
12. Knight, J.F., Baber, C., Schwirtz, A., Bristow, H.W.: The Comfort Assessment of Wearable Computers. In: Proc. of ISWC 2002 (2002)
13. Li, I., Dey, A.K., Forlizzi, J.: Understanding my data, myself: Supporting self-reflection with ubicomp technologies. In: Proc. of UbiComp 2011, pp. 405–414 (2011)
14. Meyer, J., Hein, A.: Live Long and Prosper: Potentials of Low-Cost Consumer Devices for the Prevention of Cardiovascular Diseases. J. Med. Internet Res. 2(2) (2013)

15. Philllips, L.J., Markis, N.: Using the Fitbit Motion Tracker to measure physical activity in residential care/assisted living residents. In: State of the Science Congress on Nursing Research (2012)

16. Ryoo, D.-W., Bae, C.: Design of The Wearable Gadgets for Life-Log Services based on UTC. IEEE Transactions on Consumer Electronics 53(4), 1477–1482 (2007)

17. Schmidt, M.D., Blizzard, L.C., Venn, A.J., Cochrane, J.A., Dwyer, T.: Practical considerations when using pedometers to assess physical activity in population studies. Research Quarterly for Exercise and Sport 78(3), 162–170 (2007)

18. Wilcox, L., Feiner, S.: Evaluating Physical Activity Monitors for Use in Personal Health Research. In: UbiComp 2012 Workshop on Evaluating Off- the-Shelf Technologies for Personal Health Monitoring (2012)

Towards Consent-Based Lifelogging in Sport Analytic

Håvard Johansen[1], Cathal Gurrin[2], and Dag Johansen[1]

[1] UIT The Arctic University of Norway
[2] Dublin City University, Ireland

Abstract. Lifelogging is becoming widely deployed outside the scope of solipsistic self quantification. In elite sport, the ability to utilize these digital footprints of athletes for sport analytic has already become a game changer. This raises privacy concerns regarding both the individual lifelogger and the bystanders inadvertently captured by increasingly ubiquitous sensing devices. This paper describes a lifelogging model for consented use of personal data for sport analytic. The proposed model is a stepping stone towards understanding how privacy-preserving lifelogging frameworks and run-time systems can be constructed.

1 Introduction

Wearable and ambient lifelogging technologies that individuals intentionally use to capture aspects of their own activities, promise life enriching benefits like heightened self-awareness, personalized health-care applications, and new ways of learning. This might lead to longer and more active lifespans, increased productivity in the workplace, increased independence, or increased mobility for people suffering from various memory and cognitive impairments. Lifelogging also fosters new forms of social interaction and sharing [4]. We are already seeing applications in triggering recall of recent memories. This is an application of lifelogging where the detailed lifelog acts as a memory prosthesis, thereby providing support for people with Alzheimer's or other forms of dementia [2,10,16]. Extending this concept from the care-giving domain to every-day life, there is potential for lifelogging to provide memory support to fallible human memory [6], or for logging of activities in molecular medicine [15].

The *quantified-self* movement [18] is perhaps a first mass-scale instantiation of these technologies and has lead to the emergence of pervasive lifelogging as a mainstream activity where individuals use automated digital sensors [8] to capture and permanently store a comprehensive unified digital archive with data related to their lives. An overview of the different categories of lifelogging tools that have been employed can be found in the work of Machajdik et al. [17]

In this paper, we are concerned with the specific case of using lifelogging as a tool for improving coaching and avoiding injuries in professional sport. Collecting, storing, analyzing, and correlating large volumes of personal data, known as *big-data analytic*, from teams of athletes, is important for the emerging next

X. He et al. (Eds.): MMM 2015, Part II, LNCS 8936, pp. 335–344, 2015.

generation sports analytic systems. Such systems enable coaches and medical staff to find useful performance and health indicators that might not be visible from studying single records alone [20]. This has the potential to detect individual health and performance problems at an early stage so that coaching and exercise programs can be individualized, which is clearly highly beneficial for both the athletes and the elite sports clubs.

In collaboration with Tromsø Idrettslag (TIL), a Norwegian professional soccer club, we have developed several lifelogging tools and systems specifically targeting this particular domain [13, 9, 12]. As such tools are becoming more common, comprehensive, and continuous, new privacy concerns are coming to the fore and rise important ethical and legal problems [1].

Some data, such as ambiently recorded images and audio, will naturally pose more concern than others. Indeed, privacy is an inherently fuzzy concept [19] and has had many meanings, definitions, and expectations that differ across jurisdictions, areas, and over time. Also, privacy is affected by political, social, and economic changes and by technological developments; and include psychological, social, and political aspects.

While ambiguities inevitably arise, this paper derives a simple model for reasoning about such privacy using the principles of attribution and access to captured data. This model is derived directly from our real-world experience from applying lifelogging tools for both personal and professional use in elite soccer. Although our model does not cover all privacy related issues, it forms a useful initial framework in which more complex models can be explored.

2   Lifelogging in Sports: A Use Case

Elite sport is a fiercely competitive domain where technology is currently being widely adopted as a game changer. Lifelogging has in particular surfaced as a potent tool for athlete quantification, reshaping how sports are played and how athletes are being developed.

In our case, we have developed and deployed several prototype systems for lifelogging in elite soccer clubs in Norway and for the Norwegian national soccer team. Figure 1 illustrates specific prototype deployments related to the elite soccer club TIL, enumerating the different lifelogging components. As seen in the figure, each individual athlete carries FitBit Flex armbands 24/7 (1), and he also reports perceived wellness and fitness data on a daily basis. This includes parameters like, for instance, perceived fatigue, sleep quality, and muscle soreness, and this must be manually submitted every morning before 9 AM through a smartphone app (2).

Similar reporting is also done post-practice, where perceived personal training load (RPE, perceived exertion on a category 10 scale) is submitted (3). A central server based on the Ohmage platform collects and stores this data, and results of statistical queries can be graphically depicted for coaches and physicians (4). Additionally, medical staff and physicians collect test results periodically for individual players. This includes pure medical stress tests (for instance measuring

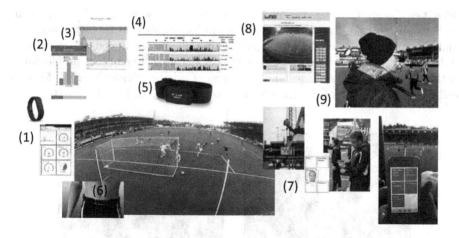

Fig. 1. Lifelogging equipment used for athlete quantification in TIL

lactate threshold predicting athletic endurance performance) and use of jumping boards measuring how high an athlete jumps.

Supplementing devices and services can be used during a practice session. One is Polar pulse belts strapped on each player (5), a wearable monitoring device wirelessly connected to a tablet carried around on the field by the club physician. This way, the individual athletes' pulse is constantly monitored and can be used for interventions (for immediate personalized lower or higher load adaption). Another system used is ZXY Sport Tracking (6), a radio based body sensor network system computing and storing positions and physical data of players on the soccer field with a resolution of up to 20 samples per second.

Bagadus [9] is a novel video processing and real-time smartphone-based notation system we have developed (7). This provides video footage of specific events tagged at run-time, which can be used for feedback purposes instantly, in the intermission break in the locker room, or for post-game analysis where players get involved through a social network service (8).

As illustrated in this use case, lifelog data originates from a wide-range of sources ranging from wearable *inward looking* sensors, like those positioned on the body to monitor heart-rate and lactate, to the new generation of *outward looking* wearable devices that has matured and come to market. These devices incorporate wearable cameras (among other sensors) and can capture detailed photo and video logs of a person's daily activities in an automated manner. Devices like the Narrative Clip wearable camera or Google Glass, enable us to record video of every waking or sleeping moment of the athletes (9). The usage of such recording devices is of great interest to sport analytic because it can be used to capture factors, like food intake and environmental effects, that might influence the athlete's restitution outside of the training and game arena.

The use of such technologies is not particularly problematic in simple usage scenarios, like when a Google Glass is being worn by a soccer coach to capture activities during training session (9). However, once worn outside of the controlled sphere of the station, such devices will inevitably capture highly personal and sensitive aspects of the lifelogger's activities, like what you are reading or toilet visits, as illustrated in scenes A & B of Figure 2. Even more problematic, is the capturing of the images and activities of other individuals in the form of colleagues, family members, friends, as illustrated in C & D of Figure 2.

Fig. 2. Examples of potentially private data (A & B) and bystanders (C & D) in a Google Glass captured lifelog from 2014

The inclusion of such data in sport-analytic performed by coaches and staff in the sport clubs therefore might infringe upon both the privacy rights of the athlete and the privacy rights of inadvertently captured bystanders.

3   Consent in Sport Analytic

We have a principled approach to *privacy-by-design* and all personal data in analytic projects should have a strong notion of being voluntarily contributed. As the particular type of lifelogging described in our use case will often capture highly sensitive and personal data, we adapt to our purpose a fundamental definition on privacy based on the one by Dodge and Kitchin [5]. We considers the privacy for athletes to be based on

1. the right to choose the composition and the usage of your lifelog data, and
2. the right to choose what happens to your representation in the lifelogs of others.

A key problem for professional athlete quantification is that data capture is not initialized and controlled by the individual himself, but rather managed by a large number of people in the form of coaches, physicians, and support staff that typically surrounds sport clubs. These people typically manage the sensor systems that collect data on the individual athletes and specify which analytic functions to run. The rights for a sports club to captured and analyze data must therefore be attested formally as an explicit stated *informed consent*. It is required that such consents are an individual's autonomous authorization of an intervention or participation in some project where private information might be disclosed and used [3].

Informed consent from subjects of study, is already a well established concept in the field of medicine. Physicians or medical researchers requiring personal data have an obligation to tell the subject about the procedure of the participation, the potential risks, and benefits to the subject. The subject must not be deceived or coerced, which implies that each subject must adequately comprehend the consent they are asked to give. Finally, the subject must intentionally sign the consent form.

Traditional consents in the form of passive paper documents, stated and signed at the time data is first collected, does however not capture well the dynamic protection and consensual agreement needed for long term usage of lifelogging data in sport analytic. Emerging data mining algorithms that can discover new sensitive personal traits from existing data or organizational changes in sport clubs, might change the mind of athletes in what they to provide and for what purpose. Using paper-based consent forms, the fine-grained consent management required to support such fine-grained control of personal data is a daunting task.

4   Modeling Lifelogging for Privacy

As highlighted in our use case in Section 2, we are deploying sensors and devices to systematically capture, in digital form, a finite set of personal attributes for each athlete in the sport club. To express this, we have adopted a simple data model that captures data from each lifelogging device d as sequence of measurement samples

$$s_i = [d, t_i, \delta_i, v_i] , \quad i = 1, 2, \ldots$$

Here t is a unique monotonically increasing number for source d denoting the time when the sample is recorded, and $d = (type, device)$ identifies data class contained in the record, like "pulse" or "step" in combination with the identity of the device that generated the data, like `zxy.belt.13` or `RunKeeper.app.78a9fac2`. By explicitly stating device names, multiple devices that provide similar types of data can be supported. For instance, both the ZXY Sports Tracking (ZXY) system and the Polar Belts provide pulse data. The value vector $v = [v_1, \ldots, v_l]$ denotes l source specific measurement points for the sample, and may contain arbitrary data like integers, text strings, or even large binary objects like images and sounds.

In addition to the time-stamp t, each sample s includes a time offset δ that indicates the time-span $[t, t + \delta]$ for when s is valid. For instance, positional data from modern $10\,\text{Hz}$ GPS device, $\delta = 0.1$. A value $\delta = 0$ indicates the end of a sequence of samples, which is used to distinguish time-spans with no samples from time-spans between two valid samples. We denote the superset of all recorded samples as $\mathcal{L}$, the aggregate lifelog of our sport club.

Recorded data samples in $\mathcal{L}$ are stored in a digital archive that spans multiple personal computers, proprietary systems, and Internet services. Moreover, each computer system storing samples may do so for multiple individuals. For instance, in the ZXY sport-tracking system, a single database table holds entries for many athletes. This is also true for most modern solipsistic lifeloggers, as many popular sensor devices, like the FitBit Flex or the Narrative Clip, are hardwired to upload data to the vendors' shared data store. As such, the lifelogs of the principals p, $\mathcal{L}_p$ cannot be defined solely by the devices and systems storing his data, but must instead be defined relative to how data is related to the individual athlete.

4.1   Attribution of Data

To find a method for modeling how individuals' lifelog can be related to the data captured in $\mathcal{L}$, we turn to the notion of data attribution. Let $I(S) = P$ be some function that maps a set of data samples S to some set of principals P. This explicit mapping function I models the fact that most data types do not directly encode information that identify individuals. Clearly, a singular heart-rate sample like $(6{:}10\,\text{pm}, 130\,\text{bpm})$ cannot by itself be attributed to a specific individual. The extra meta-data required for correct attribution, or the means to obtain it, must therefore instead be encoded as of this assignment function. Moreover, we say that I is p-*correct* if principal $p \in P$ would agree to the mapping upon manual inspection.

With this we define attribution, in the following manner:

Definition 1. *A data sample s is attributable to principal p if and only if there exists some p-correct mapping function I such that $s \in S$ and $p \in I(S)$.*

This gives us the following definition of a lifelog

Definition 2. *The lifelog $\mathcal{L}_p$ of principal p is the subset of all samples in $\mathcal{L}$ attributable to p.*

Correct attribution of data samples to individual principals is axiomatic to acquiring informed consent and therefore to privacy. To see this, consider a data sample s correctly belonging to p_1 but incorrectly attributed to p_2. This sample will not only damages integrity of $\mathcal{L}_{p_2}$, but also the confidentiality of p_1. Although correct attribution is a requirement for confidentiality it is not sufficient. This because we allow I to return more than one principal and do not require s to only be attributable to p. To see this, consider our use in Section 2 with stationary cameras on the soccer pitch. Each captured image might be attributable to multiple soccer players and therefore be included in multiple lifelogs, potentially leaking sensitive information. We can formulate this property for consent and privacy in lifelogging as the following

Property 1. $\forall s \in \mathcal{L}_{p_1}$ where there exists some attribution function such that $\{p_2, p_1\} \in I(s)$ and $p_2 \neq p_1$, then p_2 have consented to the storage of s.

4.2   Data Access

Another key requirement of granting informed consent is that the recorded athletes can know about what is captured and what the data is being used for. Indeed, any data sample s that is attributable to p, but that p does not know about, violates privacy because the p cannot consent to its acquisition and use.

A key problem granting such individual detailed insight is that many athlete quantification systems do not provide direct access to low-level sensor data. This is also true for many consumer devices used for inward looking lifelogging where the sensors, like the Fitbit Flex armband, are often hardwired to upload data directly to some online backend server owned by the hardware vendors. These hardware vendor typically only present derived values.

The privacy requirement of informed consent fortunately does not require athletes to have access to the raw sensor data. Indeed, this might even be counter productive as low-level signaling data rarely gives insight in the captured data. The purpose of these systems is to improve athlete performance and prevent injuries and athletes may gain attributable output derived from their data in the form of individualized training and exercise programs, or high-level performance reports [14].

We therefore include in our model of data access, the relaxed constraint of only having the ability to read the output derived through some analytic function T on attributable data samples. We then obtain the following:

Definition 3. *Given a set of data samples S such that $\forall s \in S, P \in I(s)$. Then S is accessible to P if there exists some transformation $S' = T(S)$ such that $\exists s' \in S'$ where $P \in I(s')$ and s' is readable by P.*

In medical research, similar usage characteristics can be found in epidemiological studies on larger population cohorts, like the Tromsø Study [11]. Here, the usage of body sensors and other lifelogging tools is gaining in popularity, enabling more accurate longitudinal data to be collected. Such studies have previously relied on anonymization and de-identification techniques, like crowd blending [7], to preserve the privacy of the subjects. However, as personalized intervention technologies are introduced based on real-time backend analytic over collected data sets, the ability to attribute high-level output back to the individual data donors becomes a key function.

Having the definitions of access we can formulate the following axiomatic property for consent and privacy in lifelogging:

Property 2. $\forall s \in \mathcal{L}_p$, s must be accessible to p.

4.3 Bystanders

Athletes wearing outward looking sensors and engaging in normal every-day interaction with other individuals, like when attending work or socializing with friends, will inadvertently lead to situations where an attributable principal $p \in I(s)$ is not granted access to s—he might not even be aware that he is being captured. This is a highly undesired situation as without access, p cannot consent to being captured and stored and thus storing and using s might damage his privacy. We denote such principals as bystanders, defined in the following manner:

Definition 4 (Bystander). *Let s be a data sample from some lifelogging device worn by principal p_i. Any principal $p_j \neq p_i$ captured, either inadvertently or on purpose, by s such that s is attributable to p_j but not accessible to p_j, is said to be a bystander to s. If s is accessible and attributable to p_i then we also say that p_j is a bystander to the lifelog L_{p_i} of p_i.*

Depending on the lifestyle and sensor devices in use, bystanders could be predominantly strangers or they could be work colleagues, family members, or friends. It is expected that bystanders will be present in a significant fraction of the data samples captured using outward looking devices. Having this definitions we can state another axiomatic property for consent and privacy in lifelogging as:

Property 3. $\forall s \in \mathcal{L}_p$, s must not contain bystanders.

5   Discussions

Publication of lifelog data with bystanders is perhaps the most obvious privacy violation. All stages of the data life cycle have their concerns. For instance, in jurisdictions such as Japan, simply sampling pixels from a worn camera without consent of a bystander can be considered a breach of privacy.

Another interesting aspect of lifelogging devices and also a criticism that has been leveled at it, is that it is primarily a Write Once Read Never (WORN) technology. The view proposed by Bell and Gemmell in Total Recall [4] is that one never knows when some piece of data could be very valuable. This view is certainly also prevalent for data analysts in the sports domain, but is certainly not without its issues. In Norway, for instance, data retention laws that restrict such desultory storage of attributable data are already in place for non-personal use. In lifelogging frameworks that only evaluate data-access policies during capture, such restriction are important for preserving privacy as a lack of an explicit plan on its usage undermines the ability of bystanders to give informed consent.

We must also consider that lifelogging is typically carried out ambient or passively without the lifelogger having to initiate recording. This causes the problems of non-curation or non-filtering, in which the individual lifelogger may not even be aware of all the data being recorded, or the implications of keeping it. Also, due to sheer data volume, there is no practical opportunity for the lifelogger to manually curate his data post-capture to remove or resolve any potential privacy concerns. For instance, the dominant visual lifelogging sensor used thus far is the wearable camera that passively can capture upwards of 4,000 images per day. We therefore propose that a lifelog can store indefinitely the complete representation of the activities of the lifelogger. Still if there at some point in time emerge an identifiable tort as a result of publishing the representation of an individual, then that individual should have the right to remedy. Hence, privacy preserving lifelogging assumes the presence of some mechanism that can postpone the acquisition of consent from the capture stage and until the later stage when derived data is accessed.

6   Conclusions

Lifelogging is a phenomenon whereby people can digitally record personal data in varying amounts of detail, for a variety of purposes. In a sense, a lifelog represents a comprehensive archive of a human's life activities and offers the potential to mine or infer knowledge on those activities using a multitude of software and sensors.

As we move away from naive implementation of lifelogging frameworks, where individuals solipsistic gather data into private life archives, to shared environments where collected data is used for purposes transcending the individual, like improving the performance of a soccer team, putting in place an appropriate privacy-preserving framework becomes an imperative. Towards this end, this paper have defined attribution as a key element of data in lifelogs, and argued that access to attributable data is a fundamental property to privacy. We also identify the presence of bystanders as a key problem, which can only be addressed by either deleting conflicting data samples or acquiring consent.

Acknowledgments. This work has been performed in the context of the iAD center for Research-based Innovation project number 174867 and FRINATEK project number 231687, funded by the Norwegian Research Council.

References

1. Allen, A.: Dredging-up the Past: Lifelogging, Memory and Surveillance. Scholarship at Penn Law. University of Pennsylvania, Law School (2007)
2. Baecker, R.M., Marziali, E., Chatland, S., Easley, K., Crete, M., Yeung, M.: Multimedia biographies for individuals with alzheimer's disease and their families. In: 2nd International Conference on Technology and Aging (2007)
3. Beauchamp, T.L., Childress, J.F.: Principles of Biomedical Ethics. Oxford University Press (2009)
4. Bell, G., Gemmell, J.: Total Recall: How the E-Memory Revolution Will Change Everything. Penguin Books (2009)
5. Dodge, M., Kitchin, R.: Outlines of a world coming into existence: Pervasive computing and the ethics of forgetting. Environment and Planning B 34(3), 431–445 (2007)

6. Doherty, A.R., Moulin, C.J., Smeaton, A.F.: Automatically assisting human memory: A SenseCam browser. Memory 7(19), 785–795 (2011)
7. Gehrke, J., Hay, M., Lui, E., Pass, R.: Crowd-blending privacy. In: Safavi-Naini, R., Canetti, R. (eds.) CRYPTO 2012. LNCS, vol. 7417, pp. 479–496. Springer, Heidelberg (2012)
8. Gurrin, C., Smeaton, A.F., Doherty, A.R.: Lifelogging: Personal big data. Foundations and Trends in Information Retrieval 8(1), 1–125 (2014)
9. Halvorsen, P., Sægrov, S., Mortensen, A., Kristensen, D.K., Eichhorn, A., Stenhaug, M., Dahl, S., Stensland, H.K., Gaddam, V.R., Griwodz, C., Johansen, D.: Bagadus: An integrated system for arena sports analytics—a soccer case study. In: Proc. of ACM MMSys, pp. 48–59 (March 2013)
10. Hodges, S., Berry, E., Wood, K.: SenseCam: A wearable camera that stimulates and rehabilitates autobiographical memory. Memory 7(19), 685–696 (2011)
11. Jacobsen, B.K., Eggen, A.E., Mathiesen, E.B., Wilsgaard, T., Njølstad, I.: Cohort profile: The Tromsø study. International Journal of Epidemiology (2011)
12. Johansen, D., Halvorsen, P., Johansen, H., Riiser, H., Gurrin, C., Olstad, B., Griwodz, C., Kvalnes, Å., Hurley, J., Kupka, T.: Search-based composition, streaming and playback of video archive content. Multimedia Tools and Applications 61(2), 419–445 (2012)
13. Johansen, D., Stenhaug, M., Hansen, R.B.A., Christensen, A., Høgmo, P.-M.: Muithu: Smaller footprint, potentially larger imprint. In: Proc. of the 7th IEEE International Conference on Digital Information Management, pp. 205–214 (August 2012)
14. Johansen, H.D., Pettersen, S.A., Halvorsen, P., Johansen, D.: Combining video and player telemetry for evidence-based decisions in soccer. In: Proc. of the 1st International Congress on Sports Science Research and Technology Support. Special Session on Performance Analysis in Soccer. INSTICC (September 2013)
15. Kumpulainen, S., Järvelin, K., Serola, S., Doherty, A.R., Smeaton, A.F., Byrne, D., Jones, G.J.F.: Data collection methods for task-based information access in molecular medicine. In: Proc. of the International Workshop on Mobilizing Health Information to Support Healthcare-related Knowledge Work, pp. 1–10 (2009)
16. Lee, M.L., Dey, A.K.: Using lifelogging to support recollection for people with episodic memory impairment and their caregivers. In: Proc. of the 2nd International Workshop on Systems and Networking Support for Health Care and Assisted Living Environments, HealthNet 2008, pp. 14:1–14:3. ACM (2008)
17. Machajdik, J., Hanbury, A., Garz, A., Sablatnig, R.: Affective computing for wearable diary and lifelogging systems: An overview. In: Machine Vision-Research for High Quality Processes and Products-35th Workshop of the Austrian Association for Pattern Recognition. Austrian Computer Society (2011)
18. Meyer, J., Simske, S., Siek, K.A., Gurrin, C.G., Hermens, H.: Beyond quantified self: Data for wellbeing. In: CHI, Extended Abstracts on Human Factors in Computing Systems, CHI EA 2014, pp. 95–98. ACM, New York (2014)
19. Spiekermann, S.: The challenges of privacy by design. Commun. ACM 55(7), 38–40 (2012)
20. Steadman, I.: IBM's Watson is better at diagnosing cancer than human doctors. Technical report, Wired magasine (February 2013), http://www.wired.co.uk

A Multi-Dimensional Data Model
for Personal Photo Browsing

Björn Þór Jónsson[1], Grímur Tómasson[1], Hlynur Sigurþórsson[1]
Áslaug Eiríksdóttir[1], Laurent Amsaleg[2], and Marta Kristín Lárusdóttir[1]

[1] CRESS, School of Computer Science, Reykjavík University, Iceland
bjorn@ru.is
[2] IRISA-CNRS, Rennes, France
laurent.amsaleg@irisa.fr

Abstract. Digital photo collections—personal, professional, or social—
have been growing ever larger, leaving users overwhelmed. It is therefore increasingly important to provide effective browsing tools for photo
collections. Learning from the resounding success of multi-dimensional
analysis (MDA) in the business intelligence community for On-Line Analytical Processing (OLAP) applications, we propose a multi-dimensional
model for media browsing, called M^3, that combines MDA concepts with
concepts from faceted browsing. We present the data model and describe
preliminary evaluations, made using server and client prototypes, which
indicate that users find the model useful and easy to use.

Keywords: Photo Browsing, Multi-Dimensional Analysis, Data Model,
Graphical User Interface, Evaluation.

1 Introduction

With the recent technological changes, photo collections have been growing very
rapidly and there appears to be no end to this growth. This calls for very effective tools for not only *finding* content in those collections, but also gaining
insights into the collections and *analyzing* them. Search and browsing tools are
ubiquitous and to some extent they do help with finding photos in collections,
although it can be argued that with the growth in collections the effectiveness
of current search and browsing tools is likely to diminish. They offer no support
for insight and analysis, however, so what is clearly needed is a new approach
to browsing collections. Such an approach has many applications in diverse domains, including professional photo management, online photo stores, digital
heritage, and personal photo browsing.

1.1 Background

For traditional databases, the multi-dimensional analysis (MDA) model used in
on-line analytical processing (OLAP) was the key to allowing analysis and understanding of large data collections. The MDA model introduced two key concepts

X. He et al. (Eds.): MMM 2015, Part II, LNCS 8936, pp. 345–356, 2015.
© Springer International Publishing Switzerland 2015

that revolutionized users' perception of data, namely *dimensions*, including hierarchies, used for specifying interesting sets of data and *facts*, or numerical attributes, which are aggregated for an easy-to-understand view of the data of interest. These simple concepts put the focus squarely on the value of data items and the relationships that exist between data items.

Based on the resounding success of OLAP applications, it is not surprising that multimedia researchers have studied the application of OLAP to multimedia retrieval for some time (e.g., see [1–3]). The fact that the MDA model is geared towards simple numerical attributes, however, is a serious limitation when it comes to multimedia collections where tags and annotations are a very important part of the meta-data.

The use of tags has been studied in faceted search, however (e.g., see [4–6]). Faceted search uses a single tag-set, but proposes to build multiple hierarchies (or even DAGs) over that tag-set, one for each aspect that could be browsed. These hierarchies are then traversed to interactively narrow the result set, until the user is satisfied. Item counts or sample queries are typically used to present the result while it is very large; when it is sufficiently small it is presented in a linear fashion. A major drawback of the faceted approach, compared to MDA, is the use of a single tag-set; although the hierarchies do help users somewhat to disambiguate the different uses of an ambiguous tag, it is more logical to categorize the tags into different tag-sets.

1.2 Contributions

The major contribution of this paper is a proposal for a new model for photo browsing, which builds on and combines aspects of MDA and faceted search to get the best of both worlds. We call this model the Multi-dimensional Media Model, or M^3; we choose to pronounce this as *emm-cube*, which also refers to the fact that the data model essentially constructs hyper-cubes of photos.

In the M^3 model, photos correspond roughly to OLAP facts, but meta-data items (we call these *tags*, but they can refer to any meta-data, including numerical data, dates and time-stamps, annotations and textual tags) can be associated with the photos; one photo may be associated with many tags while a particular tag may be associated with many photos. Tags are grouped into multiple *concepts*; each concept encapsulates a particular conceptual group of tags, such as people, objects, animals, creation dates, focal length, and so on. Borrowing from faceted search, however, *concepts* are then further organized by building (multiple) browsing *dimensions*, typically structured as hierarchies or DAGs, to facilitate browsing.

The M^3 model thus allows us to define the concepts, according to which photos can be grouped, and the organization of the tags in the concepts. OLAP operations are transformed into adding or removing selection predicates applied to the dimensions of a media collection. Predicates act as filters on tags, concepts or dimensions, and restrict the set of photos to display. A photo browsing session thus consists of: repeatedly adding or removing filters; retrieving the photos that pass through all the applied filters; and displaying the photos dynamically according to the organization of the currently visible browsing dimensions.

2 The M³ Data Model

This section develops the M³ model and contrasts it with the MDA model used to view and analyze numerical data.

2.1 Photo Description

The following components of the M³ model apply to photos and their content.

Object. An object is any entity that a user is interested in storing information about, in this case a photo. Objects correspond to *facts* in MDA, as both represent information users are interested in analysing and both have associated meta-data that further describe them. Unlike facts, however, objects such as photos represent complex content, making aggregation difficult.

Tag. A tag is any meta-data that can be associated with objects. There is no limitation on how many objects a tag can be associated with, nor is there a limitation on how many tags can be associated with a single object. Tags correspond roughly to *members* in MDA.

Location. The location refers to where a particular tag applies to a particular object. The details of the implementation of location may depend on the context, but in the case of photos a bounding box is typically sufficient. A tag that applies to the entire photo, however, does not require a bounding box. There is no counterpart in MDA, as the data items in MDA have no structure.

2.2 Multi-Dimensional Aspects

In this section we define four abstractions that concern categorization and grouping of data, and are at the heart of the multi-dimensional nature of the M³ model.

Concept. A concept is a set of tags that the user perceives to be related; a conceptual group of tags. In a concept named 'People' the tags can, e.g., be names of people or names of subcategories, say, 'Children' or 'Class Mates'. Some concepts may be entirely user generated, e.g., an 'Event' concept. Others may be based purely on the meta-data associated with each object, e.g., a 'Creation Date' concept. Yet others may partially use automated content analysis, e.g., face identification for the 'People' concept. Any implementation of the M³ model must therefore support automated media analysis methods.

Concepts are mathematical sets in the sense that tags are distinct. The order of tags, however, can be relevant; consider for example a concept containing creation date tags. Concepts are very similar to *dimensions* in MDA. In both cases the user perceives the tags as being strongly related to each other, and in both cases hierarchies can be built to organize their contents.

Dimension. A dimension adds structure and order to a subset of the tags of a concept. A dimension is typically derived from a single concept and only contains tags from that concept, but may include tags from different concepts.

Each concept has a "default" dimension which includes all the tags in the concept, but may additionally have zero, one, or more associated dimensions. The 'People' concept, e.g., could have one dimension called 'Friends' and another called 'Family', which would typically be largely disjoint but might share some tags. While concepts may or may not be ordered, the vertices of a dimension are explicitly ordered.

The dimension concept is highly similar to the hierarchy concept from MDA, although the dimension concept is more flexible. Hierarchies in the MDA model can represent either level or value based hierarchies, while a dimension may be a more freely structured hierarchy or even a DAG. In both cases, however, a node is the aggregation of its children. In MDA, aggregation is typically based on mathematical functions, such as sum or average, but in M^3 aggregation will typically take some form of grouping.

A node in a dimension may optionally have a title that applies to its children, called a *child category title*; the children are then instances of the category that the title names. To use a standard example, 'Month' could be a child category title, while the months themselves would be the children. A minor difference between the two models is thus that in MDA a column name supplies a level name in a hierarchy, but in M^3 the child category name is applied to a hierarchy node and only applies to the children of that node.

Hypercube. A hypercube is created by selecting and storing information about one or more dimensions which the user wishes to browse objects by.

In MDA implementations, the hypercube is typically stored in a specialized data structure for efficiency. This data structure stores both base facts and pre-calculated aggregations of facts. Due to the differences between objects and facts, however, it is possible that only base data can be stored for the M^3 model and the hypercube will therefore only be conceptual. The representation of complex M^3 hypercybes is a topic for future research.

Cell. A cell in the hypercube is the intersection of a single tag from each of the dimensions in a hypercube. A cell can contain zero or more photos, unlike the MDA model which simply aggregates all the facts corresponding to each cell.

2.3 Retrieval

So far, we have defined abstractions to describe and organize media objects, but now we turn to the retrieval of the objects. Unlike typical search applications, the retrieval is based on browsing, where the retrieved object sets are defined incrementally, based on the user's interest. In the M^3 model, different *filters* can be applied to the various dimensions, resulting in a *browsing state*. We now define these in more detail.

Filter. A filter is a constraint describing a sub-set of photos that the user wishes to see. Each filter applies to a single dimension, but many filters may be applied to the same dimension. All objects that are associated with tags that satisfy the constraint of a filter are said to pass through the filter.

We have defined three different filter variants:

- *Tag Filter:* The tag filter is a filter that selects a single tag from a concept. It is used to retrieve only photos associated with that particular tag.
- *Range Filter:* A range filter applies to ordered concepts and defines a value range by two boundary values, where both boundary values are included in the range. The boundary values themselves need not exist as tags in the concept the filter is applied to.
- *Dimension Filter:* A dimension filter selects a single node in a dimension. The entire sub-structure, or sub-dimension, of that node is said to pass through the dimension filter. A dimension filter on the root of a dimension thus returns the entire dimension, but excludes the tags from the underlying concept that are not part of the dimension.

Note that a filter can be applied to any dimension, regardless of whether that dimension is used for organizing photos in the user interface. Furthermore, a filter continues to restrict retrieval until it is explicitly revoked. A browsing session thus consists of repeatedly applying and/or removing filters and retrieving objects that pass through all applied filters.

The filter concept corresponds to two MDA abstractions. First, it serves the same purpose as *selection*, as both can be used to define a filter to restrict the data retrieved. Second, the filter also corresponds to *page dimension*, as both can be used to restrict retrieval using a dimension not in the cube.

Browsing State. As mentioned above, a browsing session consists of applying or removing filters and retrieving the objects that pass through all the applied filters. The browsing state therefore contains information about filters, tags or sub-dimensions that pass through the filters, and objects associated with these. Note that tags and sub-dimensions are included even when there are no corresponding objects, as the existence of such tags and sub-dimensions is highly useful information for the user. Objects are only returned, however, if they pass through *all* the applied filters. It is sufficient that one tag that the object is associated with passes through each filter, and it is possible that different tags may pass through different filters.

Informally, we can think of each filter as selecting a sub-set of the objects. The objects in the browsing state are then selected by the intersection of the sets passing through the filters. Concepts with no filters are excluded from consideration; if no filters are in effect the browsing state therefore includes all objects. The browsing state corresponds loosely to a *sub-cube* of the hypercube in the MDA model, as the browsing state contains enough data to build a sub-cube.

3 Evaluation

In this section, we discuss our evaluations so far. After briefly describing our prototypes and photo collection, we present a detailed browsing scenario followed by the results of two different user evaluations. Finally, we address one pressing issue in photo browsing, namely that of effective tag generation.

Fig. 1. A three-dimensional browsing state from the scenario

3.1 Prototypes and Photo Collection

We have developed a media server—called O^3 since it serves hypercubes of objects [7]. It implements most of the M^3 data model, but dimensions are currently restricted to hierarchies over a single concept. A *plug-in* architecture is used to generate tags from Exif metadata and extract faces from photos [8]. In [7], an extensive performance evaluation shows that the server scales well for large photo collections. We have also developed a photo browser—called P^3 since it displays hyper-cubes of photos [9]. P^3 includes a graphical user interface which grants users the access to the powerful and flexible browsing operations of the M^3 data model, such as the drill-down and roll-up operations of dimensions, dimension pivoting, and general filtering. Various aspects of these prototypes have been demonstrated at the ICMR, CBMI and MMM conferences [10–12].

The photo collection contained photos from a five day trip from 2010 along a well-known hiking trail. The hiking group consisted of 9 adults (including one of the authors) and 9 children from 5 different families. The collection consisted of 1,140 images. Aside from 19 meta-data concepts extracted from photo headers, there were 126 tags belonging to 7 concepts with a total of 8 dimensions, in addition to the default dimensions. The concepts were: 'Events'; 'Days'; 'Locations'; 'People'; 'Objects'; 'Animals'; and 'Impression' (containing the tag 'Beautiful').

3.2 Detailed Browsing Scenario

An adult user is sitting down with her children to recall a hiking trip. She first selects the family dimension (a hierarchy on top of the people concept) as a starting point, and drills down to her own family, which has four members. Then she selects the location dimension as the "up" axis, which has such nodes as "cabin" and "river". Being a photo nerd, she becomes interested in the light conditions, and selects the aperture value to the third axis. The current browsing state, shown in Figure 1, then has three dimensions, where each cell has one particular family member in one particular location type with one aperture value. Note that photos containing all four family members will show up in four cells (and, if a cabin were situated next to a river, it could show up in 8 stacks); this is an important feature of the model as the photos belong *logically* in all these cells. Then she decides to focus on photos taken at rivers, as she wants to talk about wading the glacial rivers on the hike. As there was one such river every day of the hike, she replaces the aperture value axis with a day axis. She might then go on to consider animals or objects, rotate the cube for a better view, etc.

This scenario demonstrates several common operations in the M^3 model, namely filtering, drilling down into dimensions, and pivoting browsing dimensions. It also shows how well the model uses the screen to indicate why each photo is included in the result. Finally, the scenario illustrates the suitability for storytelling and discovery, which we have experienced vividly in demonstrations.

3.3   Evaluation I: Advanced Users

The first evaluation was a pilot study with experienced computer users, in order to quickly gain insights into the pros and cons of the M^3 data model. For this study we used the same image collection as before. The subjects were five male advanced students of computing. Their experience of image browsers was very varied, but none of the participants had any knowledge of the collection. Aside from questionnaires and an open interview at the end, the focus was on solving specific tasks.

Task Performance: Each participants was asked to perform the following tasks:

1. Show images of kids by location.
2. Show images that contain a sheep.
3. Show images containing hiking shoes which have Aperture value 4 - 5.
4. Show images of people playing football.
5. Show images containing the participating author, grouped by F-number, ISO-Speed and location.
6. Show some images which the participant found to be "cool".

The purpose of the final task was to allow the users to "play around" with the prototype on their own. The performance on the tasks was noted by the experimenter, using the following performance indicators A (finished without problems), B (finished after experimentation), C (finished with minor help) and F (did not finish).

Table 1 shows the outcome of the task performance evaluation. Each row contains the task performance of one individual participant. The table shows that all participants, except for participant 3, experienced some difficulties with the first two tasks. This is most likely due to the learning curve of the model and prototype. Two users experienced significant difficulties with the first tasks and had to ask for clarifications. The instructions they received, however, consisted solely of reminding them of functionalities of the model, available operations in the browsers and information about the dataset. No direct help instructions were given; yet we do not see any F labels.

Table 1. Eval I: Task Performance

Participant	Tasks 1 2 3 4 5 6
1	B B B A A A
2	B B A A A A
3	A A B A A A
4	C B A A A A
5	C C A B A A

Table 2. Eval I: User Experience (7 = best)

User Experience Factor	Score	Range
Simple	4.8	3–7
Pliable	5.0	4–6
Easy to use	5.2	4–6
Comfortable	5.2	5–6
Encouraging	5.4	4–6
Enjoyable	6.2	6–7
Imaginative	6.4	6–7
Useful	6.4	6–7
Fascinating	6.6	6–7

Overall, participants experienced the most difficulties with Task 1, as only one user was able to finish this task without problems. In Task 2 the performance was slightly better, but still only one user was able to finish the task without any problems. In Task 3 the performance improved further, as two users were able to finish the task without problems and the others without any assistance. By Task 5 all participants were able to complete the task without any problems, even though that is a relatively difficult task. These results indicated that there was a learning curve for the browser, but that a few browsing sessions might be sufficent to overcome it.

User Experience Factors: Table 2 shows the results of a user experience questionnaire, both the average score and the range of scores. The user experience factors are ordered from the lowest score to the highest score. Overall, we observe that the scores are rather high, ranging from 4.8 to 6.6, but with notable difference between the score for different factors. The factors in the upper half, Simple, Pliable, Easy to use, Comfortable and Encouraging all have a comparably low average score, ranging from 4.8 to 5.6, with a wide range of scores. The widest range is for the Simple factor, where the average value is the lowest and the scores range from 3 to 7. On the other hand, the factors in the lower half, Enjoyable, Imaginative, Useful and Fascinating, all have comparably high average scores of 6.2 through 6.6. Furthermore, the participants all seem to agree on these values, as the range of scores is quite narrow. These results indicate that our participants find the prototype rather complicated and cumbersome to use, while finding it at the same time highly enjoyable, imaginative, and useful.

3.4   Evaluation II: Novice Users

The second evaluation was a more detailed user evaluation with the actual people from the family hike. This is actually a unique evaluation, in the sense that all participants felt as if they were browsing their private image collection. Nine users participated in all, five adults and four children/teenagers, with similar balance between genders. All the participants can be considered novice users.

The experimental protocol was similar as before, except that no predefined tasks were given. Instead, the participants were asked to browse the photo collection freely and make observations about the photos—to "show" the photo collection to the experimenter. This phase lasted 40 to 60 minutes, ending when the participant appeared tired and/or ready to quit. Due to the small size of the user group, we did not compare directly to other browsing tools, but instead asked participants to compare their experience to their regular photo browser.

AttrakDiff Results: The AttrakDiff questionnaire, a standard questionnaire for evaluating software (http://attrakdiff.de/index-en.html), was filled in by seven participants (young children were excluded). The results are interpreted along four different axes: Pragmatic Quality (PQ); Hedonic Quality - Stimulation (HQ-S); Hedonic Quality - Identity (HQ-I); and Attractiveness (ATT). The report states that the PQ and HQ-S values are average, and hence improvements are needed in terms of usability and stimulation. The HQ-I and ATT values are above average, however, meaning that the users found the prototypes attractive and they identified with it. It is possible that these results are biased, however: The fact that users were indeed browsing photos from their own experience, can positively impact the HQ-I value, and the fact that users knew one of the researchers from this trip may have positively impacted the ATT value.

Browsing Questionnaire: We also prepared a questionnaire specifically for photo browsing. The questionnaire contained six questions about the user experience, and the participants responded using a 7-point Lickert scale (1 = I agree; 7 = I disagree). The results are shown in Table 3. Overall, the participants were very positive towards the prototype. In particular, the last question of Table 3 shows that the participants enjoyed using P^3 more than using their regular photo browser. Two participants mentioned that they felt obliged to rate the software as it was at the time of testing, but expressed their expectation that the software would improve with further development and thus would probably deserve an even higher rating in the future. Similar reservations are likely made by other users as well, as the last two questions, which put the prototypes into a more general context, have less positive replies.

Result Summary: It was encouraging to see how quickly this varied set of novice users was able to start using the prototype effectively. An open interview at the end also pointed towards some improvements that could be made in the interface. The measured results from this experiments were not very conclusive, however, but we nevertheless feel that this experiment is worth reporting on as it points to a general methodology for studying personal photo browsing: Find groups with shared experiences and subject them to collections from those experiences.

3.5   Facilitating Tagging

To effectively support media browsing, the M^3 model relies on the existence of tags; when tags exist they can be used very effectively for browsing. In the past, systems relying on tagging have had problems [13] but we believe that several features of our prototypes can help make it very easy for people to *create* the tags, through automated tagging and through user-interface techniques.

This is work in progress, but we already made initial steps towards facilitating tagging. We have, e.g., designed an automated plug-in architecture for tag gen-

Table 3. Eval II: User Experience (1 = I agree; 7 = I disagree)

Questions About P³	Average	StdDev
The program helped me recall the hike	1.25	0.46
I enjoyed browsing the photos in this program	1.22	0.67
The program made it easy to browse photos	1.25	0.46
The program made it easy to find photos	1.12	0.35
I would like to show others my photos using this program	1.62	1.19
I enjoyed using this program more than the one I normally use	1.56	1.13

eration which could be used to implement any sort of analysis technique. During import of photos, tags can be selected that apply to all photos in the import. Furthermore, the new photos are assigned two random numbers that can be used to scatter them on screen, to give a fresh light-board style overview.

Finally, we have designed a drag-and-drop tagging process, which is a new tagging paradigm. For example, a photo, a part of a photo, or a set of photos, may be dragged to any tag that is visible on the screen and dropped there, resulting in a connection between that tag and the photo.

4   Related Work

Many research projects have considered photo browsing and proposed many interesting methods. While the literature is far too vast to be all cited here, we describe some of the most interesting techniques.

PhotoFinder [14] provides boolean search capabilities for finding photos based on several attributes. PhotoMesa [15] provides a zoomable interface to multiple directories of images at once, grouping the images from the folders into clusters for maximal use of the screen. Harada et al. [16] proposed a browser focusing on automatically generated event assignments of photos; this method could well be implemented as a plug-in for our prototype. Girgensohn et al. [17] provide an interface for grouping and browsing photos based on several similarity measures: visual; geological; date; and tag. The Camelis photo browser uses co-occurrences of tags in images to deduce relationships and uses those relationships to facilitate browsing [18]. Several researchers have considered photo spreadsheets; in one of the most recent works, Kandel et al. focus on a biological application [19]. A good summary of the issues and early developments is found in [20].

As mentioned in the introduction, some efforts have been made towards adapting the MDA model to photo browsing [1–3]. These systems, however, are static representations of the respective collections, and hence ill-suited for general and personal photo browsing. The approach most similar to the MDA approach is that of *faceted search* which has been applied to many domains, including photos [4]. Faceted search uses a single set of tags, but proposes to build multiple hierarchies (or even DAGs) over that tag-set, one for each aspect that could be browsed. The single tag-set is a limitation; although the hierarchies do help users somewhat to disambiguate the different uses of an ambiguous tag, it is more logical to place distinct tags in different concepts. Furthermore, faceted browsers typically employ a linear presentation, resulting a dimensionality reduction of sorts, where it is unclear why photos appear on screen. Girgensohn et al. [21] propose a system that groups photos along multiple hierarchies, corresponding to different concepts, allowing each photo to appear logically in many places. By selecting different parts of different hierarchies, filters are applied to select subsets of the photos. The resulting photo set is still presented as single-dimensional, and each hierarchy can only contribute one filter to the set.

Scenique [5] is a faceted photo browser that breaks from tradition; it is conceptually the browser most similar to our proposal. Scenique allows image browsing in 3D browsing rooms, where each dimension corresponds to a facet. In addition to tag-based facets, Scenique also offers facets based on content-based descriptors. The M^3 model does not prevent this in any way and we plan to add support for content-based browsing to the server prototype. Scenique can not, however, show different parts of the same hierarchy of different browsing dimensions and in fact it is not clear how it handles the case when constraints are given for 1–2 or 4+ dimensions. We believe that the underlying abstractions of the M^3 model, in particular the cell and the hypercube, are the key difference.

5 Conclusion

Collections of digital media are growing ever larger, leaving users overwhelmed with data but lacking insights. Looking at the literature, the interest in photo browsing peaked shortly after the turn of the century and today fewer papers are published on photo browsing. Clearly, however, it remains an unsolved problem and one that merits further work. We have therefore proposed the M^3 data model for media browsing, based on the highly successful multi-dimensional analysis model from the business intelligence community, which is a natural progression of earlier work on faceted search. We have presented a detailed browsing scenario showing the expressiveness of the data model, as well as the results from two preliminary user studies which indicate that users find the data model and prototype both useful and engaging.

There are many interesting avenues for future work. Further development of the interface is necessary, of course, as are further user studies. We plan to add support similarity metrics, e.g. based on visual similarity measures, and for dynamic browsing dimensions, e.g., based on key-word search or content-based similarity, thus integrating browsing and searching into a single framework. Extending the model and prototypes towards professional and social—big—collections will also be interesting and challenging, both in terms of data model expressiveness and not least in terms of scalability. Furthermore, adapting to haptic and tactile interface is an interesting research direction.

Acknowledgments. This work was partially supported by Icelandic Research Fund grant 70005021, Student Innovation Fund grants 110395-0091 and 131956-0091, and CNRS PICS grant 6382, "MM-Analytics".

References

1. Zaïane, O.R., Han, J., Li, Z.N., Hou, J.: Mining multimedia data. In: Proc. CASCON (1998)
2. Arigon, A.M., Miquel, M., Tchounikine, A.: Multimedia data warehouses: A multi-version model and a medical application. Multimedia Tools and Applications 35(1) (2007)
3. Jin, X., Han, J., Cao, L., Luo, J., Ding, B., Lin, C.X.: Visual cube and on-line analytical processing of images. In: Proc. CIKM (2010)
4. Yee, K.P., Swearingen, K., Li, K., Hearst, M.: Faceted metadata for image search and browsing. In: Proc. CHI (2003)
5. Bartolini, I., Ciaccia, P.: Integrating semantic and visual facets for browsing digital photo collections. In: Proc. SEBD (2009)
6. Diao, M., Mukherjea, S., Rajput, N., Srivastava, K.: Faceted search and browsing of audio content on spoken web. In: Proc. CIKM (2010)
7. Tómasson, G.: ObjectCube – a generic multi-dimensional model for media browsing. Master's thesis, Reykjavik University (2011)
8. Rúnarsson, K.: A face recognition plug-in for the PhotoCube browser. Master's thesis, Reykjavik University (2011)
9. Sigurþórsson, H.: PhotoCube: Multi-dimensional image browsing. Master's thesis, Reykjavik University (2011)
10. Tómasson, G., Sigurþórsson, H., Jónsson, B.Þ., Amsaleg, L.: Photocube: Effective and efficient multi-dimensional browsing of personal photo collections (demonstration). In: Proc. ICMR (2011)
11. Tómasson, G., Sigurþórsson, H., Rúnarsson, K., Ólafsson, G.K., Jónsson, B.Þ., Amsaleg, L.: Photocube: Effective and efficient multi-dimensional browsing of personal photo collections (demonstration). In: Proc. ICMR (2011)
12. Jónsson, B.Þ., Eiríksdóttir, A., Waage, O., Tómasson, G., Sigurþórsson, H., Amsaleg, L.: Photocube: Effective and efficient multi-dimensional browsing of personal photo collections (demonstration). In: Proc. ICMR (2011)
13. Mathes, A.: Folksonomies - cooperative classification and communication through shared metadata (2004)
14. Kang, H., Shneiderman, B.: Visualization methods for personal photo collections: Browsing and searching in the photofinder. In: Proc. IEEE ICME (III) (2000)
15. Bederson, B.B.: PhotoMesa: A zoomable image browser using quantum treemaps and bubblemaps. In: Proc. UIST (2001)
16. Harada, S., Naaman, M., Song, Y.J., Wang, Q., Paepcke, A.: Lost in memories: Interacting with large photo collections on PDAs. In: Proc. JCDL (2004)
17. Girgensohn, A., Shipman, F., Turner, T., Wilcox, L.: Flexible access to photo libraries via time, place, tags, and visual features. In: Proc. JCDL (2010)
18. Ferré, S.: Camelis: Organizing and browsing a personal photo collection with a logical information system. In: Proc. CLA (2007)
19. Kandel, S., Paepcke, A., Theobald, M., Garcia-Molina, H., Abelson, E.: Photospread: A spreadsheet for managing photos. In: Proc. CHI (2008)
20. Shneiderman, B., Bederson, B.B., Drucker, S.M.: Find that photo! interface strategies to annotate, browse, and share. Communications of the ACM 49(4) (2006)
21. Girgensohn, A., Adcock, J., Cooper, M.L., Foote, J., Wilcox, L.: Simplifying the management of large photo collections. In: Proc. INTERACT (2003)

Discriminative Regions: A Substrate
for Analyzing Life-Logging Image Sequences

Mohammad Moghimi[1,2], Jacqueline Kerr[1], Eileen Johnson[1], Suneeta Godbole[1],
and Serge Belongie[2]

[1] UC San Diego, La Jolla, CA 92093
[2] Cornell University, Ithaca, NY 14850
mmoghimi@cs.cornell.edu

Abstract. Life-logging devices are becoming ubiquitous, yet still processing and extracting information from the vast amount of data that is being captured is a very challenging task. We propose a method to find discriminative regions which we define as regions that are salient, consistent, repetitive and discriminative. We explain our fast and novel algorithm to discover the discriminative regions and show different applications for discriminative regions such as summarization, classification and image search. Our experiments show that our algorithm is able to find discriminative regions and discriminative patches in a short time and extracts great results on our life-logging SenseCam dataset.

1 Introduction

We are entering the age of wearable computing. Wearable devices are becoming more powerful and ubiquitous. Wearable cameras such as Google's Project Glass, Narrative and SenseCam are adopted more and more by people. However, indexing the enormous amount of pictures that is being captured by these devices remains a challenge. Furthermore, these visual logs of people's everyday lives provide a rich source of data for information extraction, including a variety of different Computer Vision tasks. In this paper, we a propose method to find regions of interest in the life-logging image sequences and also showcase a few applications of this representation.

In this work we are using SenseCam, a chest-mounted wearable camera, that periodically takes pictures every 20-30 seconds. This results in about two thousand images per day. Thus, there is a need for algorithms to analyze and extract information from these image sequences to facilitate the search process. The process would ideally be unsupervised or supervised with minimal human input.

We propose an algorithm that highlights regions of interest in life-logging images. We define these regions to be *salient* i.e. conspicuous regions, *consistent* i.e. reliably appearing in a few consecutive frames, *repetitive* i.e. frequently appearing in the image set, and *discriminative* i.e. are specific to a particular scene.

Saliency and consistency constraints help focus on regions that are in the foreground. We design a novel robust method to find consistent regions based on a forward-backward search in the feature space. Then those consistent regions are ranked based on their discriminative power and the number of their appearances. The motivation behind discriminative regions is that the parts that are visible everywhere are not very informative. Consider a scene where the user is working in his/her office. In this case, there are parts of the image that specifically belong to the office while others are shared with

X. He et al. (Eds.): MMM 2015, Part II, LNCS 8936, pp. 357–368, 2015.

Input Image	Consistent Regions	Discriminative Patches	Discriminative Regions

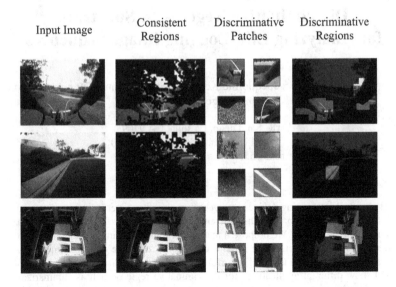

Fig. 1. Discriminative Regions. First column is the input image. Second column shows consistent regions. Third column shows four discriminative patches discovered for set a of images from the same scene as the input image and Fourth column highlights the discriminative regions.

several other scenes. The idea is that the interesting regions are those that are specific to a scene and these regions should contain features that discriminate a given scene from others. Based on this definition, the office objects such as monitor, keyboard, etc that are visible in the office become the discriminative regions for the office scene.

The organization of this paper is as follows. Sec. 2 reviews the related works. Sec. 3 and Sec. 4 explain our algorithm that finds regions of interest that satisfy the four requirements. Sec. 5 discusses how we find discriminative regions after discovering discriminative patches. Later in Sec. 6, we show a few applications we build on top of these discriminative regions. Sec. 7 discusses our experiments and results and we conclude in Sec. 8.

2   Background

Ego-centric, First Person or Wearable Computer Vision has recently emerged as an area of great interest to computer vision researchers. Recently there have been works on finding objects and their relations to activities [6], [5] and [11]. These methods usually require extensive amounts of annotation of object segments, bounding boxes or other labels for each image or video frame in the sequence.

Another relevant line of work is object discovery [12] [7] [2] [10] [14] [15]. Most of these works address category level object discovery or object category discovery. These methods are designed for cases where multiple instances of each category are available in the dataset whereas in life-logging image sequences where there is usually one instance of each object category. The challenge in discovering objects in life-logging focuses more on finding objects in a variety of imaging conditions e.g. viewing angle, illumination, and occlusion rather handling intra class variation of objects such as different shapes or sizes of objects. Among these methods, Kang et al. work [9] is the most relevant paper. Their idea is to find groups of mutually consistent image segments. They first ran segmentation on all of the images of their dataset to get many small segments and designed an algorithm find co-occuring segments. Later, those segments are joined to form object segments. In this process, since their goal was instance-level object discovery, they also imposed appearance and geometry constraints in their optimization. They tested their methods on lab controlled image sets. The difference between their work and this paper is that we are working with real life-logging images and this introduces many challenges such as the size of the dataset, uncontrolled illumination and occlusion. The other difference is that we focus more on finding candidate segment while their work is on how to connect these segments to form objects.

Our discriminative patch discovery method is similar to [3] and [13]. Singh et al. randomly subsample patches from a dataset of images with labels and apply an iterative discriminative learning method to find the patch clusters that demonstrate high discriminative power. Their iterative SVM learning is based on Ye et al. work on clustering [16]. We adapt the discriminative patch discovery idea and propose a simpler and faster algorithm to find discriminative patches.

3   Consistent Regions

As mentioned earlier, finding consistent regions is a crucial step in finding image regions containing objects of interest. We define consistent regions as parts that are visible across several consecutive frames in the image sequence. For example, in Fig. 3, the consistent regions are bike handle and hands while everything else lie on the background. The key factor is that the frame rate is variable between 20-30 as opposed to $\frac{1}{24}$ in SD videos. Due to the low frame rate, optical flow algorithms, object tracking and background subtraction methods fail to find the consistent regions.

Life-logging scenes can be classified into two groups: stationary scenes and dynamic scenes. Stationary or low motion scenes are dominant in the life-logging images. These scenes are mostly sitting cases including working with computer, watching TV, eating meals or sleeping and less occasional standing or other postures. Images of these scenes show very small change and large portions of these images are consistent. The other type of scenes are dynamic scenes in which usually the camera is translating in space. This includes biking, driving or walking. In the biking or driving scenes, there are parts of the image that are changing while some parts remain consistent. The consistent parts of the image may change their position in the next frames but they remain visible for at least a few frames. Walking or running are another types of dynamic scenes where the whole scene is significantly changing, do not have consistent regions.

There are several solutions for standard frame rate videos. Optical flow or background subtraction methods might solve this problem when the object movements are small. Fig. 3 shows a sample of three consecutive frames. In these frames, the bike's handle bar, some of parts of bike's body and hands constitute the consistent regions while the road, trees and sky compose the background. Conventional foreground-background segmentation methods fail because of two reasons: (1) the appearance of those regions change between consecutive frames due to lighting changes, shadows,

Fig. 2. Consistent Patch Detection. Two patches from the right image are selected A and B. Their nearest neighbors in HOG feature space are labeled as A' and B' in the left image. Finally A'' and B'' are the nearest neighbors of A' and B' respectively.

viewing angle and deformations. (2) the movement of these regions between these frames can be reasonably large e.g. the center of the bike handle may reside at the center of one frame and move to the far left in the next frame. Thus, we need a method that is able to handle more variations in the appearance and the movement of objects than the conventional methods do.

We propose a forward-backward search method to address these two issues which is based on searching small regions of the image in the images before or after it. To handle the location change, we search for each region in the next or previous frames in all of the possible locations. And we use HOG features to encode the structure of the region and cope with the variations in the appearance. Fig. 2 shows two consecutive frames from a biking scene. Consider a patch from the middle frame e.g. the center of the bike handle. We search for this patch in the previous frame to find its nearest neighbor. We do not constrain the search space to a local region and search throughout the previous frame for the best match. Our goal is to find another instance of the given patch in the previous frame if it is visible. One solution is to look at the distance between the given patch (A) and its nearest neighbor (A') in the feature space and accept the match as a correct match if the distance is lower than a threshold. But our experiments show that this approach is not very reliable, as the distances to the nearest neighbor for the wrong matches are sometimes lower than the correct matches. We propose not to rely on the distances in the feature space. Instead, we do another search for A' in the middle frame A''. If the starting patch A is a consistent patch, then A'' would be the same as or very close to A. And if the starting patch e.g. B is not a consistent patch, then its nearest neighbor, B', is a wrong match and B'' falls at a different location.

To implement the proposed method, first we extract dense HOG features from a grid of points from middle and left frames and create a kd-tree on the extracted features for each frame. Using the kd-trees, two searches are done for each patch in the given image. If the resulting patch is spatially close to the given patch, they the given patch is considered a consistent patch. The same procedure is done on the middle and right frames and the final consistency map is the intersection of the two maps. Fig. 3 shows threes consistency maps where the middle one is the intersection of the left and right maps. Finally the last image shows the middle image with the overlaid consistency mask.

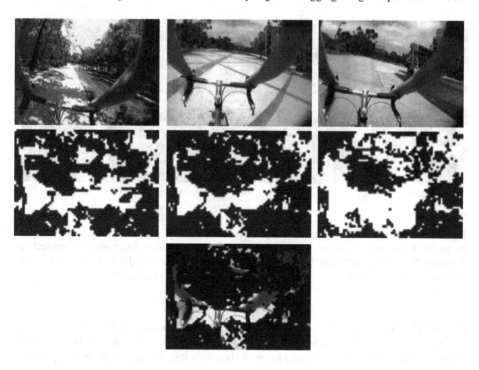

Fig. 3. Consistent Region Detection Sample. The top row show three consecutive images from a biking scene. The second row shows the consistency maps. The right binary map is generated using first two images and the left map is generated using the last two images. The middle map is the intersection of the two maps which is overlaid on top of the original image in the third row.

4 Discriminative Regions

After finding the salient and consistent regions, our goal is to group these regions to form clusters. Some of these clusters may represent meaningful regions and some clusters do not, even with very low intra cluster error. We define meaningful clusters as those who demonstrate discriminative power. For the discrimination, we use two sets (labels). Since we use image labels, this work can be considered as a weakly supervised method. Later we discuss ways to automatically extract information we need for these labels.

We assume that we have two image sets: $\mathbb{P}$ and $\mathbb{N}$. $\mathbb{P}$ contains a group of images from a particular scene. The scenes can be dynamic or static and they usually correspond to human activities such as biking, walking, driving, working in the office and watching TV. $\mathbb{N}$ is the universal or negative set and contains everything but $\mathbb{P}$. For example $\mathbb{P}$ may represents images of a biking scene while $\mathbb{N}$ is the set of all other images. Later in the Sec. 7, we discuss different combinations of $\mathbb{P}$ and $\mathbb{N}$.

$$\mathbb{U} = \mathbb{P} \cup \mathbb{N} \subset \mathbb{R}^d \tag{1}$$

More specifically $\mathbb{P}$ and $\mathbb{N}$ are the sets of extracted features from patches randomly sampled from the consistent and salient regions. We used histogram of oriented gradients

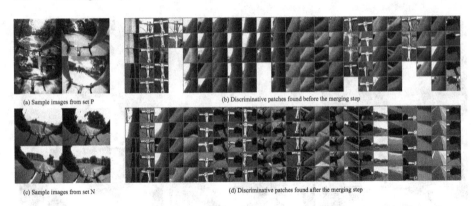

(a) Sample images from set P

(b) Discriminative patches found before the merging step

(c) Sample images from set N

(d) Discriminative patches found after the merging step

Fig. 4. Discriminative Patch Discovery. (a) and (c) show examples of $\mathbb{P}$ and $\mathbb{N}$ respectively. (b) shows the discriminative patches that are discovered by our algorithm. Each column represents a discriminative patch and the patches in each column are the nearest neighbors of the patch in the first row. Finally, (d) shows the discriminative patches after the merging step.

(HOG) [1] and color ($a^\star$ and $b^\star$ channels of $La^\star b^\star$ color space) to represent each patch. Furthermore, all of the selected patches from an image share the same labels as their image source. There is a one-to-one mapping between labels and sets. All of the patches in $\mathbb{P}$ have $+1$ label while patches in $\mathbb{N}$ have -1 label.

We propose to solve the discriminative patch discovery by looking at nearest neighbors of each feature point. $N_k(x)$ is defined as set of k nearest neighbors to x based on the distance function d. If we look at the distribution of labels in $N_k(x)$, we are able to find patches that most likely appear in either $\mathbb{P}$ or $\mathbb{N}$ and patches and appear in both sets.

$$N_k(x) = \{x'|x' \in \mathbb{U}, \forall y \in \mathbb{U} - N_k(x) : d(x,x') < d(x,y)\} \\ \text{and } |N_k(x)| = k \tag{2}$$

Specifically, we measure the label proportion in $N_k(x)$ by counting the number of element in each set (i.e. $\mathbb{P}$ or $\mathbb{N}$) and dividing the two numbers. The larger the value of division, the more discriminative the patch. The idea of iterative discriminative learning [3], [13] is to first run unsupervised clustering such as k-means and then at each iteration train an SVM classifier to separate each cluster from the rest and take the top matches of each classifier each cluster as positive in the next iteration to train another set of classifiers. Following their method, we implemented the iterative SVM learning but we did not see improvements over our nearest neighbor based discriminative patch discovery. There are two main reasons. The first reason is in the nature of our data. Since we are using life-logging images, there is inherent repetition of objects in the image sequence that help to achieve good clusters with our method. The second reason is that our results is much cleaner than k-means clustering which makes it hard for the iterations to improve the quality.

$$D(x) = \frac{|N_k(x) \cap \mathbb{P}|}{|N_k(x) \cap \mathbb{N}|} \tag{3}$$

Fig. 5. Frame selection to visualize spatial relationship among discriminative patches. Note that highlighted regions are discriminative patches found in the top ranked frames.

After calculation of the discriminativeness value $D(x)$ for all the patches in the $\mathbb{P}$, we sort patches based on their discriminativeness value in the decreasing order and pick those with highest values. In our experiments, we use $D(x) = 4$ as a cut-off value to select the discriminative patches. $D(x) >= 4$ is equivalent to have more than or equal to 80% of the nearest neighbor points in $\mathbb{P}$.

$$\mathbb{D} = \{x | D(x) > \text{threshold}\} \tag{4}$$

Consider a patch that has a high discriminativeness value. It is expected for its nearest neighbors to have high discriminativeness values as well. The reason is that in that case the discriminative patch resides in a part of the feature space that is mainly filled with patches from $\mathbb{P}$. Because of this, some of the elements of $\mathbb{D}$ become very similar. Thus, we need to remove and merge some of the clusters. The last step is to merge the clusters that have significant amount of overlap. If two clusters are to be merged, we remove the cluster center that has less discriminativeness value from $\mathbb{D}$ and add its cluster members to the nearest neighbors of the cluster with higher discriminativeness value.

5 Spatial Relations among Patches

Depending on the definition of objects and their sizes, the discriminative patches that we find may represent whole objects or object parts. When the patches are smaller than object, we need to link them together to build objects or discriminative regions. For that we need to find the spatial relationship between discriminative patches.

We experimented with different methods to find spatial relationships among patches. The best method turned out to be frame selection. In this method, we find images that contain high number of discriminative patches and deduce about the patches geometry based on their placements in the candidate frames. Specifically, we rank the frames based on the number of appearances on the discriminative patches in them. This is a very fast process since we only need to go through discriminative patches and their nearest neighbors and all of those patches have pointers to their original image frames. Fig. 5 shows four frames that very highly ranked for the biking scene.

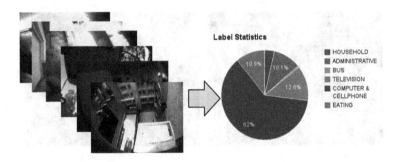

Fig. 6. Example images and a sample of histogram of daily activities

6   Applications

There are many uses for the discriminative regions we present in this paper. These regions depict information about the scene and the activity that is being performed. These regions are of importance in many applications such as classification of images, summarization and content-based image search.

Summarization. Since discriminative regions contain object and foreground information, they carry high level information about the activity that is being done from the image sequence and can be used for summarization of high level concepts. The idea is to select a minimal set of frames and cover all the discriminative patches. In this way most of the activities in the image sequences will be summarized with a small number of images.

Classification and Statistics. Discriminative Regions, by definition, are suitable for classification. Our algorithm is able to highlight regions that have discriminative power and can be used for classification. One example classifier, would be a linear classifier on the histogram of discriminative patches found in images.

The result of classification can be turned into histogram of activities or scenes e.g. Fig. 6. This will give a nice overview of a given day and can be accompanied by analysis of physical activities and recommendations about personal health.

Labeling and Search. The idea of labeling and search is similar to face labeling in Google Picasa and iPhoto. First these softwares detect faces in image galleries and cluster them into face clusters. They then ask the user to label representative faces from each cluster and using those label they label other faces in the image gallery.

In life-logging image sequences, we are not only looking for faces but anything else that are frequently appearing in the image set. After finding discriminative patches, we ask the user to label the representative patch from each cluster. Since the discriminative patches we find might be a part of an object, we show the whole image and highlight the discriminative patch and ask the user to label the enclosing object. The reason is that a cropped patch might not be informative enough to be perceived by the user but given the context it becomes easy to label. In this process, the user can choose one of the previously chosen labels or add a new label. In this way, we control the number of labels and discourage the user from adding excessive number of labels. Then, our algorithm propagate the labels to all the instances in the clusters. Fig. 7 demonstrates a few images with the propagated labels. Once we establish labels

Fig. 7. Example images with the two propagated labels: bike and hand. We manually labeled top cluster centers and ran our algorithm to propagate the labels to all of the members of those clusters. Green boxes are hand labels and red boxes are bike labels.

for images, we can use them for search and ranking. For Fig. 7, we ran our software and labeled the top 100 discriminative patches and our algorithm propagated the labels to all the candidate frames.

7 Experiments and Results

7.1 Data Acquisition and Annotation

The participants were adult cyclists recruited through a university based cycle-to-work network. Eligible participants were aged 18-70 years, were university employees, routinely bicycled for transportation. Each participant wore the SenseCam during waking hours for 35 days. They were instructed to perform their normal daily living activities, and turn off SenseCam in private time (e.g., bathroom). Excluding night time or private pictures that were removed before doing our experiments, our dataset contains over 360,000 images.

All of the images in our dataset, are labeled with position labels (i.e. Sitting, Standing Still, Walking/Running, Biking), and activity labels (such as Household Activity, Administrative Activity, Television, Other Screen Use and Eating). To the best of our knowledge, this becomes the largest SenseCam dataset with reliable labels. Our dataset is available on our lab's website[1].

7.2 Implementation

The first step is the consistency detection. For that, we randomly select 400 images from $\mathbb{P}$ and $\mathbb{N}$. Then, first we extract dense HOG from a grid of 70×50 points from all of the selected images. We chose to extract $8 \times 8 \times 31$ bin HOG from patches cropped around each point on the grid using HOG implementation of Girshick et al. [8]. Note that, throughout this paper we use a fixed patch size of 100×100 pixels. This size

---

[1] http://vision.ucsd.edu/%7emohammad/sensecam_dataset

was selected to represent a cover a good size for object parts while the images have a resolution of 640×480 pixels.

For consistency detection of each image, we need to extract HOG features from three images. After that, we search for all of the feature points in the given frame and frames before and after that and vice versa. Thus, we need to perform a total of four searches that are implemented using [4]. Assuming we have run this process for a given image, for the next image, we only need to run the feature extraction on the new image and do two searches corresponding to the new image. Using this technique, we achieved speed of about 5 seconds per image on a MacBook Pro with 2.9 GHz CPU. We also do this process for all images but with two frame distance i.e. frames $i - 2$, i and $i + 2$. The consistency maps are later combined to be used in the patch sampling process.

Having the consistency maps computed, we extract 50 patches from each image that is selected for each study. The extracted patches have to satisfy two requirements: saliency and consistency. We use a low level saliency definition which is defined as a minimum standard deviation as well as a minimum on the length of the vectorized HOG feature. For meeting the consistency requirement, we randomly sample based on the binary consistency maps. The result of this step is about 20000 HOG-color feature points.

Then we compute the distance between all pairs using [4] and pick the top 50 nearest neighbors for each point. These nearest neighbor are then used for the ranking of discriminative patches. The whole process takes about 5 minutes assuming that the consistency maps are pre-computed.

7.3 Experiments

The first experiment is to show the effect of consistency maps. In this experiment, $\mathbb{P}$ is a set of biking image of a particular person where $\mathbb{N}$ is a set of image from an office scene. Fig. 8 show the discriminative patches that are discovered in this experiment. The top row is the discriminative patch discovery without using consistency detection. For second and third rows we used three and five images for consistency detection. As it suggests, using consistency maps helps to focus more at the foreground object rather than background regions such as sky, trees or road. It also shows that using more

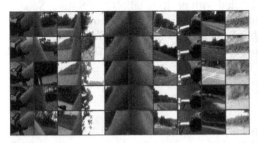 

Fig. 8. Effect of using consistency detection. The left part shows the discriminative patches without using consistency detection and the right part demonstrates the effect of consistency detection with five images. Using five images reduces the resulting discriminative patches that are mostly foreground patches.

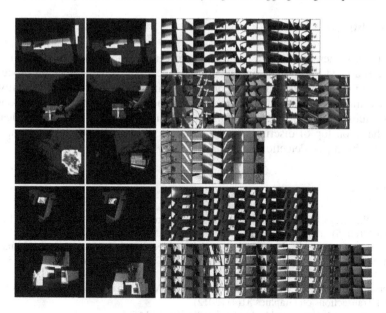

Fig. 9. Discriminative patches and discriminative regions highlights for different scenes. The scenes are Car, Biking, Watching TV, Walking/Running and Sitting from top to bottom respectively.

frames for consistency detection, results in better foreground object clusters. Additionally, since the patch sampling is more restricted and number of discriminative patches become less.

The second experiment is studying the effect of type of scene. For this experiment, we have used five different labels: Car, Biking, Watching TV, Walking/Running and Sitting. The image where chosen the image sets of two different persons. In each case $\mathbb{P}$ is defined to be the set of images having each of the five labels from one person's data and $\mathbb{N}$ is the images from the same label but from the other person. Fig. 9 shows the discovered discriminative patches and discriminative regions from top image selected by our frame selection process.

The Car and Biking scenes are similar in nature. There are some parts are visible in most of the images and some parts that are in the background. Fig. 9 shows that our algorithm is able to detect bike's parts, hands as well as parts of car's dashboard. The Watching TV and Sitting scenes are stationary scenes and our algorithm discovered monitor, TV, TV stand and well as other objects in the two rooms. Walking/Running is a completely dynamic scenes and set of discriminative patches are very small compared to other labels and our algorithm focused at fence, some parts of sky and lamps that seems to be specific for one person.

8 Conclusion

In recent years, there has been a surge of interest in wearable cameras in the industry. Yet there is a need for efficient and accurate image analysis techniques for processing life-logging images. In this paper, we presented a method to extract discriminative regions with minimal supervision. We also hinted at how these discriminative regions can serve as a substrate for life-logging applications. Finally, we showed a few applications we can build on top of discriminative regions such as summarization and improved search and object part detection.

References

1. Dalal, N., Triggs, B.: Histograms of oriented gradients for human detection. In: CVPR, vol. 1. IEEE (2005)
2. Deselaers, T., Alexe, B., Ferrari, V.: Localizing objects while learning their appearance. In: Daniilidis, K., Maragos, P., Paragios, N. (eds.) ECCV 2010, Part IV. LNCS, vol. 6314, pp. 452–466. Springer, Heidelberg (2010)
3. Doersch, C., Singh, S., Gupta, A., Sivic, J., Efros, A.A.: What makes paris look like paris? ACM Transactions on Graphics 31(4) (2012)
4. Dollár, P.: Piotr's Image and Video Matlab Toolbox (PMT),
   http://vision.ucsd.edu/~pdollar/toolbox/doc/index.html
5. Fathi, A., Farhadi, A., Rehg, J.M.: Understanding egocentric activities. In: ICCV. IEEE (2011)
6. Fathi, A., Ren, X., Rehg, J.M.: Learning to recognize objects in egocentric activities. In: CVPR. IEEE (2011)
7. Fritz, M., Schiele, B.: Decomposition, discovery and detection of visual categories using topic models. In: CVPR. IEEE (2008)
8. Girshick, R.B., Felzenszwalb, P.F., McAllester, D.: Discriminatively trained deformable part models, release 5,
   http://people.cs.uchicago.edu/~rbg/latent-release5/
9. Kang, H., Hebert, M., Kanade, T.: Discovering object instances from scenes of daily living. In: ICCV. IEEE (2011)
10. Kim, G., Faloutsos, C., Hebert, M.: Unsupervised modeling of object categories using link analysis techniques. In: CVPR. IEEE (2008)
11. Pirsiavash, H., Ramanan, D.: Detecting activities of daily living in first-person camera views. In: CVPR. IEEE (2012)
12. Russell, B.C., Freeman, W.T., Efros, A.A., Sivic, J., Zisserman, A.: Using multiple segmentations to discover objects and their extent in image collections. In: CVPR, vol. 2. IEEE (2006)
13. Singh, S., Gupta, A., Efros, A.A.: Unsupervised discovery of mid-level discriminative patches. In: Fitzgibbon, A., Lazebnik, S., Perona, P., Sato, Y., Schmid, C. (eds.) ECCV 2012, Part II. LNCS, vol. 7573, pp. 73–86. Springer, Heidelberg (2012)
14. Sivic, J., Russell, B.C., Efros, A.A., Zisserman, A., Freeman, W.T.: Discovering object categories in image collections (2005)
15. Weber, M., Welling, M., Perona, P.: Towards automatic discovery of object categories. PAMI 2, 101–108 (2000)
16. Ye, J., Zhao, Z., Wu, M.: Discriminative k-means for clustering. NIPS 20 (2007)

Fast Human Activity Recognition in Lifelogging

Stefan Terziyski, Rami Albatal, and Cathal Gurrin

Insight Centre for Data Analytics,
Dublin City University, Dublin, Ireland
{stefan.terziyski,rami.albatal,cathal.gurrin}@insight-centre.org

Abstract. This paper addresses the problem of fast Human Activity Recognition (HAR) in visual lifelogging. We identify the importance of visual features related to HAR and we specifically evaluate the HAR discrimination potential of Colour Histograms and Histogram of Oriented Gradients. In our evaluation we show that colour can be a low-cost and effective means of low-cost HAR when performing single-user classification. It is also noted that, while much more efficient, global image descriptors perform as well or better than local descriptors in our HAR experiments. We believe that both of these findings are due to the fact that a user's lifelog is rich in reoccurring scenes and environments.

1 Introduction

Recent technological development in personal and ubiquitous computing, data storage and computational power has provided the environment for lifelogging to become a normative activity [1]. At the same time, Human Activity Recognition (HAR), a well-established research field is receiving increasing attention, however much of this attention has been directed towards using fixed cameras observing people, rather than from the point of view of the wearer. HAR from lifeloggers' point of view visual context has been described by Steve Mann as "sousveillance" (different than the classic surveillance concept) in [2]. In sousveillance we do not get to observe what the human or lifelogger is doing, but rather the view that the lifelogger is seeing. This provides a visual context of the user and this visual context is a source for HAR.

The purpose of this research effort is to examine the feasibility of fully-automatic, low-cost HAR in lifelogging. When considering HAR in lifelogging we mean the process of identifying what activity the wearer is performing at the moment of capture. While HAR may be performed at any point in the lifecycle of lifelogging data (capture time, processing time, access and feedback time), our key interest here is in supporting real-time low-cost HAR using visual lifelog data. Consider the following scenario, a Google Glass wearer, upon the advice of his doctor, is interested in identifying and logging his dietary habits. Given current battery considerations, continual capture and upload to a server for analysis is not feasible. A HAR algorithm, successfully running on his device would be able to trigger capture of a series of detailed photos upon detection of and eating event.

X. He et al. (Eds.): MMM 2015, Part II, LNCS 8936, pp. 369–379, 2015.

Another scenario is where a lifelogger is interested in storing a record of his life activities, but due to privacy considerations with wearable cameras, he does not want to transmit and store wearable camera photos. Such considerations motivate this paper.

The ability to capture visual user context in real-time would be an enabling technology for many applications. Considerations of battery life and network-transport delay means that continual transmission of visual content to a server for analysis and feedback is far from ideal. In this work, we illustrate how low-cost HAR can be achieved on a per-user basis and we evaluate this by means of an experimental evaluation on data gathered from 5 users over 14 days of visual lifelog data. It is our conjecture that a low-cost implementation of HAR would be more applicable to on-device deployment than more conventional techniques such as SIFT. We show that the colour distribution combined with the global image gradient intensity performs well on low resolution images for a HAR task for an individual user.

The rest of the paper is as follows: in 2 we outline previous work done on activity recognition. In section 3 we provide a description of the activities, the visual data and the features extraced for it. Subsequently in section 4 we describe our evaluation method, then in section 5 the results we have obtained and in section 6 what conclusions we can draw from this initial exploration.

2   Related Works

Lifelogging represents a phenomenon whereby individuals can digitally record their own daily lives in varying amounts of detail and for a variety of purposes. Although there are many definitions in the literature, we define lifelogging to be *a form of pervasive computing which utilises software and sensors to generate a permanent, private and unified multimedia record of the totality of an individual's life experience and makes it available in a secure and pervasive manner.* A key aspect of this definition is that the lifelog should archive the totality of an individual's experiences, outlined in Bell and Gemmels' vision of total capture [3]. This means that lifelogging will generate a rich set of multimedia data, gathered from wearable sensors on the lifelogger.

One early attempt in activity recognition in lifelogging is by Doherty, et al. in [4], where the authors aimed at developing a trait interpreter tool in lifelogging data. Some of the target traits were activities, such as shopping and using mobile phone. An average crossover accuracy of 61% was reported. The crossover is between system output and lifelogger self-report. The authors used MPEG-7 features, namely ScalableColor and ColorLayout.

There is a research direction within the domain of HAR from visual lifelog data, where certain objects are detected and the co-occurrence of these within visual images forms the basis of activity recognition, as by Wang and Smeaton in [5], where they report an average F1-score over all activities, used in their experiments, of 90% with a baseline object detection accuracy of 65% on average. A system with such classification framework would be difficult run in real-time context, because it is necessary to determine what objects are needed to detect in order to determine the activities for a particular user. In practice human assistance would be needed for that, hence the system would not be fully automatic.

Another approach is the work by Hamm et al. in [6], where global image descriptors are combined with other sensory data to form what the authors call a multi-sensory bag-of-words. The work produces an average F1-score for the best classifier of 92%, however this is averaged from the F1-scores of the trained classifiers per activity. The authors do not propose a model, where they fuse all of the per activity classifiers to form a final activity classification model. If, for example, two activities were mutually exclusive then the system should be able to choose only one activity, but not both. The sensors that they use are: digital camera, accelerometer, audio and GPS. The authors have shown that fusing different sensors improves the accuracy and also show that discriminative models tend to perform better on global descriptors. The visual features used in [6] are colour histograms in HSV-space.

In this work we aim to provide for a fully automatic, low-cost and fast classification model for HAR from visual lifelog data that has potential to be deployed on the lifelogging device. We achieve that by examining a more direct relationship between the visual context and the activity being performed, thus rendering our classification process automatic. We employ computationally inexpensive visual features operating over low resolution images.

3   HAR Technique

Our HAR technique starts with visual feature extraction from image data. The chosen feature(s) (which can be any or a combination of Colour Histogram, HOG, SIFT) are the one(s) that yielded the best results. Different parameters (such as image resolution, colour bins, number of gradient orientations, etc.) would also be tuned at this stage. Then the feature set is used for training a per-activity and per-user SVM classifier (with radial-basis kernel). We aim to improve the performance of those classifiers. Our experimental data set is divided into training and evaluation subsets as it is explained later on in section 4.

In developing the HAR technique, we identified five research questions that we needed to address. Firstly, the experiments that we set up were to determine whether the amount of training examples affects the classification performance. This was achieved by comparing the F1-scores per activity for two data sets, where the second one contained an additional training examples for one activity. Secondly, given the same configuration, we examined whether this affected the classification performance for the rest of the activities. Thirdly, we examined various combinations of features namely colour with texture and local with global features. Fourthly, we carried out tests to see what feature scaling techniques would be most appropriate. Finally we compare the computational efficiency of extractions of the different features.

3.1   Human Activities

In selecting a set of human activities for classification, we turned to the work of Kahneman, et al. in [7], in which the 15 most enjoyable daily activities of people were identified by means of a large-scale survey. Of the 15 activities identified, we have chosen 9 for this work. We purposely excluded 5 activities, namely 'intimate relations', 'relaxing', 'pray/worship/meditate', 'napping' and 'taking care of children', because we consider those to be private or intimate and beyond the scope of our current research. One final activity, 'exercising', is excluded because it was not actually present in the lifelog of the any of 5 users. This is most probably due to the difficulties of wearing a lanyard mounted camera during exercise.

3.2   Visual Data and Annotations

As mentioned above, this paper focuses on processing the visual input. Those are lifelog images taken from an OMG Autographer wearable camera[1] which is a fish-eye camera that can be worn on a lanyard around the neck, or clipped onto clothing. In normal use, it is oriented with the viewpoint of the wearer. We had the camera configured to capture an image approximately every 20-30 seconds.

The data set consists of 41,397 images from 5 users, with 14 days of lifelog data per user. The experiments were carried out mainly on one user's data which consists of 4,315 images. The rest of the data from 4 other users with a total of 37,082 images was used for validating the cross-user generalisation of the selected features. It is to be noted that the lifelogging was not continuous and some gaps may occur, in terms of consecutive days. See Fig. 1 for examples of the visual lifelog data that we employed. The visual repetition of some human activities are clearly visible.

We note the unbalanced nature of lifelog data, due to the naturally varying lifestyle across the users. Refer to Table 1 for a detailed distribution. In Table 1 'N/A' stands for Not Available.

Table 1. Activity distributions among users' lifelog data in number of images

Activity	User 1	User 2	User 3	User 4	User 5
1.Commuting	1317	2068	2937	373	2566
2.Computer	2140	5223	3969	4961	3337
3.Eating	59	505	538	558	677
4.Housework	63	205	90	57	111
5.On the Phone	23	1087	436	N/A	227
6.Preparing Food	33	130	144	989	12
7.Shopping	N/A	98	228	13	408
8.Socialising	680	845	487	525	2800
9.Watching TV	N/A	192	386	N/A	N/A

---

[1] www.autographer.com

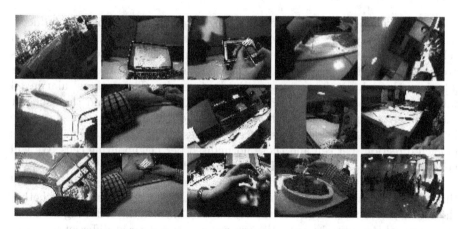

Fig. 1. Example activities showing the similar visual context of the activities. This is also shows that some activities occur in the same environment, which affects the classification process. From left column to right column: commuting, computer, on the phone, eating, socialising.

3.3   Visual Features

For fast image processing, the resolution of the lifelog images was reduced to 91x68 pixels keeping the original aspect ratio. We chose two features, namely colour histogram and Histogram of Oriented Gradients (HOG) by Dalal and Triggs in [8] and we examine the importance of colour and texture and the importance of local and global descriptors and their combination. Colour histogram is a global image descriptor and provides one colour distribution per image. The latter feature (HOG) is a local descriptor and provides texture information per image.

Colour Histogram. For each image colour histograms were extracted, where the combination of the RGB values are preserved. We extracted Colour Histograms at 16 bins, 8 bins and 4 bins per each channel. More detailed colour information such as 32, 64, 128 and 256 bins per channel, may prove more valuable, however the feature vectors would be larger, which could have a consequential impact on the computational time of the SVM classification process. Hence we compared 4, 8 and 16 bins per channel, giving 64, 512 and 4096 bins respectively. All of them have comparable performance and we can report that 512 bins and 4096 bins give similar performance on the test that we have performed, therefore we chose the 512 bin histogram due to the considerable reduction in computation time, which is due to the time complexity of the SVM. We have subsequently used the 512 bin RGB histogram for all of the experimentation in this paper, since the colour precision at this point proves to be sufficient.

Histogram of Oriented Gradients. HOG features were extracted. The format is: 9 orientation bins; 8 by 8 pixels per cell; 2 by 2 cells per block, hence

16 by 16 pixels per block. Hence a feature vector of length 2,520 is produced, where 280 cell responses are captured, giving 9 orientations each. The amount of cell responses is derived from the per-block normalisations, however the blocks are overlapping, as suggested in [8]. The parameters were also chosen as per [8], since those have shown to give the best results. The authors' aim in [8] was to improve object recognition from a detection window and they suggest that this gives comparative results to the state-of-the-art object classifiers. We have chosen HOG as it describes a scene by describing the gradient intensities in each direction, per cell and those normalised per neighbouring cells. We will henceforth refer to the original HOG as 'local' HOG.

We altered HOG's locality by constructing a 'global' HOG and also evaluated that. The evaluation of the 'global' HOG is done in order to evaluate whether the locality of a feature affects the classifier. Consider the following scenario: given several 'computer' scenes, the computer may appear at different locations in the scenes, which would affect the local HOG, but would not affect the global HOG. This hypothesis was evaluated by the introduction of 'global' HOG.

We did so by modifying the parameters, so that the cell becomes 91 by 68 pixels and there is only 1 by 1 cell per block, thus we allow the algorithm to only compute the gradient intensity for the entire image, instead of breaking it down into cells and blocks. This can be seen as global texture descriptor, similar to but not the same as an edge orientation histogram. In the latter only edges are counted per direction.

Considering Scale Invariant Feature Transform. Scale Invariant Feature Transform (SIFT) could be considered to be a widely deployed feature extraction approach. Hence we employed SIFT and we benchmarked the performance of SIFT when compared to the features employed in this paper. We used dense sampling on every 6 pixels and OpponentSIFT for a descriptor as suggested and implemented by Koen van de Sande in [9]. We used the codebook model to construct a codeword histogram per image and evaluated that with a RBF SVM. The resulting descriptor of an image is a normalised histogram of all the visual words encountered in the image, binned into their respective codewords via clustering. We chose a codebook of size 512, which reported comparable results to those obtained by RGB and HOG features when using the full sized image, but lower performance when using low resolution images, where many visual details might be lost.

For the computing performance comparison, colour histogram and SIFT were employed as executable binary files, implemented in C/C++, and were ran on the same machine over the same set of images, performing the same input-output operations. We report that the average extraction time of SIFT descriptor from one image is 0.05s, as per the implementation provided [9], whereas using an RGB histogram extraction, as provided by OpenCV takes about 0.00037s and a binary executable implementation of HOG extracts the feature in 0.00032s. Thus the total extraction time of HOG and RGB histograms together is 0.0069, which make both of these features many times quicker to extract than SIFT.

We do not consider the codebook creation also, even though it is a lengthy process, since it is only needed once, whereas the feature extraction process per image is required for all new input.

For this reason, and given the focus of this research on fast and efficient local processing that could be applied to a relatively low-power device (such as Google Glass), we decided that the overhead of implementing SIFT was too great for our use-case, so we proceed to compare and evaluate the two prior algorithms.

4   Evaluation

Evaluation was carried out via splitting the data set into 80% training examples and 20% examples for evaluation. In this instance the original distribution of data was kept, meaning that for each activity the amount of training examples is 80% and the amount of evaluation examples is 20% out of the total available examples of that activity.

The evaluation metric that we used is F1-score. We obtain the F1-score, as it is expressed in terms of the harmonic mean of precision and recall. However, we work within a classification context, hence both the precision and the recall were obtained in that context. In the classification context, TP refers to True Positives, FP refers to False Positives and FN refers to False Negatives, hence we calculate the F1-score as:

$$Precision = \frac{TP}{TP + FP}$$
$$Recall = \frac{TP}{TP + FN}$$

$$F1 = 2 * \frac{Precision * Recall}{Precision + Recall}$$

When performing the classification of the various features, we also used feature scaling, because that improved the performance of all activity classifiers over the same features. Based on a prior evaluation, the feature scaling outlined in equation (1) was selected because it had proven to be the most successful in initial evaluations on colour histograms and HOG.

$$x_j = (x_j - mean)/range$$
$$x_j = x_j/max \qquad (1)$$
$$x_j = x_j/\sum_X$$

5   Results

We noted that for the visually similar activities, the number of the positive training examples affected the F1-score, hence the performance of the classifier.

Whereas for activities whose context is more visually dissimilar that was not necessarily the case. By dissimilar we mean that the visual context is different and vice versa for visually similar.

Table 2. F1-scores of the available activities for User 1. Each activity has its own classifier.

Activity	local HOG	RGB Hist	global HOG + RGB Hist
1.Commuting	0.82	0.94	0.95
2.Computer	0.88	0.92	0.93
3.Eating	0.59	0.66	0.77
4.Housework	0.27	0.13	0.35
5.On the Phone	0.40	0.66	0.86
6.Preparing Food	0.67	0.57	0.67
7.Socialising	0.57	0.77	0.77

See Table 2 'local HOG' for results. When using global HOG, we had actually found an F1-score of 0 for all activities except for 'computer', where that was 0.66, so this did not perform well at all.

When combining both the HOG features and the colour histograms, we used both types of HOG. When using the local HOG we had no success whatsoever, in spite of the promising results from both in Table 2 - F1-scores for User 1 above. We got low F1-score throughout the activities. But when we combined the RGB histogram with the global HOG descriptor we got an improvement in the classification performance. The RGB histogram alone gives an average F1-score of 65%, whereas in combination with the global HOG the average F1-score becomes 75%. See Table 2 under 'global HOG + RGB Hist'.

Given the results in Table 2, we can say that colour provides for a greater discriminative potential than texture when comes to HAR. We can also say that a global image descriptor in our case could outperform local ones. We have obtained the best results from using a RGB histogram combined with a global HOG.

With regards to the SVM for the activity classification process, we used a radial-basis-function kernel SVM with hyper-parameters $C=1.0$ and $\gamma=1.0$ for all of the results presented, except where stated otherwise. We present a grid-search optimisation for C being 10 whose exponent ranges from -2 to 2 and γ being 10 whose exponent ranges from 2 to -4. This was used purely for exploratory data analysis in order to understand the relationship between the visual context and performed activity by the wearer. If we observe a high C, that means that those activities have similar visual contexts that are difficult to distinguish between. Respectively for γ where this hyper-parameter is low it means that the data is diverse, whereas with high values it means that those activities have a less diverse visual context.

Table 3. C and γ optimisation in the SVM Classifier

	local HOG			RGB Hist		
Activity	F1-score	C	γ	F1-score	C	γ
1.Commuting	0.89	10	0.1	0.95	1	1
2.Computer	0.92	0.10	0.1	0.93	100	1
3.Eating	0.62	0.01	0.1	0.77	100	1
4.Housework	0.47	0.01	0.1	0.29	0.01	0.1
5.On the Phone	0.66	10	0.1	0.66	0.01	1
6.Preparing Food	0.66	1	0.01	0.80	10	1
7.Socialising	0.58	10	0.1	0.81	10	1

Given Table 3, we can see that in terms of texture as a local descriptor, activities 1, 5 and 8 are difficult to distinguish from the rest of the training examples. Whereas in the case of the colour histogram, activities 2, 3, 6 and 8 are difficult to discriminate from the rest, due to the training examples overlapping with each other. All activities' classifiers whose C parameter is above 1 are considered to generalise poorly for new examples.

We validated the importance of colour for activity recognition over the rest of the lifelog data provided by the other four users. See the Table 4 for a more detailed information.

Table 4. F1-scores for the rest of the users on Colour Histograms

	User 2		User 3		User 4		User 5	
Activity	Qty	F1	Qty	F1	Qty	F1	Qty	F1
1.Commuting	2068	0.86	2937	0.92	373	0.47	2566	0.81
2.Computer	5223	0.92	3969	0.90	4961	0.96	3337	0.94
3.Eating	505	0.51	538	0.53	558	0.70	677	0.84
4.Housework	205	0.57	90	0.09	57	0.74	111	0.57
5.On the Phone	1087	0.68	436	0.63	N/A	N/A	227	0.41
6.Preparing Food	130	0.74	144	0.49	989	0.85	12	0.0
7.Shopping	98	0.5	228	0.60	13	1.0	408	0.73
8.Socialising	845	0.64	487	0.50	525	0.74	2800	0.83
9.Watching TV	192	0.59	386	0.90	N/A	N/A	N/A	N/A

It is our consideration that colour acts as a discriminator between the activities that a user can be engaged in during daily life and based on our experimentation, this has proven to be sufficient to classify the activities within the data of one user. However, the relation between the user's activity and the colour distribution of the visual context is user-dependent. We applied cross-user validation, meaning that the classifier was trained on one user's data and then used for validation on the another user's data. Training on user 3 and validating on user 2 has an average F1-score of 19% for all activities. The rest is as follows: user 3 on user 4 is 12%; user 3 on user 5 is 18%; user 2 on user 4 is 9%; user 3 on user 5 is 20%.

6   Conclusion

In this paper we examined the importance of texture and colour as well local versus global image descriptors in order to determine the features with the best classification performance and low-cost extraction time. We evaluated the performance of HOG, RGB Histograms and combinations thereof and used SIFT as a benchmark for time tests.

There are several conclusions that we can derive from our results. First, local HOG descriptors do not always provide the best classification performance in our scenario. Although what we named local HOG outperformed significantly what we proposed as global HOG, it (the local HOG) was outperformed by the colour histogram. Second, we have verified the importance of colour on the rest of the lifeloggers' data and we can show that the colour distribution of the visual context of a lifelogger is related to the activity, which the lifelogger is performing. Third, texture information is important, and we can confirm that it affects the performance of the classifier positively as with the global HOG. Lastly, the trained classifiers are user-dependent, therefore classifier trained on one lifelogger cannot used for inference on other lifeloggers.

We also demonstrated how real-time HAR is possible with computational and storage efficient processing, due to the low resolution that still provides good results. Finally, we cannot state that the visual aspect for HAR is alone sufficient. The performance of HAR on low-power devices could potentially be significantly enhanced by the integration of additional sensors, such as accelerometers, location, etc.

Future Work
With regard to limitations of this work, we have noted different luminance conditions in the images and we believe that this may lead to misclassification. Considering alternative colour models, such as HSV, is a further step, where the idea is to neutralise any shadows that may appear in the visual context due to different luminance conditions in the lifelog.

We also recognise the need for a larger sample of people in order to validate our findings. We also note the necessity for more lifestyle activities. This is due to a possible direction of the research into fully characterising the day of a lifelogger as opposed to identifying occasional scenes where the lifelogger is performing a certain activity.

A further direction of the research may go into a multisensory approach as that will inevitably increase the classification performance. We also confirm the need, as suggested by previous research that an ontology for an activity needs to be defined as that may have a direct impact on the classification framework and hence the classification performance.

Acknowledgements. This publication has emanated from research conducted with the financial support of Science Foundation Ireland (SFI) under grant number SFI/12/RC/2289.

Third-party libraries were used for feature extraction, specifically [10] for HOG extraction, [11] for SVM classification and [9], [12] for SIFT comparison.

References

[1] Gurrin, C., Smeaton, A.F., Doherty, A.R.: LifeLogging: Personal Big Data. Foundations and Trends® in Information Retrieval 8, 1–125 (2014)

[2] Mann, S.: Continuous lifelong capture of personal experience with Eye-Tap. In: Proceedings of the the 1st ACM Workshop on Continuous Archival and Retrieval of Personal Experiences, pp. 1–21. ACM, New York (2004)

[3] Bell, G., Gemmell, J.: Total Recall: How the E-Memory Revolution Will Change Everything. Dutton (2009)

[4] Caprani, N., Conaire, C., Kalnikaite, V., Gurrin, C., Smeaton, A.F., Doherty, A.R., O'Connor, N.E.: PassivelyRecognising Human Activities Through Lifelogging. Comput. Hum. Behav. 27(5), 1948–1958 (2011)

[5] Wang, P., Smeaton, A.F.: Using Visual Lifelogs to Automatically Characterize Everyday Activities. Inf. Sci. 230, 147–161 (2013)

[6] Hamm, J., Stone, B., Belkin, M., Dennis, S.: Automatic Annotation of Daily Activity from Smartphone-Based Multisensory Streams. In: Uhler, D., Mehta, K., Wong, J.L. (eds.) MobiCASE 2012. LNICST, vol. 110, pp. 328–342. Springer, Heidelberg (2013)

[7] Krueger, A.B., Schkade, D.A., Schwarz, N., Kahneman, D., Stone, A.A.: A Survey Method for Characterizing Daily Life Experience: The Day Reconstruction Method. Science 306(5702), 1776–1780 (2004)

[8] Dalal, N., Triggs, B.: Histograms of oriented gradients for human detection. In: IEEE Computer Society Conference on Computer Vision and Pattern Recognition, CVPR 2005, vol. 1, pp. 886–893 (2005)

[9] van de Sande, K.E.A., Gevers, T., Snoek, C.G.M.: Empowering Visual Categorization with the GPU. IEEE Transactions on Multimedia 13(1), 60–70 (2011)

[10] Schönberger, J.L., Nunez-Iglesias, J., Boulogne, F., Warner, J.D., Yager, N., Gouillart, E., van der Walt, S., Yu, T.: Scikit-image: Image processing in Python. PeerJ 2, e453 (2014)

[11] Varoquaux, G., Gramfort, A., Michel, V., Thirion, B., Grisel, O., Blondel, M., Prettenhofer, P., Weiss, R., Dubourg, V., Vanderplas, J., Passos, A., Cournapeau, D., Brucher, M., Perrot, M., Pedregosa, F., Duchesnay, E.: Scikit-learn:Machine Learning in Python. Journal of Machine Learning Researc 12, 2825–2830 (2011)

[12] van de Sande, K.E.A., Gevers, T., Snoek, C.G.M.: Evaluating ColorDescriptors for Object and Scene Recognition. IEEE Transactions on Pattern Analysis and Machine Intelligence 32(9), 1582–1596 (2010)

Iron Maiden While Jogging, Debussy for Dinner?

An Analysis of Music Listening Behavior in Context

Michael Gillhofer and Markus Schedl

Johannes Kepler University, Linz, Austria
http://www.cp.jku.at

Abstract. Contextual information of the listener is only slowly being integrated into music retrieval and recommendation systems. Given the enormous rise in mobile music consumption and the many sensors integrated into today's smart-phones, at the same time, an unprecedented source for user context data of different kinds is becoming available.

Equipped with a smart-phone application, which had been developed to monitor contextual aspects of users when listening to music, we collected contextual data of listening events for 48 users. About 100 different user features, in addition to music meta-data have been recorded.

In this paper, we analyze the relationship between aspects of the *user context* and *music listening preference*. The goals are to assess (i) whether user context factors allow predicting the song, artist, mood, or genre of a listened track, and (ii) which contextual aspects are most promising for an accurate prediction. To this end, we investigate various classifiers to learn relations between user context aspects and music meta-data. We show that the user context allows to predict artist and genre to some extent, but can hardly be used for song or mood prediction. Our study further reveals that the level of listening activity has little influence on the accuracy of predictions.

1 Introduction

Ever increasing amounts of music available on mobile devices, such as smart-phones, demand for intelligent ways to access music collections. In particular mobile music consumption, for instance, via audio streaming services, has been spiraling during the past couple of years. However, accessing songs in mobile music collections is still performed either via simple meta-data filtering and search or via standard collaborative filtering, both ignoring important characteristics of the users, such as their current activity or location. Searching by meta-data performs well when the user has a specific information or entertainment need in mind, collaborative filtering when the user wants to listen to music judged similar by like-minded users. However, these methods do not encourage serendipitous experiences when discovering a music collection.

Integrating the user context in approaches to music retrieval and recommendation has been proposed as a possible solution to remedy the aforementioned

X. He et al. (Eds.): MMM 2015, Part II, LNCS 8936, pp. 380–391, 2015.
© Springer International Publishing Switzerland 2015

shortcomings [15,19]. Building user-aware music access systems, however, first requires to investigate which characteristics of the listeners (both intrinsic and external) influence their music taste. This paper hence studies a wide variety of user context attributes and assesses how well they perform to predict music taste at various levels: artist, track, genre, and mood. The dataset used in this study has been gathered via a mobile music player that offers automated adaptation of playlists, dependent on the user context [9].

In the remainder, related work is reviewed (Section 2) and the data acquisition process is detailed (Section 3). Subsequently, the experimental setup is defined and classification results are presented, for individual users, for groups of users, and using different categories of features (Section 4). To round off, conclusions are drawn and future work is pointed out (Section 5).

2 Related Work

Context-aware approaches to music retrieval and applications for music access, which take into account the user in a *comprehensive* way, have not been seen before the past few years, to the best of our knowledge. Related work on context-aware music retrieval and recommendation hence differs considerably in how the user context is defined, gathered, and incorporated [19]. Some approaches rely solely on one or a few aspects, such as temporal features [3], listening history and weather conditions [14], while others model the user context in a more comprehensive manner.

The first available **user-aware music access systems** monitored just a particular type of user characteristics to address a specific music consumption scenario. A frequently targeted scenario was to adapt the music to the pace of a jogger, using his pulse rate [2,17,16]. However, almost all proposed systems required additional hardware for context logging [6,7,8].

A few recent approaches model the user via a larger variety of factors, but address only a particular listening scenario. For instance, Kaminskas and Ricci [12] propose a system that matches tags describing a particular place or point of interest with tags describing music. Employing text-based similarity measures between the lists of tags, they target location-based music recommendation. The approach is later extended in [13], where tags for unknown music are automatically learned via a music auto-tagger, from input of a user questionnaire. Baltrunas et al. [1] propose an approach to context-aware music recommendation while driving. The authors take into account eight different contextual factors, such as driving style, mood, road type, weather, and traffic conditions, which they gather via a questionnaire and use to extend a matrix factorization model. In contrast to these works, the mobile music player through which the data analyzed here has been collected logs the listening context in a comprehensive and unobtrusive manner.

Other recently proposed systems for user-aware music recommendation include "NextOne" and "Just-for-me", the former proposed by Hu and Ogihara [11], the latter by Cheng and Shen [5]. The NextOne player models the music recommendation problem under five perspectives: music genre, release year, user's favorite music, "freshness" referring to old songs that a user almost forgot and that should be recovered, and temporal aspects per day and week. These five factors are then individually weighted and aggregated to obtain the final recommendations. In the Just-for-me system, the user's location is monitored, music content analysis is performed to obtain audio features, and global music popularity trends are inferred from microblogs. The authors then extend a topic modeling approach to integrate the diverse aspects and in turn offer music recommendations based on audio content, location, listening history, and overall popularity.

For what concerns **user studies on the relation of user-specific aspects and music taste,** the body of scientific work is quite sparse. Cunningham et al. [6] present a study that investigates if and how various factors relate to music taste (e.g., human movement, emotional status, and external factors such as temperature and lightning conditions). Based on the findings, the authors employ a fuzzy logic model to create playlists. Although related to the study at hand, Cunningham et al.'s work has several limitations, foremost (i) the artificial setting because a stationary controller is used to record human movement and (ii) the limitation to eight songs. The study at hand, in contrast, employs a far more flexible setup that monitors music preference and user context in the real world and in an unobtrusive way.

Another study related to the work at hand was performed by Yang and Liu [21], who investigate the interrelation of user mood and music emotion. To this end, Yang and Liu identify user moods from blogs posted on LiveJournal[1] and relate them to music mentioned in the same posting. They show that user mood can be predicted more accurately from the user context, assumed to be reflected in the textual content of the postings, than from audio features extracted from the music mentioned in the postings. While their study focuses on predicting mood from music listening events, our goal is to predict music taste from a wide range of user characteristics, including mood.

3   Data Acquisition

A recently developed smart-phone application called "Mobile Music Genius" [18] allows to monitor the context of the user while listening to music. We analyze the dataset which has been recorded by this application from January to July 2013, foremost for students from the Johannes Kepler University Linz, Austria. It consists of 7628 individual samples from 48 unique persons. We managed to identify 4149 different tracks from 1169 unique artists. As genre and mood data has not been directly recorded by the application, we queried the Last.fm API[2] to obtain this additional information. Unfortunately, the Last.fm data turned out to be quite noisy or not available at all. We were nevertheless able to identify 24 different genres and 70 different moods by matching the Last.fm tags to a dictionary of genres and moods gathered from Freebase[3]. This matching resulted

---

[1] http://www.livejournal.com/

[2] http://www.lastfm.at/api/

[3] http://www.freebase.com/

Table 1. Monitored user attributes and their type (N=numerical, C=categorical)

Category	Attributes
Time	day of week (N), hour of day (N)
Location	provider (C), latitude (C), longitude (C), accuracy (N), altitude (N)
Weather	temperature (N), wind direction (N), wind speed (N), precipitation (N), humidity (N), visibility (N), pressure (N), cloud cover (N), weather code (N)
Device	battery level (N), battery status (N), available internal/external storage (N), volume settings (N), audio output mode (C)
Phone	service state (C), roaming (C), signal strength (N), GSM indicator (N), network type (N)
Task	up to ten recently used tasks/apps (C), screen on/off (C), docking mode (C)
Network	*mobile network*: available (C), connected (C); *active network*: type (C), subtype (C), roaming (C); *Bluetooth*: available (C), enabled (C); *Wi-Fi*: enabled (C), available (C), connected (C), BSSID (C), SSID (C), IP (N), link speed (N), RSSI (N)
Ambient	mean and standard deviation of all attributes: light (N), proximity (N), temperature (N), pressure (N), noise (N)
Motion	mean and standard deviation of acceleration force (N) and rate of rotation (C); orientation of user (N), orientation of device (C)
Player	repeat mode (C), shuffle mode (C), automated playlist modification mode (C), *sound effects*: equalizer present (C), equalizer enabled (C), bass boost enabled (C), bass boost strength (N), virtualizer enabled (C), virtualizer strength (N), reverb enabled (C), reverb strength (N)
Activity	activity (C), mood (N)

in 4246 and 2731 samples, respectively, for genre and mood. The most frequent genres in the dataset are rock (1183 instances), electronic (392), folk (274), metal (224), and hiphop (184). The most frequent moods are party (319), epic (312), sexy (218), happy (154), and sad (153). Arguably, not all of the Freebase mood tags would be considered as mood in a psychological interpretation, but we did not want to artificially restrict the mood data from Freebase and Last.fm. In cases where an artist or song was assigned several genre or mood labels, we selected the one with highest weight according to Last.fm, since we consider a single-label classification problem.

Table 2 summarizes the basic statistics of our dataset for different meta-data levels: the number of instances or data points, the number of unique classes, and the number of users for whom data was available. Table 3 additionally shows per-user-statistics. Notably, the average number of genres per user is quite high (5.14). This means that participants in the study showed a diverse music taste. Figure 1 shows the different activity levels of users. We see a few users have recorded lots of samples. However, compared to them, the majority have been fairly inactive.

Table 2. Basic properties of the recorded dataset: number of different data instances, number of unique classes, and number of unique users

	Instances	Classes	Users
Artists	7628	1169	48
Genres	4246	24	45
Moods	2731	70	45
Tracks	7628	4149	48

Table 3. Arithmetic mean, median, standard deviation, minimum and maximum, per user and class

Property	Mean	Med.	SD	Min.	Max.
Artists per user	27.88	13	33.68	1	158
Genres per user	5.14	4	3.84	1	16
Moods per user	9.91	9	9.03	1	36
Titles per user	89.16	46	96.66	1	387

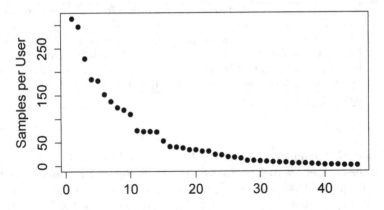

Fig. 1. Distribution of number of data instances per user, in descending order

4   Predicting the User's Music Taste

Addressing the first research question of whether user context factors allow to predict song, artist, genre, or mood, we performed classification experiments, using standard machine learning algorithms from the *Weka* [10] environment. These were *IBk* (a k-nearest neighbor, instance-based classifier), *J48* (a decision tree learner), *JRip* (a rule learner), *Random Forests*, and *ZeroR*. The last one just predicts the most frequent class among the given training samples, and is therefore used as a baseline. Optimizing the classifiers' parameters has been investigated, but we could not make out a single setting which yielded a sub-

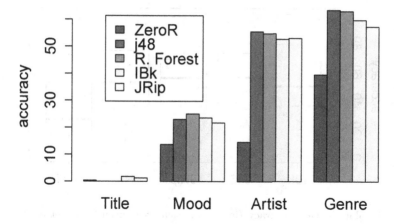

Fig. 2. Accuracy (in %) of classifications using all features

stantially better classification accuracy across multiple experiments, hence we used the default configurations in the experiments reported in the following. By performing 10-fold cross validation, we estimated the average accuracy of the classifiers' predictions.

The results evidence differences between classifiers. But no single classifier was able to outperform all others in multiple tasks (cf. Figure 2, which shows accuracies for the different classifiers in %). We also could not make out a classifier besides *ZeroR* which yields worse results than the others. Except for that, results vary only up to 10% in accuracy, depending on the experiment.

The average performance of the four non-baseline classifiers vary strongly, however, for different classification tasks: predicting genre, mood, artist, and track. Although our dataset consists of 1169 unique classes for the *artist* classification task, the classifiers managed to correctly predict about 55% of the samples, a remarkable result considering the many classes and 13% accuracy when using majority voting. The *genre* prediction results are quite good as well, since all classifiers obtained a decent accuracy of about 61% correctly predicted samples. Even given the 39% accuracy achieved by the *ZeroR* baseline, this result is remarkable. Predicting the *mood* of music succeeded on average for only about 23% of the samples. It seems that information required to accurately relate user context to music mood labels is not included in the recorded aspects. The last classification task was *title* prediction, which did not work at all. Only about 1.5% of samples have been assigned the correct title. This is not a surprise as the average playcount per title is only 1.83, thus rendering the training of classifiers almost impossible for a large number of users.

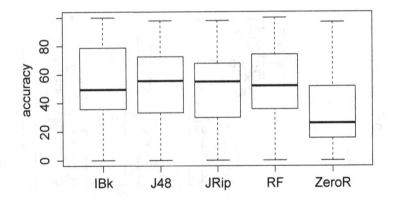

Fig. 3. Boxplot showing accuracy (in %) for each user-specific dataset on the artist prediction task

To investigate whether prediction accuracy varies for different groups of users and categories of features, we created subsets of the data in different ways:

1. for each user individually,
2. for groups of users according to their activity, and
3. for categories of features.

4.1   Individual Users

We prepared datasets in a way which required each included user to have listened to a minimum of four different tracks. Seven users did not meet this requirement and have been sorted out. We then ran experiments, using as training set only the individual user's data. Experiments were conducted again using 10-fold cross validation. For users for whom the number of samples were below 10, we performed leave-one-out cross-validation.

Figure 3 shows the distribution of the classification results for individual users in the *artist* prediction task, for each used classifier. In this boxplot, the central thick line marks the median, the upper and lower edges of the box mark the 0.25 and 0.75 percentiles, respectively, and the whiskers extend to the highest and lowest values which are not consideres outliers. We see that on average classification works considerably well, but the accuracy varies substantially between different users. We found this behavior for all four classification tasks, but investigate only the *artist* prediction task further, because results were most significant here.

By investigating the type of users for which the number of correct predictions is low, we found that they seem to have a fairly static context while listening to music. The users showing better predictability tend to listen to music in many different contexts. Recommendation systems should thus distinguish between these groups. Separating these two groups may be performed by computing the entropy of users' context features.

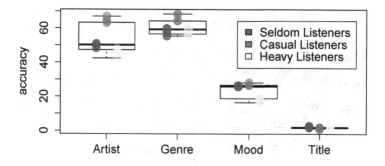

Fig. 4. Accuracies (in %) of all three user groups and all four non-baseline classifiers, for the four classification tasks. Boxplots show the aggregates of the results over all user groups, for each classifier.

4.2 User Groups with Respect to Listening Activity

Assuming that not only the diversity of the user context influences the quality of prediction results, as indicated above, but also the number of listening events recorded play an important role, we compared different types of users. To this end, we first sorted the users according to their number of listening events, in descending order. We then divided the dataset into three groups of users: *heavy listeners*, *casual listeners*, and *seldom listeners*. Each group was constructed to cover about one third of all available samples. Hence, the *heavy* group only contains 4 different users, the *casual* group 8, and the *seldom* group the remaining 36 users. The choice of using three groups and accumulated numbers of data instances to separate them was motivated by earlier work on assessing differences in activity or popularity, respectively, between users or artists. To this end, artists or users are typically categorized into three disjoint groups [4,20].

The classification results for each task are illustrated in Figure 4. We see relatively narrow boxplots for *genre*, *mood*, and *title* predictions, contrasting the results of the *artist* task. We looked deeper into the data and found a cluster of a single artist which corresponds to 18% of all samples within the casual listener group. Therefore, classification of this group seems easier, which results in a higher average accuracy of about 65% with non-baseline classifiers. A similar pattern was found in the genre prediction task, again for the casual listener group. Here, a single genre corresponds to 41% of all samples, which simplifies classification, although the impact is less pronounced. The remaining variability in each classification task can partly be explained by differences of the used classifiers. We conclude that the user's listening activity has only a small influence on the classification results, as long as the user context data is diverse enough.

4.3 Feature Categories

Table 1 displays all user aspects under consideration. Each feature was categorized already in [18] into one of the following 11 groups: *Time, Location, Weather,*

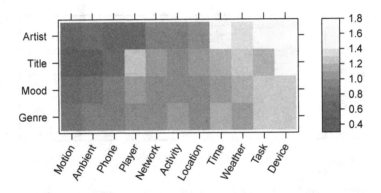

Fig. 5. The relative importance of each feature group *compared to the mean classification result* (achieved over all individual feature categories), per classification task

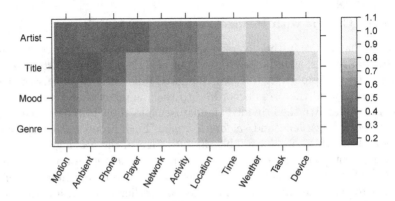

Fig. 6. The relative importance of each feature group *compared to the results obtained including all features*, per classification task

Device, Phone, Task, Network, Ambient, Motion, Player, and *Activity.* For example, the features *day of week* and *hour of day* both belong to category *Time.* By using only one category for predicting the music listening behavior in our classification tasks, it becomes possible to estimate the importance of the respective kinds of features.

We trained all classifiers for each feature group and classification task. The results are shown in Figures 5 and 6. We ordered the feature categories from left to right in increasing order according to their value for classification. Each colored box in the matrix represents the average relative performance of the respective category and class, among all four used non-baseline classifiers. Performance is measured in terms of accuracy. In Figure 5, performance values for a particular combination of feature group and classification task (one box) are relative to the mean of the achieved accuracy over all feature groups for that classification task (mean of the respective row of boxes). Performance values reported in Figure 6 for a particular feature group and classification task represent the relative accuracy of that combination, when compared to accuracy obtained by a classifier that exploits all available features.

Therefore, a neutral shade of orange in Figure 5 represents an average importance, whereas darker shades of red indicate a less important group. Consequently, the brighter the shade, the more useful information is contained within this feature group. We see that there are significant differences in the importance of groups. Interestingly, the *Player* feature category can be considered an outlier when it comes to song prediction. Although this feature category might be presumed to be a rather weak indicator, it seems to hold quite valuable information about the title. This could mean that listeners adjust player settings, such as the repeat mode, on certain songs more frequently than on others.

Figure 6 on the other hand shows the relative importance of feature groups compared to the classification accuracy using all features. Hence, a red box indicates an accuracy of only 20-30% of the accuracy achievable using all features, while a bright yellow shade indicates high performance. Therefore, we observe that *Device*, *Task*, *Weather*, and *Time* features contain almost the same amount of information as all features combined. By adding more features, we are not able to increase classification accuracy. Being in line with other research on context-aware systems, the good performance of temporal and weather features is expected. However, also the other tasks running on the user's device while using the music player seem to play a crucial role. In particular, users may prefer certain genres and artists when running a fitness app, but others when checking mails or writing instant messages. Quite surprisingly, device-related aspects are overall most important. A possible explanation is that they typically change very slowly, thus capture the general music taste of the user better than any other aspect.

5   Conclusion and Future Work

We presented a detailed analysis of user context features for the task of predicting music listening behavior, investigating the classes track, artist, genre, and mood. We found substantial differences in classification accuracy, depending on the class. *Genre* classification yielded a remarkable 60% accuracy. *Artist* classification achieved 55% accuracy. Significantly worse results were obtained in the *mood* classification task (25% accuracy) and in particular for the *track* class (1.5% accuracy). Analyzing different groups of users, we found that accuracy is not stable across users, in particular, varies with respect to diversity in user context features. Furthermore, no strong evidence for a correlation between listening activity (number of listening events of a user) and prediction accuracy, for any of the classification tasks, could be made out. We also managed to identify an importance ranking of user context features. Features related to applications running on the device, weather, time, and location turned out to be of particular importance to predict music preference. We further plan to investigate more sophisticated feature selection techniques.

Based on these results, we will elaborate context-aware music recommendation approaches that incorporate the findings presented here. In particular, this study evidences that the diversity of situations or contexts in which a user consumes music has a high impact on the performance of the predictions, and likely in turn also on the performance of corresponding music recommenders. Approaches that incorporate this knowledge along with information about the importance of particular context features should thus be capable to improve over existing solutions.

A possible limitation of the study at hand is the user data it is based upon. In particular, we cannot guarantee that the recruited participants from which we recorded data do correspond to the average music listener, as we required them to have an *Android* device and listen to local music. The user set is also heavily biased towards Austrian students. Although we believe that results are representative, a larger dataset of more and more diverse participants should be created to base future experiments on.

Acknowledgments. This research is supported by the European Union Seventh Framework Programme FP7 / 2007-2013 through the project "Performances as Highly Enriched aNd Interactive Concert eXperiences" (PHENICX), no. 601166, and by the Austrian Science Fund (FWF): P22856 and P25655.

References

1. Baltrunas, L., Kaminskas, M., Ludwig, B., Moling, O., Ricci, F., Aydin, A., Lüke, K.-H., Schwaiger, R.: InCarMusic: Context-Aware Music Recommendations in a Car. In: Huemer, C., Setzer, T. (eds.) EC-Web 2011. Lecture Notes in Business Information Processing, vol. 85, pp. 89–100. Springer, Heidelberg (2011)

2. Biehl, J.T., Adamczyk, P.D., Bailey, B.P.: DJogger: A Mobile Dynamic Music Device. In: CHI 2006: Extended Abstracts on Human Factors in Computing Systems, Montréal, Québec, Canada (2006)

3. Cebrián, T., Planagumà, M., Villegas, P., Amatriain, X.: Music Recommendations with Temporal Context Awareness. In: Proceedings of the 4th ACM Conference on Recommender Systems, Barcelona, Spain (2010)

4. Celma, O.: Music Recommendation and Discovery – The Long Tail, Long Fail, and Long Play in the Digital Music Space. Springer, Heidelberg (2010)

5. Cheng, Z., Shen, J.: Just-for-Me: An Adaptive Personalization System for Location-Aware Social Music Recommendation. In: Proceedings of the 2014 ACM International Conference on Multimedia Retrieval (ICMR), Glasgow, UK (April 2014)

6. Cunningham, S., Caulder, S., Grout, V.: Saturday Night or Fever? Context-Aware Music Playlists. In: Proceedings of the 3rd International Audio Mostly Conference of Sound in Motion, Piteå, Sweden (October 2008)

7. Dornbush, S., English, J., Oates, T., Segall, Z., Joshi, A.: XPod: A Human Activity Aware Learning Mobile Music Player. In: Proceedings of the IJCAI 2007 Workshop on Ambient Intelligence (2007)

8. Elliott, G.T., Tomlinson, B.: Personalsoundtrack: Context-aware playlists that adapt to user pace. In: CHI 2006: Extended Abstracts on Human Factors in Computing Systems, Montréal, Québec, Canada (2006)

9. Breitschopf, G.: Personalized, context-aware music playlist generation on mobile devices. Master's thesis, JKU (August 2013)
10. Hall, M., Frank, E., Holmes, G., Pfahringer, B., Reutemann, P., Witten, I.H.: The WEKA data mining software: An update. SIGKDD Explorations Newsletter 11(1), 10–18 (2009)
11. Hu, Y., Ogihara, M.: NextOne Player: A Music Recommendation System Based on User Behavior. In: Proceedings of the 12th International Society for Music Information Retrieval Conference (ISMIR), Miami, FL, USA (October 2011)
12. Kaminskas, M., Ricci, F.: Location-Adapted Music Recommendation Using Tags. In: Konstan, J.A., Conejo, R., Marzo, J.L., Oliver, N. (eds.) UMAP 2011. LNCS, vol. 6787, pp. 183–194. Springer, Heidelberg (2011)
13. Kaminskas, M., Ricci, F., Schedl, M.: Location-aware Music Recommendation Using Auto-Tagging and Hybrid Matching. In: Proceedings of the 7th ACM Conference on Recommender Systems (RecSys), Hong Kong, China (October 2013)
14. Lee, J.S., Lee, J.C.: Context Awareness by Case-Based Reasoning in a Music Recommendation System. In: Ichikawa, H., Cho, W.-D., Satoh, I., Youn, H.Y. (eds.) UCS 2007. LNCS, vol. 4836, pp. 45–58. Springer, Heidelberg (2007)
15. Liem, C.C., Müller, M., Eck, D., Tzanetakis, G., Hanjalic, A.: The Need for Music Information Retrieval with User-centered and Multimodal Strategies. In: Proceedings of the 1st International ACM Workshop on Music Information Retrieval with User-centered and Multimodal Strategies, Scottsdale, AZ, USA (November 2011)
16. Liu, H., Rauterberg, J.H.M.: Music Playlist Recommendation Based on User Heartbeat and Music Preference. In: Proc. 4th Int'l Conf. on Computer Technology and Development (ICCTD), Bangkok, Thailand (2009)
17. Moens, B., van Noorden, L., Leman, M.: D-Jogger: Syncing Music with Walking. In: Proceedings of the 7th Sound and Music Computing Conf. (SMC), Barcelona, Spain (2010)
18. Schedl, M., Breitschopf, G., Ionescu, B.: Mobile Music Genius: Reggae at the Beach, Metal on a Friday Night. In: Proceedings of the 2014 ACM International Conference on Multimedia Retrieval (ICMR), Glasgow, UK, April 02-04 (2014)
19. Schedl, M., Flexer, A., Urbano, J.: The neglected user in music information retrieval research. Journal of Intelligent Information Systems 41, 523–539 (2013)
20. Schedl, M., Hauger, D., Urbano, J.: Harvesting microblogs for contextual music similarity estimation — a co-occurrence-based framework. Multimedia Systems (May 2013)
21. Yang, Y.-H., Liu, J.-Y.: Quantitative Study of Music Listening Behavior in a Social and Affective Context. IEEE Transactions on Multimedia 15(6), 1304–1315 (2013)

Travel Recommendation via Author Topic Model Based Collaborative Filtering

Shuhui Jiang[1], Xueming Qian[1,*], Jialie Shen[2], and Tao Mei[3]

[1] SMILES LAB, Xi'an Jiaotong University, China
[2] Singapore Management University, Singapore
[3] Microsoft Research, USA
jsh.0531.smiling@stu.xjtu.edu.cn, qianxm@mail.xjtu.edu.cn,
jlshen@smu.edu.sg, tmei@miscrosoft.com

Abstract. While automatic travel recommendation has attracted a lot of attentions, the existing approaches generally suffer from different kinds of weaknesses. For example, sparsity problem can significantly degrade the performance of traditional collaborative filtering (CF). If a user only visits very few locations, accurate similar user identification becomes very challenging due to lack of sufficient information. Motivated by this concern, we propose an Author Topic Collaborative Filtering (ATCF) method to facilitate comprehensive Points of Interest (POIs) recommendation for social media users. In our approach, the topics about user preference (e.g., cultural, cityscape, or landmark) are extracted from the textual description of photos by author topic model instead of from GPS (geo-tag). Consequently, unlike CF based approaches, even without GPS records, similar users could still be identified accurately according to the similarity of users' topic preferences. In addition, ATCF doesn't predefine the category of travel topics. The category and user topic preference could be elicited simultaneously. Experiment results with a large test collection demonstrate various kinds of advantages of our approach.

Keywords: Multimedia, Travel Recommendation, Author Topic Model.

1 Introduction

In our daily lives, travel planning is always a tedious and difficult task. Gaining useful information from the fussy raw materials via manual analysis of travel guide website like IgoUgo (www.igougo.com) could be very time consuming, especially when travelers face a new city. Personalized travel recommendation techniques [1-9], [10-14], which can effectively integrate user preferences (e.g., cultural, cityscape or landscape), are gaining more and more attentions due to various potential applications in real world [11],[12].

Users' photos on social media record their travel history and much information about daily life. As shown in Fig. 1, a typical Flickr user's photo contains metadata like "User Id", "tags", "Taken data" "Latitude" and "Longitude".

---

* Corresponding author.

X. He et al. (Eds.): MMM 2015, Part II, LNCS 8936, pp. 392–402, 2015.

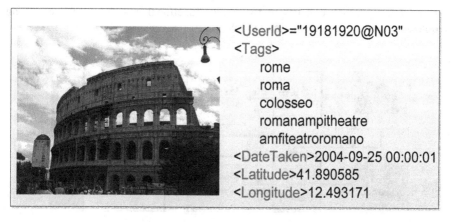

Fig. 1. Example of Flickr image information

Except the GPS trajectory information, the textual descriptions (such as tags and comments) that user gave when sharing the photos to social media networks (e.g. Flickr), is an important clue to infer user's latent interests [12-14]. For example, if a user visits a gym, the information about where he has gone can be identified and extracted from the GPS trajectory data. However, more detailed information about his interest such as "football" or "vocal concert" can be gained via visual analysis over the images and related tags.

Collaborative filtering (CF) is a well-known personalized travel recommendation approach [1]. However, it generally suffer from well-known "sparsity problem" in recommendation. Travel related data from real domain can be is very "sparse" and it makes accurate similar user identification very difficult if the user has only visited very few POIs. Recently, topic model (TM) learning method is introduced to solve the "sparsity problem" in travel recommendation [2]. Basic idea of TM is to infer users' travel topic preferences from the POIs that user has visited previously. Then the user preferred POIs can be recommended with similar topics. Usually the topic is determined by the naive category information from recommender system [2]. Unfortunately, for the community photo sharing websites like Flickr and Panoramio, it is difficult to define the category of travel topics due to lack of the accurate topic classification.

Motivated by these concerns, we develop an ATM based approach to model social users to carry out personalized travel recommendation. Due to data complexity, effective user travel topic preference mining with only textual description is challenging. Natural language models such as PLSA [10], LDA [4] and Author Topic Model (ATM) [3] are often utilized to cluster words to discover the latent topics that are combined to form documents in a corpus. LDA robustly discovers multinomial word distributions of these topics [4]. However, they cannot model authors and documents simultaneously. ATM directly annotates the user's interest with automatically divided semantic classes with respect to the distribution of the labels.

In this paper, we present an ATCF based personalized POI recommendation method by effectively extracting and integrating user travel topic preference from their tags of photo sets on social media. As illustrated in Fig 2, our ATCF based travel recommendation approach consists of two major functional modules - offline mining module and online recommendation module.

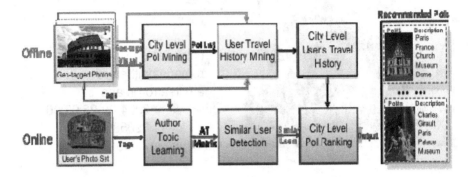

Fig. 2. A detail illustration of our travel recommendation system

In offline mining module, firstly, POIs of each city are mined using coarse-to-fine method from geo-tagged community-contributed photos by exploring both visual feature and geo-tags. Secondly, for each user, we mine each user's travel history (the POIs that user has visited) from geo-tags of user's community-contributed photos by a coarse-to-fine mapping method.

In online recommendation module, an ATM based approach is proposed to learn category of the topics and user travel topic preferences from the tags simultaneously. Secondly, similar users are mined based on the similarities of their topic preferences. Then, POIs in the new city are ranked based on the history of similar users' POIs visiting histories and the top ranked POIs are recommended to the user.

The main contributions of the research can be summarized as follows:

1)   In this paper, we propose author topic collaborative filtering (ATCF) method based personalized travel recommendation systems. We utilize users' topic preferences as the law for collaborative filtering instead of location co-occurrences. It improves the sparsity problem of classical location based collaborative filtering (LCF).

2)   We introduce an author topic model to adaptively elicit the topic category from tags associated with the Flickr photos. Using the scheme, topics about user preferences can be accurately extracted and applied to personalized travel recommendation. We also carry out large scale empirical study and the results show our approach enjoys great recommendation effectiveness.

The rest of paper is organized as follows: Section 2 is the introduction of related work. Section 3 and Section 4, the offline and online systems are described in detail. Section 5 presents experimental result and analysis. Finally, we conclude the paper in Section 6.

2   Related Work

In recent years, many different techniques have recently been developed to support travel recommendation based on different kinds of data. They include blogs [7], GPS trajectory [8], check in [5] and geo-tags [1,6]. Particularly, collaborative filtering algorithm shows its promising effectiveness in travel recommendation. The scheme is based on a Gaussian density estimation of co-occurrence space to cluster related geo-tags. They measure the similarity between two landmarks based on the similarity of travelers. By using the found similar user, a new trip plan can be made to a new location for a user [1]. While CF based recommendation methods demonstrate promising results, it suffers from the "data sparsity" problem. To solve this problem, topic model based methods are introduced to facilitate effective personalized travel recommendation [2, 9]. In [9], the authors conduct a study on exploiting online travel information by developing a tourist-area-season topic model. Bao et al., present a location-based and preference-aware recommender system that offers venues within a geospatial range [2]. They model each individual's personal preferences with a weighted category hierarchy using an iterative learning model in their offline system. However, in travel recommendation, it is difficult to find the authoritative category definition. Even though for check-in data, we could apply the original classification on website like Foursquare [2], for Flickr dataset with photos and textual descriptions [12-14] as shown in Fig.1, POIs are difficult to be categorized.

Distinguished from the existing POIs recommendation methods using CF, tags of photos on social media are used to represent user travel history in our system to mine user latent interest. Also, we use topic distribution to find similar users in a new city instead of the location co-occurrence. So accurate similar users could still be mined even the user has only visited very few POIs. Different from these mentioned topic model based methods, pre-defined the categories about travel topics are not required in our approach. By the author topic model, latent topics of travel could be mined adaptively.

3   Offline Mining Module

The offline mining module aims at mining POIs and all the users' travel history in the dataset from geo-tagged community-contributed photos. We propose coarse-to-fine POIs mining method when mining city-level POIs for each city. In "User Travel History Mining" part, we propose a coarse-to-fine-mapping method to mine user travel history for all the users using their geo-tagged photos.

3.1   City Level POIs Mining

The input is geo-tagged community-contributed photos with visual feature, tags and geo-tags. We have collected about 7 million social images from Flickr as Fig.1. In order to ensure that the noise photos of the dataset for each city are as less as possible, we use both the tags of city name and geo-tags of the location to double restrict the data of each city. After filtering, we get the geo-tagged photos of each city.

The input of **coarse-to-fine** method of POI mining is the geo-tag collection and visual feature collection of the city. First mean-shift clustering is used towards all the geo-tags of the photos in a city at a very small bandwidth as 0.0005, which is smaller than the radius of a landmark [6]. Each cluster contains a specified view. In this paper, we only use clusters containing at least 20 photos in the city and the number of users is no less than 10. Thus we get a list of clusters of a city denoted as CL.

We merge these clusters belonging to the same landmark (POI) by visual feature matching. First, we extract the 128D SIFT features for each image. Then we use bag of words (BoW) to present the SIFT descriptor. The size of the codebook is 61,944. Each image I_i is represented by BoW histogram. For a cluster C_n, the images belonging to it are represented by BoW histograms H_n, which is obtained by averaging the histograms of all the J images belonging to this cluster C_n. Their visual similarity $S(C_i, C_j)$ is measured by Euclidean Distance of H_i and H_j. For simplicity, we introduce a vector $AL=[A_1, A_2, ..., A_N]$ to record the assignment for the N clusters C_n to POI P_k as follows. If $S(C_i, C_j)$ is smaller than the threshold ThV, then the clusters C_i and C_j are belonging to the same landmark and we merge them together.

3.2   User Travel History Mining

This step aims at mining travel history for all the users. We make full use of the coarse layer clusters CL and assignment vector AL to get accurate travel history rather than directly compare the distances of a given image I_k to the centroids of the refined POIs PL, which is named as **coarse-to-fine mapping.**

Firstly, we determine the assignment of I_k to P_k by mapping user photos to cluster according to the geo-distances between user photos and the center of clusters. Then we determine which POIs the user has visited by mapping from CL to PL according to the assignment vector AL.

4   Online Recommendation

Online recommendation module aims at recommending POIs to a new user who has travel history about one city (city#a) and wants to visit a new city (city#b). We propose author topic learning based approach to mine user travel topic preference. Then POI recommendation is based on the history of similar user detection by the similarity of topic preference.

4.1   Author Topic Learning

Thus in this paper, we propose an ATM based approach to model social users to carry out personalized travel recommendation. The ATM is a generative model for document collections, which is able to extract information about authors and topics from large-scale text collections [3]. The input of this step contains two parts. The first part is a photo set I^u of user u_u with tag set τ_u. The second part is community users' photo sets with tags of each city. The output is topic preference distribution for each user.

In this section, first, the terminologies of ATM are introduced by combining the travel data. Then data processing of ATM procedure describes how to process the data as the input of ATM. At last the algorithm of ATM is shown.

4.1.1   Terminologies of ATM

In order to describe the model in our proposed method, we use the original terms (i.e., words, vocabulary, authors, and documents) to define the terminologies ATM in this paper as follows:

1) The **vocabulary** $V = \{1, 2, \ldots, N_d\}$ is the set of different tags of all the photos in a certain city. N_d is the size of V that represents the number of the different tags.

2) The **word** $w_i \in \{1, 2, \ldots, N_d\}$ represents the label of one tag of the photo, which can be considered as a representation of same tags of the photos. Note that each tag of an image is mapped to vocabulary V whose size is N_d through character matching and each tag can be represented by corresponding word w_i.

3) The document $d \in \{1, 2, \ldots, D\}$ corresponds to a tag set τ_j of the image I_j. So a user with NI images in the photo set has NI documents. So each photo with tags could be regard as one document.

4) The **authors** $a_d \in \{1, 2, \ldots, B\}$ is the label of the user who uploads the document d. $\{1, 2, \ldots, B\}$ is the set of labels of the B users in the city. In our paper, each a_d has only one element, as each photo could only be uploaded by one user. a_d is only used ATM.

4.1.2   Data Processing for ATM

First, to construct the vocabulary V, we filter all the tags with both "stop words" and "Flickr-style words". A stop word can be identified as a word that has the same likehhood of occurring in those documents not relevant to a query as in those documents relevant to the query like "his", "on" and etc. "Flickr-style words" is a list of words frequently appear in Flickr tags but not in ordinary "stop words" like "Canon". We define these words manually after rank the words according to the frequency. After tag filtering, N_d tags without repetition construct the vocabulary V. Each tag in V has a label $w_i \in \{1, 2, \ldots, N_d\}$.

Second, for each user $a_d \in \{1, 2, \ldots, B\}$ who has upload document d (corresponds to an image), we map all the tags of the image to the V to get the label w. Thus all the tags of the city have been mapped to corresponding labels.

Third, we record the relationship between each document d with author ad. We also record the relationship between each word w with document.

4.1.3   Algorithm of ATM

The ATM is a Bayesian network as LDA. However, each author's interest is modeled with a mixture of topics by ATM, ATM is a hierarchical generative model in which each word w_i in a document d is associated with two latent variables, i.e., an author x_i and a topic z_i.

The generative process of ATM mainly consists of two steps: first an author x_i and a topic z_i are picked, and then, a word is generated according to the probability distributions. The details are as follows:

1) For each author $a_d \in \{1, \ldots, B\}$, choose a dimensional Dirichlet random variable $\theta_a \sim$ Dirichlet (α). For each topic $t \in \{1, \ldots, T\}$, choose $\phi_t \sim$ Dirichlet (β).

2) For each document $d \in \{1, 2, \ldots, D\}$, given the vector of authors a_d, for each word w_i, indexed by $i \in \{1,\ldots,N_d\}$, do

   (a) Conditioned on a_d, choose an author $x_i \sim \mathrm{Uniform}(a_d)$.

   (b) Conditioned on x_i, choose a topic $z_i \sim \mathrm{Discrete}(\theta_{x_i})$.

   (c) Conditioned on z_i, choose a word $w_i \sim \mathrm{Discrete}(\phi_{z_i})$.

As a result, we get AT matrix (Author Topic matrix) for all the users. AT is a sparse $A \times T$ matrix, where A is the number of authors, and T is the number of topics. $AT(x,z)$ contains the times that a word w token associated with author x has been assigned to topic z.

4.2   POI Recommendation

In this section, we recommend POIs for a new user u_u in a new city#b according to his or her travel topic distribution AT_u mined by ATM in the city#a that has been visited before.

First, normalize each AT_i. Then we calculate the similarity between u_u and u_i from their author topic vectors using Cosine distance.

$$Sim(u,i) = \frac{AT_u \times AT_i{}'}{\parallel AT_u \parallel \bullet \parallel AT_i \parallel} \tag{1}$$

And then we rank the users in the city according to their AT distribution similarity. The top ranked NS users are selected as the set of similar users U^s.

Secondly, POIs in city#b are ranked according to the similar users' travel histories in city#b and then top ranked POIs are recommended to the user u_u.

5   Experimentation

We compare our proposed method (ATCF) with different approaches including recommendation by Popularity (PO), Collaborative Filtering (CF) and recommendation by LDA (LDA) to check the robustness of ATCF. The performance of these three methods and our proposed method are evaluated by criteria of MAP on Flickr dataset crawled by Flickr API on Flickr Website. MAP is one of the most well known criteria for measuring the relevance of recommendation. The descriptions of four compared methods are described as follows:

PO: First, POIs of the city are ranked according to how many users have uploaded photos of this POI.

CF: Location-based Collaborative Filtering is the most common way that can be most easily realized [1]. This baseline utilizes the users' location histories in a city to detect similar users.

LDA: To test the robustness of Author-Topic Model in the ATCF method, we replace the ATM with LDA model to mine user travel topic preference. Different from ATM, in LDA, we need to carry out an additional step to get user's topic distribution AT. In the first step, all tags of the photos in the city are allocated to different topics using LDA [4]. Each observed word is generated from a multinomial word distribution, specific to a particular topic. However in this step the relationship between authors and words, and authors and documents are not considered yet. Therefore, in the second step, we calculate the proportion of user's tags allocated to each topic mined in first step as *AT*. The other steps of LDA based approach are the same as ATCF.

5.1   Dataset

To facilitate comprehensive empirical study, we collected 7 million Flickr photos by open API. These photos are uploaded by 7,387 users and the heterogeneous metadata are associated with the photos.

After crawling Flickr photos, we only retain the photos with both tags and geo-tags from the original Flickr dataset. Though only tags are used to mine user's topic preference, geo-tags are also important to the recommendation system and evaluate experiments. On one hand, in offline system, geo-tags are involved in city level POIs mining and community users' travel history mining. On the other hand, in the evaluate experiments, the geo-tag that user labeled originally are regarded as ground truth of what the user have actually visited.

We select nine top popular cities to evaluate the performance of the five methods like that utilized in [1]. These nine selected cities are **Barcelona, Berlin, Chicago, London, Los Angeles, New York, Pairs, Rome** and **San Francisco.** We use the coarse-to-fine method in Part A of Section IV to mine POIs of these nine cities. Table II shows the corresponding number of users, POIs and photos in each city. There are 2,892 users, 307 POIs and 150,101 photos in total.

5.2   Performance Evaluation

In test data (Part A in this section), all the user photos retained to test contain both tags and geo-tags. And geo-tags, which record which POIs the user actually visited, are regarded as the **Ground Truth.** For example, if we recommend POIs in London to a user u_b, what he/she's travel history in London would be the Ground Truth. In the offline system, we mine travel history of u_b as $Q_b = \{q_1, ..., q_i, ...q_M\}$. To evaluate the performance, we compare the recommended POIs with POIs the user actually visited by his or her geo-tagged photos

We use MAP@n [1] to evaluate the performance of our method and the four comparative methods. It is one of the most well known criteria of the evaluation of recommendation system. In these two criteria, n denotes the number of POIs that we recommend to the user. We also provide the performance under MAP (without @n). In MAP, the number of recommended POIs is the same as the number that the user actually visited. The equation of MAP@n is as follows:

MAP@n: Mean average **precision** for a set of m users in the test data is the mean of the average precision scores for each user as follows:

$$MAP @ n = (\sum_{i=1}^{m} AP_i) / m \qquad (2)$$

where AP_i is Average Precision of each user as follows

$$AP @ n = (\sum_{i=1}^{n} \sum_{j=1}^{i} rel_j / i) / n \qquad (3)$$

where rel_i is a relevance value. Suppose that we recommend n POIs in city#b to the user. To the i-th POI, we calculate how many POIs from 1-th to i-th POIs which we recommend are within the list of POIs user have actually visited in city#b. $rel_i =1$ if the user has actually visited the recommended POI, otherwise, $rel_i =0$. Then we average the results of n POI to get AP@n for the user.

5.3   Performance Comparison

Table 1 shows the recommendation results of ATCF (ours) on MAP in comparison with PO, CF and LDA. Similar users NS=40, and distance metric is Cosine distance. In LDA and ATCF, the number of topics is set to be K=50.

Table 1. Performance of POI recommendation on MAP of PO, CF, LDA and ATCF

Perf.	PO	CF	LDA	ATCF
MAP	0.3408	0.4137	0.4166	**0.4225**
MAP@1	0.4861	0.5595	0.5678	**0.5876**
MAP@5	0.3557	0.4312	0.4361	**0.4483**
MAP@10	0.3076	0.4059	0.4005	**0.4115**
MAP@20	0.2642	0.3519	**0.3545**	**0.3545**
MAP@30	0.2438	0.3151	0.3163	**0.3184**

Table 1 shows the recommendation results of ATCF (ours) on MAP in comparison with PO, CF and LDA. Similar users $NS=40$, and distance metric is Cosine distance. In LDA and ATCF, the number of topic is set to be $K=50$.

The performance on MAP of ATCF is 0.4225, which outperforms PO, CF and LDA by 8.17%, 0.88% and 0.59% respectively. Table 1 also shows the performance of MAP@n by $n=1,5,10,20$ and 30. We could see the performances of ATCF and LDA are higher than PO and CF. ATCF is the best when $n=1,5,10$ and 30.

5.4   Discussion

We conducted two experiments in order to evaluate the robustness of ATCF in "sparsity" condition. In the first experiment, we randomly sample the POIs from user travel history we mined in the baseline experiment. In Fig.3 (a), the x-coordinate means the proportion of POIs that we sampled. In the second experiment, we select the users whose travel histories are much sparser to construct the experimental data. In Fig.3 (b), the x-coordinate means users whose number of visited POIs is less than the certain value. In both experiments, $NS=40$, $K=50$.

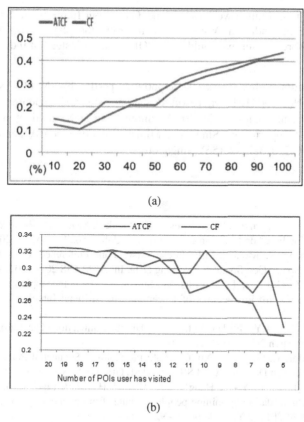

(a)

(b)

Fig. 3. (a) MAP curves of CF and ATCF under sampled user travel history of different percentage. (b) MAP curves of CF and ATCF under users with sparser travel history.

The results of CF and ATCF under MAP criteria are shown in Fig.3 we could see that the performances of both CF and ATCF decrease when the data becomes sparser. Under the "sparsity" condition, the performance of ATCF is higher than CF in these two experiments. In the first experiment, the largest improvement of ATCF to CF is more than 0.5 when we sample user history at around 30%. In Fig. 3(b), in most cases, ATCF is 0.2 higher than CF. Only when the number of POIs is between 11 and 13, CF is higher than ATCF. When the number of POI is set to be 6, ATCF is 0.8 higher than CF.

6   Conclusion

This paper presents a novel author-topic collaborative filtering (ATCF) based personalized travel recommendation approach for social users. We model user travel preference and detect similar user simultaneously using author topic learning. User topic preference can be mined from the textual descriptions of photos by author-topic model (ATM) instead of history of locations from GPS (geo-tag) as most previous works.

However, there is still much work to be done. One of our future works is personalized travel route recommendation. We continue to crawl the photos from social media website. With more dataset, we could mine POI sequence instead of individual POIs.

Acknowledgments. This work is supported partly by NSFC No.61173109, No.61128007, No.60903121, Microsoft Research Asia, and Fundamental Research Funds for the Central Universities for Xueming Qian at Xi'an Jiaotong University. Jialie Shen was supported by Singapore Ministry of Education Academic Research Fund Tier 2 (MOE2013-T2-2-156), Singapore.

References

[1] Clements, M., Serdyukov, P., de Vries, A.P., Reinders, M.J.T.: Personalized Travel Recommendation based on Location Co-occurrence. CoRR, abs/1106.5213 (2011)

[2] Bao, J., Zheng, Y., Mokbel, M.F.: Location-based and Preference-Aware Recommendation Using Sparse Geo-Social Networking Data. In: ACM GIS 2012 (2012)

[3] Michal, R.Z., Griffiths, T., Steyvers, M., Smyth, P.: The author-topic model for authors and documents. In: Proceedings of Uncertainty in Artificial Intelligence, pp. 487–494 (2004)

[4] Blei, D.M., Ng, A.Y., Jordan, M.I.: Latent Dirichlet allocation. The Journal of machine Learning research 3(5), 993–1022 (2003)

[5] Sang, J., Mei, T., Sun, J.-T., Xu, C., et al.: Probabilistic Sequential POIs Recommendation via Check-In Data. In: ACM SIGSPATIAL GIS, USA (November 2012)

[6] Cheng, A.-J., Chen, Y.-Y., Hang, Y.-T., Hsu, W.H., Mark Liao, H.-Y.: Personalized travel recommendation by mining people attributes from community-contributed photos. In: ACM MM 2011, NY, USA, pp. 83–92 (2011)

[7] Kori, H., Hattori, S., Tezuka, T., Tanaka, K.: Automatic Generation of Multimedia Tour Guide from Local Blogs. In: Cham, T.-J., Cai, J., Dorai, C., Rajan, D., Chua, T.-S., Chia, L.-T. (eds.) MMM 2007. LNCS, vol. 4351, pp. 690–699. Springer, Heidelberg (2006)

[8] Zheng, Y., Zhang, L.Z., Ma, Z.X., Xie, X., Ma, W.-Y.: Recommending Friends and Locations Based on Individual Location History. ACM Transactions on the Web 5(1), Article 5 (February 2011)

[9] Liu, Q., Ge, Y., Li, Z.M., Chen, E., Xing, H.: Personalized Travel Package Recommendation. In: Proceedings of 11th ICDM (2011)

[10] Hofmann, T.: Unsupervised learning by probabilistic latent semantic analysis. Machine Learning 42(1/2), 177–196 (2001)

[11] Shen, J., Cheng, Z., Shen, J., Mei, T., Gao, X.: The Evolution of Research on Multimedia Travel Guide Search and Recommender Systems. In: Gurrin, C., Hopfgartner, F., Hurst, W., Johansen, H., Lee, H., O'Connor, N. (eds.) MMM 2014, Part II. LNCS, vol. 8326, pp. 227–238. Springer, Heidelberg (2014)

[12] Qian, X., Feng, H., Zhao, G., Mei, T.: Personalized Recommendation Combining User Interest and Social Circle. IEEE Trans. Knowledge and Data Engineering 26(7), 1487–1502 (2014)

[13] Feng, H., Qian, X.: Mining User-Contributed Photos for Personalized Product Recommendation. Neurocomputing 129, 409–420 (2014)

[14] Qian, X., Liu, X., Zheng, C., Du, Y., Hou, X.: Tagging photos using users' vocabularies. Neurocomputing 111, 144–153 (2013)

Robust User Community-Aware Landmark Photo Retrieval

Lin Wu[1,2], John Shepherd[1], Xiaodi Huang[2], and Chunzhi Hu[3]

[1] School of Computer Science and Engineering, The University of New South Wales, Australia
[2] School of Computing and Mathematics, Charles Sturt University, Albury, Australia
[3] Tianjin Aviation Electro-Mechanical Co., Ltd, Tianjin, China
{linw,jas}@cse.unsw.edu.au, xhuang@csu.edu.au,
chunzhi_hu@126.com

Abstract. Given a query photo characterizing a location-aware landmark shot by a user, landmark retrieval is about returning a set of photos ordered in their similarities to the photo. Existing studies on landmark retrieval focus on exploiting location-aware visual features or attributes to conduct a matching process between candidate images and a query image. However, these approaches are based on a hypothesis that a landmark of interest is well-captured and distinctive enough to be distinguished from others. In fact, distinctive landmarks may be biasedly taken due to bad viewpoints or angles. This will discourage the recognition results if a biased query photo is issued. In this paper, we present a novel approach towards landmark retrieval by exploiting the dimension of user community. Our approach in this system consists of three steps. First, we extract communities based on user interest which can characterize a group of users in terms of their social media activities such as user-generated contents/comments. Then, a group of photos that are recommended by the community to which the query user belongs, together with the query photo, can constitute a set of multiple queries. Finally, a pattern mining algorithm is presented to discover regular landmark-specific patterns from this multi-query set. These patterns can faithfully represent the characteristics of a landmark of interest. Experiments conducted on benchmarks are conducted to show the effectiveness of our approach.

Keywords: Landmark Photo Retrieval, User Community, Query Expansion.

1 Introduction

The explosion of personal digital photography, together with the Internet, has led to the phenomenal growth of landmark photo sharing in many websites such as panoramio. com and Picasa Web Album[1]. Tourists usually take photos of local landmarks which are easy-recognizable and well-known to the public such as monuments, churches, and towers. The study of landmark recognition has opened up a possibility of extensive Web applications including GPS estimation [13], media organization [6], mobile image retrieval [28], and online tour guide developing [30]. In this work, we are interested in the problem of landmark retrieval in the context of social media networks. Given a query image issued by a query user (called a q-user for short) in which a particular

---

[1] picasa.google.com

X. He et al. (Eds.): MMM 2015, Part II, LNCS 8936, pp. 403–414, 2015.
© Springer International Publishing Switzerland 2015

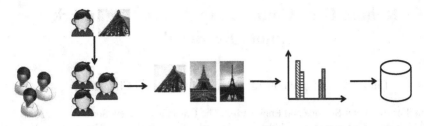

User community discovery   Multiple query expansion   Pattern mining & landmark retrieval

Fig. 1. The flow chart of our framework. Our framework comprises three main phases: 1) discover user groups; 2) select multiple landmark photos from user groups together with a query photo to form a multi-query set; and 3) Obtain feature representations of the multi-query set to encode the photo database

landmark has been captured, our system returns from its corpus a set of representative images in which the landmark appears.

Many approaches [7,8,12,25,26,30,3] have been proposed to tackle this problem from different perspectives. A variety of landmark-aware representation features, such as patch level region features [7], semantic-level attributes [8], and the combination of low-level features [12], is used to retrieve a set of photos with landmarks that are visually similar to those in the query under the landmark-aware feature space. In [12], Hays *et al.* estimated the geographic information from an image at the world scale. Zheng *et al.* [30] inherited this vision and built an engine of planet-scale landmark recognition, so as to facilitate landmark recognition, modeling, and 3D reconstruction. In [8] an approach is presented to discover discriminative attributes for location recognition. One common assumption in the relevant literature is that the landmark in a query photo from a q-user is accurately captured. However, this cannot be always possible in practice, especially when there are man-made errors made by the q-user such as bad viewpoints, occlusions. To address this problem, in this paper we propose to complement the biased query by recommending more photos so as to augment the one query to be a multi-query set. This robust query set can faithfully represent the landmark of interest.

In this paper, we utilize user communities to complement the photo query issued by a user in social media networks, and form a multi-query set to achieve a robust landmark retrieval. The information of users as well as photos associated with textual descriptions are collected from social media, e.g., Flickr. In our framework, we first discover user communities by applying a graph based community detection model [18,5] in which users are divided into clusters based on user similarities. The metric of user similarities is defined based on shared interests, which can be obtained from user social activities or textual description for photos they have uploaded. Second, with user communities detected, we re-organize the collection of users and photos in which each community, denoted as user-photo community, contains a set of similar users and corresponding photos uploaded by these users. Then, a matrix factorization is conducted on a user-photo matrix associated with each community to which the q-user belongs. This can facilitate the selection of retrieved photos by enriching the query photo in a form of a multi-query set. The key observation underlying our approach is that if people have taken photos at the same locations, there is a high likelihood that they share similar interests and come from the same community. Third, a pattern mining based method is developed to discover the common feature representation of all photos in the multi-query set. Such a representation is also utilized to efficiently retrieve the landmark photos with respect to the query photo. The flow chart of our framework is illustrated in Fig. 1. Our main contributions can be summarized as follows:

- We introduce a robust multi-query expansion approach on landmark retrieval by exploring social user communities, which complements the issued photo and faithfully represents the landmark of interest to facilitate the retrieval.
- We mine regular patterns among the photos in a multi-query set so as to find representative landmark features for effective and efficient retrieval.
- We conducted extensive experiments on landmark retrieval against datasets collected from social media networks.

Roadmap. The rest of this paper is organized as follows. We briefly review related literature in Section 2. Then, in Section 3, we describe user community discovery scheme, multiple query recommendation, and landmark-specific pattern mining. We validate the performance of our approach in Section 4, and conclusions are made in Section 5.

2 Related Work

In this section, we briefly review literature on social media search and understanding.

Extensive research efforts have been dedicated to tag-based social image search. In [10], Gao *et al.* propose to learn the joint relevance of visual content and tags, which is achieved by a hypergraph learning approach with vertices corresponding to images and hyperedges for tags. The relevance scores of images are calculated by iteratively updating them and hyperedge weight. Another study to classify tags is presented by Li *et al.* [15] where the relevance between image and manual tags is estimated by exploiting both relative positive examples and negative examples rather than only positive examples in conventional methods. Meanwhile, all examples can be automatically selected without extra manual annotations. Instead of using the pre-defined features for these applications in social media network, a novel method of learning latent feature based on deep learning is proposed in [29], which can effectively leverage the collective effect of social media data objects into the learned latent feature, while keeping discriminative to characterize their heterogeneity. This can lead to superior performance than approaches relying on pre-defined features. Motivated by the quality specified image retrieval on improving user experiences in mobile searching, a novel prior-assessment scheme for evaluating the quality of the user generated images (UGI) is proposed by defining four attribute dimensions [16], each of which contributes a quality score of UGI. Then, these scores are leveraged to yield a final score of UGI. This scheme benefits mobile search due to its great bandwith savings.

3 Approach

Given a landmark photo query q which might be biasedly captured by a user u_q, and a stack of landmark photos with user information collected from social media networks, our goal is to handle this noisy query q and robustly retrieve a set of photos from database ranked by their similarity to q. We systematically decompose our framework into three stages with important notations defined in Table 1. Specifically, in the first stage, we construct graph G to model topic similarity, and segment users into clusters C based on their interest. Secondly, we recommend top-K photos j from the community C_u ordered by the score $S(u_q, j, C_u)$ to form a robust query set Q. Finally, a pattern mining process is conducted to find out model H over Q, and retrieve each photo y ranked by $score(y) = \frac{1}{D(h_q, h_y)}$. We summarize notations used throughout this paper in Table 1.

Table 1. Table of notations

Notation	Definition
q	query photo
u_q	the user issues query q
$G = (V, E)$	a graph to model the similarity of users
V	a set of users
E	edges that connect users
A	user-user similarity matrix
$w[i, j]$	the similarity weight between user i and j
C	user-photo community set
C_u	the community that user q_u falls into
M_c	user photo matrix
$S(u_q, j, c)$	possibility score between user u_q and photo j from community c
$Q = \{q, q_i\}_{i=1}^{K}$	a multi-query set
$H = \{h_i\}$	a model contains a set of pattern h_i
$h_i = \{e_k\}$	a pattern contains a set of visual words e_k
$D(h_q, h_y)$	distance between query q and photo y in pattern representation h_q and h_y

3.1 User Interest Based Community Discovery

Generally, users in the same community are assumed to share same interests in terms of their social media activities such as their travel trajectory in history and comments in social media websites. In this aspect, a landmark query issued by a particular user can be naturally augmented by multiple photos recommended from the community this query user falls into. However, it is difficult to cluster users into communities, which is usually related to user's hidden interest [14]. In our approach, we explicitly measure user similarities in terms of user interest, and then a graph based algorithm is applied to cluster users into different communities. Each community is naturally comprised of user with similar interest as well as photos take by those users. This community is referred as user-photo community. Next, we will present a graph based approach to extract sensible user communities from social media networks.

User Similarity Based on User Interest. In our framework, we collect photos, user comments on photos, and user information from Flickr through user interface of Flickr

API [2]. From gathered data, we explicitly define user similarities in terms of these tags offered by users for photos they uploaded. This is because tags are highly semantics to reflect user interest and discriminative to identify users. Then, we choose the probability distribution of 100 LDA topics one these text tags [2] as feature vector to represent tags. As a result, users who contribute tags into photos can be encoded as a 100-dimensional LDA topic vector. Thereafter, we represent the user-user relationship as a symmetric square matrix, A, with dimension equal to the number of users. Each diagonal entry of A, i.e., $A[i, i]$ is set as to be zero, and other entries $A[i, j]$, $i \neq j$, denotes the similarity between user i and j. For example, given two users i and j with their feature vector l_i and l_j, the value of $A[i, j] = \langle l_i, l_j \rangle$ where $\langle, \rangle$ denote the inner product. Many community extraction algorithms take graphs as input, thus, we convert the matrix to an undirected graph $G = (V, E)$, where V represents the users and E represents the edges between them. We assign an edge between two users if there exist common tags. The weight between two vertices i and j, denoted by $w[i, j] = A[i, j]$.

Community Discovery. The community discovery problem is typically formulated as finding a partition $C = \{c_1, \ldots, c_k\}$ of a simple graph $G = (V, E)$ according to some meaningful criteria. C can also be referred as a clustering of G. Clauset, Newman, and Moore (CNM) is a efficient and widely used greedy community extraction algorithm [5]. It defines a metric of modularity, which is a quantitative measure of the quality of a partition of a graph. Let e_{ij} denote one-half of the fraction of edges in a graph that connects vertices in community i to those in community j. Therefore, $e_{ij} + e_{ji}$ is the total fraction of such edges for communities i and j. Let e_{ii} be the fraction of edges that fall into the group i, then $\sum_i e_{ii}$ is the total fraction of edges that fall within individual groups and be $a_i = \sum_j e_{ij}$ the total fraction of all ends of edges that are attached to vertices in group i. Thus, the modularity M of a clustering C can be defined as:

$$M(C) = \sum_i (e_{ii} - a_i^2). \tag{1}$$

Maximizing Eq.(1) is to promoting the higher number of intra-community edges compared to inter-community edges, reflecting the concept of communities clearly.

3.2 Multiple Query Expansion and Pattern Mining

After obtaining $|C|$ user-photo communities [3], we can select K queries from a particular community, together with the original single query, to form a robust multi-query set for landmark retrieval. Formally, denote c as a user community that contains a number of $|C_u|$ users and $|C_i|$ images, such that C_u is the user set, C_i is the photo set, and the user-photo matrix is $M_c \in \mathbb{R}^{|C_u| \times |C_i|}$, where $M_c(u, j) = 1$ if a user $u \in C_u$ uploaded a photo j into C_i, and $M_c(u, j) = 0$ otherwise. Then we perform the matrix factorization over M_c by minimizing the objective function below.

$$\arg \min_{P_c, Q_c} \|M_c - P_c Q_c\|_F, \tag{2}$$

---

[2] https://www.flickr.com/services/api/
[3] In our setting, each user-photo community contains a large amount of photos displaying the same landmark of interest.

where $||\cdot||_F$ denotes the Frobenius norm. Two factorized matrices $P_c \in R^{|C_u| \times L}$ and $Q_c \in R^{L \times |C_i|}$ are obtained, where P_c models the mapping of users in the reduced latent space with L dimensions, and Q_c defines the mappings of photos to the same reduced latent space. That is, each user u is represented as the u-th row vector of P_c denoted by p_u, while each photo i is by j-th column vector Q_c denoted by q_j. For a q-user u_q, we compute a score as the possibility that photo j will be recommended to u_q by user u from community c. The possiblity is calculated by the inner product of p_u and q_j as follows

$$S(u_q, j, c) = < p_u, q_j > . \tag{3}$$

We finalize the top-K list of photos as a multi-query set by ranking all $S(u_q, j, c)$ scores in descending order. This set augments the original single query landmark to be a more robust query set. The landmark retrieval is then performed against a landmark database with respect to this multi-query set. A naive way of generating a top-K list is to issue each query in the set individually, and then to combine their retrieved results. However, such a late fusion fashion is highly computational demanding [22], especially when K is large. By contrast, we present a pattern mining method that mines a landmark-specific mid-level representation to best represent the information in the query set. The presentation is further utilized to encode and retrieve photos in a landmark database.

Given a pool of landmark queries $Q = \{q, q_i\}_{i=1}^{K}$ where q is the original query and $q_i (i = 1, \ldots, K)$ are queries recommened from the user-photo community, we aim to use pattern mining techniques to discover regular patterns that can serve as a landmark-specific representation. A pattern, denoted as h, is a generalization of a set of transactions and consists of a set of items, $h = \{e_1, e_2, e_9\}$, for example. In our case, each item $e_j \in E$ is a label of the visual word where E is a set of items. Each transaction t is a set of items ($t \in PowerSet(E)$), e.g., $t_1 = \{e_2, e_3, e_5\}$. A single pattern h can only describes part of the query location and a pattern h is considered to be matched/mapped to transaction t if $h \subseteq t$. A set of patterns that together describe a query landmark is referred to as a model H. In our framework, we use root-SIFT descriptors extracted from Hessian-affine regions [17] and a visual vocabulary created by K-means clustering. Each SIFT descriptor extracted from a key point is assigned to a weighted combination of r nearby visual words from r cluster centers (soft assignment [19]) instead of its single nearest-neighbor center (hard assignment). Intuitively, we assign weights to each cell proportionally to the value of $\exp -\frac{d^2}{2\sigma^2}$, where d is the distance from the cluster center to the descriptor point. Thus, there are two essential parameters: the spatial scale σ and the number of nearest neighbors r. After computing the weights to the r nearest neighbors, the descriptor is represented by an r-vector, which is then L_1 normalized.

We use local-bag-of-words (LBOW) as a representation of a local region surrounding a key point [23,24]. These LBOWs are constructed as histograms over the visual words in the spatial neighborhood of a detected key point. Finally, LBOWs are transformed into transactions by considering each non-zero bin as an item. As a result, we obtain a transaction database B, containing all images encoded by visual transactional words.

KRIMP Algorithm. To learn a model that best explains the query landmark of interest, we employ the KRIMP algorithm [20] that uses the Minimum Description Length

(MDL) principle [11] to discover a set of patterns that best explain the landmark. The obtained patterns can be generalized to image archive, because MDL produces patterns that are the most representative of the landmark. The KRIMP algorithm proceeds as follows. First, we compute a cover function that functions a model H and transaction t, which is defined as:

$$cover : \{H \times t\} \to PowerSet(PowerSet(E)). \tag{4}$$

This cover function indicates which patterns in the model contribute to "cover" the transaction. Intuitively, it selects the model patterns that are used to encode all items in the transaction. We omit the details on how to compute a cover function, which can be found in [20]. To measure how many times each pattern $h \in H$ is used to encode the query image, we deploy the **usage metric** of h based on B that can be computed as $U(h|B) = |\{t \in B : h \in cover(H,t)\}|$.

Given a set of models $\mathbb{H}$, the best model H^* that we seek is the one that minimizes

$$H^* = \arg\min_{H \in \mathbb{H}} L(H) + L(B|H), \tag{5}$$

where B is the database containing all transactions, $L(H)$ is the length of the model in bits and $L(B|H)$ is the length of the query image once it is encoded with the model H. Mathematically, $L(H)$ is computed as $L(H) = \sum_{h \in H}[L(h|H) + L(h|\hat{H})]$, where $\hat{H}$ is the standard model consisting of only singleton items, and $L(h|H) = -\log(P(h|B))$ where the quantity $P(h|B)$ is computed using the query image information as $P(h|B) = \frac{U(h|B)}{\sum_{\forall y \in H} U(y|B)}$. The second term in (5), $L(B|H)$, is the summation of all the lengths of transactions in B once encoded by the model H. Thus, we have $L(B|H) = \sum_{t \in B} L(t|H)$, and $L(t|H) = \sum_{h \in cover(H,t)} L(h|H)$. We refer to [20] for more details on how to compute the $L(H)$ and $L(B|H)$. By optimizing Eq.(5), we find the set of patterns in the optimal model (H^*) that are considered as a new set of mid-level features. With these patterns, we construct a histogram (bag-of-pattern) for each image such that a queried unknown location can be understood and recognized by labeled images from an established database.

3.3 Landmark Photo Retrieval Rule

Once the bag-of-patterns are learnt, we can use tf-idf weighting on each bin of a histogram. For recognizing a candidate photo y in the database, we first represent it by counting bins of histograms based on codewords (patterns) in the vocabulary of landmark patterns. Then we have its histogram vector $h_y = [h_y(1), \ldots, h_y(N)]$ where N is the number of bins. Note that h_y is normalized, i.e., $\sum_{i=1}^{N} h_y(i) = 1$. Likewise, the query photo can be coded as h_q, and the distance between image q and y is computed by a chi-square (χ^2) distance,

$$D(h_q, h_y) = \sum_{i=1}^{N} \frac{(h_q(i) - h_y(i))^2}{h_q(i) + h_y(i)}. \tag{6}$$

Henceforth, the ranking score for a candidate y is defined as $score(y) = \frac{1}{D(h_q, h_y)}$.

4   Experimental Studies

In this section, we first present our experimental settings, and then report the experimental results to verify the effectiveness and efficiency of our approach.

4.1   Datasets

Two datasets are constructed by collecting photos from social media websites of **Flickr** and **Picasa Web Album**. They are suitable for location retrieval because they both contain user information and corresponding landmark photos.

Flickr. We use the **Flickr** API to retrieve landmark photos taken at a city posted by a large number of users. We sort out 11 cities: London, Paris, Barcelona, Sydney, Singapore, Beijing, Tokyo, Taipei, Cairo, New York city, and Istanbul. In each city, e.g., Paris, we obtained images by querying the associated text tags for famous Paris landmarks such as "Paris Eiffel Tower" or "Paris Triomphe". In summary, Flickr dataset contains 49,840 photos in total from 11 cities, and 7,332 users.

Picasa Web Album. Providing rich information about interesting tourist attractions, this source contains a vast amount of GPS-tagged photos uploaded by users along with their text tags. We manually download a fraction of photos and their user information on 6 cities: London, Paris, Beijing, Sydney, Chicago, and Barcelona. Thus, Picasa album datasets consists of 41,000 photos from 6 cities, and 5,772 users in total.

4.2   Settings

Evaluation Metric. We choose averaged precision for the top 100 retrieved photos as the evaluation metric due to two reasons: (1) in practice users are usually concerned about the top ranked results; (2) computing the averaged precision of full list is time-consuming. For a query q, the average precision (AP) is defined as $AP(q) = \frac{1}{L_q} \sum_{z=1}^{l} P_q(z) \varpi_q(z)$, where L_q is the number of ground-truth neighbors of q in database, l is the number of entities in database, $P_q(z)$ denotes the precision of the top z retrieved entities, and $\varpi_q(z) = 1$ if the z-th retrieved entity is a ground-truth neighbor and $\varpi_q(z) = 0$, otherwise. Ground truth neighbors are defined as items which share at least one semantic label. Given a query set of size F, the MAP is defined as the mean of the average precision for all queries: $MAP = \frac{1}{F} \sum_{i=1}^{F} AP(q_i)$. In each city, we issue 50 queries and an average precision score is computed for each of the 50 queries, and these are averaged to obtain a mean Average Precision (mAP) for the city category [27,21].

We also report the performance of our method with varied values of parameters of K (the number of queries in the multi-query set), and results are given in Section 4.3. This is because a larger size of query set might yield better pattern mining results, whilst result in high computational burden. Thus, it desires an optimal K as a trade-off. Throughout experiments in this paper, we use $r = 3$, $\sigma^2 = 6,250$ in the construction of visual words, and the dimension of latent space L is set to be 32.

Competitors. Several state-of-the-art methods are implemented for comparison: (1) **K-NN** [12]: A data-driven photo matching method, setting nearest neighbors K as 20 to match its best performance; (2) **LF+SVM**: Low-level features [31] combined with

SVM; (3) **DRLR** [7]: A region based location recognition method that detects discriminative regions at patch-level; (4) **AQE** [4]: Average Query Expansion method proceeds as follows: given a query region, it ranks a list of photos using tf-idf scores. Bag-of-Word vectors corresponding to these regions are averaged with BoW vectors of the query, resulting in an expanded vector used to re-query the database; (5) **DQE** [1]: Discriminative Query Expansion enriches the query in the exactly same way as AQE. Then it considers images with lower tf-idf scores as negative data to train a linear SVM for further rankings and retrievals; (6) **PQE** [9]: A Pattern based Query Expansion algorithm that combines top-K retrieved images with the query to find a set of patterns.

4.3 Parameters Learning

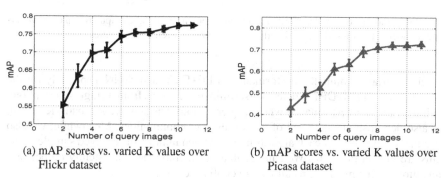

(a) mAP scores vs. varied K values over Flickr dataset

(b) mAP scores vs. varied K values over Picasa dataset

Fig. 2. mAP scores by varying the number of images recommended from user communities

Fig.2 depicts the mAP scores varying with a range of top-K images selected from the user group. It can be seen that when the number of query photos increases, our method can obtain higher values of mAP. This is because a larger query set comprehensively provides more informative features about the queried landmark. As such, our approach is more robust against occlusion or bad-viewpoint captures. However, a large K will result in a forbidden computational cost, as shown in Fig.3 (a) where the running time (sec) is reported as per each K value for two databases of Flickr and Picasa. Therefore, we set the value of K to 7 as a tradeoff value.

4.4 Landmark Retrieval and Ranking

To evaluate the performance, we randomly sample 50 landmark photos per city in **Flickr** and **Picasa Web Album** and compute the average precision. Upon each city, we repeat this retrieval process 5 times and use mAP (mean Average Precision) as the evaluation metric by averaging these values over all test cities. In this experiment, by default we use 10 spatial neighbors of a key-point to construct the LBOWs. Multiple queries are composed of top-7 ranked images (see the detailed analysis in the section 4.3 for how we set up this parameter). Note that each time we only sort out top-10 items and compute the precision values.

The compared landmark retrieval results are shown in Table 2 and 3. We can observe that: (1) Our method outperforms all competitors, due to the effectiveness of exploiting user communities as well as the robust pattern mining technique; (2) The large

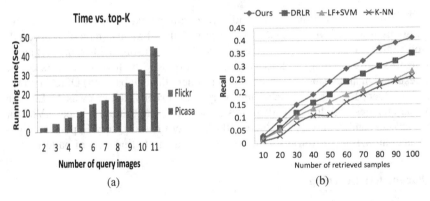

Fig. 3. Computational cost and performance comparison on recall. (a) Computational costs versus varied numbers of top-Ks. (b) Recall with varied numbers of retrieved samples.

intra-class variance limits the performance of **LF+SVM** and **K-NN**, especially for the **Flickr** dataset; and (3) **DRLR** detects discriminative regions from a single query photo, which degrades its performance when a query photo was shot from a bad viewpoint. This also demonstrates the need of exploiting user communities. Meanwhile, we also report the recall values of each algorithm with varied number of retrieved samples. The results are plot in Fig.3 (b). We can see that our method can consistently outperform competitors with a marginal recall gain over all baselines.

Table 2. The mAP values (%) of landmark retrieval on examined approaches over Flickr dataset

Method	London	Paris	Barcelona	Sydney	Singapore	Beijing	Tokyo	Taipei	Cairo	NYC	Istanbul	mAP
K-NN	46.75	32.03	33.55	38.93	29.83	54.33	17.21	32.39	44.37	22.56	74.58	38.78
LF+SVM	40.52	52.35	45.78	36.09	29.17	61.23	38.96	15.99	34.55	33.58	62.33	40.96
DRLR	53.45	35.65	51.48	34.58	40.33	62.45	42.33	40.24	44.58	38.34	57.35	45.52
Ours	57.03	61.57	54.33	44.29	61.48	63.88	58.98	57.32	68.74	52.58	71.38	60.39

Table 3. The mAP values (%) of landmark retrieval on examined approaches over Picasa dataset

Method	London	Paris	Beijing	Sydney	Barcelona	Chicago	mAP
K-NN	58.37	61.24	44.04	70.26	24.87	35.48	49.03
LF+SVM	50.03	61.05	71.28	33.48	12.37	48.38	46.10
DRLR	64.65	56.77	72.57	56.88	49.67	43.76	57.38
Ours	73.49	67.77	75.08	72.54	53.13	51.44	61.59

4.5   Comparing Query Expansion Approaches

Unlike **AQE** and **DQE**, our method for selecting a multi-query set exploits groups of users to complement a q-user. While **PQE** uses a similar approach to expanding a single query into a multiple one in which a pattern mining can be used, its hard assignment in descriptor quantization definitely loses feature information to some extent.

To demonstrate the superiority of our method over existing multiple query methods that can be adopted for landmark recognition, we compare our method against the above three state-of-the-art approaches of query expansion. They are **AQE**, **DQE** and **PQE**. Results are shown in Fig.4.We conclude that our soft-assignment based pattern mining approach outperforms **AQE**, **DQE** and **PQE** for location retrieval by a large margin.

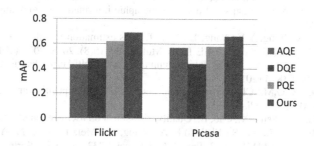

Fig. 4. Comparison of query expansion methods

5 Conclusion

In this paper, we study the problem of landmark retrieval in the context of social media networks. A robust multi-query expansion with the mining approach has been presented that makes use of user community information. Automatically selecting multiple landmark photos, we expand an unideal single landmark query photo into a set of multi-query photos. A pattern mining method is presented to discover a new representation of landmark-specific features of photos in the multi-query set, which is further utilized to efficiently search over location-aware landmark databases. Our experimental results on real-world datasets have validated the effectiveness and efficiency of our method.

References

1. Arandjelovic, R., Zisserman, A.: Three things everyone should know to improve object retrieval. In: CVPR (2012)
2. Blei, D.M., Ng, A.Y., Jordan, M.I.: Latent dirichlet distribution. J. Mach. Learn. Res. 3, 993–1022 (2003)
3. Cheng, Z., Ren, J., Shen, J., Miao, H.: Building a large scale test collection for effective benchmarking of mobile landmark search. In: Li, S., El Saddik, A., Wang, M., Mei, T., Sebe, N., Yan, S., Hong, R., Gurrin, C. (eds.) MMM 2013, Part II. LNCS, vol. 7733, pp. 36–46. Springer, Heidelberg (2013)
4. Chum, O., Philbin, J., Sivic, J., Isard, M., Zisserman, A.: Total recall: automatic query expansion with a generative feature model for object retrieval. In: ICCV (2007)
5. Clauset, A., Newman, M., Moore, C.: Finding community structure in very large networks. Physical Review E 70(6), 066111 (2004)
6. Crandall, D., Backstrom, L., Huttenlocher, D., Kleinberg, J.: Mapping the world's photos. In: WWW (2009)
7. Doersch, C., Singh, S., Gupta, A., Sivic, J., Efros, A.A.: What makes paris look like paris? ACM Trans. Graph. 31(4), 101 (2012)

8. Fang, Q., Sang, J., Xu, C.: Giant: Geo-informative attributes for location recognition and exploration. In: ACM Multimedia (2013)
9. Fernando, B., Tuytelaars, T.: Mining multiple queries for image retrieval: on-the-fly learning of an object-specific mid-level representation. In: ICCV (2013)
10. Gao, Y., Wang, M., Zha, Z.-J., Shen, J., Li, X., Wu, X.: Visual-textual joint relevance learning for tag-based social image search. IEEE Trans. Image Processing 22(1), 363–376 (2013)
11. Grunwald, P.D.: The minimum description length principle. The MIT press (2007)
12. Hays, J., Efros, A.A.: im2gps: estimating geographic information from a single image. In: CVPR (2008)
13. Li, J., Qian, X., Tang, Y.Y., Yang, L., Mei, T.: Gps estimation for places of interest from social users' uploaded photos. IEEE Trans. Multimedia 15(8), 2058–2071 (2013)
14. Li, Q., Gu, Y., Qian, X.: Lcmkl: latent-communtiy and multi-kernel learning based image annotation. In: CIKM (2013)
15. Li, X., Snoek, C.G.M.: Classifying tag relevance with relevant positive and negative examples. In: ACM Multimedia (2013)
16. Liu, Q., Yang, Y., Wang, X., Cao, L.: Quality assessment on user generated image for mobile search application. In: Li, S., El Saddik, A., Wang, M., Mei, T., Sebe, N., Yan, S., Hong, R., Gurrin, C. (eds.) MMM 2013, Part II. LNCS, vol. 7733, pp. 1–11. Springer, Heidelberg (2013)
17. Perdoch, O.C.M., Matas, J.: Efficient representation of local geometry for large scale object retrieval. In: CVPR (2009)
18. Palsetia, D., Patwary, M.M.A., Zhang, K., Lee, K., Moran, C., Xie, Y., Honbo, D., Agrawal, A., Keng Liao, W., Choudhary, A.: User-interest based community extraction in social networks. In: SNA-KDD Workshop (2012)
19. Philbin, J., Chum, O., Isard, M., Sivic, J., Zisserman, A.: Lost in quantization: improving particular object retrieval in large scale image databases. In: CVPR (2008)
20. Vreeken, J., Leeuwen, M., Siebes, A.: Krimp: mining itemsets that compress. Data Min. Knowl. Discov. 23, 169–241 (2011)
21. Wang, Y., Lin, X., Wu, L.: Exploiting correlation consensus: Towards subspace clustering for multi-modal data. In: ACM Multimedia (2014)
22. Wang, Y., Lin, X., Zhang, Q.: Towards metric fusion on multi-view data: a cross-view based graph random walk approach. In: ACM CIKM (2013)
23. Wang, Y., Lin, X., Zhang, Q., Wu, L.: Shifting hypergraphs by probabilistic voting. In: Tseng, V.S., Ho, T.B., Zhou, Z.-H., Chen, A.L.P., Kao, H.-Y. (eds.) PAKDD 2014, Part II. LNCS, vol. 8444, pp. 234–246. Springer, Heidelberg (2014)
24. Wang, Y., Pei, J., Lin, X., Zhang, Q., Zhang, W.: An iterative fusion approach to graph-based semi-supervised learning from multiple views. In: Tseng, V.S., Ho, T.B., Zhou, Z.-H., Chen, A.L.P., Kao, H.-Y. (eds.) PAKDD 2014, Part II. LNCS, vol. 8444, pp. 162–173. Springer, Heidelberg (2014)
25. Wu, L., Cao, X.: Geolocation estimation from two shadow trajectories. In: CVPR (2010)
26. Wu, L., Cao, X., Foroosh, H.: Camera calibration and geo-location estimation from two shadow trajectories. Computer Vision and Image Understanding 114(8), 915–927 (2010)
27. Wu, L., Wang, Y., Shepherd, J.: Efficient image and tag co-ranking: A bregman divergence optimization method. In: ACM Multimedia (2013)
28. Xue, Y., Qian, X., Zhang, B.: Mobile image retrieval using multi-photos as query. In: ICME (2013)
29. Yuan, Z., Sang, J., Liu, Y., Xu, C.: Latent feature learning in social media network. In: ACM Multimedia (2013)
30. Zheng, Y., Zhao, M., Song, Y., Adam, H., Buddemeier, U., Bissacco, A., Brucher, F., Chua, T.-S., Neven, H.: Tour the world: building a web-scale landmark recognition engine. In: CVPR (2009)
31. Zhu, J., Hoi, S.C.H., Lyu, M.R., Yan, S.: Near-duplicate keyframe retrieval by nonrigid image matching. In: ACM Multimedia (2008)

Cross-Domain Concept Detection with Dictionary Coherence by Leveraging Web Images

Yongqing Sun, Kyoko Sudo, and Yukinobu Taniguchi

NTT Media Intelligence Laboratories,
1-1 Hikarinooka Yokosuka-shi Kanagawa, 239-0847, Japan
{yongqing.sun@lab.ntt.co.jp}

Abstract. We propose a novel scheme to address video concept learning by leveraging social media, one that includes the selection of web training data and the transfer of subspace learning within a unified framework. Due to the existence of cross-domain incoherence resulting from the mismatch of data distributions, how to select sufficient positive training samples from scattered and diffused social media resources is a challenging problem in the training of effective concept detectors. In this paper, given a concept, the coherent positive samples from web images for further concept learning are selected based on the degree of image coherence. Then, by exploiting both the selected dataset and video keyframes, we train a robust concept classifier by means of a transfer subspace learning method. Experiment results demonstrate that the proposed approach can achieve constant overall improvement despite cross-domain incoherence.

Keywords: Visual concept detection, Web image mining, Sparse representation, Dictionary learning, Transfer learning.

1 Introduction

Nowadays, the explosive growth of visual contents on the Internet presents a challenge in how to manage the ever-growing size of the multimedia collections, particularly in how to extract sufficiently accurate semantic metadata (concepts) to make them searchable [15]. Typical machine leaning methods gather training dataset of a concept manually and a classifier is estimated based on them. Then the semantic concept of a new image is recognized based on the classifier. In order to learn effective concept detectors, a critical step is to acquire a sufficiently large amount of training samples, especially positive training samples [15]. Fortunately, with the explosive growth of visual contents on the Internet, large amounts of training samples have become available through Web searching [12, 25]. Consequently, how to utilize these abundant web images to improve concept detection has been the subject of intensive research by a large multimedia research community, since it has offered promising ways to automatically annotate the contents at relatively low cost [12, 25].

X. He et al. (Eds.): MMM 2015, Part II, LNCS 8936, pp. 415–426, 2015.

Fig. 1. Web Image Example of "Airplane-flying"

However, the online web images are very noisy, cover a wide range of unpredictable contents, and have quite different data distributions with any close dataset such as TREC-Vid dataset [19, 23]. As shown in Figure 1, for example, the content of web images searched from Google Image with the keyword "Airplane-flying" varies greatly. Obviously, the images in the top row of the figure are incoherent from the concept "Airplane-flying" in the TRECVid dataset. Thus these images can not facilitate the training of the concept and may even harm it. Only the images in the bottom row are consistent with the dataset and hence helpful. Therefore the challenges of using web images for building video concept detectors can be briefly broken down into two problems. First, how to select coherent positive training samples from diffused web images. Second, the approach of switching knowledge between different domains is crucial since there are large differences in distribution between web images and video keyframes.

In this paper, we propose a new scheme to address these problems in a unified framework. It exploits the properties of both web images and $TRECVID$ keyframes for video concept learning. First, we propose to measure the coherence in terms of how dictionary atoms are shared since shared atoms represent common features with regard to a given concept and are robust to occlusion and corruption. Thus, two kinds of dictionaries are learned through online dictionary learning methods: one is the concept dictionary learned from key-point features of all the positive training samples while the other is the image dictionary learned from those of web images. Intuitively, the coherence degree is then calculated by the Frobenius norm of the product matrix of the two dictionaries. Next, exploiting both the selected web data and $TRECVID$ video keyframes, transfer learning subspace learning is adopted to build a robust classifier, which can also alleviate the over-fitting problem since it simultaneously conducts dimensionality reduction.

1.1   Related Work

Our work is related to several research topics, including web image learning, canonical image selection from the web, and transfer learning. Due to limited space, we give only a brief overview here.

Existing work on visual detector learning using web images has mainly focused on how to leverage compact features, such as region-based features [22] or image salience [21], to alleviate the visual differences. Since an image is greatly reduced to a very compact feature vector, the effect of these approaches is not evident. As mentioned above, the main purpose of the existing web image learning methods is to exploit web image properties to boost web image search results by current search engines, it is not appropriate to apply them directly to the application of video concept learning.

Recently, cross-domain learning has been applied to solve the problem of training data scarcity in video concept detection in $TRECVID$. By leveraging the knowledge of a large amount of labeled training data from other domains called source domains, cross-domain learning methods make efforts to learn robust classifiers for the target domain. In [9], adaptive SVM is proposed to enhance the prediction performance of video concept detection, which adapts one or more existing classifiers trained from the source domain(program data in one TV channel data) to the target domain(other TV channel program data). Note that this method does not utilize the properties of the target domain, which is also useful in improving the classification performance. An instance-transfer approach is proposed in [10] to solve the cross-domain problem, in which the weights of the source domain data are estimated and a concept model is then trained based on the reweighted data. However, this approach is based on a strong assumption that both training and test data are drawn from the same feature distribution. Note that the above methods may perform desirably because they focus on cross learning between different TV channels or different TRECVID datasets, in which the distribution may be similar. However, in the context of web-based learning, since the difference between the properties of web images and TRECVID video keyframes is large, it is necessary to include information on both source and target domain data to build a robust classifier.

2   Training Data Selection from Web

In this paper, we propose a novel sampling approach on how to exploit bundles of local key-point features to measure how coherent a web image is with a given concept, from the aspects of sparse coding and dictionary learning. Before the presentation of our method, an overview of sparse coding is as the bellow.

2.1   Sparse Coding and Dictionary Learning

Recently, modeling data or signals as sparse linear combinations of a few elements (atoms) of some redundant bases (dictionary), sparse coding or sparse representation has been widely applied to classification problems where the data on multiple subspaces relies on the notion of sparsity due to its robustness to occlusion and corruption [18].

Formally, given a dictionary $\mathbf{D} = [d_1, \ldots, d_k] \in \mathbf{R}^{l \times k}$, and the i-th data instance vector $x_i \in \mathbf{R}^l$ from the observed data matrix $\mathbf{X} = [x_1, \ldots, x_n] \in \mathbf{R}^{l \times n}$, where d_i is the dictionary atom, l is the data dimensionality, k is the size of the dictionary, and n is the number of data instances ($l < k \ll n$), sparse representation solves the following non-convex program to seek the sparsest solution for the coefficient vector (i.e., sparse code) $\alpha_i \in \mathbf{R}^k$:

$$\min_{\alpha_i} \|\alpha_i\|_0 \ s.t. \ \mathbf{D}\alpha_i = x_i \tag{1}$$

where $\|\alpha_i\|_0$ denotes the l_0 pseudo-norm of the coefficient vector $\alpha_i \in \mathbf{R}^k$, i.e., the number of non-zero elements.

Since minimizing l_0 is NP-hard, a common approximation is to replace it with the l_1-norm according to theories from compressive sensing [13]. Taking noise into consideration, the equality constraint must be relaxed. Hence, an alternative is to solve the unconstrained problem after using the Lagrange multiplier method:

$$\min_{\alpha_i} \frac{1}{2} \| x_i - \mathbf{D}\alpha_i \|^2 + \lambda \| \alpha_i \|_1, \tag{2}$$

where λ is a regularization parameter that balances the tradeoff between reconstruction error and sparsity induced by the alternative l_1-norm constraint. This is a convex problem called Lasso in statistics and can be efficiently solved by the LARS-Lasso algorithm [14].

Here, how to determine the dictionary $\mathbf{D}$ is very important for sparse representation. It has been shown that dictionaries learned from data can significantly outperform off-the-shelf ones such as wavelets [18]. Given the observed data matrix $\mathbf{X}$, the goal is to seek an optimal $\mathbf{D}$ so that all the data instances can be represented as a sparse linear combination of their atoms. There are many dictionary learning methods such as the method of optimal directions (MOD), the K-SVD algorithm , and the Generalized Principal Component Analysis (GPCA) [18]. All the methods using classical optimization alternate between the dictionary and sparse code, and can obtain good results, but are too slow to scale up to large data sets [18]. Recently, efficient online learning methods were proposed in [18], which can handle large scale, potentially infinite, or dynamic data sets.

2.2   Sampling from Web Images

Inspired by the observation that dictionary atoms representing common features in all categories tend to appear to be repeated almost exactly in dictionaries corresponding to different categories, [20] promotes incoherence between the dictionary atoms to improve the speed and accuracy of sparse coding.

Motivated by this work, since the shared dictionary atoms learned from data can represent common features with regard to a given concept (represented by the set of positive training samples) and are robust to occlusion and corruption [18], we propose to use dictionary coherence in terms of how an image and a given concept share dictionary atoms to measure the degree of image coherence with the concept. That is, the more atoms they share, the higher the dictionary coherence is, which means it is more probable that the web image is coherent with the concept.

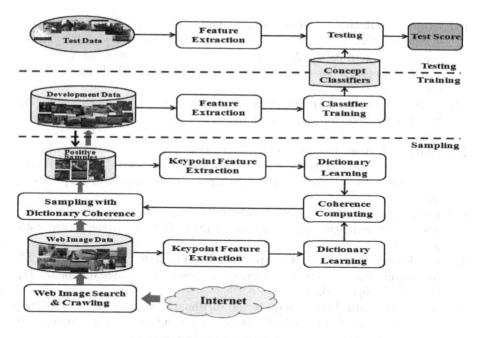

Fig. 2. Proposed Framework

In order to compute the dictionary coherence, we learn two kinds of dictionaries through the online dictionary learning method [18]: one is the concept dictionary learned from key-point features of all the positive training samples while the other is the image dictionary learned from those of web images. Intuitively, the coherence degree is then calculated by the Frobenius norm of the product matrix of the two dictionaries since it reflects the sum of the absolute values of inner products between dictionary atoms.

On the basis of the dictionary coherence, we propose a novel adaptive sampling approach to select coherent positive samples from diffused web images for further concept learning.

2.3   Algorithm

As shown in the framework of Figure 2, for each concept, the algorithm of the proposed sampling principally consists of the following steps:

(1) **Construction of Concept Set:** Select all the positive training samples from a development dataset such as TRECVid development set to represent the concept.

(2) **Feature Extraction of Concept Set:** Extract local key-point features, such as SIFT [17] or SURF [11], and collect each key-point feature $x_i \in \mathbf{R}^l$ of all the images in the concept set to form the data matrix $\mathbf{X}_c = [x_1, \ldots, x_n] \in \mathbf{R}^{l \times n}$. Here, l is the feature dimensionality, and n is the total number of keypoints.

(3) **Concept Dictionary Learning:** Adopt the efficient online dictionary learning methods [18] to learn the concept dictionary $\mathbf{D}_C \in \mathbf{R}^{l \times k}$ from the concept data matrix $\mathbf{X}_C$, where k is the size of the dictionary, i.e., the number of atoms. For the SIFT feature, we set $k = 192$ about 1.5 to 2.0 times of the feature size $l = 128$ [24].

(4) **Collection of Web Image Set:** After query construction or mapping [12] based on the concept name, search the web images and crawl the top-ranked ones.

(5) **Feature Extraction of Web Image:** For each image in the web image set, extract the same local key-point features as the second step, and form the image data matrix $\mathbf{X}_i \in \mathbf{R}^{l \times m}$, where m is the number of keypoints in the image.

(6) **Image Dictionary Learning:** Adopt the same dictionary learning methods [18] to learn the image dictionary $\mathbf{D}_i \in \mathbf{R}^{l \times k}$ from the image data matrix $\mathbf{X}_j$.

(7) **Dictionary Coherence Computing:** Use Equation (4) in subsection 2.5 to compute the dictionary coherence C_i between the image dictionary $\mathbf{D}_i$ and the concept dictionary $\mathbf{D}_C$.

(8) **Adaptive Sampling:** Compare the dictionary coherence C_i of the current web image with the adaptive threshold in subsection 2.6 to determine whether to add the current web image to the training set.

As shown in Figure 2, after adding the selected coherent positive web samples (a manual check is advised to ensure it is positive) to the training set, we can do further concept learning for training more effective concept detectors. We will detail the key procedures in the following subsections.

2.4 Dictionary Learning

In our study, we use the efficient online learning methods [18] to learn the dictionary. Due to the advantage of non-negativity constraints in learning part-based representations [24], which is helpful for object-oriented concept learning, we impose the positivity constraints on both dictionary D and sparse code α_i in solving the optimization problem as below:

$$\min_{\mathbf{D}, \alpha_i} \sum_{i=1}^{n} \left(\frac{1}{2} \parallel x_i - \mathbf{D}\alpha_i \parallel^2 + \lambda \parallel \alpha_i \parallel_1 \right), \ s.t., \ \mathbf{D} \geq 0, \ \alpha_i \geq 0. \tag{3}$$

while restricting the atoms to have a norm of less than one. The optimization is achieved through an iterative approach consisting of two alternative steps: the sparse coding step on a fixed $\mathbf{D}$ and the dictionary update step on fixed α_i [18]. As mentioned above, we learn two types of dictionaries: (1) a concept dictionary $\mathbf{D}_C$; (2) an image dictionary $\mathbf{D}_j$.

2.5 Dictionary Coherence Computing

The natural way to measure the degree of coherence C_i between the image dictionary $\mathbf{D}_j$ and the concept dictionary $\mathbf{D}_C$, is to inspect the product matrix: $\mathbf{D}_i^{\mathbf{T}} \mathbf{D}_c$, where the superscript T denotes the matrix transposition. This is because the element d_{ij} of the product matrix represents the inner product between a pair of the two dictionary atoms, i.e., $d_{ij} = d_i \cdot d_j$, here, $d_i \in \mathbf{D}_j$, $d_j \in \mathbf{D}_C$. Therefore, as shown in Equation (4), we compute dictionary coherence C_i through a Frobenius norm defined as the square root of the sum of the absolute squares of the matrix's elements d_{ij}:

$$C_i = \parallel \mathbf{D}_i^{\mathrm{T}} \mathbf{D}_C \parallel_F = \sqrt{\sum_{i=1}^{k} \sum_{j=1}^{k} |d_{ij}|^2} \tag{4}$$

where the subscript F denotes the Frobenius norm.

2.6   Adaptive Sampling

After computing the dictionary coherence C_i between the current web image and the concept, we can easily determine whether to add the current web image to the training set by simply comparing the C_i with a pre-given threshold C_{th}. If $C_i \geq C_{th}$, meaning that the web image is coherent with the concept, then we accept it. Otherwise, we discard it.

Here, we propose an adaptive off-line method through automatic calculation of the threshold C_{th} from the distribution of the coherence degrees of all the positive train samples. According to the theory of hypothesis testing, the threshold C_{th} can be adaptively determined by:

$$C_{th} = \mu - \eta \sigma, \tag{5}$$

where μ and σ are the mean and standard deviation of all the coherence degrees C_{Pos} between each positive training sample and the concept, and η is an empirical parameter that can be determined universally. In our experiments, we set $\eta = \sqrt{3}$.

3   Cross-Domain Video Concept Learning

In many cases, the number of positive samples of TRECVID data(*target data*) is much lower than that of negative samples. Therefore, we want to find a cross-domain learning method which not only improves the classification performance by leveraging the knowledge of *source data*(web data), but also conducts dimensionality reduction that is expected to solve the overfitting problem due to the insufficiency of positive samples. The method proposed in [5],in which the age estimation problem is solved by a cross-domain subspace learning process, fits our goal well. In our work, by applying [5] with the web training images denoted as *source data* and video keyframes denoted as *target data*, a common subspace is learned based on a regularization subspace learning method by considering both distributions of both domains. For each image, a global visual feature or *BoW* is used; this will be described further in the Experiment section.

First, we adopted a mixture of Gaussian distributions to estimate the positive images in the combination of *source data* and *target data* referred to as CD and their common subspace. According to our empirical experience, the Gaussian component K is set from 15–50. The objective function of the cross-domain subspace learning is modeled by finding an optimal projection or common subspace W, in which the projected features of both CD and its corresponding subspace share the same distribution. Using a regularization method the optimization can be carried out iteratively by the following function:

$$W_{k+1} = W_k + \kappa \left\langle \frac{\delta F(W)}{\delta W} + \lambda \frac{\delta H(W)}{\delta W} \right\rangle \tag{6}$$

Where $F(W)$ is the distribution of CD, $H(W)$ is a regularization term to adjust the likelihood function in subspace, and λ is the regularization weight. The above gradient of the distribution function of CD can then be obtained based the conventional discriminant subspace learning method, which is defined as:

$$\frac{\delta F(W)}{\delta W} = 2tr^{-1}(W^T S_w W)S_b W$$
$$-2tr^2(W_T S_w W)tr(W^T S_b W)S_w W$$

Here S_b and S_w denote the between-class and within-class scatter matrices, respectively. In [5], S_b and S_w are calculated according to the LDA algorithm. In our work, however, they are calculated according to the BDA proposed in [3], in which positive images are estimated by mixed Gaussian distribution and any distribution assumptions on negative images are relaxed. For actual applications this is more reasonable since negative images are very sparse within the span of feature space. With BDA, positive samples are clustered closely and the negative samples are pulled far away from the positive ones.

Let μ_k and Σ_k be the mean and covariance of the $k-th$ Gaussian component in CD and π_k be the mixing coefficient. Let μ'_k and Σ'_k be the projected mean and covariance in the subspace. That is, $\mu' = W^T \mu_k$ and $\Sigma'_k = W^T \Sigma_k W$. Then the gradient of $H(W)$ follows the function, here $y_i = W x_i$, G is the Gaussian probability density function:

$$\frac{\delta H(W)}{\delta W} = \sum_{i \in CD} \sum_{k=1}^{K} \gamma \Sigma'^{-1}(y_i - \mu'_k)\mu_k^T \tag{7}$$

Where γ is:

$$\gamma = \frac{\pi_k G(y_{ij}\mu'_k, \gamma'_k)}{\sum_{k=1}^{K} \pi_k G(y_{ij}\mu'_k, \gamma'_k)} \tag{8}$$

Next using the above W, each image x_i of labeled *target data(target development data)* is then represented as a low-dimensional y_i. A SVR regression is then conducted on the *target data* to find the classification function $f(y)$. Here f can also be carried out by other linear or non-linear processes.

Finally, based on the above W and regression function f, for an unlabeled image from TRECVID keyframes(*target test data*),the classification of a video concept can be achieved.

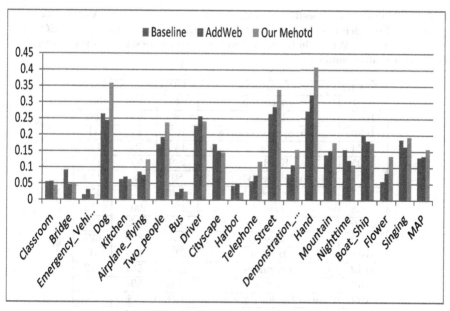

Fig. 3. Comparison results

4   Experiment

We tested the proposed method on the TRECVid 2008 [19, 23]. TRECVid is now widely regarded as the actual standard for performance evaluation of concept based video retrieval systems [24]. The number of positive training samples for each concept in the TRECVid 08 development set is shown in the column "#DPos" of Table 1 [24].

First, we used the Google API to search and download the top 1000 web images for each concept by constructing a query with the concept name. Then we annotated the images manually; the number of positive samples for each concept in the initial web image set is shown in the column "#WPos" of Table 1. Finally, we used our proposed sampling method to select the positive samples for each concept; the number of positive samples for each concept selected from the web images is shown in the column "#SPos" of Table 1. To test the effectiveness of our proposed method, we performed three runs for each concept:

- **[Baseline]**: Use only positive training samples in the TREC-Vid 08 development set ("#DPos" in Table 1).
- **[AddWeb]**: Use positive training samples of the TREC-Vid 08 development set and the initial positive web image set ("#DPos+#WPos" in Table 1).
- **[Sampling]**: Use positive training samples of the TREC-Vid 08 development set and the web image set after the proposed sampling("#DPos+#SPos" in Table 1).

Table 1. The number of positive samples for 20 concepts in TRECVid08. Note: The column "#DPos" denotes the number of positive training samples in the TRECVid 08 development set, "#WPos" in the initial positive web image set, "#WPos" in the final web image set after sampling.

1.0pt ID	Concept	#DPos	#WPos	#SPos
1001	Classroom	241	790	347
1002	Bridge	186	420	235
1003	Emergency-Vehicle	103	151	11
1004	Dog	136	795	123
1005	Kitchen	289	537	174
1006	Airplane-flying	80	395	113
1007	Two-people	4140	729	458
1008	Bus	106	902	312
1009	Driver	302	489	157
1010	Cityscape	331	879	623
1011	Harbor	217	261	76
1012	Telephone	203	557	412
1013	Street	1799	693	508
1014	Demonstration-Or-Protest	159	68	25
1015	Hand	1879	384	302
1016	Mountain	265	507	284
1017	Nighttime	490	594	229
1018	Boat-Ship	506	783	215
1019	Flower	620	948	513
1020	Singing	441	646	187

In the above runs, we used the SIFT features [17] for dictionary learning during sampling, and the well-known BoW feature [16] based on soft-weighting of SIFT, due to its widely reported effectiveness [24].

Figure 3 shows the comparison results of AP for each concept and mean AP (MAP) of the three runs. As shown, the proposed run [Sampling] achieved the highest MAP of 0.144, which is 9.92% higher than the run [Baseline] (MAP 0.131), and 6.67% higher than the run [AddWeb] without sampling(MAP 0.135). In particular, the proposed method outperformed the others on 9 out of 20 concepts, including Airplane-flying, Dog, Telephone, Demonstration-Or-Protest, Hand, and Flower, which had been selected with sufficient visually-coherent positive samples, while little was gained with the concepts such as Harbor, Kitchen, Bridge, and Emergency-Vehicle because these concepts on the old documentary TRECVid videos may be too outdated for enough positive web samples to be obtained. On the other hand, the run [AddWeb] achieved only a 3.05% improvement in MAP compared with the run [Baseline].

Compared with the best runs in TRECVid 2008 [16], significant improvement was obtained in handling concepts with few TRECVID positive training samples. The experimental results show that the proposed approach can achieve constant overall improvement despite cross-domain incoherence.

5   Conclusion

In this paper, we proposed a novel sampling of web images for cross-domain concept detection based on coherence between an image dictionary and concept dictionary. Experimental results on the TRECVid 08 benchmark show the effectiveness and necessity of the proposed sampling method.

References

1. Wessel, K.: TRECVID-2007 High-Level Feature task:Overview, http://wwwnlpir.nist.gov/projects/tvpubs/tv7.slides/
2. Jinqiao, W., et al.: IVA-NLPR-IA-CAS TRECVID 2009: High LevelFeatures Extraction, http://www-nlpir.nist.gov/projects/tvpubs/tv.pubs.org.html#2009
3. Sun, Y.: A Novel Region-based Approach to Visual Concept Modeling using Web Images. In: ACM Multimedia (2008)
4. Jinhui, T.: To construct optimal training set for video annotation. In: ACM Multimedia (2006)
5. Su, Y.: Cross-database age estimation based on transfer learning. In: ICASSP 2010, pp. 1270–1273 (2010)
6. Fergus, R.: Learning Object Categories from Google's Image Search. In: ICCV, vol. 2, pp. 1816–1823 (2005)
7. Kennedy, L.S.: Generating diverse and representative image search results for landmarks. In: WWW, pp. 297–306 (2008)
8. Baudat, G., et al.: Feature vector selection and projection using kernels. Neurocomputing 55(1-2), 21–38 (2003)
9. Yang, J.: Cross-Domain Video Concept Detection using Adaptive SVMs. In: ACM Multimedia 2007, pp. 188–197 (2007)
10. Chang, S.: Columbia University/VIREO-CityU/IRIT TRECVID2008 High-Level Feature Extraction and Interactive Video Search, http://www-nlpir.nist.gov/projects/tvpubs/tv8.papers/
11. Bay, H., Ess, A., Tuytelaars, T., Gool, L.V.: Surf: Speeded up robust features. Computer Vision and Image Understanding 110(3), 346–359 (2008)
12. Borth, D., Ulges, A., Breuel, T.M.: Automatic concept-to-query mapping for web-based concept detector training. In: ACM Multimedia 2011, New York, NY, USA, pp. 1453–1456 (2011)
13. Donoho, D.: For most large underdetermined systems of linear equations the minimal l1-norm solution is also the sparsest solution. Comm. Pure and Applied Math. 59(6), 797–826 (2006)
14. Efron, T.J.I., Bradley, H., Tibshirani, R.: Least angle regression. Annals of Statistics 32(2), 407–499 (2004)
15. Huiskes, M.J., Thomee, B., Lew, M.S.: New trends and ideas in visual concept detection: the mir flickr retrieval evaluation initiative. In: MIR 2010, New York, NY, USA, pp. 527–536 (2010)
16. Jiang, Y.-G., Yang, J., Ngo, C.-W., Hauptmann, A.G.: Representations of keypoint-based semantic concept detection: A comprehensive study. IEEE Transactions on Multimedia 12, 42–53 (2010)
17. Lowe, D.G.: Distinctive image features from scale-invariant keypoints. International Journal of Computer Vision 60(2), 91–110 (2004)

18. Mairal, J., Bach, F., Ponce, J., Sapiro, G.: Online learning for matrix factorization and sparse coding. J. Mach. Learn. Res. 11, 19–60 (2010)
19. Over, P., Awad, G., Rose, R.T., Fiscus, J.G., Kraaij, W., Smeaton, A.F.: Trecvid 2008 - goals, tasks, data, evaluation mechanisms and metrics. In: TRECVID Workshop (2008)
20. Ramirez, I., Sprechmann, P., Sapiro, G.: Classification and Clustering via Dictionary Learning with Structured Incoherence and Shared Features. In: CVPR 2010, pp. 3501–3508 (June 2010)
21. Sun, Y., Kojima, A.: A novel method for semantic video concept learning using web images. In: ACM Multimedia 2011, New York, NY, USA, pp. 1081–1084 (2011)
22. Sun, Y., Shimada, S., Taniguchi, Y., Kojima, A.: A novel region-based approach to visual concept modeling using web images. In: ACM Multimedia 2008, New York, NY, USA, pp. 635–638 (2008)
23. Tang, S., Li, J.-T., Li, M., Xie, C., Liu, Y.-Z., Tao, K., Xu, S.-X.: TRECVID 2008 High-Level Feature Extraction By MCG-ICT-CAS. In: Proc. TRECVID 2008 Workshop, Gaithesburg, USA (November 2008)
24. Tang, S., Zheng, Y.-T., Wang, Y., Chua, T.-S.: Sparse ensemble learning for concept detection. IEEE Transactions on Multimedia 14(1) (2012)
25. Zhu, S., Wang, G., Ngo, C.-W., Jiang, Y.-G.: On the sampling of web images for learning visual concept classifiers. In: CIVR 2010, New York, NY, USA, pp. 50–57 (2010)

Semantic Correlation Mining between Images and Texts with Global Semantics and Local Mapping

Jiao Xue, Youtian Du, and Hanbing Shui

Ministry of Education Key Lab for Intelligent Networks and Network Security,
Xi'an Jiaotong University 710049, China
duyt@mail.xjtu.edu.cn

Abstract. This paper proposes a novel approach for the modeling of semantic correlation between web images and texts. Our approach contains two processes of semantic correlation computing. One is to find the local media objects (LMOs), the components composing text (or image) documents, that match the global semantics of a given image(or text) document based on probabilistic latent semantic analysis (PLSA); The other is to make a direct mapping among LMOs with graph-based learning, with those LMOs achieved based on PLSA as a part of inputs. The two cooperating processes consider both dominant semantics and local subordinate parts of heterogeneous data. Finally, we compute the similarity between the obtained LMOs and a whole document of the same modality and then get the semantic correlation between textual and visual documents. Experimental results demonstrate the effectiveness of the proposed approach.

Keywords: semantic correlation, web images, heterogeneous data, graph-based learning.

1 Introduction

There has been a massive explosion of multimedia contents on the web which includes the multi-modality of texts, images and videos. Different types of modality data always co-exist in a variety of ways and describe the same or similar semantic concepts. This leads to challenges of semantic correlation mining between two types of medias due to the difficulty in computing the similarity across two heterogeneous feature spaces [1]. The research is significant to many applications such as image tagging [2–4] and cross-media retrieval [5, 6].

Three categories of methods were focused on in the past research. The first category is linear/nonlinear mapping, which builds a closed-form transformation between two heterogeneous spaces [5, 7]. The main challenge of these methods is to choose nonlinear models with a suitable function capacity. The second category is based on probabilistic models such as probabilistic latent semantic analysis (PLSA) [8, 9], which focuses on the dominant semantics of medias but ignores the direct corresponding relationship among the components of documents such as textual words and local visual regions. The third type of methods

X. He et al. (Eds.): MMM 2015, Part II, LNCS 8936, pp. 427–435, 2015.
© Springer International Publishing Switzerland 2015

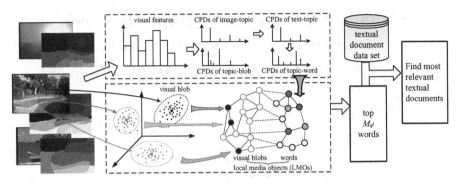

Fig. 1. Our proposed framework (taking the task to return relevant text documents for the image query for instance)

is to make correlation propagation over a graph with documents as vertices and correlation as the weight of edges [6, 10]. The graph-based method [11] is a kind of local learning which has been demonstrated with better performance than global learning [12], while its computation cost increases rapidly for a large scale of data. In addition, there are some other work about the social images including both textual and visual information. Gao et al. [13] proposed a method for social image searching based on both visual and textual information to estimate the relevance of user tagged images.

In this paper, we propose a new semantic correlation modeling approach that contains two processes. One is to find the local media objects (LMOs), the components composing text (or image) documents, that match the global semantics of a given image(or text) based on PLSA. This process focuses on analyzing the global semantics of documents; The other is to make a direct mapping among LMOs based on graph-based learning, with the relevant LMOs achieved in the first process as a part of inputs. By combining the two processes, our method considers both the dominant semantics and the local subordinate parts of heterogeneous data. Finally, we compute the similarity between the obtained LMOs and a whole document of the same modality and then get the semantic correlation between different modalities of documents.

2   Our Framework

We denote the dataset by $D = \{d_1, d_2, \cdots, d_N\}$, where $d_i =< d_i^{\mathcal{I}}, d_i^{\mathcal{T}} >$ and the superscripts $\mathcal{I}$ and $\mathcal{T}$ indicate images and texts, respectively. We suppose that $d_i^{\mathcal{I}}$ and $d_i^{\mathcal{T}}$ share the same semantics. In general, images are composed of multiple local regions (generally called visual blobs), each with uniform textures or colors, and texts consist of a number of words. We define a visual blob or a word as a *local media object (LMO)*.

Without loss of generality, we introduce our framework with the task to find the relevant texts for a given image. As shown in Fig.1, the proposed framework includes two channels of correlation computing. One is to find the most relevant textual words for a whole image with PLSA technique (see upper part of Fig.1). The topic-based technique tends to obtain the textual words relevant to the dominant parts of images that reflect the main semantics. Further, we present a local media object mapping between the visual and textual modality based on the information propagation over graphs, and fuse the output of correlation computing in channel 1 to the information propagation (see lower part of Fig.1). The graph builds the relation directly among LMOs instead of on the level of documents of images and texts, which results in the advantage that the size of graph is not so tremendous because the amount of LMOs building up the documents is approximately finite and in control compared to image and text documents. In addition, the direct LMO mapping emphasizes the correspondence between each pair of LMOs contained in images and texts. Thus, the proposed framework considers the correlation of both the dominant semantics and the local subordinate parts of heterogeneous data.

3   The Proposed Method

3.1   Image and Text Representation

Images are represented with the bag-of-visual-words model. First, images are segmented into a number of local regions. Each region is represented by 128 original features including 96-dimensional color histogram for three independent components of RGB space with 32 bins for each component, and 32-dimensional texture feature vector based on the Gabor filter with 4 scales and 8 directions. Finally, k-means clustering algorithm is applied to separate all the local regions into M_I clusters that generally called visual blobs, and quantize each feature vector into the mean of the cluster that it belongs to. Each image can thus be described as $(\#(I_i, b_1), ..., \#(I_i, b_{M_I}))$, where $\#(I_i, b_j)$ denotes the number of regions in image I_i that belong to the blob b_j.

Textual documents are represented based on the bag-of-words model with tf-idf weighted features, and each text is represented as an M_T-dimensional feature vector.

3.2   Image-Text Correlation Computing

The proposed approach includes two types of semantic correlation computing. We first find the LMOs of one modality (i.e. textual words) that relevant to the global semantics of a document of the other modality (i.e. an image), and then use the result as an input of local media object mapping to further adjust the relevant LMOs for the given document. Finally, we find the documents that are similar with, and of the same modality as the achieved LMOs by computing their distance.

Correlation Transferring with Global Semantics. PLSA[14] introduces a latent aspect and models the joint probability of the co-occurrence of words and documents (w, d) as a mixture of conditional probability distributions (CPDs):

$$\Pr(w_j, d_i) = \Pr(d_i) \sum_{z_k} \Pr(z_k | d_i) \Pr(w_j | z_k) \tag{1}$$

where z_k denotes latent topics. $\Pr(w_j | z_k)$ and $\Pr(z_k | d_i)$ can be estimated by Expectation-Maximization (EM) algorithm.

As is supposed above, one image shows the same semantics as its corresponding texts and then $\Pr(z_k^\mathcal{T} | d_i^\mathcal{T}) = \Pr(z_k^\mathcal{I} | d_i^\mathcal{I})$. This builds a bridge between images and texts. Therefore, we can learn $\Pr(z_k | d_i)$ from all training documents of one modality (visual or textual) and then keep fixed for the other modality as introduced in [8]. It should be noticed that the distributions $\Pr(w^\mathcal{I} | z_k^\mathcal{I})$ and $\Pr(w^\mathcal{T} | z_k^\mathcal{T})$ are possibly different due to different low-level representation.

Without loss of generality, we learn the semantic aspect distribution from visual modality and keep it fixed for the textual modality. The inference from visual modality to textual modality includes two steps: 1) N_z conditional distributions $\Pr(w | z_k)$ and N conditional distributions $\Pr(z_k | d_i)$ are estimated based on EM algorithm over the training set of visual modality; 2) Keep the achieved $\Pr(z_k | d_i)$ fixed for the textual modality, and maximize the joint probability in Eq.(1) to obtain N_z conditional distributions $\Pr(w^\mathcal{T} | z_k^\mathcal{T})$ for texts.

To the image from a new document d_{new} that consists of no texts, the corresponding textual words can be predicted based on the following conditional probability:

$$\Pr(w^\mathcal{T} | d_{new}^\mathcal{I}) = \sum_{z_k} \Pr(w^\mathcal{T} | z_k) \Pr(z_k | d_{new}^\mathcal{I}) \tag{2}$$

where $\Pr(z_k | d_{new}^\mathcal{I})$ and $\Pr(w^\mathcal{T} | z_k)$ are achieved in above steps 1 and 2, respectively. The top K_g words, denoted by $O_g^\star$, with the largest conditional probability are the most relevant to the given image in the view of global semantics.

Local Media Object Mapping. In many cases, the LMOs of different modalities shows direct correspondence. Therefore, directly mapping the LMOs including visual blobs and textual words is important to mine the semantic correlation of images and texts.

We denote the total n LMOs by $\Omega = \{o_i^z\}_{i=1}^n$, where $z \in \{\mathcal{T}, \mathcal{I}\}$ and o_i^z is actually the alternative notation of $w^\mathcal{I}$ or $w^\mathcal{T}$. In this subsection, the direct mapping between LMOs is performed with graph-based methods. Two types of LMOs are uniformly served as vertices and the correlation between them is the weight of the edges. The correlation between each pair of LMOs is described as an $n \times n$ matrix W with $W_{k,l} = dis(o_k^{z_1}, o_l^{z_2})$ defined as follows:

$$dis(o_k^{z_1}, o_l^{z_2}) = \begin{cases} \|o_k^{z_1} - o_l^{z_2}\|_2 + \alpha(1 - f_{occ}(o_k^{z_1}, o_l^{z_2})), & if\, z_1, z_2 = \mathcal{I} \\ \beta(1 - f_{occ}(o_k^{z_1}, o_l^{z_2})), & else \end{cases} \tag{3}$$

where $f_{occ}(\cdot, \cdot)$ denotes the percentage that two LMOs occur in one document at the same time, α and β are parameters. We then form the affinity matrix

C with $C_{kl} = \exp(-W_{kl}^2/2\sigma^2)$ if $k \neq l$ and $C_{kk} = 0$, and normalize it by $\tilde{C} = D^{-1/2}CD^{-1/2}$, where D is a diagonal matrix with its (k,k)-element equal to the sum of the k-th row of C. Finally we only keep the edges with K largest weights for each vertex and construct a K-nearest neighbor (K-NN) graph.

The task to find the relevant LMOs of the other modality for a query of document d_q^z can be accomplished by propagating the correlation information on the graph, and obtain a proper matching value function $f : \Omega \to R$. We define an input vector $\mathbf{y} = [y_1, y_2, \cdots, y_n]^T$ with

$$y_i = \begin{cases} 1, & i : o_i^z \ belongs \ to \ d_q^z \\ \lambda \cdot \Pr(o_i^{z_1}|d_q^z), & i : o_i^{z_1} \in O_g^\star, z_1 \neq z \\ 0, & otherwise \end{cases} \quad (4)$$

where λ is a factor that linearly normalizes $\Pr(o_i^{z_1}|d_q^z)$ (obtained with Eq.(2)) for $o_i^{z_1} \in O_g^\star$ so that the largest $\lambda \cdot \Pr(o_i^{z_1}|d_q^z)$ equals 1. Eq.(4) fuses the local mapping and the global semantic mining introduced in above section together. We define $\mathbf{f} = [f_1, f_2, \cdots, f_n]^T$, where $f_i = f(o_i^z)$. The correlation information spreads over graphs by iterating the process $\mathbf{f}^{(t+1)} = \eta\tilde{C}\mathbf{f}^{(t)} + (1 - \eta)\mathbf{y}$ and converges to

$$\mathbf{f}^\star = (1 - \eta)(I - \eta\tilde{C})^{-1}\mathbf{y} \quad (5)$$

where $\eta \in (0,1)$ is a small constant and we let $\eta = 0.98$ here.

To a given image $d_q^\mathcal{I}$, the multiple LMOs $o_i^\mathcal{I}$ in $d_q^\mathcal{I}$ are used as the visual query, and the textual LMOs $o_j^\mathcal{T}$ belonging to some texts with top K_{gl} scores of f function, denoted by $O_{gl}^\star$, can be considered the most correlated. If $d_q^\mathcal{I}$ is an *out-of-sample image* (i.e. not included in the graph), a simple solution is to find the neighbors of the LMOs belonging to $d_q^\mathcal{I}$ on the graph, and use these neighbors as visual query to find the relevant local media objects of the other modality.

Computing Similarity between LMOs and Documents. In cross-media retrieval, the task is to find the relevant documents $d^\star$ for a query d_q of the other modality. We consider that the documents $d^\star$ are those with high similarity to the sequence $O_{gl}^\star$ of K_{gl} ranked LMOs that are most relevant to d_q and are achieved above. Intuitively, the LMOs belonging to $d^\star$ should be at the front of the sequence $O_{gl}^\star$. Thus, the similarity is defined based on Average Precision as

$$sim(d, O_{gl}^\star) = \frac{1}{K_{gl}} \sum_{n'=1}^{K_{gl}} p_{n'}\Delta_{n'} \quad (6)$$

where $\Delta_{n'} = 1$ when the n'-th LMO in $O_{gl}^\star$ occurs in d and $\Delta_{n'} = 0$ otherwise, and $p_{n'}$ is the precision at cut-off rank n' and defined as: $p_{n'} = \#(d, O_{gl}^\star, n')/n'$, where $\#(d, O_{gl}^\star, n')$ denotes the number of LMOs in d that occur in the first n' places of $O_{gl}^\star$. Finally, we can obtain the most relevant document $d^\star$ for the query document d_q of the other modality:

$$d^\star = \arg\max_d sim(d, O_{gl}^\star) \quad (7)$$

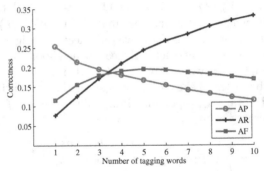

Fig. 2. The average precision(AP), average recall(AR) and average F1-score (AF) of image tagging for test images

4 Experimental Results

This paper conducts the experiments on Corel dataset, in which each example consists of an image and a segment of text. There are 50 categories in this dataset and each category is made up by 100 images. In the experiment, the dataset is divided into the training set of 4000 examples and the test set of 1000 examples.

The performance of the proposed method is evaluated by precision, recall and F-score defined as follows. Precision is the fraction of the documents retrieved that are relevant to the user's information need.

$$precision = \frac{|\{relevant\ documents\} \cap \{retrieved\ documents\}|}{|\{retrieved\ documents\}|} \tag{8}$$

Recall is the fraction of the documents that are relevant to the query that are successfully retrieved.

$$recall = \frac{|\{relevant\ documents\} \cap \{retrieved\ documents\}|}{|\{relevant\ documents\}|} \tag{9}$$

$F1$ score that combines the precision and recall and is defined in the form of

$$F1 = \frac{2 \cdot precision \cdot recall}{precision + recall} \tag{10}$$

4.1 Application of Image Tagging

We use the top K_{gl} textual words $O_{gl}^{\star}$ as the image tagging results, and evaluate the performance with recall, precision and F1-score. The parameters α, β and σ are determined by 3-fold cross validation in our paper, and the size of neighborhood in K-NN graph is $K = 10$.

As is shown in Fig.2, the average precision (AP) declines while the average recall (AR) improves as the number of tagging words increases from 1 to 10 for an image. The results show the precision-recall tradeoff, a well known phenomenon in object detection and information retrieval. The average F1-score

Table 1. The ranked tagging words for images, $K_{gl} = 5$

No.	Input images	Method 1	Method 2	Our method
1	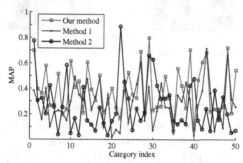	Sky Snow Clouds Mountain Plane	Aerial Runway Frost Ice Jet	Sky Aerial Clouds Plane Jet
2		Sky Snow Clouds Bear Polar	Moss Elk Ice Polar Tundra	Moss Snow Clouds Polar Bear
3		Water Tree Ocean Beach Coral	Moss Fish Waves Coral Reefs	Coral Water Ocean Beach Fish

Fig. 3. MAP results of retrieval for each category

(AF), which fuses precision and recall, is up to 19.57% based on the simple visual and textual features when the number of tagging words is $K_{gl} = 5$. In table 1, we also compare the proposed method to the PLSA-based method [8] (method 1) and graph-based method (called method 2 and introduced in section 3.2.2), and find that our method is superior to them. For example, the tagging words "Sky", "Aerial", "Clouds", "Plane" and "Jet" are more consistent of the image 1.

4.2   Application of Cross-Media Retrieval

In the cross-media retrieval, we use texts as query and return the relevant images back.

Fig.3 shows the MAP performance of cross-media retrieval for 50 categories of test data in Corel dataset. For each category, we select the first 30% of words occurring the most frequently in the category, and use each of them as the

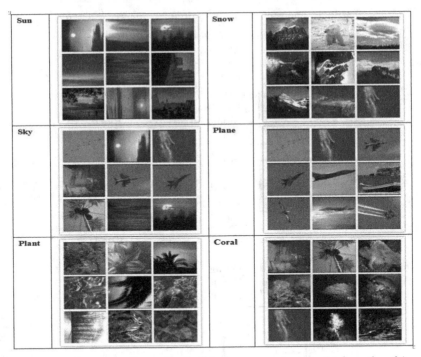

Fig. 4. Results of cross-media retrieval with the query of one textual word and images as returned results

query (one word query). In Fig.3, the result for each category is the average of MAP over the selected words. From the figure, we find the proposed method has better performance than two compared methods. The average MAPs for the total categories of methods 1 and 2 are 0.287 and 0.264, respectively, and the result of our method is 0.389. Fig.4 illustrates some cross-media retrieval results, in which we use one textual word as the query and return images. From the figure we find that the returned images are very relevant to the word query. It is noticed that some returned images are very similar for different query words, e.g. "sky" and "plane". The result is mainly caused by the high correlation of the queries.

5   Conclusions

In this paper, we propose a novel approach to the modeling of semantic correlation between images and texts. Different from the past research, this work includes two processes of semantic correlation computing based on global semantics and local mapping, respectively, and fuses the results of the former process of semantic correlation computing to the correlation propagation of the latter. The cooperation makes our method consider both dominant semantics and local subordinate parts of heterogeneous data. Experimental results on image tagging and cross-media retrieval show the proposed method is superior to the methods based on PLSA and graphs, and demonstrate its effectiveness.

Acknowledgement. This work is supported in part by the National Natural Science Foundation (61375040, 60905018, 61221063) and 111 International Collaboration Program, of China.

References

1. Wu, X., Qiao, Y., Wang, X., Tang, X.: Cross matching of music and image. In: ACM International Conference on Multimedia, pp. 837–840 (2012)
2. Jeon, J., Lavrenko, V., Manmatha, R.: Automatic image annotation and retrieval using cross-media relevance models. In: Annual ACM SIGIR Conference, pp. 119–126 (2003)
3. Wu, L., Jin, R., Jain, A.K.: Tag Completion for Image Retrieval. IEEE Transactions on Pattern Analysis and Machine Intelligence 35, 716–727 (2013)
4. Wang, M., Ni, B., Hua, X., Chua, T.: Assistive tagging: a survey of multimedia tagging with human-computer joint exploration. ACM Computing Surveys 44, 25–25 (2012)
5. Rasiwasia, N., Perieira, J.C., Cobiello, E., Doyle, G., Lanckriet, G.R.G., Levy, R., Vasconcelos, N.: A new approach to cross-modal multimedia retrieval. In: MMM, pp. 251–260 (2010)
6. Zhuang, Y., Yang, Y., Wu, F.: Mining semantic correlation of heterogeneous multimedia data for cross-media retrieval. IEEE Transaction on Multimedia 10, 221–229 (2008)
7. Jiang, T., Tan, A.: Learning image-text associations. IEEE Transactions on Knowledge and Data Engineering 21, 161–177 (2009)
8. Monay, F., Perez, D.G.: Modeling semantic aspects for cross-media image indexing. IEEE Transactions on Pattern Analysis and Machine Intelligence 29, 1802–1817 (2007)
9. Zhou, Y., Liang, M., Du, J.: Study of cross-media topic analysis based on visual topic model. In: 24th Chinese Control and Decision Conference, pp. 3467–3470 (2012)
10. Zhai, X., Peng, Y., Xiao, J.: Effective heterogeneous similarity measure with nearest neighbors for cross-media retrieval. In: Schoeffmann, K., Merialdo, B., Hauptmann, A.G., Ngo, C.-W., Andreopoulos, Y., Breiteneder, C. (eds.) MMM 2012. LNCS, vol. 7131, pp. 312–322. Springer, Heidelberg (2012)
11. Zhou, D., Bousquet, O., Lal, T.N., Weston, J., Scholkopf, B.: Learning with local and global consistency. In: NIPS, pp. 237–244 (2003)
12. Vapnik, V.: Local learning algorithms. Neural Computation 4, 888–900 (1992)
13. Gao, Y., Wang, M., Zha, Z., Shen, J., Li, X., Wu, X.: Visual-textual joint relevance learning for tag-based social image search. IEEE Transactions on Image Processing 22 (2013)
14. Hofmann, T.: Unsupervised learning by probabilistic latent semantic analysis. Machine Learning 42, 177–196 (2001)

Image Taken Place Estimation via Geometric Constrained Spatial Layer Matching

Yisi Zhao, Xueming Qian[*], and Tingting Mu

SMILES LAB, Xi'an Jiaotong University, China
zyswhy0203@stu.xjtu.edu.cn, qianxm@mail.xjtu.edu.cn,
T.Mu@liverpool.ac.uk

Abstract. In recent years, estimating the locations of images has received a lot of attention, which plays a role in application scenarios for large geo-tagged image corpora. So, as to images which are not geographically tagged, we could estimate their locations with the help of the large geo-tagged image set by visual mining based approach. In this paper, we propose a global feature clustering and local feature refinement based image location estimation approach. Firstly, global feature clustering is utilized. We further treat each cluster as a single observation. Next we mine the relationship of each image cluster and locations offline. By cluster selection online, several refined locations likely to be related to an input image are pre-selected. Secondly, we localize the input image by local feature matching which utilizes the "SIFT" descriptor extracted from the refined images. In this process, "spatial layers of visual word" (SLW) is built as an extension of the unorganized bag-of-words image representation. Experiments show the effectiveness of our proposed approach.

Keywords: Location Estimation, Spatial Layer Matching, Bag-of-Words.

1 Introduction

Given a query image, in this paper, our goal is to estimate its location by mining image content. Automatic location estimation for an image is possible with the help of the large scale geo-tagged photos shared by millions of worldwide users. State-of-the-art large scale image retrieval systems have relied on local SIFT descriptors [5]. Traditionally, a visual vocabulary is trained by clustering a large number of local feature descriptors. The exemplar descriptor of each cluster is called a visual word, which is then indexed by an integer. However, experimental results of existing work show that the commonly generated visual words are still not as expressive as the text words. Spatial information of visual words should be exploited for better performance. Moreover, we find that although purely using global features is not so efficient, some images can be recognized well via global feature matching.

Therefore, we propose image visual mining based image geographic location estimation approach. In our work, firstly, the clusters are mined to generate refined locations for an input image using global features. Secondly, we exploit sufficient

---

[*] Corresponding author.

X. He et al. (Eds.): MMM 2015, Part II, LNCS 8936, pp. 436–446, 2015.

information by mining spatial information of visual words. "Spatial layers of visual word" (SLW) is proposed, which plays a significant role for image location estimation. SLW is generated by involving one visual word and its spatial relationships with its neighbor visual words. Unlike what is introduced in [11], their "spatial pyramid" is generated by partitioning an image into increasingly fine sub-regions and computing histograms of local features found inside each sub-region. We go deep into each word whose multiple neighbors are taken into consideration in sequence.

The contributions of this paper are as follows: (1) Refined locations of an input image are generated via cluster selection based on cluster location estimation. (2) Spatial layer matching is proposed to improve the estimation accuracy for an input image. (3) Useful local features selection is utilized, on the basis of which our proposed SLW shows better results.

The rest of the paper is organized as follows: Firstly, related works on location estimation are reviewed. Secondly, we provide the system overview. Finally, we give a description on our approach in section 4 and 5. Experiments containing the comparison with the recently popular method and parameters discussions are shown in Section 6. In Section 7, the conclusion is drawn.

2   Related Work

Many methods are intended to estimate the geographic location of images. An approach which is based purely on visual features is presented by Hays and Efros in [9]. They characterize each image using a number of image features. Then they compute the distances on different feature spaces and use the k-nearest-neighbor technique to estimate the GPS of an input image. Finally, cluster with the highest cardinality is selected and its GPS is assigned as GPS of the input image

Bag-of-words image representation has been utilized for many multimedia and vision problems. Li et al. utilize multi-class SVM classifiers using bag-of-words for large scale image location estimation [2]. They also show that through adding textual features such as tags, they can improve the performance. Han et al. propose an object-based image retrieval algorithm. They combine a novel feature descriptor based on context-preserving bag-of-words and a two-stage re-ranking technique to measure the similarity between the query image and each image in the dataset [16]. Zhang et al. propose a spatial coding based image retrieval approach by building the contextual visual vocabulary [1]. The spatial coding encodes the relative positions between each pair of features in an image. They focus on user traces across the micro-blogging platform Twitter. Chum et al. also propose an approach for estimating the location of the image by using local feature matching [10]. And user interaction is required to confine the locations of the input image to really small ranges. In [17,18], the GPS information is served as an important clue to improve tag recommendation performances for social user shared photos.

Researchers have proposed many works e.g. visual synonyms [7, 14-15], embed geometry constraint [3, 12-13], etc. Spatial information can reinforce the discriminative power of single word. Wu et al. [12] employed the detector of Maximally Stable Extremal Regions (MSER) to bundle point features (SIFT) into groups instead of taking all of them individually. Moreover, the database can be constructed with a 3D model [6]. Liu et al. propose an approach which is capable of providing a complete set of more accurate parameters about the scene geo—including the actual locations of both the mobile user and perhaps more importantly the captured scene along with the viewing direction [6]. They firstly perform joint geo-visual clustering in the cloud to generate scene clusters, with each scene represented by a 3D model. The 3D scene models are then indexed using a visual vocabulary tree structure.

3   System Overview

The system of our proposed approach is shown in Figure 1. It consists of two systems: the online system and the offline system.

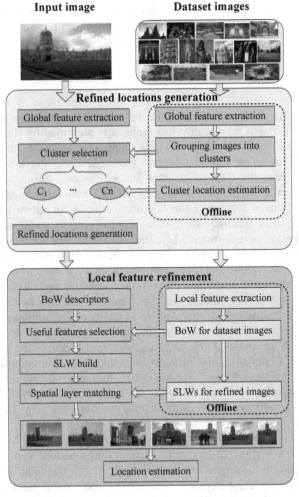

Fig. 1. Block diagram of the location estimation system

Firstly, we obtain refined locations related to an input image offline. In this process, global feature clustering is utilized. We further treat each cluster as a single observation to mine the relationship of each image cluster and locations. Then several refined locations likely to be related to an input image are pre-selected by cluster selection in our online system. Secondly, we estimate image GPS by local feature refinement by making full use of the images in refined locations. In our work, "spatial layers of word" (SLW) is proposed as an extension of bag-of-words image representation. SLWs for dataset images are built offline. We estimate the location of an input image by spatial layer matching.

4   Refined Locations Generation

Generating refined locations is the first step of our framework in our offline system. In this section, we introduce how to generate image clusters, and how to select the refined location candidates.

4.1   Grouping Images into Clusters

We propose to cluster the dataset images using their global features, such as color feature and texture feature. Similar to our previous work [4], color moment (CM) and hierarchical wavelet packet descriptor (HWVP) [19] are utilized here. The global feature clustering is carried out on the 215d vector including 45d CM and 170d HWVP. K-means clustering is utilized to divide dataset images into M clusters $C_i (i = 1,...,M)$. In this paper, we set M to be 50, according to the suggestions in [4].

4.2   Cluster Location Estimation

Due to the fact that our dataset images are geo-tagged, each image has one geo-tag. The geo-tag indicates the taken place of the input image. In our offline system, before the cluster selection, we first mine the relationship of clusters and locations. Our approach consists of the following steps:

Assume that the cluster C_n has g images $I_{nj} (j = 1,...,g)$. Firstly, for each image in the cluster C_n, we gather its R most similar images across the entire dataset images based on the similarities of the global visual features (We will discuss the situation that using local features instead of global features in our experiments.). We select the top ranked $K = R \times g$ neighboring images $I_i (i = 1,...,K)$.

Secondly, through analysis of the K geo-tagged neighboring images, we can predict the probable locations for the cluster C_n. We divide the geo-tags of the K images into L ($L \leq K$) sets according to their true locations. Each set corresponds to a unique location. Let $D_i (i = 1,...,L)$ denote the L sets. By ranking the L locations according their frequencies (i.e. the numbers of images belonging to the locations) in the descending orders, we can get the probabilities that the cluster C_n belonging to. As shown in Figure 2, among the K neighbor images, $I_1, I_2, I_3, I_4, I_6, I_{K-1}$ belong to the same location D_l. So, the score (frequency) of location D_l is 6. We select $V\%$ of the L locations as location candidates related to the cluster.

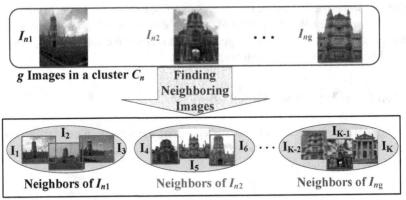

Fig. 2. Finding neighboring images for a cluster

Finally, we select candidate clusters for an input image in our online system. Let F_x denote the 215d global features of the input image. The candidate cluster selection is based on the distances between F_x and M centers $C_i\,(i=1,...,M)$. In this paper, the top ranked fifteen clusters are selected, i.e. $g=15$. Based on the found neighboring images for each cluster we can get the refined locations.

5   Local Feature Refinement

After the global feature clustering and refined location generation, we can determine the candidate locations for the input image. In order to improve image location estimation performances, we further conduct local feature refinement. In this section, to capture some unique and representative details in images, we utilize SIFT to carry out spatial layer matching. In our work, we first quantize the SIFT points into visual words by using a hierarchical K-means clustering approach [4].

5.1   Useful Features Selection

Given a query image, its visual words have different discrimination power for location estimation. Some of them are useful for location estimation, and some of them may be noise. To mine useful features, we compute the score of the visual word while considering the frequency and the weight of word by employing a term-frequency inverse-document-frequency (tf-idf) weighting scheme. For an image, the score of each visual word is computed as follows:

$$S_w = \frac{f_w}{\sum_w f_w} \times \log \frac{N}{n_w} \tag{1}$$

where f_w is the frequency of w-th visual word in the image, n_w is the number of images containing the w-th visual word. In Figure 3 (a) the raw SIFT points are shown, and in Figure 3 (b) the useful feature are kept. We find that by useful feature mining many non-discriminative visual words are removed.

(a) All features　　　　　　　　(b) useful features

Fig. 3. (a)All features of an image and (b) the useful features

5.2　Spatial Layer Matching

After the above-mentioned steps, each refined image is represented by a set of useful visual words. In this section, we build SLW for each useful visual word of the refined images, which is generated by integrating a visual word and its neighboring visual words. In our feature extraction, we represent each SIFT point by a 128-D descriptor vector and a 4-dimensional DoG key-point detector vector (x, y, scale, and orientation). In this part, the coordinates (x, y) are utilized to calculate the distance of visual words, according to which we build SLW.

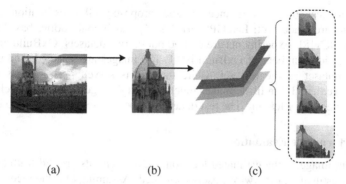

(a)　　　　　　　(b)　　　　　　　(c)

Fig. 4. Spatial layers of visual word (SLW), (a) local region, (b) the neighboring visual words around a visual word, (c) spatial layer representation for a visual word

Therefore, from a visual word w, we build its spatial layers as shown in Figure 4. The enlarged location region for Figure 4 (a) is as shown in Figure 4(b), its spatial layer representation for the visual word is shown in Figure 4(c). Let $SLW(w)$ to record both the word and its neighbor visual words. We define each layer of $SLW(w)$ as:

$$SLW_n(w) = \left\{ w, \left(NW_n \right)_{n=0}^k \right\} \tag{2}$$

where NW is neighbors of word w, and n is one of the neighbors of w. k is the number of layers. $SLW_0(w)$ denotes the first layer of w, which only contains the visual word itself. The second layer is composed of the visual word and its nearest neighbor. The third layer consists of the visual word, and its two nearest neighbors, and so on. During our experiments, we build 3 layers for each visual word. The value of n will be discussed in section 6.3.

Then, we score for each image of refined locations like this: For each visual word q_m of a query image, we build its $SLW(q_m)$. For a refined image r, SLW of its any useful visual word g is denoted as $SLW(g)$. If $SLW_n(q_m)$ is found in $SLW(g)$, the score of refined image r (denoted as $Score_r$) is accumulated by one. We assume that query image has Z visual words. Then, we iterate over all the visual words of the query image to calculate the score of the image as follows:

$$Score_r = \sum_{m=1}^{Z} f_m \tag{3}$$

where f_m is an indicator, it records whether the visual word q_m belonging to image r.

$$f_m = \begin{cases} 1, if\ SLW_n(q_m) \in SLW(g),\ g \in r \\ 0, if\ SLW_n(q_m) \notin SLW(g),\ g \in r \end{cases}, n = 0,1,...,k \tag{4}$$

So, we obtain scores of all candidate images. Then we rank all the candidate images according to their scores. At last, we use K-NN based approach to estimate the location of input image.

6   Experimentation

In order to test the performance of the proposed GPS estimation approach, comparisons are made with IM2GPS [9], CS [4] and spatial coding based approach (denoted as SC) [8]. Experiments are carried out on two datasets: OxBuild and GOLD [4]. The location numbers of OxBuild is 11. 100 images are selected randomly from the whole dataset as the test set, while the rest is served as training set. GOLD contains more than 3.3 million images together with their geo-tags. 80 travel spots are randomly selected for testing. The test dataset for the 80 sites contains 5000 images.

6.1   Performance Evaluation

For an input image, if the estimated location is exact with its ground-truth location, it is correctly estimated, otherwise falsely estimated. Assuming that the recognition rate of the i-th spot (RR_i) is the correct, then average recognition rate (AR) is utilized to evaluate the performance which is given as follows:

$$AR = \frac{1}{G} \sum_{i=1}^{G} RR_i \tag{5}$$

$$RR_i = \frac{NC_i}{NI_i} \times 100\%, i \in \{1,...,G\} \qquad (6)$$

where NC_i is the correct estimated image number, NI_i is the test image number. G is the number of locations, 11 and 80 for OxBuild and GOLD respectively.

6.2   Performance Comparison

As for IM2GPS, Spatial Coding (SC) and Cosine Similarity (CS), we choose the best parameters provided in [9], [8] and [4]. From Figure 5 we find that our method SLW outperforms the other methods. The results of IM2GPS in the two test datasets are 39.67% and 53.06%. The results of spatial coding (SC) in the two test datasets are 59.48% and 70.39%, while the results of Cosine Similarity (CS) in the two test datasets are 89.27% and 84.86% respectively. Those of ours for the two datasets are 90.15% and 86.03% respectively. The performance of CS is better than IM2GPS and SC. We can conclude that both global and local visual features are contributive in image location estimation. Our SLW further gets some improvement over our previous work CS [4]. This shows that the spatial layer information information is worth exploiting for improving the image location estimation performance.

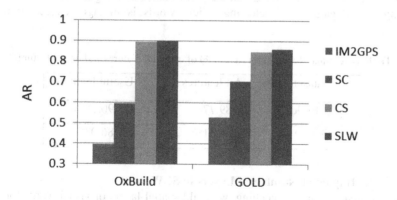

Fig. 5. ARs of IM2GPS, SC, CS and SLW

6.3   Discussion

The performance of our approach is influenced by several main factors. Hereinafter, we discuss their impacts respectively by carrying out a set of experiments.

6.3.1   The Impact of Using Global Features or Local Features
In our experiments, when mining a cluster, the global features are utilized. For each image in a cluster, we gather its R most similar images across the entire training set based on the similarity of the global visual features. We conduct an experiment that in this process, local feature is used instead of global features. We extract local features scale-invariant feature transform (SIFT). A SIFT feature consists of a 128-D

descriptor vector and a 4-dimensional DoG key-point detector vector (x, y, scale, and orientation). We can see from Table 1 that performance improvement is not obvious. The reason is that the clusters are mined for location estimation. We further obtain refined locations of input image via cluster selection. After this process, we just want to generate refined locations of the input image to narrow the scope of retrieval. Moreover, the time cost of local features is certainly more. So, global features are utilized in our work.

Table 1. Average Recognition Rates (%) of using global features and local feature in cluster mining

Dataset	Global features	Local feature
OxBuild	90.15%	90.17%
GOLD	86.03%	86.29%

6.3.2   The Impact of Using Useful Features or All Features

In the part of local feature refinement, salient features selection of images is carried out. For different images, their visual words have different weights during location estimation. The performances of using all features and useful features are discussed here. It can be seen from Table 2 that the performance of using all features is interior to using useful features. So selecting salient words is of significance for image retrieval.

Table 2. Average Recognition Rates (%) of using all features and useful features

Dataset	All features	Useful features
OxBuild	89.77%	90.15%
GOLD	85.51%	86.03%

6.3.3   The Impact of Number of Layers in SLW n

In the part of spatial layer matching, we build spatial layers of visual words for each useful word of those refined images. In our experiments, we build n layers for each visual word. The impact of layer number n to location estimation is discussed here. The AR values of SLW on GOLD are 70.56%, 86.03%, 86.97% and 63.44% respectively when $n=\{1,3,5,7\}$. It can be seen from Table 3 that with the increase of n the AR is first increasing and then into decline. When n is in the range of [3, 5], better performance can be achieved. If the distance of two visual words is larger than the proper value, their correlation is obviously weaker. During our experiments, n is set to be 3.

Table 3. Average Recognition Rates (%) of different values of n

Dataset	$n=1$	$n=3$	$n=5$	$n=7$
OxBuild	76.82%	90.15%	91.21%	70.58%
GOLD	70.56%	86.03%	86.97%	63.44%

6.3.4   The Impact of Percentage of Cluster Location Candidates V

In cluster location estimation, for a cluster, we obtain the ranking list of L locations. We select V percent of the L locations as location candidates related to the cluster. Here, we discuss the impact of V to image GPS estimation performances on both GOLD and OxBuild as shown in Figure 6. The AR values of SLW on GOLD is 39.41%, 60.39%, 75.19%, 86.03%, 86.73% and 61.32% with the increase of V. V is set 60 in our experiments. If the V is too large, more unrelated locations will be taken into consideration. If the V is too small, the related location will be cut out, which has a worse impact on performance. It can be seen from Figure 6 that with the increase of V, the AR is first increasing and then into decline.

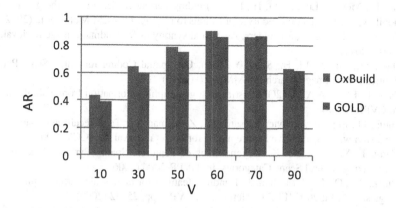

Fig. 6. Impact of cluster location candidates V

7   Conclusion

In this paper, we present a method for image location estimation. In our work, first the images are clustered relying on global features. And each cluster is seen as a single observation so as to mine the relationship among those clusters and locations. Then refined locations are further selected via a cluster selection strategy online. Afterwards, spatial information of visual words is mined. We build spatial layers of visual words (SLW) for further matching, which are generated by involving visual words and the neighboring visual words. The final location estimation is yielded via an online spatial layer matching process. Experiments show that our proposed SLW has better results.

Acknowledgments. This work is supported partly by NSFC No.61173109, No.61128007, No.60903121, Microsoft Research Asia, and Fundamental Research Funds for the Central Universities.

References

[1]   Zhang, S., Huang, Q., Hua, G., Jiang, S., Gao, W., Tian, Q.: Building Contextual Visual Vocabulary for Large-scale Image Applications. In: MM 2010, October 25–29 (2010)

[2]   Li, Y., Crandall, D.J., Huttenlocher, D.P.: Landmark Classification in Large-scale Image Collections. In: ICCV 2009 (2009)

[3]   Zhang, Y., Jia, Z., Chen, T.: Image retrieval with Geometry-Preserving visual phrases. In: CVPR (2011)

[4]   Li, J.J., Qian, X.X., Tang, Y.Y., Yang, L.L., Mei, T.T.: GPS estimation for places of interest from social users' uploaded photos. IEEE Trans. Multimedia (2013)

[5]   Lowe, D.G.: Distinctive Image Features from ScaleInvariant Keypoints. In: HCV (2004)

[6]   Liu, H., Mei, T., Luo, J., Li, H., Li, S.: Finding perfect rendezvous on the go: accurate mobile visual localization and its applications to routing. In: ACM Multimedia (2012)

[7]   Gavves, E., Snoek, C., Smeulders, A.: Visual synonyms for landmark image retrieval. In: CVIU (2011)

[8]   Zhou, W., Lu, Y., Li, H., Song, Y., Tian, Q.: Spatial Coding for Large Scale Partial-Duplicate Web Image Search. In: MM 2010 (2010)

[9]   Hays, J., Efros, A.A.: IM2GPS: estimating geographic information from a single image. In: CVPR (2008)

[10]   Chum, O., Philbin, J., Sivic, J., Isard, M., Zisserman, A.: Total recall: automatic query expansion with a generative feature model for object retrieval. In: ICCV (2007)

[11]   Wang, J., Yang, J., Yu, K.: Beyond Bags of Features: Spatial Pyramid Matching for Recognizing Natural Scene Categories. In: CVPR 2006 (2006)

[12]   Wu, Z., Ke, Q., Isard, M., Sun, J.: Bundling features for large scale partial-duplicate web image search. In: 2009 IEEE Conference on CVPR, pp. 25–32 (2009)

[13]   Chen, J., Feng, B., Zhu, L., Ding, P., Xu, B.: Effective near-duplicate image retrieval with image-specific visual phrase selection. In: ICIP 2010 (2010)

[14]   Gavves, E., Snoek, C., Smeulders, A.: Visual synonyms for landmark image retrieval. In: CVIU (2011)

[15]   Xue, Y., Qian, X., Zhang, B.: Mobile image retrieval using multi-photo as query. In: ICMEW (2013)

[16]   Han, J., Xu, M., Li, X., Guo, L., Liu, T.: Interactive Object-based Image Retrieval and Annotation on iPad. Multimedia Tools and Applications 72, 2275–2297 (2014)

[17]   Qian, X., Liu, X., Zheng, C., Du, Y., Hou, X.: Tagging photos using users' vocabularies. Neurocomputing 111, 144–153 (2013)

[18]   Liu, X., Qian, X., Lu, D., Hou, X., Wang, L.: Personalized Tag Recommendation for Flickr Users. In: Proc. ICME, pp. 1–6 (2014)

[19]   Qian, X., Liu, G., Guo, D., Li, Z., Wang, Z., Wang, H.: Object Categorization using Hierarchical Wavelet Packet Texture Descriptors. In: Proc. ISM 2009, pp. 44–51 (2009)

Recognition of Meaningful Human Actions for Video Annotation Using EEG Based User Responses

Jinyoung Moon[1], Yongjin Kwon[1], Kyuchang Kang[1], Changseok Bae[1], and Wan Chul Yoon[2]

[1] Human Computing Research Section, SW·Content Research Laboratory, ETRI, 218 Gajeong-ro, Yuseong-gu, Daejeon, 305-700, South Korea
{jymoon,scocso,k2kang,csbae}@etri.re.kr
[2] Department of Industrial and System Engineering, KAIST, 291 Daehak-ro, Yuseong-gu, Daejeon 305-701, South Korea
wcyoon@kaist.ac.kr

Abstract. To provide interesting videos, it is important to generate relevant tags and annotations that describe the whole video or its segment efficiently. Because generating annotations and tags is a time-consuming process, it is essential for analyzing videos without human intervention. Although there have been many studies of implicit human-centered tagging using bio-signals, most of them focus on affective tagging and tag relevance assessment. This paper proposes binary and unary classification models that recognize actions meaningful to users in videos, for example jumps in the figure skating program, using EEG features of band power (BP) values and asymmetry scores (AS). As a result, the binary and binary classification models achieved the best balanced accuracies of 52.86% and 50.06% respectively. The binary classification models showed high specificity on non-jump actions and the unary classification models showed high sensitivity on jump actions.

Keywords: classification, human action recognition, video, EEG, signal processing, video annotation, and implicit tagging.

1 Introduction

To provide users with exactly what they want to view, it is important to generate relevant annotations and tags for video clips as well as the whole videos. However, generating annotations and tags with an editing tool manually is a cumbersome and time-consuming process. In addition, some people generate irrelevant tags intentionally for the purpose of increasing the number of video views or just for fun. Therefore, there have been numerous studies of generating metadata for automatic annotation and tagging on videos or images.

Among them, the human-centered approaches for implicit tagging [1] generate metadata for annotations or tags from the viewpoint of users by analyzing their

X. He et al. (Eds.): MMM 2015, Part II, LNCS 8936, pp. 447–457, 2015.
© Springer International Publishing Switzerland 2015

responses while the users viewed the target multimedia or while the users determined the topic relevance of a suggested tag after viewing the target multimedia. The human-centered approaches assume that users show their responses on the viewed multimedia or the suggested tags through facial expressions, body movements, eye gaze, peripheral physiological signals, and brain signals. By analyzing the user responses, the human-centered approaches obtain user states, such as emotion and preference.

Joho and others [2] analyzed the pronounced levels of facial expressions and their change rate. Although facial expressions are collected from a camera in an unobtrusive way, it is difficult even for people to interpret facial expressions because people have neutral ones in general. Peng and others [3] proposed an interest meter component for measuring interest scores of a user by analyzing a captured video containing the user's upper body. Their framework utilizes facial expressions for its emotion model and detects head and eye movements for its attention model. Money and Agius [4] analyzed various peripheral physiological signals including electrodermal response, respiration amplitude, respiration rate, blood volume pulse, and heart rate. Although the peripheral physiological signals delicately reflect changes of user states, compared to their facial expressions and body movements, the proposed method asks subjects to wear four sensor devices for collecting the signals. In addition, their method is useful for videos in the comedy and horror genres, which arouse noticeably significant responses.

There have been many studies of analyzing user responses on videos and images using EEG, which is an electrical signal collected from the brain, because the brain has been considered as the center of cognitive activities. For emotion recognition, the studies provided users with emotion-inducing visual stimuli, which are images or videos [5]-[10]. The studies classified the collected EEG signals into primitive emotion types, which are distributed over the valance-arousal space in the two-dimensional emotion model [11]. The studies of measuring preference proposed the classification method for binary preference, that is like and dislike, on images [12] for four levels of preference classes on videos [13].

Contrary to the most studies that focused on how users feel while viewing images and videos for affective annotation, some of human-centered approaches analyzed their content [10][14][15]. They collected EEG signals while they provided a tag, which was either relevant or irrelevant to the topic of images or videos, to a subject after viewing images or videos. They assumed that tag relevance and irrelevance induce different EEG user responses. Koelstra and Patras [15] showed that assessment task of tag relevance on video also arouse the occurrence of N400 event-related potential (ERP) for irrelevant tags. Soleymani and others achieved tag relevant assessment task using only EEG signals in [10] and using various non-verbal bodily signals including EEG, facial expressions, and eye gaze in [14]. Although the studies on tag relevance assessment using EEG signals can be applied to filtering irrelevant tags, they cannot suggest a tag relevant to the topic of a video.

Therefore, we focus on recognizing meaningful human actions in a video, which are parts important to users and worth to be annotated, by analyzing EEG user responses. Although there have been many studies of recognizing human actions in video sequences using global or local video features in computer vision, they have concentrated on determining the action occurred in the video among candidate actions and there were no significant difference on importance among the actions. Niebles and others proposed unsupervised learning methods of action categories and classified three types of spins, stand-spin, sit-spin, and camel-spin for the figure skating dataset. [16] Contrastively, our approach focuses on *jump* actions in a video for a figure skating program, as shown in Fig. 1, which is a critical element to everyone because most people know the difficulty and execution of the element influence the total scores though they do now know its specific type, grade, and deductions.

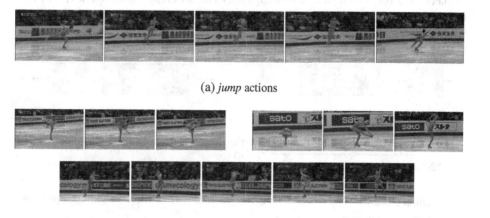

(a) *jump* actions

(b) *non-jump* actions including spins and step sequences

Fig. 1. The short and free programs for ladies single skating competition consist of four elements including jumps, spins, step sequences and spiral sequences. (a) shows the *jump* actions, which are more significant to users than the other actions because most of them know high scores are determined by their perfect executions. The short and free programs for ladies single skating competition is comprised of three and seven *jump* actions respectively. (b) The *non-jump* actions mean all actions performed by a skater except *jump* actions.

To our best knowledge, this is the first work to recognize meaningful actions in a video that arouse different EEG user responses without human's intentional control. We propose binary and unary classification models for recognizing *jump* actions using band power (BP) and asymmetry score (AS) features. In binary classification, the best and second best balanced accuracies of 52.86% (±4.24%) and 52.47% (±1.46%) were achieved by the models based on NB and SVM with RBF kernel using BP features respectively. Most binary classification models showed high specificity on *non-jump* actions and the one class classification models showed high sensitivity on *jump* actions.

2   Materials and Methods

The proposed methods follow the typical procedure for EEG signal analysis including EEG data acquisition, feature extraction, and classification.

2.1   EEG Data Acquisition

EEG data were collected from seven subjects, four males and three females. They we all right-handed and healthy people. Their ages ranged from 26 to 40. They were workers for a research institute. Although they had different knowledge levels of figure skating program, all of them knew the good execution of jumps is directly related with high scores basically.

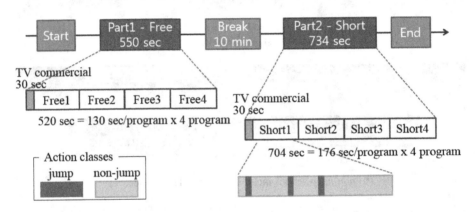

Fig. 2. Video stimuli provided to subjects. The video stimuli were divided into two parts, four free programs and four short programs of ladies single figure skating. The former part included the four latter halves of the free programs. For the former part, each video was extracted from each four-minute free programs. The latter part contained the whole 176-second short programs. Action classes were categorized into *jump* and *non-jump* actions.

In this study, we selected figure skating program videos for recognizing meaningful human actions because most subjects basically know the good execution of jumps is directly related with high scores without learning the specific rules of figure skating programs. In addition, the importance of jump actions is definitely distinguishable from the other actions in the figure skating program. This study chose four short and four free program videos of ladies single performed by four most famous figure skaters in World Championships 2013 as the video stimuli.

While wearing an EEG headset, a subject was instructed to try to maintain a nearly static position during video viewing for reducing artifacts in the collected EEG signals. After an experimenter checked out good signals from all electrodes, the experimenter let the subject view the video stimuli. As shown in Fig. 2, the video stimuli were divided into two parts, four free programs and four short programs. For the former part, the latter halves of four free programs, which took 520 seconds for four

130-second program videos, were provided to the subject successively immediately after 30-second TV commercials. During ten-minute intermediate break, the subject could move freely in the experiment room. After the 30-second TV commercials, four short programs, which took 704-seconds for four 176-second short program video, were provided in a row as the second part of video stimuli.

As shown in Fig. 3, the EEG device collected the EEG signals from the fourteen channels of electrodes placed on the scalp as follows: AF3, AF4, F7, F8, F3, F4, FC5, FC6, T7, T8, P7, P8, O1, and O2. EEG signals were recorded at the sampling rate of 128 Hz. The bandwidth of the recorded EEG signals was from 0.2 Hz to 45 Hz using a 0.16-Hz high-pass filter, an 83-Hz low-pass filter, and 50-Hz and 60-Hz notch filters for removing the power line artifacts. The environmental artifacts caused by the amplifier and aliasing were removed from the hardware of the EEG device.

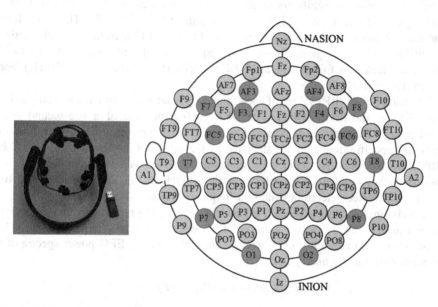

Fig. 3. EEG device and its electrodes positions of the EEG device in international 10-20 system. This figure shows the EEG device consists of a mobile headset and a USB receiver. The EEG signals from electrodes were collected in a desktop through wireless transmission between the EEG headset and the USB receiver. The EEG device used in the experiment collects EEG signals from fourteen electrodes, which represent orange circles in international 10-20 system [17].

2.2   EEG Feature Extraction

As shown in Fig. 2, for both the former and latter part, 30-second TV commercials and the one-second start and end of each video were excluded and 1208 instances for each subject (512 from the former part and 696 from the latter part) were extracted. For classifying jump actions, each second of a program video was assigned to one of two action classes, *jump* and *non-jump*, according to the action class performed by the skater.

For preprocessing, the mean of the EEG signal from each channel was subtracted from the EEG signal for each channel in the baseline removal so that all EEG signal values were distributed to around zero.

The EEG values of each channel were filtered by a Butterworth filter with a pass-band between 4 Hz and 40 Hz, which is an artifact-resilient frequency range. It was required to minimize the side effects of the biological artifact from the live subjects after recording their EEG signals. This study rejected artifact-influenced frequency bands, around 12 Hz by heartbeat, below 4 Hz by eyeball movement, and above 5 Hz by muscle movements [18].

This study adopted one of the most common features for investigating EEG signals, power values at significant frequency bands, which are delta (<4 Hz), theta (4-7 Hz), alpha (8-13 Hz), beta (13-30 Hz), and gamma (>30 Hz) [19]. There is a little variation on the range of each frequency band in the EEG literature. By eliminating the artifact-resilient frequency range, four frequency bands in the range of 4 Hz to 40 Hz were selected as follows: theta (4-7 Hz), alpha (8-13 Hz), beta (13-30 Hz), and gamma (31-40 Hz).

The band power (BP) values of each frequency band were extracted by using a discrete Fourier transform (DFT). The DFT coefficient $X(k)$ of a one-second non-overlapping EEG segment is obtained by fast Fourier transform. The spectral power value at a specific frequency is calculated by the square of the absolute value of $X(k)$ [20]. Each BP value is obtained by summing all power values within the range of each frequency band on a logarithmic scale. Therefore, this study extracted 56 BP features for fourteen electrodes and four frequency bands.

In addition, this study extracted asymmetry scores (AS) resulting from the spectral difference between the power values of EEG electrodes in a symmetrical pair. The features were extracted from (1), wherein the right and left EEG power spectra of a symmetric pair are P_R and P_L in [21].

$$\text{Asymmetry Score} = \ln(P_R) - \ln(P_L) . \tag{1}$$

Therefore, this study extracted 28 AS features for seven symmetrical pairs and four frequency bands.

2.3   Classification

This study investigated on the EEG signals by adopting two classification approaches, binary and one class classification. The binary classifiers separate video stimuli into specific actions and the other actions, for example, *jump* and *spin* actions, by learning from the training set including both of the action instances. Contrary to the binary classifiers, one class classifiers distinguish the specific actions, such as *jump* actions, from the others, such as *non-jump* actions, by learning the training set including only the specific action instances. The one class classification, which has been applied to novelty detection, anomaly detection, and outlier detection, identifies instances of a specific class.

In binary classification, this study employed four classifiers for recognizing *jump* and *non-jump* actions, which included the k nearest neighbors(k-NN), a neural network (NN), a naive Bayesian classifier (NB), and a support vector machine (SVM) with a radial basis function (RBF) kernel. We compared the eight classification models based on the four classifiers using either BP or AS features.

The k nearest neighbor algorithm (k-NN) is an intuitive method for assigning a test instance to the dominant class among its k nearest neighbors in Euclidean distance through the majority vote [22]. This study selected a value of 4 for k, which is a small positive integer, because the curse of dimensionality in k-NN is considered. The artificial neural networks are computational models, which were inspired by brains, presented as systems of interconnected neurons, which can compute values form inputs [23]. This study adopted a feed-forward neural network (NN) with a single hidden layer. The number of units in the hidden layer was set to ten and the maximum number of iterations was set to 300. The naive Bayesian classifier (NB) is one of probabilistic classifiers based on Bayes' theorem with strong independence assumption between the features [24]. An SVM is one of the most popular supervised learning techniques for classification and regression [25]. An SVM constructs with an N-dimensional hyperplane for minimizing the classification errors on the test set and maximizing its margins. This study set 50 and 0.01 to cost and gamma parameters respectively.

In one class classification, this study employed one class SVM (OCSVM) [26], which finds the smallest sphere including all training data, using linear and BRF kernels. The value for nu parameter was set to 0.2 by tuning the values ranged from 0.1 to 0.5 for balanced accuracies. The gamma parameter was set to 0.01 for both linear and RBF kernels, which is the same value used in the binary SVM. We compared the four unary classification models based on the two classifiers based on SVMs with linear and RBF kernels using BP and AS features.

In both the binary and unary classification, the classification accuracy of a model based on each classifier using each feature set was measured by leave-one-subject-out (LOSO) cross validation. The LOSO cross-validation scheme partitions instances into *k* subsets according to *k* subjects. One subset for a subject is used to validate the generated model trained by the other *k–1* subsets for the other subjects. The process is repeated *k* times. The accuracy is obtained by averaging the accuracies from *k* models.

Table 1. Number of instances used in the training and testing for binary classification

Target actions	# of instances			
	Training		Testing	
	Positive class	Negative class	Positive class	Negative class
Jump	924	6324	154	1054

Total 8456 instances from seven subjects (1208 instances/subject x 7 subjects) were used both for training and predicting, as shown in Table 1. In the experiment, 1208 instances, which consisted of 512 and 696 instances extracted from the former and latter part, were obtained from each subject respectively. As shown in Table1, the instances for jump and non-jump actions were unbalanced because the meaningful jump actions occupy a small portion of the whole video stimuli of figure skating programs. This study, therefore, measured sensitivity, specificity, balanced accuracy as well as accuracy of each classification model.

3   Experiment Results

In binary classification, the classification result of recognizing *jump* actions is shown in Table 2. The best and second best balanced accuracies of 52.86% (±4.24%) and 52.47% (±1.46%) were achieved by the models based on NB and SVM-RBF using BP features respectively. However, considering the classifiers used, there was no significant difference ($p < 0.05$) among balanced accuracies of the models. Considering the type of features, the models using AS features achieved less balanced accuracies than the models using BP features.

Compared the balanced accuracies of all the models, their accuracies were high because the models achieved high specificity on non-jump actions, which occupy a large portion of the whole video stimuli. This was the reason the accuracy of the model based on NB classifier using BP features showed the worst accuracy. Most binary classification models showed high specificity and low sensitivity irrespective of both classifiers and features because they were trained by the unbalanced training set biased to *non-jump* actions.

Table 2. Result of eight classification models classifying *jump* and *non-jump* actions in binary classification

Classifiers		Sensitivity	Specificity	Bal. Accuracy	Accuracy
k-NN	BP	10.02(±2.73)	90.92(±2.16)	50.47(±1.60)	80.61(±1.87)
	AS	8.07(±3.47)	90.96(±1.09)	49.52(±1.80)	80.39(±1.03)
NN	BP	14.01 (±5.71)	88.11(±3.63)	51.06(±3.26)	78.67(±3.20)
	AS	8.63(±2.04)	92.30(±2.62)	50.46(±1.93)	81.63(±2.40)
NB	BP	31.73(±11.23)	74.00(±13.78)	**52.86**(±4.24)	68.61(±10.93)
	AS	7.14(±3.09)	94.33(±2.23)	50.74(±0.92)	83.22(±1.65)
SVM-RBF	BP	11.22(±5.24)	93.72(±3.33)	**52.47**(±1.46)	83.21(±2.35)
	AS	5.66(±2.53)	94.05(±2.47)	49.85(±1.54)	82.78(±2.10)

In one class classification, the classification result of identifying *jump* actions is shown in Table 3. the models based on OCSVM with a linear kernel using AS features and OCSVM with an RBF kernel using BP features achieved the best and second best balanced accuracies of 50.06% (±2.45%) and 49.79% (±1.84%). Although there were no significant difference ($p < 0.05$) among the models based on OCSVMs with RBF and linear kernels considering the classifiers, the model based on

OCSVM with an RBF kernel showed the big gap between sensitivities and specificities contrary to the model based on OCSVM with a linear kernel.

Contrary to the result of binary classification, the balanced accuracies of all the models were better than their accuracies because the models achieved high sensitivities on *jump* actions. However, their low specificities deteriorated the accuracies. Compared to the binary classification models, one class classification models performed similar balanced accuracies and low accuracies because their speficitieied high sensitivities and low specificities irrespective of both classifiers and features because they were trained by the training set including only *jump* actions. Although the one class classification models showed high sensitivities,

Table 3. Result of four classification models identifying *jump* actions in one class classification

Classifiers		Sensitivity	Specificity	Bal. Accuracy	Accuracy
OCSVM	BP	74.86($\pm$7.45)	24.72($\pm$7.18)	**49.79**($\pm$1.84)	31.11($\pm$5.46)
-RBF	AS	76.72($\pm$9.06)	20.47($\pm$5.73)	48.59($\pm$2.45)	27.64($\pm$4.02)
OCSVM	BP	52.41($\pm$7.85)	45.68($\pm$6.22)	49.04($\pm$1.84)	46.54($\pm$5.46)
-Linear	AS	57.51($\pm$8.37)	42.60($\pm$6.30)	**50.06**($\pm$2.45)	44.50($\pm$4.02)

4   Conclusion

This paper proposed the classification models for recognizing meaningful actions in videos annotations and tags. To the best of our knowledge, it is the first study to classify meaningful actions performed in videos using band power (BP) and asymmetry score (AS) features extracted from EEG user responses. As a result, the best and second best balanced accuracies of 52.86% (±4.24%) and 52.47% (±1.46%) were achieved by the models based on NB and SVM-RBF using BP features respectively in binary classification by learning from the training set including *jump* and *non-jump* actions. Most binary classification models showed high sensitivities because they were trained by the training set balanced to the *non-jump* actions. In addition, the models based on OCSVM with a linear kernel using AS features and OCSVM with an RBF kernel using BP features achieved the best and second best balanced accuracies of 50.06% (±2.45%) and 49.79% (±1.84%). Most one class classification models showed high sensitivity compared to the binary classification models but low specificity because they were trained by the training set including only *jump* actions. Considering the balanced accuracies of the models for the both binary and one class classification, the classification performance for recognizing *jump* actions was not satisfactory.

As a future work, we are going to devise an ensemble method to use high sensitivity from one class classification and high specificity from binary classification. In addition, we are going to extended experiment with more subjects to extinguish the accuracies of binary and one class classification models based on different classifiers and using different features.

Acknowledgments. This work was supported by the IT R&D program of MSIP/IITP, [2014-044-020-001, Development of High Performance Visual BigData Discovery Platform for Large-Scale Realtime Data Analysis].

References

1. Soleymani, M., Pantic, M.: Human-centered implicit tagging: Overview and perspectives. In: IEEE International Conference on Systems, Man, and Cybernetics, pp. 3304–3309. IEEE Press, New York (2012)
2. Joho, H., Jose, M.J., Valenti, R., Sebe, N.: Exploiting Facial Expressions for Affective Video Summarisation. In: ACM International Conference on Image and Video Retrieval, Article No. 31, pp. 1–8. ACM, Santorini (2009)
3. Peng, W.-T., et al.: Editing by Viewing: Automatic Home Video Summarization by Viewing Behavior Analysis. IEEE Transaction on Multimedia 13(3), 539–550 (2011)
4. Money, A.G., Agius, H.: ELVIS: Entertainment-Led Video Summaries. ACM Trans. Multimedia Computing, Communications, and Applications, 17:1–17:30 (2010)
5. Schaaff, K., Schultz, T.: Towards Emotion Recognition from Electroencephalographic Signals. In: 3rd IEEE International Conference on Affective Computing and Intelligent Interaction, pp. 1–6. IEEE Press, New York (2009)
6. Liu, Y., Sourina, O., Nguyen, M.K.: Real-Time EEG-Based Human Emotion Recognition and Visualization. In: International Conference on Cyberworlds, pp. 262–269. IEEE Press, New York (2010)
7. Nie, D., Wang, X.W., Shi, L.C., Lu, B.L.: EEG-based Emotion Recognition during Watching Movies. In: 5th International Conference on Neural Engineering, pp. 186–191. IEEE Press, New York (2011)
8. Koelstra, S., et al.: Single Trial Classification of EEG and Peripheral Physiological Signals for Recognition of Emotions Induced by Music Videos. In: Yao, Y., Sun, R., Poggio, T., Liu, J., Zhong, N., Huang, J. (eds.) BI 2010. LNCS, vol. 6334, pp. 89–100. Springer, Heidelberg (2010)
9. Yazdani, A., et al.: Affect Recognition Based on Physiological Changes during the Watching of Music Videos. ACM Trans. Interactive Intelligent System 2(1), 7:1–7:26 (2012)
10. Soleymani, M., Pantic, M.: Multimedia Implicit Tagging using EEG Signals. In: IEEE International Conference on Multimedia and Expo. IEEE Press, New York (2013)
11. Russell, J.A.: A Circumplex Model of Affect. Journal of Personality and Social Psychology 39(6), 1161–1178 (1980)
12. Aurup, G.M.M.: User Preference Extraction from Bio-signals: An Experimental Study. Master's thesis. Concordia University (2011)
13. Moon, J., Kim, Y., Lee, H., Bae, C., Yoon, W.C.: Extraction of User Preference for Video Stimuli Using EEG-Based User Responses. ETRI Journal 35(6), 1105–1114 (2013)
14. Soleymani, M., Kaltwang, S., Pantic, M.: Human behavior sensing for tag relevance assessment. In: The 21st ACM International Conference on Multimedia, pp. 657–660. ACM, New York (2013)
15. Koelstra, S., Patras, I.: EEG analysis for implicit tagging of video data. In: Workshop on Affective Computing and Intelligent Interaction, pp. 1–6 (2009)
16. Niebles, J.C., Wang, H., Fei-Fei, L.: Unsupervised Learning of Human Action Categories Using Spatial-Temporal Words. International Journal of Computer Vision 79(3), 299–318 (2008)

17. Malmivou, J., Plonsey, R.: Bioelectromagnetism: Principles and Applications of Bioelectric and Biomagnetic Fields, 1st edn. Oxford University Press, New York (1995)
18. Petrantonakis, P.C., Hadjileontiadis, L.J.: Emotion Recognition from Brain Signals Using Hybrid Adaptive Filtering and Higher Order Crossings Analysis. IEEE Trans. Affective Computing 1(2), 81–97 (2010)
19. Sörnmo, L., Laguna, P.: Bioelectrical Processing in Cardiac and Neurological Applications, 1st ed. Elsevier Academic Press, Waltham (2005)
20. Iscana, Z., Dokura, Z., Demiralp, T.: Classification of Electroencephalogram Signals with Combined Time and Frequency Features. Expert Systems with Applications 38(8), 10499–10505 (2011)
21. Allen, J.J., et al.: The Stability of Resting Frontal Electroencephalographic Asymmetry in Depression. Psychophysiology 41(2), 269–280 (2004)
22. Altman, N.S.: An introduction to kernel and nearest-neighbor nonparametric regression. The American Statistician 46(3), 175–185 (1992)
23. Artificial neural networks,
    http://en.wikipedia.org/wiki/Artificial_neural_networks
24. Naive Bayes classifier,
    http://en.wikipedia.org/wiki/Naive_Bayes_classifier
25. Support vector machine,
    http://en.wikipedia.org/wiki/Support_vector_machine
26. Schölkopf, B., Platt, J.C., Shawe-Taylor, J., Smola, A.J., Williamson, R.C.: Estimating the support of a high-dimensional distribution. Neural Computation 13(7), 1443–1471 (2001)

Challenging Issues in Visual Information Understanding Researches

Kyuchang Kang, Yongjin Kwon, Jinyoung Moon, and Changseok Bae

Human Computing Research Section, SW · Content Research Laboratory,
ETRI, 218 Gajeongno Yuseong-gu Daejeon, 305-700, South Korea
{k2kang,scocso,jymoon,csbae}@etri.re.kr

Abstract. Visual information understanding is known as one of the most difficult and challenging problems in the realization of machine intelligence. This
paper presents research issues and overview of the current state of the art in the
general flow of visual information understanding. In general, the first stage of
the visual understanding starts from the object segmentation. Using the saliency
map based on human visual attention model is one of the most promising methods for object segmentation. The next step is scene understanding by analyzing semantics between objects in a scene. This stage finds description of image
data with a formatted text. The third step requires space understanding and context awareness using multi-view analysis. This step helps solving general occlusion problem very easily. The final stage is time series analysis of scenes and a
space. After this stage, we can obtain visual information from a scene, a series
of scenes, and space variations. Various technologies for visual understanding
already have been tried and some of them are matured. Therefore, we need to
leverage and integrate those techniques properly from the perspective of higher
visual information understanding.

Keywords: Object segmentation, scene understanding, space understanding,
context awareness, time series analysis.

1 Introduction

Recent progresses of mobile computing devices and network infrastructure are changing aspects of our daily lives considerably. These environment of progresses enable to
produce and consume visual data, such as images or videos, very conveniently. We
can easily take pictures with our mobile phones, and share them with our friends using social network services. Further, we are living with plenty of CCTV cameras
which are watching most of our daily lives. Recent studies have reported image or
video data will occupy about 60% of annual digital data production in 2020 [1, 2].

Now, we can call this trend of explosion of image or video data as we are living in
the age of visual bigdata. As we all well know, visual data carries lots of information
on it. There are lots of researches to understand visual data on our computing
environment. However, the understanding of visual information in images or videos

X. He et al. (Eds.): MMM 2015, Part II, LNCS 8936, pp. 458–469, 2015.
© Springer International Publishing Switzerland 2015

still has lots of challenging issues. Recent advances in computing and networks technologies give great promises in high level image or video understanding.

In this paper, we present our studies about current research issues and state of the art in each step of the general visual data understanding flow. Fig. 1 illustrates basic steps required in visual data understanding. First, we apply preprocessing algorithms to extract bags of features which are useful in the next stages. SIFT (Scale-Invariant Feature Transform) [3] is a widely used algorithm for detecting and describing local features. Second, we need to separate objects from backgrounds. Saliency map based on a human visual attention model is one of the promising method in this step [4]. In third step, we identify what kinds of objects are in a scene. There are plenty of classifiers and machine learning algorithms for recognizing objects in a scene. In the next stage, we can describe the situation or context of a scene by analyzing semantics between objects within the scene. There are lots of recent progresses including I2T (Image Parsing to Text Description) [5]. In the fifth stage, we can understand and describe a space by analyzing the understanding results of multi-view scenes. A research to understand structure of world famous hot spots through the world's photos is a good example of this trial [6]. In the final stage, we can understand a space and its changes in time series. After this stage, we can find and describe high level knowledge from images or videos.

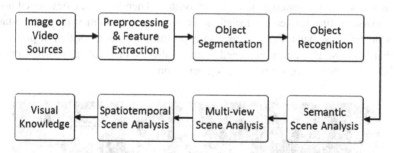

Fig. 1. The general flow of visual data understanding process

Recent progresses in every step of visual data understanding make it possible to develop a method to process whole steps comprehensively. According to this stream, there are big projects such as VIRAT (Video and Image Retrieval and Analysis Tool) [7] or Mind's Eye [8]. This paper presents recent progresses and prospect of every stages in video data understanding. This paper is organized as follows. Section 2 addresses research issues in visual data understanding such as object segmentation, scene understanding, space understanding and context awareness, time series analysis of scenes and a space. Section 3 summarize the recent progresses.

2   Research Issues in Visual Data Understanding

This section describes each challenging issues and their related previous works in every step of visual data understanding.

2.1   Object Segmentation

The first stage of the visual understanding starts from the object segmentation. Using the saliency map based on human visual attention model is one of the most promising method for object segmentation. The saliency map model generates the most salient and interesting region in the visual scene. This model conventionally leverages one of the cognitive characteristics of the human visual system such as selective perceptron [9], which means that human eyes pay more attention to the regions with visual saliency.

Currently various application of image processing models the selective perceptron of the human visual system and leverages it in image segmentation [10], object recognition [11], and content-aware image resizing [12].

J. S. Kim, *et al.* [13] proposed a graph-based multi-scale saliency-detection algorithm by modeling eye movements as a random walk on a graph. First, this algorithm extracts intensity, color, and compactness features from an input image. Then, it constructs a fully connected graph by employing image blocks as the nodes. If the two connected nodes have dissimilarities in intensity or color features and the ending node is more compact than the starting node, a high edge weight is assigned in the algorithm. Based on this assignment, the algorithm computes the stationary distribution of the Markov chain on the graph as the saliency map. However, the relative block size in an image influences the performance of the saliency detection algorithm. Therefore, they developed and applied a coarse-to-fine refinement technique based on the random walk with restart for multi-scale saliency maps. Fig. 2 shows an application of this algorithm in proto-object extraction [14], which can be extracted from a saliency map, are primitive forms in a scene before the object recognition and segmentation.

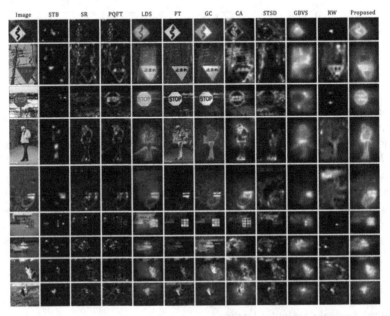

Fig. 2. Comparison of the proto-object extraction performance [13]

H. Lee *et al.* [4] proposed an adaptive extraction of focus of attention (FOA) region for saliency-based visual attention. This method determine the most salient point by checking every value in saliency map, and expand the neighborhood of the point until the average value of the neighborhood is smaller than 75% value of the most salient point, and then find the contour of the neighborhood.

From the input images, the visual feature maps that contribute to attentive selection of stimulus are derived and combined into one single topographically oriented map, the saliency map which integrates the normalized information from the individual feature maps. Then, they find the maximum point from the saliency map and expand the neighborhood of point until the average value of the neighborhood is expandable up to 75% of the maximum point. Finally, the contour of the neighborhood is an area of FOA

Fig. 3 describes the overall architecture of adaptive FOA region extraction for saliency-based visual attention. An input image is decomposed into a set of feature map which extract local spatial discontinuities in the modalities of color, intensity and orientation [15]. All the feature maps are then combined into a unique scalar saliency map which is topographically encoded for conspicuity at every location in the input images [15]. Finally the FOA is determined through adaptive FOA extraction algorithm.

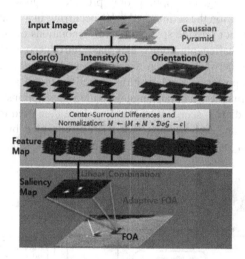

Fig. 3. Overall architecture of adaptive FOA region extraction for saliency-based visual attention [4]

In addition, various studies related to the salient region detection have been developed. Itti *et al.* [16] introduced a visual attention model based on center-surround concept. Costa [17] and Harel *et al.* [18] proposed a graph-based saliency detection algorithm.

2.2   Scene Understanding

Scene understanding requires analysis of semantics between objects in a scene. This stage finds description of image data in a text format which is structured as extended markup language (XML) or Web ontology language (OWL).

If the image space is transformed as a text space and its relation, we can apply the conventional and popular text processing techniques used in text mining area.

Scene understanding technology is the most important component for the visual information understanding research area. Many previous works have been performed for scene understanding, especially, activity recognition, in computer vision research area.

On the other hand, in the point of scene characteristics, scene understanding means action, activity, or interaction embedded in the scene. Complex events can be thought of as compositions of atomic actions performed by people holding different roles. The use of action and/or role models trained with extensive spatiotemporal annotations have shown to boot event recognition performance in videos [19, 20].   Such detailed annotations require expensive human efforts and severely restrict the scalability with the inclusion of more actions and roles.

V. Ramanathan *et al.* [21] proposed a method to learn such models based on natural language descriptions of the training videos. There are two challenging points. One is that natural language descriptions only provide a high-level summary and often do not directly mention the actions and roles occurring in a video. The other is that these descriptions do not provide spatiotemporal annotations of actions and roles. To overcome these two challenging points, they have introduced a topic-based semantic relatedness measures between a video description and an action and role label, and incorporate it into a posterior regularization objective. Fig. 4 shows an overview of their system. They first use natural language video descriptions to train action and role models. The prediction scores from the model are then used to train event recognition models. In setup, each training video is accompanied by a natural language description, which might or might not contain the action label present in the video.

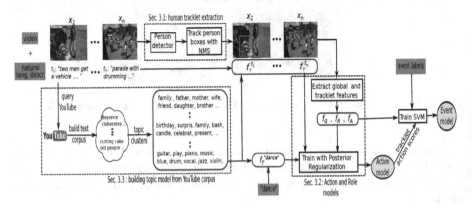

Fig. 4. An overview of a method to learn models based on natural language descriptions of the training data [21]

B. J. Yao *et al.* [5] proposed an image parsing framework that generates text description of image and video content based on image understanding. They called it as image to text (I2T) framework, consisting of four key components: (1) an image parsing engine that converts input images or video frames into parse graphs. (2) An And-or Graph visual knowledge representation that provides top-down hypotheses during image parsing and serves as an ontology when converting parse graphs into semantic representations in RDF format. (3) A general knowledge base embedded in the Semantic Web that enriches the semantic representations by interconnecting several domain specific ontologies. (4) A text generation engine that converts semantic representations into human readable and query-able natural language descriptions. Fig. 5 illustrates a diagram of the I2T framework and operational flows.

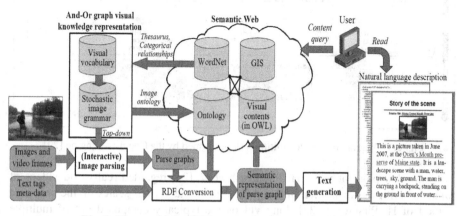

Fig. 5. Diagram of the I2T framework and operational flows [5]

A. Oltramari *et al.* [22] proposed an integrated cognitive engine for automatic video surveillance. They focused on the task of classifying the actions occurring in a scene. To do this, they developed a semantic infrastructure on top of a hybrid computational ontology of actions. They aimed to approximate human visual intelligence in making effective and consistent detections: humans evolved by learning to adapt and properly react to environmental stimuli, becoming extremely skilled in filtering and generalizing over perceptual data, taking decisions and acting on the basis of acquired information and background knowledge.

In contrast with the conventional machine vision technology which focused on progressing in recognizing a wide range of objects and their properties such as nouns in the description of a scene, they focused on the perceptual and cognitive underpinnings for recognizing and reasoning about the verbs in those scenes, enabling a more compete narrative of action in the visual experience.

Fig. 6 shows a diagram of the recognition task performed by the Cognitive Engine. The horizontal black arrow represents the sequence time framing while the vertical one represents the interconnected levels of information processing. The Cognitive Engine finds semantic disambiguation of the scene elements, and figures out them with predefined schema. The light-green box displays the results of semantic disambiguation of the scene elements, while the gray box contains the schema of the output, where importance reflects the number of components in a detected pattern and observed is a boolean parameter whose value is 1 when a verb matches an immediate activity recognition detection and 0 when the verbs is an actual result of extended activity reasoning processing.

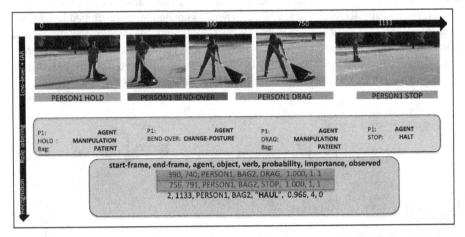

Fig. 6. A diagram of the recognition task performed by the Cognitive Engine [22]

From other point of view, H. Pirsiavash *et al.* [23] regarded real-world videos of human activities as exhibition of temporal structure at various scales. In the proposed method of H. Pirsiavash [23], long videos are typically composed out of multiple action instances, where each instance itself is composed of sub-actions with variable durations and orderings. They describe simple grammars that capture hierarchical temporal structure while admitting inference with a finite-state-machine. This makes parsing linear time, constant storage, and naturally online.

A. Gupta *et al.* [24] considered that analyzing videos of human activities involves not only recognizing actions, typically based on their appearances, but also determining the story or plot of the video. In the proposal of A. Gupta [24], the storyline of a video describes causal relationships between actions. This approach is to learn a visually grounded storyline model of videos directly from weakly labeled data. The storyline model is represented as an And-Or graph. They formulate an Integer Programming framework for action recognition and storyline extraction using the storyline model and visual groundings learned from training data.

2.3   Space Understanding and Context Awareness

The third step requires space understanding and context awareness using multi-view analysis. This step helps to solve general occlusion problem very easily through multi-view geometry analysis such as epipolar constraints, calculating fundamental matrix and finding matching points from different viewpoints.

S.Y. Bao *et al.* [25] proposed a framework for robustly understanding the geometrical and semantic structure of a cluttered room from a small number of images captured from different viewpoints. They addressed the tasks such as estimating the 3D layout of the room and identifying and localizing all the foreground objects in the room. In the analysis, they use multi-view geometry constraints and image appearance to identify the best room layout configuration.

Fig. 7 shows multi-image room layout understanding framework summarized as the flow chart. The feature points play a role to create the 3D reconstruction of the points in rooms and help estimate camera parameters. And a structure-from-motion (SFM) pipeline is used to estimate a set of 3D points in the scene. The region segments are critical for evaluating the possibility of a room hypothesis. Therefore, the region segments should be matched across images.

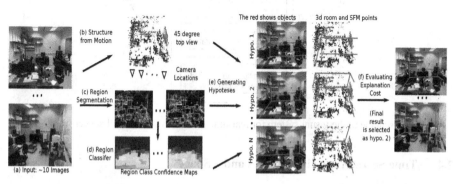

Fig. 7. Multi-image room layout understanding framework based on the geometrical and semantic structure of a cluttered room [25]

M. Leo *et al.* [26] proposed real-time multi-view analysis of soccer matches. In general, soccer video analysis requires different challenging tasks which ball and players have to be localized in each frame, tracked over time and, their interactions have to be detected and analyzed. Frequent occlusion of ball and players is one of main problems in soccer video analysis. In this proposal, 3D ball trajectories are extracted by triangulation from multiple cameras and used to detect the interactions between the players and the ball.

Fig. 8 shows the scheme of the visual system. All data are collected and synchronized by the central supervisor. Each node performs motion detection, players and ball detection and tracking and players classification. Extracted information are then sent to the supervisor that retrieves the 3D ball trajectories by triangulation from multiple sources, projects ball and players positions on a virtual play-field and detects interactions among players and ball by temporally analyzing variations in ball trajectories onto the virtual play-field. Moreover the supervisor unit infers temporal and spatial localization of the detected interactions and determines which players interacted with the ball.

These results are very encouraging for building an automatic system for complex event detection if the rules for action or event are provided in sports, where it is strategic the exact localization of the players in the field, but also the precise detection of the frame in which the shot was recognized.

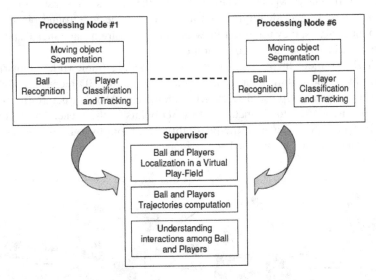

Fig. 8. The scheme of the visual system for real-time multi-view analysis of soccer matches [26]

2.4   Time Series Analysis of Scenes and a Space

The final stage is time series analysis of scenes and a space. This step helps to fine understanding of a scene. Several studies [27 - 29] based on vision-based trajectory learning allowing for time series analysis of scenes and space. The idea of these is that trajectory data can define general set of activities, applicable to a wide range of scenes.

U. Gaur *et al.* [27] proposed analysis of activities in low-resolution videos or far fields which is a research challenge not receiving much attention. In this application scenario, it is often the case that the motion of the objects in the scene is the only low-level information available, other features like shape or color being unreliable. However, this proposal [27] provided a method to classify activities of multiple interacting objects in low-resolution video by modeling them through a set of discriminative features which rely only on the object tracks. The core of the proposed method involves a transformation of the noisy tracks of the multiple objects into a motion feature space that encapsulates the individual characteristics of the tracks, as well as their interactions.

Each individual track is represented by its gradients as a function of time. Each pair of tracks is represented by the relative distance between the components as a function of time. By considering multiple pairs, interactions between more than two objects can be modeled in an iterative manner. Thus each pair of tracks is now represented by a multi-dimensional feature vector that, at each time instant, consists of the gradients of each track and the distance between the tracks.

B.T. Morris *et al.* [30] summarized trajectory-based activity analysis for visual surveillance. Generally, high-level activity analysis requires large amounts of domain knowledge while low-level analysis assumes very little. In this point, trajectory dynamics analysis provides a medium between low- and high-level analysis. Trajectory analysis can predict the next activity and detect abnormality if the some kinds of rules are provided.

Fig. 9 shows the example of the abnormality detection and prediction of proper action.

(a) Trajectory analysis for abnormality detection

(b) Trajectory analysis for left turn prediction at intersection showing the probability of the top three best paths

Fig. 9. Examples of trajectory-based activity analysis for visual surveillance [30]

3   Concluding Remarks

Video data understanding is one of the hottest issues in the data processing researches and there are lots of challenging problems. In this paper, we have reviewed recent progresses in visual data, such as images or videos, understanding processes including segmentation, recognition, semantic analysis, multi-view analysis, and spatiotemporal analysis. We can easily find an excellent research result in a step leads progress of the neighbor steps in visual data understanding. This kind of consecutive good effect will make it possible to understand visual information on the visual data in a very high level.

In addition, we can consider today's tremendous volume of visual data as video bigdata. In order to find out useful meanings from this video bigdata, lots of studies are under way for trying to analyze it on the bigdata analytics platform.

Acknowledgement. This work was supported by the IT R&D program of MSIP/IITP, [2014-044-020-001, Development of High Performance Visual BigData Discovery Platform for Large-Scale Realtime Data Analysis].

References

1. IDC, Digital Universe Study: Big Data, Bigger Digital Shadows, and Biggest Growth in the Far East. IDC (December 2012)
2. Riding the multimedia big data wave, SIGIR 2013, Keynote (July 2013)
3. Lowe, D.G.: Object recognition from local scale-invariant features. In: Proceedings of the International Conference on Computer Vision, vol. 2, pp. 1150–1157 (1999)
4. Lee, H., Bae, C., Lee, J., Sohn, S.: Adaptive FOA Region Extraction for Sailency-Based Visual Attention. IJIPM: International Journal of Information Processing and Management 3(3), 36–43 (2012)
5. Yao, B.Z., Yang, X., Lin, L., Lee, M.W., Zhu, S.C.: I2T: Image Parsing to Text Description. Proceedings of the IEEE 98(8), 1485–1508 (2010)
6. Snavely, N., Garg, R., Seitz, S.M., Szeliski, R.: Finding Paths through the World's Photos. In: Proceedings of ACM Sigraph 2008, vol. 27(3) (August 2008)
7. DARPA, Video and Image Retrieval and Analysis Tool (VIRAT), Broad Agency Announcement (March 3, 2008)
8. DARPA, Mind's Eye Program, Broad Agency Announcement (March 25, 2010)
9. Treisman, A.M., Gelade, G.: A Feature-Integration Theory of Attention. Cognitive Psychology 12(1), 97–136 (1980)
10. Ko, B.C., Nam, J.-Y.: Object-of-Interest Image Segmentation based on Human Attention and Semantic Region Clustering. JOSA A 23(10), 2462–2470 (2006)
11. Rutishauser, U., Walther, D., Koch, C., Perona, P.: Is Bottom-up Attention Useful for Object Recognition. In: Proceedings of the 2004 IEEE CVPR, pp. II-37–II-44 (2004)
12. Kim, J.S., Kim, J.H., Kim, C.S.: Adaptive Image and Video Retargeting Technique based on Fourier Analysis. In: Proceedings of the 2009 IEEE CVPR, pp. 1730–1737 (2009)
13. Kim, J.S., Sim, J.Y., Kim, C.S.: Multiscale Saliency Detection Using Random Walk with Restart. IEEE Transactions on Circuits and Systems for Video Technology 24(2), 198–210 (2014)
14. Walther, D., Koch, C.: Modeling Attention to Salient Proto-Objects. Neural Network 19(9), 1395–1407 (2006)
15. Caduff, D., Sabine, T.: On the Assessment of Landmark Salience for Human Navigation. Cognitive Processing 9(4), 249–267 (2008)
16. Itti, L., Koch, C., Niebur, E.: A Model of Saliency-Based Visual Attention for Rapid Scene Analysis. IEEE Transactions on Pattern Analysis and Machine Intelligence 20(11), 1254–1259 (1998)
17. Costa, L.D.F.: Visual Saliency and Attention as Random Walks on Complex Networks. arXiv preprint physics/0603025 (2006)
18. Harel, J., Koch, C., Perona, P.: Graph-Based Visual Saliency. Advances in Neural Information Systems, 545–552 (2006)
19. Izadinia, H., Shah, M.: Recognizing Complex Events using Large Margin Joint Low-Level Event Model. In: Fitzgibbon, A., Lazebnik, S., Perona, P., Sato, Y., Schmid, C. (eds.) ECCV 2012, Part IV. LNCS, vol. 7575, pp. 430–444. Springer, Heidelberg (2012)

20. Lan, T., Sigal, L., Mori, G.: Social Roles in Hierarchical Models for Human Activity Recognition. In: Proceeding in Computer Vision and Pattern Recognition (CVPR), pp. 1354–1361 (2012)
21. Ramanathan, V., Liang, P., Fei-Fei, L.: Video Event Understanding using Natural Language Descriptions. In: Conference on Computer Vision (ICCV), pp. 905–912 (2013)
22. Oltramari, A., Lebiere, C.: Using Ontologies in a Cognitive-Grounded System: Automatic Action Recognition in Video Surveillance. In: Conference on Semantic Technology for Intelligence, Defense, and Security (2013)
23. Pirsiavash, H., Ramanan, D.: Parsing videos of actions with segmental grammars. In: Conference on Computer Vision and Pattern Recognition (CVPR) (2014)
24. Gupta, A., Srinivasan, P., Shi, J., Davis, L.S.: Understanding Videos, Constructing Plots: Learning a Visually Grounded Storyline Model from Annotated Videos. In: Conference on Computer Vision and Pattern Recognition (CVPR), pp. 1063–6919 (2009)
25. Bao, S.Y., Furlan, A., Fei-Fei, L., Savarese, S.: Understanding the 3D Layout of a Cluttered Room From Multiple Images. In: IEEE Winter Conference on Applications of Computer Vision (WACV), pp. 690–697 (2014)
26. Leo, M., Mosca, N., Spagnolo, P.: Real-Time Multiview Analysis of Soccer Matches for Understanding Interactions between Ball and Players. In: Conference on Content-based Image and Video Retrieval, pp. 525–534. ACM (2008)
27. Gaur, U., Song, B., Roy-Chowdhury, A.K.: Query-based Retrieval of Complex Activities using Strings of Motion-Words. In: IEEE Workshop on Motion and Video Computing (WMVC), pp. 1–8 (2009)
28. Bashir, F.I., Khokhar, A.A., Schonfeld, D.: Object Trajectory-based Activity Classification and Recognition using Hidden Markov Models. IEEE Transactions on Image Process 16(7), 1912–1919 (2007)
29. Piciarelli, C., Foresti, G.L.: On-line Trajectory Clustering for Anomalous Events Detection. Pattern Recognition Letters 27(15), 1835–1842 (2006)
30. Morris, B.T., Trivedi, M.M.: A Survey of Vision-Based Trajectory Learning and Analysis for Surveillance. IEEE Transactions on Circuits and Systems for Video Technologies 18(8), 1114–1127 (2008)

Emotional Tone-Based Audio Continuous Emotion Recognition

Mengmeng Liu, Hui Chen, Yang Li, and Fengjun Zhang

Beijing Key Lab of Human-Computer Interaction
Institute of Software, Chinese Academy of Sciences, Beijing, China
liumengmeng12@mails.ucas.ac.cn, liyang19880204@163.com,
{chenhui,fengjun}@iscas.ac.cn

Abstract. Understanding human emotions in natural communication is still a challenge problem to be solved in human-computer interaction. Emotional tone that people feel within a period of time can affect the way people communicate with others or environment. In this paper, a new emotional tone-based two-stage algorithm for continuous emotion recognition from audio signals is presented. Gaussian mixture models of hidden Markov models (GMM-HMMs) are employed to infer the dimensional emotional tone and affect labels. Two emotional tones, positive or negative, which represent the overall emotion state over an audio clip are first obtained. Then, based on that emotional tone, corresponding positive or negative GMM-HMM classifier is refined to finish the continuous emotion recognition. The experimental results show that our method outperforms the GMM-HMM and SVR in baseline for the Audio-Visual Emotion Challenge (AVEC 2014) database [1].

Keywords: Audio Emotion Recognition, Human-Computer Interaction.

1 Introduction

A widespread consensus about the future human-computer interaction (HCI) is that the interaction will need to be human-centered, build for humans, instead of computer-centered [2]. In the traditional HCI, computer initiate communications simply responding to the user's commands while ignoring implicit information about the user, such as changes in the emotional state. Yet, a change in the user's emotional states is a fundamental component of human-human communication. Some emotional states inspire human actions, and others enrich the meaning of human communication. This means that the traditional HCI, which ignores the user's emotional state, leaves out a large portion of information originated in the communication process. Thereby, a system which can detect and track a user's emotional states is of great significance to improve the human-computer interaction.

Audio signals convey a lot of emotional information, being an important communicative modality in human-human interaction. Emotion recognition based on audio signals is of great importance in human-computer interaction. Earlier work of emotion recognition often focused on recognition of a subset of basic emotions, e.g., anger, disgust, fear, happiness, sadness, and surprise. However, this kind of categorical

X. He et al. (Eds.): MMM 2015, Part II, LNCS 8936, pp. 470–480, 2015.

approach cannot describe the non-basic, subtle and rather complex affective states like thinking, embarrassment or depression. In other words, a small number of discrete categories (e.g., happiness and sadness) may not reflect the complexity and subtlety of the affective state [3]. Consequently, there is a trend towards analyzing emotions along a set of emotional dimensions, which models continuous emotion as a combination of several dimensional values [1]. A well-accepted and commonly used emotion dimensional description is the 3D emotional space of pleasure, arousal and dominance, which is usually referred to as the PAD emotion space [4]. An emotion dimensional representation, which can encode small changes in affect over time and distinguish between many more subtly different displays of affect, allows a more complete description of the human emotional state in daily life [5].

In this paper, we present our efforts toward audio continuous emotion recognition. An emotional tone-based two-stage GMM-HMM method is proposed to infer the continuous emotion labels from audio cues. In this methodology, the continuous emotion recognition problem is solved through a two-stage automatic pattern recognition system with the Hidden Markov Model (HMM) framework. Two emotional tones, positive and negative, are set for each affective dimension. In the first stage, we employ the classifier to obtain the emotional tone of one audio clip. And then in the second stage, according to different emotional tone, corresponding positive or negative classifier is chosen to finish the recognition task. The emotional dimensions analyzed in our work are Arousal, valence and dominance. The work is evaluated on the Audio/Visual Emotion Challenge (AVEC 2014) [1] database. Our method has got 0.458 average correlations on development dataset, higher than traditional GMM-HMM 0.426 and SVM baseline 0.362. The results show the effectiveness of the proposed method.

The remainder of this paper is structured as follows. Section 2 provides a review on related works. Section 3 describes the proposed emotional tone-based two-stage algorithm. Section 4 presents the experimental results. And section 5 presents the conclusions.

2   Related Work

Audio signals convey affective information through explicit (linguistic) messages and implicit (acoustic and prosodic) messages that reflect the way the words are spoken. The popular features are prosodic features (e.g., pitch-related feature, energy-related features, and speech rate) and spectral features (e.g., MFCC and cepstral features) in speech based emotion recognition [2]. Many studies show that speech energy and pitch among these features contribute the most to affect recognition (e.g., [6, 7]). Besides, approaches combining acoustic features and spoken words, as well as approaches using linguistic features have also been proposed [8]. In addition, in terms of dimensional models, a most reliable finding is that pitch appears to be an index into arousal [9]. Another finding is that mean of the fundamental frequency (F0), mean intensity, speech rate, as well as pitch range and high-frequency energy [10] are positively correlated with the arousal dimension. There is relatively less evidence on the relationship between certain acoustic parameters and other emotion dimensions such as valence and dominance. It is worth noting that deciding the optimal audio feature set is still an open research problem.

Most of the previous work on the audio emotion recognition has focused on classifying human emotional states into discrete labels such as happy, angry or surprised. Researchers usually model this problem as pattern classification, HMM, GMM, ANN and SVM have been widely used. Nwe *et al.* [11] used a four-state fully connected HMM to recognize six archetypical emotions from speech, obtaining recognition performance comparable to subjective observers' ratings. In [12], GMM was used as the classifier for an infant-directed KISMET database. GMM can be seen as a HMM with only one hidden state. They built the GMM with spectral and energy related features and the classification accuracy is 78.77%. Many other researchers attempted to detect coarse affective states, i.e. positive, negative, and neutral (e.g., [13], [14]). However, the emotion state is not always one or the other, but has a level that indicates how strong the expressed feeling is.

As for continuous dimensional emotion recognition from audio signals, the most commonly employed strategy is to reduce the recognition problem of each dimension to a two-class problem, such as positive *vs.* negative or active *vs.* passive [15], or a four-class problem (e.g., four quadrants of 2D arousal-valence space [16]). In the work of Ozkan *et al.* [17], they present the emotion in a way of quantisation, in which continuous emotion values were modeled by a set of discrete classes. As far as the actual continuous dimensional emotion prediction, there are several methods (e.g., [18], [19], [20]). Grimm *et al.* [20] used the SVR, and compared their results to that of the distance-based fuzzy k-NN and rule-based fuzzy-logic estimators, and they found that the SVR get a better predict performance. The work by Wöllmer *et al.* used the Long Short-Term Memory neural networks and SVR [18]. Espinosa *et al.* used the Vera am Mittag database [21] and examined the importance of different acoustic features with continuous PAD dimensions [19]. However, how to model dimensional affect space (continuous *vs.* quantized) and which classifier is better suited for automatic, continuous emotion analysis with dimensional representation is still no consensus.

In this paper, we present an emotional tone-based two-stage GMM-HMM method for quantized emotion recognition that first model emotion dimension into a step-wise representation, and then use the GMM-HMM to infer the dimensional emotional tone and affect labels.

3   Emotional Tone-Based Two-Stage Algorithm

In this section, the framework of emotional tone-based two-stage algorithm is first described. Then audio feature extraction, training set construction and GMM-HMM classifier construction are described in details separately.

3.1   Overall Framework

During communication, emotions originated by a speaker can also be affected by the other speakers or the surrounding environment. This implies that emotional tone over a period of time reflects the potential predisposing of emotion expression in some way.

We hypothesize that although emotion ratings are fluctuating, one's emotional tone remains unchanged over a relatively short period of time. Based on this assumption, an emotional-tone based two-stage algorithm is proposed to detect continuous emotion values. The main idea behind this scheme is to build a multistage classification system that can achieve the most targeted emotion-specific recognition that is, using different classifier under different emotion tone. Fig.1 shows the overall framework of the proposed method. After feature preprocessing, the GMM-HMM in the first stage is employed to obtain the emotional tone. Specifically, the GMM-HMM outputs a set of emotional predictions for each segment feature of one audio sequence; the mean value of this set of predictions is used as the emotional tone. According to this mean value, two emotional tones, positive and negative, are predicted. Then in the second stage, corresponding positive or negative classifier is chosen based on the emotional tone to complete the final continuous emotion recognition. The positive and negative classifiers are modeled with GMM-HMMs as well.

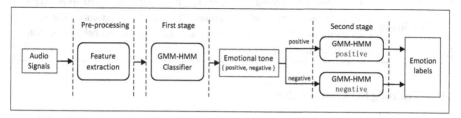

Fig. 1. Overall framework of the emotional tone-based two-stage algorithm

3.2 Feature Extraction

We adopted the audio features from the AVEC 2014 dataset, a detailed description of the dataset can be found in subsection 4.1. The   audio feature vector consists of 2268 dimensions, comprising 32 features that are related to the overall energy and spectrum (e.g., energy per band, entropy, and harmonicity) and 6 features that are voice-related (e.g., F0, jitter, and harmonics–to-noise ratio). As different features range in different values, all the features were normalized to have zero mean and unit variance.

To reduce the high feature dimensions, a connected correlation based feature subset selection and principal component analysis (CFS-PCA) method is applied. First, correlation based feature subset selection (CFS) [22] with a best first search strategy is employed to extract the most significant functionals for arousal, valence and dominance from the training set. 49 dimensions for arousal, 48 dimensions for valence and 49 dimensions for dominance are extracted. Unfortunately, the extracted dimensions by CFS from training set and development set are of large difference. For example, we extract 49 dimensions with training set and 43 dimensions with development set for arousal, but there exist only 9 same dimensions among all 92 dimensions. Thus, the Principle Component Analysis (PCA) method with 85% energy reserved is applied to the rest dimensions of each feature excepting the ones selected by CFS. Table 1 listed the selected features respectively for three emotional dimensions. Concatenating the CFS results and PCA results, an integrated vector is formed to jointly represent audio signal. Those entire integrated vector are concatenated to form the whole integrated vector. Then, PCA is applied again to reduce the dimensions of the whole integrated vector.

Finally, the reduced dimension for arousal, valence and dominance is 85, 85 and 83 respectively.

Table 1. Set of features for Arousal, Valence, Dominance dimension

Affective dimension	Features
Arousal	Loudness, zero crossing rate, energy in 250-650Hz, 1kHz-4kHz, spectral variance, kurtosis, flatness, MFCC(1-7, 9-14, 16), probability of voicing, jitter, shimmer(local), jitter(delta: "jitter of jitter")
Valence	Loudness, energy in bands from 250-560, spectral variance, skewness, kurtosis, flatness, MFCC(1-2, 4-7, 9-14,16), probability of voicing, shimmer(local)
Dominance	Loudness, zero crossing rate, energy in 250-650Hz, 1kHz-4kHz, 25%, 50%, 90% spectral roll-off points, spectral flux, variance, skewness, harmonicity, flatness, MFCC(1, 3-4, 6-12, 14), F0, probability of voicing, Jitter, logHNR

3.3   Training Set Construction

In our system, Gaussian mixture models of hidden Markov models (GMM-HMMs) are employed to infer the dimensional emotional tone and emotion labels. Different training sets are constructed for classifiers with different targets.

At first, we perform a step-wise representation of the continuous emotion dimension, in which the continuous emotion value is assigned to a discrete value that can be seen as a class label. The label discretization is performed by a percentile approach. For each emotion dimension, we use the labels from the training set to determine the range of continuous emotion levels of each discrete label class. The range relies on the percentage of the continuous levels that have similar values. For instance, if we want to represent the data with 4 discrete labels, then we automatically find 5 thresholds such that each label class contain 1/4 samples from the training set that have similar continuous levels. These thresholds can then be used to determine the class labels for development   dataset.

For the first stage, multiple sub-datasets containing only sub-sequences with the same class labels are organized. Each GMM is learned for each discrete label with corresponding sub-dataset. All sub-datsets are employed to train the GMM-HMM classifier for emotional tone prediction. While in the second stage, positive and negative GMM-HMM classifiers are trained seperately. The construction of sub-datasets for both above two classifiers is similar with the way in the first stage. But, the difference is that in the first stage sub-datasets are created with all sub-sequences in traing set, but in the second stage the taining set is divided into positive emotional tone sub-dataset and negative emotional tone sub-dataset. Apparently, the positive emotional tone sub-dataset contains only audio sequences with positive emotional tone, similarly for negative emotional tone sub-dataset.

3.4   GMM-HMM Classifier

We choose the GMM-HMMs as the classifiers in both two stages. A Hidden Markov Model learns a probability distribution over a sequence of observations of X={x_1, x_2 ..., x_m}, where each segment observation x_j is represented by a feature vector in R^d. In HMM, these observations are associated with a set of hidden states H= {h_1, h_2, ..., h_m}, where the state of h_t at time t depends only on the previous state h_{t-1} at time t-1 and the observation x_t at time t. And the joint distribution of state variables and the sequence observations can be found by the followin:

$$p(h, x) = p(h_1) \, p(x_1|h_1) \prod_t p(h_t|h_{t-1}) \, p(x_t|h_t) \qquad (1)$$

The first two terms are the model priors, i.e., distribution over the initial state. Training of a HMM involves finding the probability distribution over $N \times N$ state transition matrix that defines $p(h_t|h_{t-1})$, and the output model that defines $p(x_t|h_t)$, which can be modeled in different ways. We use mixtures of gaussians, since our observations are real values. As shown in Fig. 2, with each sub-datasets containing only sub-sequences with the same class label, one GMM model is learned for each step-wise class label, which represents the probability distribution of the corresponding feature. Count the number of transitions between class labels to get the $N \times N$ state transition matrix that defines $p(h_t|h_{t-1})$. Thus, we get the GMM-HMM.

In our method, HMM is implemented as a fully-connected network with four states. The number of hidden states is determined according to the training set. The observation x_i is associated with a hidden state h_i, and each hidden state h_i directly corresponds to a class label l_i. For the HMM testing, the classification problem is converted into a best path-finding problem for the class label sequence. The Viterbi algorithm [23] is used to produce the best match label sequence.

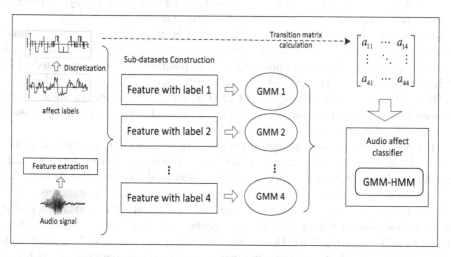

Fig. 2. Training procedure for GMM-HMM model

4   Experiments

To evaluate the performance of our approach, experiments are performed on the AVEC 2014 audio dataset. In the following, we first describe the dataset in 4.1. Then, the setup of these experiments is described in 4.2. And the results of the experiments are presented in 4.3.

4.1   AVEC 2014 Dataset

The Audio/Visual Emotion Challenge (AVEC) 2014 dataset is a subset of the AVEC 2013 audio-visual depression corpus [24]. The dataset contains 150 video clips from 84 subjects performing a human-computer interaction task in a number of quiet settings. The audio was recorded using a headset connected to the built-in sound card of a laptop at a variable sampling rate, and was resampled to a uniform audio bitrate of 128kbps using the AAC codec. The speakers were recorded between one and four times, with a period of two weeks between the measurements. There is only one person in every recording and some subjects feature in more than one recording. 18 subjects appear in three recordings, 31 in 2, and 34 in only one recording. The length of the full recordings is between 50 minutes and 20 minutes (mean = 25 minutes). The total duration of all clips is 240 hours. The mean age of subjects was 31.5 years, with a standard deviation of 12.3 years and a range of 18 to 63 years.

The behavior within the clips consisted of 2 different human-computer interaction tasks which were power point guided. The two tasks selected are as follows:

- Northwind – Participants read aloud an excerpt of the fable "Die Sonne under Wind" (The North Wind and the Sun), spoken in the German language

- Freeform – Participants respond to one of a number of questions such as: "What is your favourite dish? "; "What was your best gift, and why?"; "Discuss a sad childhood memory", again in the German language

The unit of classification for the audio dataset is an audio segment. The feature vector for each segment consists of 2268 components, composed of 32 energy and spectral related low-level descriptors (LLD) ×42 functionals, 6 voicing related LLD ×32 functionals, 32 delta coefficients of the energy / spectral LLD ×19 functionals, 6 delta coefficients of the voicing related LLD ×19 functionals, and 10 voiced / unvoiced durational features. The set of LLD covers a standard range of commonly used features in audio signal analysis and emotion recognition.

The data is continuously annotated per frame by human raters for three affective dimensions: arousal, valence, and dominance. These dimensions form a well-established basis for emotion analysis in the psychological literature [25]. The valence dimension indicates the over positive or negative feeling of an individual toward the object at the focus of his/her emotional state. Arousal indicates the individual's global level of dynamism or lethargy. It subsumes mental activity, and physical preparedness to act as well as overt activity. Dominance is an individual's sense of how much they feel to be in control of their current situation. The original continuous label traces were binned in temporal units of the same duration as a single video frame (i.e., 1/30 seconds). The raw joystick data for arousal, valence, dominance lies in the range [-1000, 1000] labels, which is scaled by a factor 1/1000 to the range [-1, 1].

4.2   Experimental Setup

Two sets of data, "Northwind" and "Freeform" are used in our experiments. Each data is divided into 3 subsets: training, development and testing. The short segment features, extracted using a sliding window of 3s and an overlap of 1s, were used, which are intended for the emotion task. Owing to the fact that the data is continuously labeled per frame, and there is no emotion labels specific to the audio segment features. Hence, we take the way of averaging all the frame labels falling into the sliding window of 3s to get the audio emotion labels. The frame sample rate is 30 frames per second. So, we average 90 (30x3) frame labels to get an affect label for an audio short segment feature. The proposed method was trained on the training dataset and the performance was evaluated on development set.

The quality of the predicted target is measured via its Pearson's correlation coefficient with the ground-truth targets. High correlations are achieved when we get the same emotional trends with the ground-truth label. The correlations are computed per clip on each one of the three emotion dimensions, subsequently, averaged over all clips in the development data to obtain the final quality measure.

4.3   Results

Table 2 and 3 provide an overview of the results on the Northwind dataset and the Freeform dataset, obtained separately using the single GMM-HMM and our emotional tone-based two-stage algorithm. The comparison for these two methods presented in the tables reveals that our method outperforms the GMM-HMM in all the three emotion dimensions. This demonstrates the effectiveness of the emotional tone-based classify scheme. Nevertheless, as for the Freeform dataset, our method only gain a bit higher improvement than the GMM-HMM (i.e., 0.394 vs 0.4 for Valence, 0.496 vs 0.501 for Dominance), this may be due to the property of this dataset. It contains the freeform question about something general about the participator's life, which means that it may not embody obvious emotion color or the emotion state tends to neutral. As shown in Table 4, on the average of the development set, the mean correlation we obtain is 0.458, which is quite a bit better than the performance of the baseline SVR result of 0.362 in [1].

Table 2. Correlations performances of Northwind dataset

Method	Arousal	Valence	Dominance	Average
GMM-HMM	0.379	0.435	0.462	0.425
Our method	0.442	0.469	0.475	0.462

Table 3. Correlations performances of the Freeform dataset

Method	Arousal	Valence	Dominance	Average
GMM-HMM	0.390	0.394	0.496	0.427
Our method	0.46	0.40	0.501	0.454

Table 4. Average correlations performances on the AVEC 2014 development set

Method	Arousal	Valence	Dominance	Average
GMM-HMM	0.385	0.414	0.479	0.426
Our method	0.451	0.435	0.488	0.458
Baseline system	0.412	0.355	0.319	0.362

5 Conclusion

With an automatic emotion recognizer, a computer can respond appropriately to the user's emotional state rather than user commands. In this way, the human-computer interaction would become more meaningful and intelligent. We investigate audio emotion recognition along three affective dimensions in this paper. An emotional tone-based two-stage algorithm was proposed to tackle this problem. Our approach relied on the assumption that although emotion ratings are fluctuating, one's emotional tone remains relatively fixed within a period of time. Based on this, we proposed this multistage classification scheme, in which corresponding positive or negative classifier was chosen based on the emotional tone to complete the emotion recognition task.

We have evaluated our approach on the audio set of the AVEC 2014 database. The results show considerable improvement over the baseline SVR results. This classification algorithm can be generalized to other kind of features. In future work, we will improve the proposed emotional tone-based scheme to achieve affective recognition with audio-visual information. In addition, a more complex version of GMM-HMM will be used to explore the intrinsic dynamics within each class label.

Acknowledgements. This work was supported by the National Natural Science Foundation of China (NSFC: 61135003, 61173059) and the National Fundamental Research Grant of Science and Technology (973 Project: 2013CB329305).

References

1. Valstar, M., Schuller, B., Smith, K., et al.: AVEC 2014 – 3D Dimensional Affect and Depression Recognition Challenge. In: Proc. 4th ACM International Workshop on Audio/Visual Emotion Challenge (2014)
2. Zeng, Z., Pantic, M., Roisman, G.I., et al.: A survey of affect recognition methods: Audio, visual, and spontaneous expressions. IEEE Transactions on Pattern Analysis and Machine Intelligence 31(1), 39–58 (2009)
3. Ozkan, D., Scherer, S., Morency, L.P.: Step-wise emotion recognition using concatenated-HMM. In: Proceedings of the 14th ACM International Conference on Multimodal Interaction, pp. 477–484. ACM (2012)
4. Jia, J., Zhang, S., Meng, F., et al.: Emotional audio-visual speech synthesis based on PAD. IEEE Transactions on Audio, Speech, and Language Processing 19(3), 570–582 (2010)
5. Gunes, H., Schuller, B., Pantic, M., Cowie, R.: Emotion representation, analysis and synthesis in continuous space: A survey. In: 2011 IEEE International Conference on Automatic Face & Gesture Recognition and Workshops (FG 2011), pp. 827–834. IEEE (2011)
6. Kwon, O.W., Chan, K., Hao, J., Lee, T.W.: Emotion recognition by speech signals. In: INTERSPEECH (2003)
7. Scherer, K.R., Johnstone, T., Klasmeyer, G.: Handbook of Affective Sciences - Vocal expression of emotion, Affective Science, ch. 23, pp. 433–456. Oxford University Press (2003)
8. Savran, A., Cao, H., Shah, M., et al.: Combining video, audio and lexical indicators of affect in spontaneous conversation via particle filtering. In: Proceedings of the 14th ACM International Conference on Multimodal Interaction, pp. 485–492. ACM (2012)
9. Calvo, R.A., D' Mello, S.: Affect detection: An interdisciplinary review of models, methods, and their applications. IEEE Transactions on Affective Computing 1(1), 18–37 (2010)
10. Gunes, H., Schuller, B., Pantic, M., et al.: Emotion representation, analysis and synthesis in continuous space: A survey. In: 2011 IEEE International Conference on Automatic Face & Gesture Recognition and Workshops (FG 2011), pp. 827–834. IEEE (2011)
11. New, T.L., Foo, S.W., De Silva, L.C.: Speech emotion recognition using hidden Markov models. Speech Communication 41(4), 603–623 (2003)
12. Breazeal, C., Aryananda, L.: Recognition of affective communicative intent in robot-directed speech. Autonomous Robots 12(1), 83–104 (2002)
13. Lee, C.M., Narayanan, S.S.: Toward detecting emotions in spoken dialogs. IEEE Transactions on Speech and Audio Processing 13(2), 293–303 (2005)
14. Neiberg, D., Elenius, K., Laskowski, K.: Emotion Recognition in Spontaneous Speech Using GMM. In: Proc. Int'l Conf. Spoken Language Processing (ICSLP 2006), pp. 809–812 (2006)
15. Schuller, B., Vlasenko, B., Eyben, F., et al.: Acoustic emotion recognition: A benchmark comparison of performances. In: IEEE Workshop on Automatic Speech Recognition & Understanding, ASRU 2009, pp. 552–557. IEEE (2009)
16. Wöllmer, M., et al.: Combining long short-term memory and dynamic bayesian networks for incremental emotion-sensitive artificial listening. IEEE Journal of Selected Topics in Signal Processing 4(5), 867–881 (2010)
17. Ozkan, D., Scherer, S., Morency, L.P.: Step-wise emotion recognition using concatenated-HMM. In: Proceedings of the 14th ACM International Conference on Multimodal Interaction, pp. 477–484. ACM (2012)

18. Wöllmer, M., Eyben, F., Reiter, S., et al.: Abandoning emotion classes-towards continuous emotion recognition with modelling of long-range dependencies. In: INTERSPEECH 2009, pp. 597–600 (2008)
19. Espinosa, H.P., Garcia, C.A.R., Pineda, L.V.: Features selection for primitives estimation on emotional speech. In: Proc. IEEE ICASSP, pp. 5138–5141 (2010)
20. Grimm, M., Kroschel, K.: Emotion estimation in speech using a 3d emotion space concept (2005)
21. Grimm, M., Kroschel, K., Narayanan, S.: The vera am mittag german audio-visual emotional speech database. In: Proc. IEEE ICME, pp. 865–868 (2008)
22. Sánchez-Lozano, E., Lopez-Otero, P., Docio-Fernandez, L., et al.: Audiovisual three-level fusion for continuous estimation of Russell's emotion circumplex. In: Proceedings of the 3rd ACM International Workshop on Audio/Visual Emotion Challenge, pp. 31–40. ACM (2013)
23. Zeng, Z., Tu, J., Pianfetti, B.M., et al.: Audio–visual affective expression recognition through multistream fused HMM. IEEE Transactions on Multimedia 10(4), 570–577 (2008)
24. Valstar, M., Schuller, B., Smith, K., et al.: AVEC 2013: The continuous audio/visual emotion and depression recognition challenge. In: Proceedings of the 3rd ACM International Workshop on Audio/Visual Emotion Challenge, pp. 3–10. ACM (2013)
25. Fontaine, J.R., Scherer, K.R., Roesch, E.B., Ellsworth, P.C.: The world of emotions is not two-dimensional. Psychological Science 18(2), 1050–1057 (2007)

A Computationally Efficient Algorithm
for Large Scale Near-Duplicate Video Detection

Dawei Liu[1] and Zhihua Yu[1,2]

[1] Instititue of Network Technology, Institute of Computing Technology(Yantai),
CAS, Shandong, P.R. China
[2] Instititue of Computing Technology, CAS, Beijing, P.R. China
`liudw@int-yt.com, yzh@ict.ac.cn`

Abstract. Large scale near-duplicate video detection is very desirable for web
video processing, especially the computational efficiency is essential for practical applications. In this paper, we present a computationally efficient algorithm based on multi-layer video content analysis. Local features are extracted
from key frames of videos and indexed by an novel adaptive locality sensitive
hashing scheme. By learning several parameters, fast retrieval in the new hashing structure is performed without high dimensional distance computations and
achieves better real-time retrieving performance compared with other state-of-
the-art approaches. Then a descriptor filtering method and a two-level matching
scheme is performed to generate a relevance score for detection. Experiments
on near-duplicate video detection tasks including various transformed videos
demonstrate the efficiency gains of the proposed algorithm.

Keywords: Near-duplicate Detection, Locality sensitive hashing, SURF,
Multimedia content analysis.

1 Introduction

With the rapid growth of digital video content production on the web, near-duplicate
video detection for protecting and managing video content has received growing attention over the last decade. There are two general techniques for near-duplicate video
detection: digital watermarking and content-based copy detection. Compared with
digital watermarking, which embeds hidden data information called watermarking in
an image or video, the content-based techniques, which employs video content analysis and detects video copies by video signatures or key frame features, leads to better
efficiency and effectiveness.

Recently, most approaches focus on the content-based near-duplicate video detection. The general procedure of existing work can be summarized as three stages. First,
using shot detection methods, videos are segmented into clips, which then represented
by one or more key frames. Second, a set of high dimensional feature vectors are
extracted by feature detector and descriptor. Finally, the similarity between videos is
computed from the feature vectors under spatial and/or temporal sequence matching
schemes [7][11][14]. While the feature representation stage can be further classified

X. He et al. (Eds.): MMM 2015, Part II, LNCS 8936, pp. 481–490, 2015.
© Springer International Publishing Switzerland 2015

into two categories: global feature and local feature, each has different design of video content representations and similarity metrics between feature sequences. Yeh et al. proposed a global frame-level descriptor [6], which is a compact 16-dimensional feature vector based on computing the spectral properties of a graph built from partitioned blocks of a frame, and a fast sequence matching scheme: dot plot[5]. Chiu et al. [13] combines both global and local feature descriptors and integrates min-hashing and spatiotemporal matching to detect video copies. Shang et al. [8] introduced a binary spatiotemporal feature which is global and fast to compute using an indexing structure based on inverted file. Liu et al. [9] described a framework which used SIFT as local feature and employed locality sensitive hashing (LSH) [3] and random sample consensus (RANSAC) techniques to index and detect. Avrithis et al. [12] quantized local features to visual word and used a RANSAC-like matching method. Comparative study of state-of-the-art techniques [10] concluded that local feature-based methods are more robust but more computational expensive.

In this paper, we consider the content-based near-duplicate video detection based on local features. We employ SURF [2] as local visual features and we design an LSH-based indexing structure with parameterizations to reduce the computational cost, while maintaining the scalability and robustness. After the feature vectors retrieval, a descriptor filtering method is applied to further reduce the number of candidate feature sets and then a two-level matching scheme to generate a relevance score for detection. To the best of our knowledge, few framework has yet been present to merge LSH with SURF to design video content indexing and retrieval. The only work similar to ours is [1] in which Zhang et al. used SURF features and LSH indexing separately without optimization. In contrast, our algorithm leverages the characteristics of both feature vectors and indexing structures to reduce computational cost and boost retrieval performance.

The rest of this paper is organized as follows. In Section 2, we present our adaptive locality sensitive hashing for SURF indexing. Section 3 introduces the descriptor filtering and two-level matching scheme. Section 4 gives the experimental results and performance analysis of our proposed algorithm. Finally we conclude this paper and give some future work in Section 5.

2   Adaptive Locality Sensitive Hashing for SURF Indexing

Fig.1. illustrates the near-duplicate video detection framework of our algorithm. The processing consists of two parts: Indexing and Retrieval. Indexing videos are processed by shot detection, key-frame and SURF extraction to generate a set of 64-dimensional feature vectors. Then a video database is built using an indexing structure. In the retrieval parts, the same local features extraction is performed and by retrieving in the database a candidate result set is generated and filtered, then two-level matching methods are applied to get the final near-duplicate video results. We focus on the adaptive LSH indexing, feature filtering, frame-level and video-level matching, as indicated by shades.

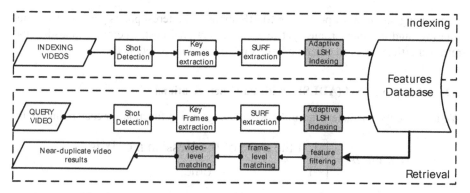

Fig. 1. Overview of Near-Duplicate Video Detection Framework

2.1   Insights into SURF and LSH

SURF (Speeded-Up Robust Features) detector and descriptor [2] is based on calculating approximate Hessian response for image points and is efficiently implemented on the basis of integral images. SURF is proved to be equal or superior to performance and significantly better computational efficiency in comparison with other local feature methods, such as SIFT, PCA-SIFT. Moreover, some intermediate results, such as the sign of the Laplacian (i.e. the trace of the Hessian matrix) and the position of each interest points, can be used to accelerate the feature vector matching process with no extra computational cost.

LSH (Locality Sensitive Hashing) is introduced in [3] for approximate nearest neighbors search in high dimensions. LSH function families have the property that objects that are close to each other have a higher probability of colliding than objects that are far apart. For different distance measures, different LSH families have been proposed. We consider the LSH for p - norms based on p -stable distributions [4] since SURF descriptors are designed to measured by the Euclidean distance. In the basic LSH scheme, a query point is hashed into several buckets in different hash tables to retrieve all points in these buckets, then the distances to each point is computed. We argue that the Euclidean distance computing in high dimensions (i.e.64-dimensions for SURF) has a high computational complexity and becomes the performance bottleneck for existing LSH-based methods [9][13]. While simply without distance computing, retrieving all points in several buckets to candidate sets also results in high computational cost in the next multi-level matching process due to the large number of points contains in these buckets. We introduce the Adaptive LSH with parameterizations to solve the problem. The key idea of our adaptive LSH scheme is that the accuracy of near-duplicate video detection based on video content analysis depends on multiple factors and the fine granularity of SURF is enough for frame-level matching. In our scenario, speed is more important than "local accuracy", more effective filtering and less unnecessary computing is demanded.

In comparison with global feature based methods, which use one compact feature vector to describe one key frame, local feature based methods characterize one key frame using hundreds of points (for SURF about 100 interest points generated without filtering) and are more robust yet more computational cost. Therefore, boosting the retrieval efficiency is an essential issue for local feature based methods.

2.2   Adaptive LSH (ADLSH) with Parameterizations

We leverage the characteristics of SURF to design a sign function for each point p as:

$$sign(p) = \begin{cases} 1 & \text{if the trace of the Hessian metrix is 1} \\ -1 & \text{others} \end{cases}$$

Merging the sign function with the basic LSH function, each of our adaptive LSH function is represent as:

$$h_{a,b}(p) = \left\lfloor \frac{a \cdot p + b}{W} \right\rfloor \cdot sign(p) \tag{1}$$

where p is a 64-dimensional SURF feature descriptor in our algorithm. Accordingly a is a 64-dimensional random vector with entries chosen independently from a 2-stable distribution (i.e. here is the Gaussian distribution for the Euclidean distance) and b is a real number chosen uniformly from the range $[0,W]$. Each hash function $h_{a,b}(p)$ maps a 64-dimensional vector v onto the set of signed integers. To implement the indexing structure, each point p is hashed into L buckets in L hash tables: $g_j(p)$, for $j = 1,...,L$, where $g_j(p) = \left(\left| h_{1,j}(p) \right|,..., \left| h_{k,j}(p) \right| \right)$. We use the absolute value $\left| h_{a,b}(p) \right|$ of each function to generate the k-dimensional label of each bucket and then split each bucket into two parts considering the sign of point. The number of buckets in our scheme is about twice the basic LSH, accordingly the number of points in each buckets is averagely reduced by 50%..

One challenge with LSH-based methods is that there are several parameters: W, k, L to be tuned, while existing approaches determine parameters by experimental results, which are uncertain and impractical. We analyze parameter restrictions and propose a sample learning approach to set parameters automatically during indexing and at the same time to further reduce the average amount of points in each buckets. We design our adaptive LSH for R-near neighbor search in 64-dimensional SURF descriptor space. To process a query point q, we retrieve all points in buckets $g_1(q),..., g_L(q)$ without distance computing while guarantee that the resulting candidate set contains all R-near neighbors v of q (i.e. $\|q - v\|_2 \le R$). As proved in [4], basic LSH solve the R-near neighbor problem with a probability at least $1 - \delta$, where δ is the fail probability (0.1% in our implement) and for two points p_1, p_2, let $c = \|p_1 - p_2\|_2$, the probability of the two points collide under a hash function chosen uniformly at random from LSH family is:

$$p(c) = \Pr_{a,b}\left[h_{a,b}(p_1) = h_{a,b}(p_2)\right] = \int_0^W \frac{1}{c} f_2(\frac{t}{c})(1 - \frac{t}{W})dt \tag{2}$$

where $f_2(t)$ denote the probability density function of the absolute value of the Gaussian distribution ($p = 2$):

$$f_2(t) = \begin{cases} \dfrac{2}{\sqrt{2\pi}} e^{-t^2/2} & \text{if } t \geq 0 \\ 0 & \text{if } t < 0 \end{cases}$$

For the R-near neighbor problem in our scenario (without loss of generality we assume that $R = 1$), to retrieve all points with the Euclidean distance less than 1, the following condition is necessary:

$$p(c) \geq p(1) = 1 - 2\Phi(-W) - \frac{2}{\sqrt{2\pi}W}(1 - e^{-(W^2/2)}) \tag{3}$$

Note that each bucket is labeled by a k-dimensional vector, the label collision probability is:

$$\Pr_g\left[g(q) = g(p)\right] = \prod_{i=1}^{k} p_i(c) \geq p(1)^k \tag{4}$$

For all L functions, a query point q finds a 1-nearest neighbor with the following probability:

$$\Pr_{NN}\left[\|q - p\|_2 \leq 1\right] = 1 - (1 - p(1)^k)^L \geq 1 - \delta \tag{5}$$

For a fixed $p(c)$, the optimal value of W is a function of c and decreasing W decreases the $p(c)$ for any two points. Moreover, increasing k or decreasing L decreases the probability to find nearest neighbors. We design the following three steps to determine the parameters:

    1)   sample learning:

We randomly choose m pairs of total n SURF feature points extracted from the indexing videos to be samples, we estimate $p(c)$ with:

$$c = \frac{1}{m}\sum_m \|p_i - p_j\|_2$$

and the average number of collisions in each bucket to be:

$$N_{bucket} = \sum_n p(c_n)^k \approx \sum_n p(c)^k .$$

    2) W:

We adopt the minimum power of 2 under restriction (3) to be the proper value for parameter W.

3) k and L

Our aim is to reduce the number of points retrieved by one query point from L buckets to about one percent of the total number for small database or about a constant (we choose 10000 in experiments of this paper) for large database, that is: the parameter k and L are determined by:

$$N_{bucket} \cdot L \approx \min(1\% \cdot n, 10000) \text{ while under restriction (5).}$$

Fig. 2. An example of filtering false matching pairs. Left: a key frame extracted from a indexing video; Right: a key frame extracted from a query video with transformations of resizing and subtitles. All the lines denote matching pairs of original SURF descriptors. Red lines denote the false matching pairs which are filtered by adaptive LSH using bucket splitting; Blue lines denote the false matching pairs which are filtered by Descriptor Filtering method based on relative distances; Yellow lines denote the right matching pairs which compose the candidate sets after pruning in our algorithm.

2.3   Descriptor Filtering (DF)

After splitting buckets according to (1) and three steps parameterizations, for each point in each key frame, a predictable size of candidate set is generated without high dimensional distance computations. Note that the SURF descriptor matching by Euclidian distance is not robust to noisy transformation, some geometrical verification techniques, such as RANSAC, are used in most recent studies. Since our candidate set is well pruned, we simply introduce a distance-based filter to remove most false matched points. For each pair of descriptor, we leverage the position coordinates, which recorded during SURF feature extraction without additional computation, to calculate a relative distance in 2-dimensional space. Mean and standard deviation are computed for distances of all descriptors in each key frame (image) and those pairs whose distances have a large difference (exceeds standard deviation greatly) to mean are considered to be noisy points. Fig. 2. as an example illustrates the filtering effect of our algorithm.

3   Two-Level Matching Scheme

In Section 2, all matched SURF descriptors is detected using adaptive LSH indexing structure, a two-level matching scheme is proposed in this section. During indexing and retrieving the 64-dimensional SURF feature descriptors, corresponding video identifiers, key frame identifiers and the numbers of points in the same bucket are recorded with each descriptor. We hash each descriptor based on key frame identifiers and after filtering the candidate set, a linear scan is performed for each frame to select the frames that have a number of matching descriptors upon a threshold (for about 100 SURF descriptors per frame in our scenario, we choose the threshold to be 60). Therefore we got the near-duplicate results at the frame-level.

To get the near-duplicate video results of a query video, for each key frame f_i^q in the query video, the similarity between f_i^q and a key frame f_j^c of an indexing video is defined as:

$$sim(f_i^q, f_j^c) = \sum_{N_{descriptor}} \sum_L w_{i,j} \cdot N_{matching} \qquad (6)$$

where $N_{descriptor}$ is the number of descriptors extracted from the query frame, $N_{matching}$ is the number of matching descriptors between the two frames in one bucket and $w_{i,j} = 1 / N_{bucket}$ is the weight of the corresponding bucket. We use the weight $w_{i,j}$ to reduce the impact of large buckets, in which many descriptors of the same frame match to one query descriptor of the query frame.

The relevance score for one indexing video v_c to the query video v_q is defined as:

$$score_c = \frac{\sum_{N_{frame}} sim(f_i^q, f_j^c)}{N_{frame}} \qquad (7)$$

where N_{frame} is the number of key frames of the query video. An indexing video v_c is considered to be a near-duplicate video of the query video if $score_c$ exceed a threshold S_t. The selection of S_t requires a trade-off between recall and precision during detection and is data dependent. Finally, the videos with first several highest scores are returned as the results of near-duplicate video detection.

4   Experiments

To evaluate the computational efficiency and robustness of our proposed algorithm, we conducted experiments using the MUSCLE-VCD benchmark [15] , which is an evaluation set of the TRECVID 2008 copy detection task. The dataset consists of 101 videos with a combined length of about 100 hours. In comparison with method proposed in [1], we tested precision, recall rates and computation time using the provided

15 query videos with different transformations such as blur, color adjustment, resizing, subtitles, encoding and crops. Let ADLSH, ADLSH+DF and BLSH represent the proposed adaptive locality sensitive hashing scheme, the adaptive LSH with descriptor filtering method and the basic LSH scheme used in [1], respectively.

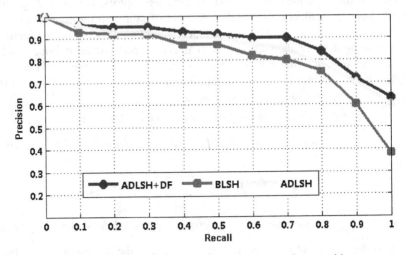

Fig. 3. Average Precision/Recall Curves across 15 query videos

Fig. 3. shows the comparison results of ADLSH+DF, ADLSH and BLSH evaluated by Precision-Recall curves. We can see that for recalls less than 0.8, all the three methods achieve high precisions larger than 0.8. At recalls larger than 0.8, our methods ADLSH+DF and ADLSH have higher precisions than BLSH. Moreover, ADLSH+DF slightly outperforms ADLSH. These results illuminate the impact of filtering. Matching SURF features under Euclidean distance without any filtering introduces noise in candidate sets and therefore cause the precision degrades. Filtering descriptors using the sign function and the relative distance contributes to robustness of detection.

Table 1. Efficiency of Indexing and Retrieval

Methods	Storage(MB)	Pruned Points	Time(s)
ADLSH+DF	19.6	6536	0.095
ADLSH	19.6	5198	0.068
LSH in [1]	18.5	-	0.347

To evaluate the computational efficiency gains of our algorithm, we focus on the index structure and retrieval speed. About 1.83×10^6 SURF descriptors (64-dimensional vectors) are extracted from 101 indexing videos and about 1.26×10^5 SURF descriptors are extracted from 15 query videos. We tested the average storage for each indexing video in the indexing structure, the number of pruned points during process and the average retrieval time cost for each query point (Note that time cost for shot detection and key frame extraction in Fig. 1. are not considered here). Table 1 shows the results. Since our proposed algorithm (ADLSH+DF, ADLSH) maintain additional information for the use of filtering, about 5% more storage space is needed. While compared with BLSH, which applies no filtering method, ADLSH pruned 5198 points per query and ADLSH+DF pruned about 1400 more points. Both algorithms achieved 3.6-5.1 times faster than BLSH. (ADLSH+DF is slower due to the descriptor filtering method). The results also illustrate that the high dimensional distance computing in BLSH seriously degrades the time efficiency, while our algorithm builds indexing structure with proposed parameterizations to avoid the inefficient computation and therefore accelerates the retrieval process.

5   Conclusions and Future Work

In this paper, we presented a computationally efficient algorithm for large scale near-duplicate video detection. An adaptive LSH scheme with parameterizations is proposed to index SURF feature vectors. Detection queries are retrieved in the proposed hashing indexing structure without high dimensional distance computations. The candidate set is then pruned by indexing buckets splitting and descriptor filtering methods. Relevance scores are generated using two-level matching methods and detection results are determined. The experimental results show the robustness preservation and computational efficiency gains of the proposed algorithm. Our future work will focus on: 1) the combination of our adaptive LSH indexing and different sequence matching algorithms, and 2) enhancing the parameterizations in an incremental way.

Acknowledgments. This work is supported by National Grand Fundamental Research 973 Program of China (No. 2013CB329602) and National Science Supported Planning (No. 2014AA015204).

References

1. Zhang, Z., Cao, C., Zhang, R., Zou, J.: Video Copy Detection Based on Speeded Up Robust Features and Locality Sensitive Hashing. In: Proc. IEEE Int. Conf. Automation and Logistics, pp. 13–18 (2010)
2. Bay, H., Tuytelaars, T., Van Gool, L.: Speeded-up Robust Features (SURF). Comput. Vis. Image Underst. 3(110), 404–417 (2008)
3. Gionis, A., Indyk, P., Motwani, R.: Similarity Search in High Dimensions via Hashing. In: Proc. Int. Conf. Very Large Data Bases, pp. 518–529 (1999)
4. Datar, M., Immorlica, N., Indyk, P., Mirrokni, V.: Locality-sensitive hashing Scheme Based on p-stable Distributions. In: Proc. ACM Symposium on Computational Geometry (2004)
5. Yeh, M.,, K.: T Cheng: Fast Visual Retrieval Using Accelerated Sequence Matching. IEEE Trans. Multimedia 13(2), 320–329 (2011)
6. Yeh, M., Cheng, K.T.: A Compact, Effective Descriptor for Video Copy Detection. In: Proc. ACM Int. Conf. Multimedia, pp. 633–636 (2009)

7. Caspi, Y., Irani, M.: Spatio-Temporal Alignment of Sequences. IEEE Trans. Pattern Anal. Mach. Intell. 24(11), 1409–1424 (2002)
8. Shang, L., Yang, L., Wang, F., Chan, K., Hua, X.: Real-time Large Scale Near-duplicate Web Video Retrieval. In: Proc. ACM Int. Conf. Multimedia, pp. 531–540 (2010)
9. Liu, X., Liu, T., Gibbon, D., Shahraray, B.: Effective and Scalable Video Copy Detection. In: Proc. ACM Int. Conf. Multimedia Information Retrieval, pp. 119–128 (2010)
10. Law-To, J., Chen, L., Joly, A., Laptev, I., Buisson, O., Gouet-Brunet, V., Boujemaa, N., Stentiford, F.: Video Copy Detection: A Comparative Study. In: Proc. ACM Int. Conf. Image and Video Retrieval (2007)
11. Kim, C., Vasudev, B.: Spatiotemporal Sequence Matching for Efficient Video Copy Detection. IEEE Trans. Circuits Syst. Video Technol. 15(1), 127–132 (2005)
12. Avrithis, Y., Tolias, G., Kalantidis, Y.: Feature Map Hashing: Sub-linear Indexing of Appearance and Global Geometry. In: Proc. ACM Int. Conf. Multimedia, pp. 231–240 (2010)
13. Chiu, C., Wang, H., Chen, C.: Fast Min-hashing Indexing and Robust Spatio-temporal Matching for Detection Video Copies. ACM Trans. Multimed. Comput. Comm. Appl. 6(2), Article 10 (2010)
14. Poullot, S., Buisson, O., Crucianu, M.: Scaling Content-based Video Copy Detection to Very Large Databases. Multimed. Tools Appl. 47, 279–306 (2010)
15. Law-To, J., Joly, A., Boujemaa, N.: Muscle-VCD-2007: A Live Benchmark for Video Copy Detection (2007)

SLOREV: Using Classical CAD Techniques for 3D Object Extraction from Single Photo

Pan Hu, Hongming Cai, and Fenglin Bu

School of Software, Shanghai Jiao Tong University

Abstract. While the perception of an object from a single image is hard for machines, it is a much easier task for humans since humans often have prior knowledge about the underlying nature of the object. Considerable work has recently been done on the combination of human perception with machines' computational capability to solve some ill-posed problems such as 3D reconstruction from single image. In this work we present SLOREV (Sweep-Loft-Revolve), a novel method for modeling 3D objects using 2D shape snapping and traditional computer-aided design techniques. The user assists recognition and reconstruction by choosing, drawing and placing specific 2D shapes. The machine then snaps the shapes to the automatically detected contour lines, calculates their orientations in 3D space, and constructs the original 3D objects following classical CAD methods.

Keywords: Image Processing, 3D Acquisition, User Interaction.

1 Introduction

The task of extracting three dimensional objects from a single photo is of great practical relevance since 1) photos are easy to obtain especially when only one photo for a particular object is needed and 2) photos can serve as a guideline in the modeling process, reducing the workload of professional 3D modelers. However, this problem is, in most cases, very difficult as some information is lost from 3D scene to 2D image and can hardly be recovered.

In this paper we introduce an interactive technique to extract 3D objects from a single photograph. SLOREV stands for sweeping, lofting and revolving, which are all classical computer-aided design techniques. Our approach takes advantage of human's perceptual abilities and prior knowledge on the object to be extracted. The user helps to recognize and roughly describe the geometric features of the object to constrain the solution space while the machine performs the rest of the task which requires high accuracy or complex computations. The information the user provides bridges the gap between what is given in the image and what is actually needed to reconstruct the object in 3D. Textures can be derived from the photograph and preserved once the object's geometry is recovered.

As mentioned earlier, our SLOREV modeling technique is composed of three submethods, namely sweeping, lofting and revolving. After the edge pixels are

X. He et al. (Eds.): MMM 2015, Part II, LNCS 8936, pp. 491–501, 2015.

automatically detected and linked to contour lines, the user can choose to apply appropriate submethods to model and extract desired 3D parts in the photo. To do sweeping, the user needs to define the 2D profile of the object, as well as its main axis; For lofting, the user can define both ends of the object in order to rebuild it; Our approach also allows the user to depict the side contour and the symmetric axis of the object, so as to extract it through revolving. Throughout the sweeping/lofting/revolving process, the machine dynamically carries out optimizations and adjustments, snapping the projection of the 3D model to its 2D counterpart in the photo.

Our approach allows the user to assist the recognition and the extraction of 3D objects by explicitly defining some of their geometric features. Meanwhile, the rest of the task is accomplished by an automatic, computational process. Using this technology makes it possible for both professionals and non-professionals to efficiently extract 3D objects from photographs. Despite the non-linearity and non-convexity of the fitting problem, the optimization process is indeed quite fast.

The remainder of this paper is organized as follows. We first briefly survey some related work and some current systems that can be used to extract 3D objects from photographs (section 2). We next investigate the issue of information loss from 3D scenes to 2D photos and explain the necessity of using human perceptual abilities to solve the problem (section 3). These observations and reasoning lead to our novel technique, SLOREV, which we describe at length in section 4. We then present our experimental results and discuss some possible future directions in section 5. Section 6 summarizes our contributions and presents our conclusions.

2   Related Work

Classical CAD Techniques. Sweeping, lofting and revolving are all well-known techniques in computer aided design [1, 2] and a large body of work has been done in this area. To create a 3D model through sweeping, one needs to first define a 2D area (usually referred to as a sweep template) and then sweep it along a specific trajectory. Early computer-aided-design systems [3] usually used boundary representations for modeling sweep surfaces. However, these sweep solids lacked a robust mathematical foundation, making self-intersecting sweeps simply invalid. Angelidis et al. proposed a volume preserving modeling technique called Swirling-Sweepers [4], which allowed arbitrary stretching while avoiding self-intersection. An implicit modeling method with no topological limitations on the 2D sweep profile was proposed by Schmidt and Wyvill [5], allowing surfaces to be blended smoothly. These methods all model 3D objects with sweeps but none of them uses images or sketches to guide the modeling process.

The lofting process first specifies a network of curves that approximates the desired shape and then interpolates them with a smooth surface [6]. Tiller [7] and Woodward [8] gave a detailed description of constructing a lofted surface with B-splines as well as the necessary background about interpolation. Filip et al. [9] pointed out the difficulty in lofting non-planar curves and provided a feasible solution to the problem. In spite of extensive work in this area, few studies have explored the possibility of using lofting techniques to extract 3D models from photographs.

Image-Based Modelng. Methods for interactive modeling from multiple photographs or video sequences have been proposed in [10–12]. These methods require users to explicitly mark edges or polygons or place 3D primitives in multiple photographs or frames before the systems could automatically align them and extract models, whereas in our problem setting there is only one single image for one particular object. Therefore, these methods could not be applied in this circumstance.

Extracting a 3D object from a single image requires a certain degree of semantic understanding of the object. To this effect, [13, 14] are proposed. In these methods, the user has to explicitly place primitives and add semantic annotations to them in order to model a 3D shape. Unlike our work, these techniques require explicit annotations and extensive manual operations.

Semantic understanding could also be integrated into the modeling system in the form of prior knowledge. Given a photograph of an object and a database of 3D models in the same class, Xu et al. [15] presented a system that creates a 3D model to match the photograph requiring only minimal user interaction. Rother and Sapiro [16] proposed a probabilistic framework of 3D object reconstruction from a single image that encodes prior knowledge about the object class in a database and projects it to 2D when it is used, casting the 3D reconstruction problem to a statistical inference problem. In contrast, our method does not rely on any database of similar models.

Our work is inspired by the recent work of Chen et al. [17] that uses three strokes provided by the user to generate a 3D component that snaps to the shape's outline in the photograph with each stroke defining one dimension of the component. However, this approach only works for a limited number of primitives such as generalized cylinders and cuboids. In our work a larger range of 2D profiles, including irregular ones are supported. Also, with the lofting method applied, our system supports modeling of non-uniformly scaled objects.

Sketch-Based Modeling. Sketches can be used to assist 3D modeling process. Igarashi et al. introduced Teddy [18], a sketching interface for 3D freeform design. This inspired a large body of follow-up work; refer to [19] for a detailed survey. These methods used a so-called sketch-rotate-sketch workflow and allowed users to build 3D objects by sketching.

Our approach is inspired by that of Shtof el al. [20], which models 3D objects from sketches rather than by sketching. In their work, the user assists recognition and segmentation by choosing and placing specific 3D geometric primitives on the relevant parts of the sketch. However, our approach tries to snap 2D profiles instead of 3D primitives. Moreover, we do not fit shapes to sketches, but to contour lines automatically detected in real images. Another difference is that [20] supports a limited number of primitives such as generalized cylinders, spheres and boxes whereas our approach is capable of more complex shapes.

The D-Sweep system proposed by Hu et al. [23], on which we build, uses sketching and real-time profile snapping techniques to extract 3D parts from a single photo. Unlike this method which utilizes only the sweeping process, our approach explores also the possibility of using another two well-known techniques, lofting and revolving, to address the problem. We provide users with the flexibility of choosing among all three submethods the most appropriate one to perform modeling. Therefore, compared with D-Sweep, our method can be used to extract a much larger range of 3D objects.

3   Analysis

When a photo is taken by a camera, a perspective projection occurs, mapping three-dimensional points to a two-dimensional plane. As the dimensionality reduces, some important information is lost. For example, parallelism and perpendicularity might not be preserved. In another word, angles might change after the projection. The extent to which a specific angle changes, relies on many factors including the positions of the object and the camera, as well as their orientations. Consequently, multiple different shapes could be projected to one same 2D shape on the resulting photo under certain conditions.

For the reasons mentioned above, it is hard for a machine to automatically determine the actual location and orientation of a projected shape in 3D space. This makes the task of extracting 3D models from single photos extremely difficult for machines. On the contrary, the cognitive task of understanding the geometry and structure of a shape, its projection and their relations is often simple for humans. This is because humans often have prior knowledge on the objects that we want to extract and can thus draw some inference based on their experience. Upon seeing a photo, we humans can often rapidly identify what the object in the photo is. Now what remains to be done is to pass the information we perceive to the machine and let it do the rest of the work.

The 3-sweep system proposed by Chen et al. [17] implicitly defines different user-perceived geometrical constraints on 2D profiles by using different gestures. Two kinds of simple 2D shapes are supported, namely circles and squares. Unlike this method, our approach allows the user to directly define geometric features of the object to be extracted and thus supports a larger range of 3D objects. We next discuss in detail our SLOREV technique.

4   SLOREV Method

Our interactive modeling approach takes as input a single photo. Our goal is to extract 3D component whose projection matches (a part of) the object in the given photo. Creating multiple 3D components one by one and using them to construct more complex 3D models is, however, beyond the scope of our paper (see, for instance, [17] and [20]). We focus on single component fitting. By explicitly describing geometric features such as 2D profiles, main axes and side contours, the user can efficiently construct 3D components that are consistent with their counterparts in the photograph.

Although the user interacts with the photo by describing some geometric features of the objects in it, SLOREV also relies on automatically snapping shapes to object outlines created from image edges. To detect object contours and build candidate object outlines from them, we use the hierarchical edge feature extraction method proposed by Arbelaez et al. [21]. Each pixel of the image is associated with a numerical value, indicating the probability of this pixel being part of a contour. We then apply the technique proposed by Cheng [22] to link the detected contour pixels into continuous point sequences. We introduce below our sweeping, lofting and revolving methods, respectively.

4.1   Sweeping

The workflow of the sweeping method is presented in Figure 1. First, an image is loaded into the system and automatic contour detection takes place, showing the user multiple candidate object outlines. The user then marks the profile contour, which will later be used for geometric snapping, as shown in Figure 1(b). This semantic task is very simple for the user to accomplish while machines can hardly distinguish bottom contours from others. We provide a smart interface to help users accomplish this task in a simple and straightforward manner. First, the user clicks somewhere on the image and the system iterates over all contour lines to find the nearest contour point, as well as the contour line on which it resides. The contour point found in this operation is marked as the starting point of the final contour line. Second, the user casually traces over the chosen contour line and marks vertex points of the profile using left clicks. As the cursor moves, the point that 1) is closest to the current cursor position and 2) resides on the previously determined contour line is marked as the ending point of the current line. Third, the user ends the whole contour marking process by using a simple right click. It is possible that the chosen bottom outline is incomplete with the other part of it sheltered by the object itself, which is indeed the case in Figure 1. To handle this issue, we check the completeness of the chosen curve by simply comparing its starting and ending position. If they are far away the system asserts that the contour is incomplete. Under the assumption of centrosymmetry and ignoring perspective distortion, the system will carry out an auto-completion to the existing contour, rotating it by 180 degrees around the center of the bottom surface. The center lies at the midpoint of the line segment connecting starting and ending point of the existing contour. Alternatively, the user can manually complete its contour using line segments in cases where the bottom shape is asymmetric.

Once the appropriate bottom contours are chosen (and completed), the user picks up a 2D shape from the pre-defined primitive list or he can draw one by himself, as shown in Figure 1(d). The user then approximately scales the 2D shape to the size of the object's profile, and drags it to the appropriate position (Figure 1(e)). When the user releases the 2D shape, an optimization using the

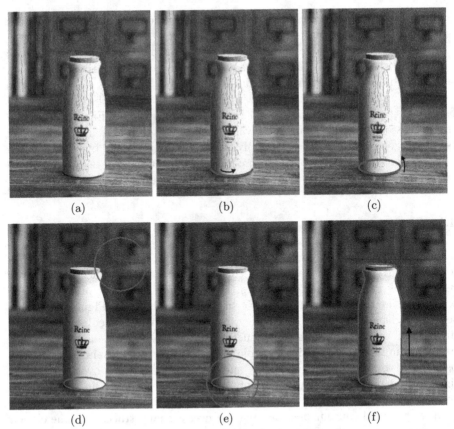

(a)   (b)   (c)

(d)   (e)   (f)

Fig. 1. Workflow of the sweeping method

following fitting objective function is performed so as to determine both the orientation and the size of the 2D profile.

$$\phi(c,n,d) = \sum_{i=1}^{N} \left[(n_z(c_{xy} - s_i))^2 + (n_{xy} \cdot (c_{xy} - s_i))^2 - (dr_i n_z)^2 \right]^2 \quad (1)$$

where $S = \{s_i\}$ represents the set of vertices on the contour line. For the user-defined polygon (a circle can also be approximated with a polygon), $n = \{n_x, n_y, n_z\}$ denotes its normal direction; $c = \{c_x, c_y, c_z\}$ defines the geometric center of its vertices; $R = \{r_i\}$ represents the distances from its vertex points to the center c (which are constants once the polygon is drawn); d defines the actual size of a polygon (which is to be optimized). This objective function is based on the observation that a 2D point lies on the projection of the ith 3D vertex point if, for a certain z coordinate, it resides on the plane defined by the center of all 3D vertices and the normal direction of the 3D polygon, and its distance from the center equals r_i. This optimization problem is essentially an unconstrained one and we simply use BFGS method to solve it. Refer to [23] for more details.

Finally the user defines the main axis of the 3D component using a straight or curved stroke. Our method for this step largely follows that of Chen et al. [17]. The user sweeps the 2D profile along a curve that approximates the main axis of the 3D object. During the curve drawing procedure, copies of the 2D profile are placed along the curve and most of them are snapped to the object's side outline. Side contours are not always correctly detected. In places where no appropriate side outline can be found, we simply clone the last profile and translate it to the current position. Figure 1(f) shows the result of this final operation. The 3D model for a milk bottle is successfully built.

4.2 Lofting

Figure 2 shows the workflow of our lofting method. As shown in Figure 2(a), this method has exactly the same pre-processing step as the sweeping method when candidate contour lines are automatically detected. Next, for each end of the object, the user applies the marking-describing-snapping technique introduced in the previous section to define the original 2D shape, as well as its size and its orientation in 3D space (Figure 2(b) and Figure 2(c)).

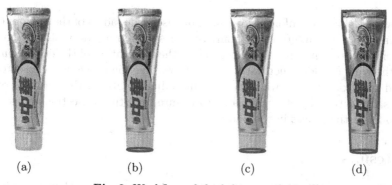

(a)   (b)   (c)   (d)

Fig. 2. Workflow of the lofting method

After both ends of the 3D part are defined, we apply the skinning technique proposed in [8] to interpolate the surface that pass through the two curves in 3D space. Note that the projection of the resulting 3D surface does not essentially coincide with the object in the photo. To address this problem, we check, at regular intervals, along the axis of the object and perform necessary adjustments to let the projection fit the corresponding side contour lines (Figure 2(d)).

4.3 Revolving

Figure 3 provides an overview of our revolving technique for object extraction. First, the user needs to pick up the side contour of the object, with which he wants to perform revolving (Figure 3(b)). During this contour marking process, the user helps to fill up the curve in places where automatic contour detection fails. This setting is extremely useful and makes revolving an ideal supplement to the sweeping method, since the sweeping method relies heavily on the availability of automatically detected and linked side contours. Second, the user places the symmetric axis to the desired place, as is shown in Figure 3(c).

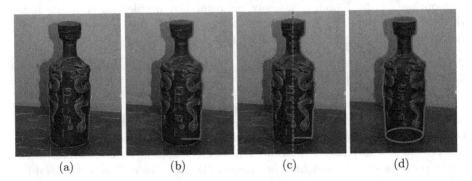

(a)	(b)	(c)	(d)

Fig. 3. Workflow of the revolving method

Now with the given information, more precisely the radius of the bottom (top) end circle and the intersection point of the symmetric axis with the bottom (top) end contour, it is possible to calculate the orientation of the object (in this case, the wine bottle). Then we can start to revolve the previously chosen curve around the axis. At last we perform some adjustments on the radiuses of 2D profiles along the object's main axis to ensure that the projection of the final 3D model fit the object in the photo.

5 Results

Modeling from Single Image. Figure 4 shows several modeling results. In this figure we show 3D models represented by lines that are extracted with sweeping, lofting and revolving techniques, respectively. Textures can then be mapped using a similar strategy to that introduced in [17]. For a certain 3D point of the object, the 2D coordinates of its projection on the image plane can be directly used as its texture coordinates. As there is no texture information regarding the back side of the extracted object, we may simply mirror the front texture to the back.

3-Sweep method proposed in [17] can only handle 3D components with square or circle bottoms. The method introduced in [20], similarly, is only capable of generalized cylinders and spheres. The previous work of Hu et al. [23] uses only sweeping to extract 3D models and thus does not allow non-uniform scale of the object. In reality, many objects have complex geometry and thus can not be well modeled with these methods. The pyraminx in Figure 4(a) has a triangular bottom while the vase in Figure 4(b) has a hexagonal bottom. These objects, and many

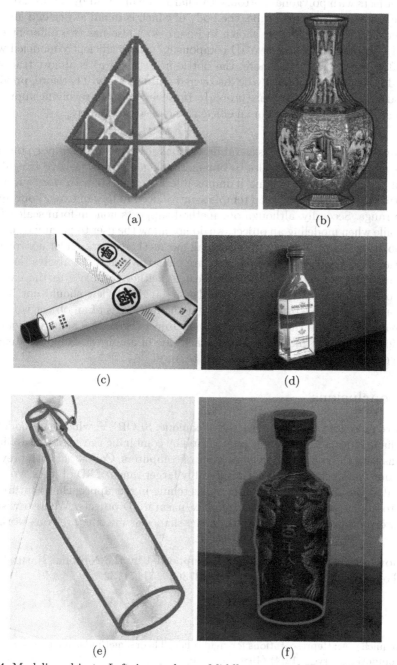

(a)

(b)

(c)

(d)

(e)

(f)

Fig. 4. Modeling objects. Left: input photos. Middle: extracted 3D models represented by lines. Right: extracted 3D models with their textures.

other objects with polygonal bottoms, can hardly be modeled by [17] or [20]. Figure 4(c) shows a toothpaste tube, the body of which is round at one end and flat at the other; The bottle neck shown in Figure 4(d) also has two different ends (round/rectangular). These two 3D components can be efficiently modeled with our lofting method. Furthermore, the bottle in Figure 4(e) is almost transparent and its contours can hardly be discovered automatically. Therefore, previous methods can not handle this case properly. However, with our revolving approach, users can model such objects in an efficient and elegant manner.

Limitations. Our work has several limitations. Firstly, although polygons and circles are supported as profiles in our method, many objects have irregularly curved profile contours, making it impossible for our system to work properly. Adopting the generalized sweep template introduced in [5] can extend our application range. Secondly, although our method supports non-uniform scale of the 2D profile when modeling an object, we do not allow the user to parameterize the interpolation (lofting) process. Thirdly, all our methods work under a strong assumption that the objects' end surfaces are planar, which is not always the case.

In our work we assume that the object is neither too close to the camera, nor too far away from it. In other words, the angle of view should not be too large or too small. We further assume that the main axis of an object is mainly visible and parallel to the projection plane. Objects extruding along the viewing direction are hard to model. Also, the back side of an object cannot be texture mapped if it does not have a symmetric texture.

6   Conclusions

We have presented a novel interactive technique, SLOREV, which can model 3D man-made objects from a single photograph by combining the cognitive ability of humans with the computational accuracy of computers. Compared with previous work, our approach supports a considerably larger range of 3D objects. We have integrated the concept of modeling by sketching in our approach. In future, we hope to allow modeling of more complex/natural 3D objects. We also wish to add inter-part semantic constraints on the shapes as in some previous work.

Acknowledgments. This research is supported by the National Natural Science Foundation of China under No. 61373030, 71171132.

References

1. Requicha, A.: Representations for rigid solids: Theory, methods, and systems. ACM Computing Surveys 12(4) (1980)
2. Foley, J.D., Van Dam, A., Feiner, S.K., Hughes, J.F., Phillips, R.: Introduction to Computer Graphics. Addison-Wesley (1993)
3. Requicha, A., Voelcker, H.: Solid modeling: A historical summary and contemporary assessment. Computer Graphics and Applications 2(2), 9–24 (1982)

4. Angelidis, A.: Swirling-sweepers: Constant-volume modeling. In: Proceedings of the 12th Pacific Conference on Computer Graphics and Applications, pp. 10–15 (2004)

5. Schmidt, R., Wyvill, B.: Generalized sweep templates for implicit modeling. In: Proceedings of the 3rd International Conference on Computer Graphics and Interactive Techniques in Australasia and South East Asia, pp. 187–196 (2005)

6. Schaefer, S., Warren, J., Zorin, D.: Lofting curve networks using subdivision surfaces. In: Proceedings of the 2004 Eurographics/ACM SIGGRAPH Symposium on Geometry Processing, pp. 103–114 (2004)

7. Tiller, W.: Rational B-splines for curve and surface representation. IEEE Computer Graphics and Applications 3(10), 61–69 (1983)

8. Woodward, C.D.: Skinning techniques for iterative B-spline interpolation. Computer Aided Geometric Design 20(8), 441–451 (1988)

9. Filip, D.J., Ball, T.W.: Procedurally representing lofted surfaces. IEEE Computer Graphics and Applications 9(6), 27–33 (1989)

10. Debevec, P., Taylor, C., Malik, J.: Modeling and rendering architecture from photographs: A hybrid geometry-and image-based approach. In: Proceedings of ACM SIGGRAPH, pp. 11–20 (1996)

11. Sinha, S., Steedly, D., Szeliski, R.: Interactive 3D architectural modeling from unordered photo collections. ACM Transactions on Graphics (TOG) 27(159) (2008)

12. Van Den Hengel, A., Dick, A.: VideoTrace: Rapid interactive scene modelling from video. ACM Transactions on Graphics (TOG) 26(3) (2007)

13. Gingold, Y., Igarashi, T., Zorin, D.: Structured annotations for 2D-to-3D modeling. ACM Transactions on Graphics (TOG) 28(148) (2009)

14. Lau, M., Saul, G., Mitani, J., Igarashi, T.: Modeling-in-context: User design of complementary objects with a single photo. In: Proceedings of the Seventh Eurographics Workshop on Sketch-Based Interfaces and Modeling, pp. 17–24 (2010)

15. Xu, K., Zheng, H., Zhang, H., Cohen-Or, D., Liu, L., Xiong, Y.: Photo-inspired model-driven 3D object modeling. ACM Transactions on Graphics (TOG) 30(80), 1–10 (2011)

16. Rother, D., Sapiro, G.: 3D Reconstruction from a Single Image. Submitted to IEEE Transactions on Pattern Analysis and Machine Learning, IMA Prepr International (2009)

17. Chen, T., Zhu, Z., Shamir, A., Hu, S., Cohen-Or, D.: 3-Sweep: Extracting Editable Objects from a Single Photo. ACM Transactions on Graphics (TOG) 32(195) (2013)

18. Igarashi, T., Matsuoka, S., Tanaka, H.: Teddy: A sketching interface for 3D freeform design. In: Proceedings of ACM SIGGRAPH, pp. 409–416 (1999)

19. Olsen, L., Samavati, F.F., Sousa, M.C., Jorge, J.A.: Sketch-based modeling: A survey. Computers Graphics 33(1), 85–103 (2009)

20. Shtof, A., Agathos, A., Gingold, Y., Shamir, A.: Cohen-or D.: Geosemantic snapping for sketch-based modeling. Computer Graphics Forum 32(2.2), 245–253 (2013)

21. Arbelaez, P., Maire, M.: Contour detection and hierarchical image segmentation. Pattern Analysis and Machine Learning 33(5), 898–916 (2011)

22. Cheng, M.: Curve structure extraction for cartoon images. In: Proceedings of the 5th Joint Conference on Harmonious Human Machine Environment, pp. 13–25 (2009)

23. Hu, P., Cai, H., Bu, F.: D-Sweep: Using Profile Snapping for 3D Object Extraction from Single Image. In: Christie, M., Li, T.-Y. (eds.) SG 2014. LNCS, vol. 8698, pp. 39–50. Springer, Heidelberg (2014)

Hessian Regularized Sparse Coding
for Human Action Recognition

Weifeng Liu[1], Zhen Wang[1], Dapeng Tao[2,3], and Jun Yu[4]

[1] China University of Petroleum (East China), Qingdao, 266580, China
[2] Shenzhen Institutes of Advanced Technology, Chinese Academy of Science, Shenzhen, China
[3] The Chinese University of Hong Kong, Hong Kong, China
[4] Hangzhou Dianzi University, Hangzhou, 310018, China
liuwf@upc.edu.cn

Abstract. With the rapid increase of online videos, recognition and search in videos becomes a new trend in multimedia computing. Action recognition in videos thus draws intensive research concerns recently. Second, sparse representation has become state-of-the-art solution in computer vision because it has several advantages for data representation including easy interpretation, quick indexing and considerable connection with biological vision. One prominent sparse representation algorithm is Laplacian regularized sparse coding (LaplacianSC). However, LaplacianSC biases the results toward a constant and thus results in poor generalization. In this paper, we propose Hessian regularized sparse coding (HessianSC) for action recognition. In contrast to LaplacianSC, HessianSC can well preserve the local geometry and steer the sparse coding varying linearly along the manifold of data distribution. We also present a fast iterative shrinkage-thresholding algorithm (FISTA) for HessianSC. Extensive experiments on human motion database (HMDB51) demonstrate that HessianSC significantly outperforms LaplacianSC and the traditional sparse coding algorithm for action recognition.

Keywords: Action recognition, sparse coding, Hessian regularization, manifold learning.

1 Introduction

Due to the development of Internet technology and smart devices, explosive growth of social videos are produced and spread on the Internet frequently. For example, every day YouTube streams more than 1 billion videos most of which are unlabeled. It is impractically expensive to manually annotate this huge volume of videos. Thus there is an emergent demand for effective methods which can help to organize this increasing visual media data including video summarizing, indexing, retrieval, classification and annotation [1]. And human action recognition is one of the most attractive research topic very recently.

Given an unknown video sequence, human action recognition aims to automatically classify ongoing actions including gestures, movement, interactions, and group

X. He et al. (Eds.): MMM 2015, Part II, LNCS 8936, pp. 502–511, 2015.
© Springer International Publishing Switzerland 2015

activities. Most action recognition methodologies employ spatio-temporal features to describe action in videos by concatenating video frame along time to form a 3-D space-time representation [2]. Briefly speaking, it can be divided into three categories: (1) action recognition with space-time volumes [3][4]; (2) action recognition with space-time trajectories [5][6][7] and (3) action recognition with space-time local features [8][9][10][11][12][14][15][16]. The methods with space-time volumes [3][4] recognize human actions by measure the similarity between the test video volume and template video volume. The methods with space-time trajectories [5][6][7] interpret an human action as a set of space-time trajectories which consist of a set of 2-dimensional or 3-dimensional points corresponding to human joint positions. The methods with space-time local features represent and recognize human actions using local features extracted from 3-D space-time volumes including concatenation local features at every frame [8][9][12][13] or on interest points [10][11][14][15].

Considering the redundancy of the space-time features for action representation, it is essential to employ a proper representation to reveal the underlying process of these observations. Sparse coding has received growing attentions because of its promising performance in machine learning, signal processing, neuroscience and statistics. Sparse coding aims to learn a dictionary and simultaneously the sparse coordinates w.r.t. the dictionary to represent the observations. It yields an easier interpretation because each data point is represented as a linear combination of a small set of dictionary atoms. And also sparse coding has some considerable connection with biological vision mechanism [17]. Hence a lot of variant algorithms and applications of sparse coding have been developed in recent years. In brief, sparse coding algorithms can be categorized into the following groups: (1) reconstructive sparse coding [18][19][20][21][22], (2) structured sparse coding [23][24][25] and (3) manifold regularized sparse coding [26][27][28][29][30]. Reconstructive sparse coding minimizes the data reconstruction error by different optimization algorithms including matching pursuit [19] and basis pursuit [18]. Structured sparse coding exploits the structure sparsity for a certain purpose such as group sparsity [23], hierarchical sparsity [24] and latent space [25]. Manifold regularized sparse coding exploits the local geometry of the data distribution by graph regularization including graph Laplacian [26][27] and Hessian [28][29].

In this paper, we propose Hessian regularized sparse coding (HessianSC) for action recognition. In contrast to Laplacian, Hessian has richer null space and favors the solution varying linearly w.r.t. geodesic distance [31]. Then Hessian regularization can better preserve the local geometry and lead to better extrapolation capability. Hence, HessianSC can achieve smoother sparse coding that preserves local similarity and result more excellent performance than traditional sparse coding algorithms. We also present the fast iterative shrinkage-thresholding algorithm (FISTA) [32] for the optimization of HessianSC. In the sense of Nemirovsky and Yudin [33], FISTA is one "optimal" first order method for sparse coding [32] with an $O(1/k^2)$ complexity. Finally, we carefully implement HessianSC for action recognition and conduct experiments on the HMDB51 database [34]. To evaluate the performance of HessianSC, we also compare HessianSC with some baseline algorithms including traditional sparse coding and Laplacian regularized sparse coding (LaplacianSC). The experimental results verify the effectiveness of HessianSC by comparison with the baseline algorithms.

The rest of this paper is assigned as follows. Section 2 provide a brief description of the proposed Hessian regularized sparse coding (HessianSC) algorithm. Section 3 introduces the optimization scheme of HessianSC using FISTA. Section 4 reports some experimental results followed with conclusions in section 5.

2   Hessian Regularized Sparse Coding

Suppose we are given N examples $S = \{x_i\}_{i=1}^N$, sparse coding aims to learn the sparse representation w_i of each example x_i simultaneously with a dictionary D. In the following section of this paper, we use $X = [x_1, \cdots, x_N] \in R^{m \times N}$ to denote the data matrix of examples and $W = [w_1, \cdots, w_N] \in R^{d \times N}$ to denote the sparse codes matrix w.r.t. to the dictionary $D = [D_1, \cdots, D_d] \in R^{m \times d}$. Then sparse coding can be formulated as follows:

$$\min_{D,W} \frac{1}{2N} \|X - DW\|_F^2 + \lambda_1 \sum_{i=1}^N \|w_i\|_1 \, , \text{s.t.} \left\|D_j\right\|_2 \leq 1, 1 \leq j \leq d. \tag{1}$$

Under manifold assumption [35], it is crucial to explore the local geometry because the sparse codes w_i, w_j of two examples x_i and x_j respectively are close to each other if the two examples are close in the intrinsic geometry of the data distribution. Hence in this paper, we integrate Hessian regularization into the objective function of sparse coding and reformulate HessianSC as below:

$$\min_{D,W} \frac{1}{2N} \|X - DW\|_F^2 + \lambda_1 \sum_{i=1}^N \|w_i\|_1 + \lambda_2 Tr(WHW^T), \text{s.t.} \left\|D_j\right\|_2 \leq 1, 1 \leq j \leq d.$$

$$\tag{2}$$

Here H is the Hessian computed from the data matrix.

The objective function in (2) is convex w.r.t. D or W separately, but it is not convex w.r.t. both variables together. In this paper, we employ alternating optimization to solve the problem by optimizing one variable while keeping the other one fixed. Thus the solution of (2) can be generally divided into two parts: sparse coding and dictionary updating. In the following section, we detail the optimization algorithm of (2).

3   Optimization of HessianSC

The optimization of HessianSC contains two steps: (1) learning sparse codes W given fixed dictionary D and (2) updating dictionary D given fixed sparse codes W. In particular, given fixed dictionary D, the problem (2) can be written as the follow subproblem:

$$\min_W \frac{1}{2N}\|X - DW\|_F^2 + \lambda_1 \sum_{i=1}^N \|w_i\|_1 + \lambda_2 Tr(WHW^T). \qquad (3)$$

Given fixed sparse codes W, the problem (2) can be written as the follow sub-problem:

$$\min_D \frac{1}{2N}\|X - DW\|_F^2, \text{s.t.} \|D_j\|_2 \le 1, 1 \le j \le d. \qquad (4)$$

In the following, we present the optimization of subproblem (3) and (4) in detail.

3.1 Learning Sparse Codes W with Fixed D

In this section, we describe the optimization of subproblem (3) using FISTA [32]. Subproblem (3) can be expressed as the general form:

$$\min_W \{F(W) \equiv f(W) + g(W)\}, \qquad (5)$$

where $f(W) = \frac{1}{2N}\|X - DW\|_F^2 + \lambda_2 Tr(WHW^T)$, $g(W) = \lambda_1 \sum_{i=1}^N \|w_i\|_1$. $f(W)$ and $g(W)$ are both convex functions.

Adopting gradient algorithm, subproblem (5) leads to the iterative scheme:

$$W_k = \operatorname{argmin}_W \left\{ Q_L(W, W_{k-1}) \equiv f(W_{k-1}) + \langle W - W_{k-1}, \nabla f(W_{k-1})\rangle + \frac{L}{2}\|W - W_{k-1}\|^2 + \lambda_1 \sum_{i=1}^N \|w_i\|_1 \right\} \quad (6)$$

where $\langle A, B\rangle = Tr(A^T B)$, and L is the Lipschitz constant of ∇f. Ignoring constant terms, (6) can be rewritten as:

$$W_k = p_L(W_{k-1}) \equiv \operatorname{argmin}_W \left\{ \frac{L}{2}\left\| W - \left(W_{k-1} - \frac{1}{L}\nabla f(W_{k-1})\right)\right\|^2 + \lambda_1 \sum_{i=1}^N \|w_i\|_1 \right\}. \qquad (7)$$

Since l_1 norm is separable, subproblem (7) can then be solved using the shrinkage operator as follows:

$$w_i = \mathcal{T}_{\frac{\lambda_1}{L}}\left(w_{k-1} - \frac{1}{L}\nabla f(w_{k-1})\right),$$

where $\mathcal{T}_\alpha(x)_j = (|x_j| - \alpha)_+ sgn(x_j)$.

Then we can state the optimization of subproblem (3) using FISTA with back-tracking stepsize in Table 1.

3.2 Update Dictionary D with Fixed W

Subproblem (4) is a l_2-constrained least squares problem and can be equally rewritten as:

$$\min_D \|X - DW\|_F^2, \text{s.t.} \|D_j\|_2 \le 1, 1 \le j \le d \qquad (8)$$

In this section, we describe the optimization of (8) using Largrange dual [36].

Table 1. FISTA optimization for subproblem (3)

— **Input**: $X, D, H, \lambda_1, \lambda_2$
— **Output**: W
— **Step 0**: chose $W_0, Z_1 = W_0, L_0, \eta > 1, t_1 = 1$
— **Step k**:
— 1. set $\bar{L} = L_{k-1}$
— 2. repeat $\bar{L} = \eta \bar{L}$, until $F\big(p_{\bar{L}}(Z_k)\big) \leq Q_{\bar{L}}(p_{\bar{L}}(Z_k), Z_k)$
— 3. set $L_k = \bar{L}$
— 4. update
— $W_k = p_{L_k}(Z_k)$
— $t_{k+1} = \dfrac{1 + \sqrt{1 + 4 t_k^2}}{2}$
— $Z_{k+1} = W_k + \left(\dfrac{t_k - 1}{t_{k+1}}\right)(W_k - W_{k-1})$

Consider $\beta = [\beta_1, \cdots, \beta_d]$ as the Lagrange multiplier, the Lagrange dual function of subproblem (8) can be written as follows:

$$g(\beta) = \min_D \mathcal{L}(D, \beta) = \min_D \|X - DW\|_F^2 + \sum_{j=1}^{d} \beta_j \big(D_j^T D_j - 1\big)$$

$$= \min_D Tr\big((X - DW)^T (X - DW)\big) + Tr(D^T DB) - Tr(B)$$

$$= \min_D Tr(X^T X - 2D^T XW^T + W^T D^T DW + D^T DB - B) \qquad (9)$$

where $B = diag(\beta)$ is $d \times d$ diagonal matrix with diagonal entry $B_{jj} = \beta_j$ for all j.

Set the first order derivative of $\mathcal{L}(D, \beta)$ w.r.t. D to zero, we have
$$D^* WW^T - XW^T + D^* B = 0.$$

Then, we have
$$D^* = XW^T (WW^T + B)^{-1}. \qquad (10)$$

Substituting (10) into (9), the Lagrange dual of subproblem (4) can be written as:

$$g(\beta) = Tr(X^T X - XW^T(WW^T + B)^{-1}WX^T - B). \qquad (11)$$

After solving the maximization of (11) w.r.t β by using Newton's method, we obtain the optimal dictionary D^* as $D^* = XW^T(WW^T + B^*)^{-1}$.

4   Experiments

To evaluate the effectiveness of the proposed HessianSC, we apply support vector machines as classifier to the sparse codes obtained by HessianSC for action recognition. We conduct the experiments on the HMBD51 database [1]. HMDB51 contains 6849 video clips of 51 distinct action categories, each containing at least 101 clips. Each clip was validated by at least two human observes to ensure consistency. The 51 actions categories can be grouped in five types: (1) general facial actions, (2) facial actions with object manipulation, (3) general body movements, (4) body movements with object interaction and (5) body movements for human interaction (see Table 2). Figure 1 shows some sample frames of different categories from the HDMB51 database.

Fig. 1. Some sample frames from HDMB51 database

We implement bag-of-words on a concatenation of the HOG and HOF features to obtain action descriptors. The HOG and HOF features have been shown to be state-of-the-art descriptors for action representation. In this paper, we use the HOG and HOF features around 3D Harris corners which are provided by Kuehne et al. [1]. In particular, we cluster a subset of 100000 features sampled from the training videos with the k-means algorithm and form a set of 2000 visual words. By matching every local point descriptors to the nearest visual words, each action clip can be represented with a 2000-dimensional feature vector which is a histogram over the index of the matched code book entries. Sequently, these action descriptors are used to obtain sparse codes by HessianSC.

Table 2. Actions categories in five types

Action groups	Action labels
General facial actions	smile, laugh, chew, talk
Facial actions with object manipulation	smoke, eat, drink
General body movements	cartwheel, clap hands, climb, climb stairs, dive, fall on the floor, backhand flip, hand-stand, jump, pull up, push up, run, sit down, sit up, somersault, stand up, turn, walk, wave
Body movements with object interaction	brush hair, catch, draw sword, dribble, golf, hit something, kick ball, pick, pour, push something, ride bike, ride horse, shoot ball, shoot bow, shoot gun, swing baseball bat, sword exercise, throw
Body movements for human interaction	fencing, hug, kick someone, kiss, punch, shake hands, swordfight

According to [1], we select 70 training and 30 testing clips from each action class to form our experiment dataset including a training set that contains 3570 clips and a test set that contains 1530 clips. We compare the proposed HessianSC with Laplacian regularized sparse coding (LaplacianSC) and the traditional sparse coding algorithms. In our experiments, The number of dictionary atoms is set to 200, the parameters λ_1 and λ_2 are tuned from the candidate set $\{1 \times 10^e | -30, \cdots, 10\}$, and the number of the neighbors for computing Hessian and Laplacian is set to 100 empirically. Considering multiple action category recognition, we adopt one vs. one method that selects two action categories each time for classification.

Figure 2 shows the confusion matrix on selected 20 action categories. Although errors look like to be spread across category labels randomly, HessianSC performs significantly better than LaplacianSC and the traditional sparse coding methods.

Figure 3 illustrates the accuracy performance on the selected 20 single action categories. From Figure 3, we can see that HessianSC outperforms than the baseline algorithms including LaplacianSC and traditional sparse coding in most cases.

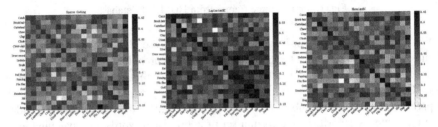

Fig. 2. Confusion matrix on selected 20 action categories

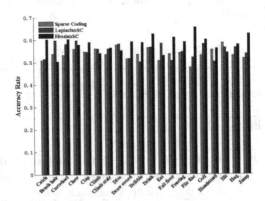

Fig. 3. Accuracy on single action category

5   Conclusion

Human action recognition have received intensive research attentions with the explosively growing of online videos. Although there are a lot of action representation methods, sparse coding has achieved stat-of-the-art performance in many computer vision applications. In this paper, we employed Hessian regularized sparse coding for human action recognition. The proposed HessianSC can well preserve local similarity benefitting from Hessian regularization. We also present a fast iterative shrinkage thresholding algorithm (FISTA) for efficient solving HessianSC. We apply HessianSC to support vector machines for action recognition. Extensive experiments on the HMBD51 database demonstrate that the proposed HessianSC significantly outperforms LaplacianSC and the traditional sparse coding algorithm for action recognition.

Acknowledgement. This paper is supported partly by National Natural Science Foundation of China (61301242, 61271407), Natural Science Foundation of Shandong Province (ZR2011FQ016), Fundamental Research for the Central Universities (13CX2096A).

References

1. Kuehne, H., Jhuang, H., Garrote, E., Poggio, T., Serre, T.: HMDB: A large video database for human motion recognition. In: IEEE International Conference on Computer Vision (ICCV), pp. 2556–2563 (2011)
2. Aggarwal, J.K., Ryoo, M.S.: Human activity analysis: A review. ACM Computing Surveys (CSUR) 43(3), 16 (2011)
3. Ke, Y., Sukthankar, R., Hebert, M.: Spatio-temporal shape and flow correlation for action recognition. In: IEEE Conference on Computer Vision and Pattern Recognition (CVPR), pp. 1–8 (2007)
4. Rodriguez, M., Ahmed, J., Shah, M.: Action MACH: A spatio-temporal maximum average correlation height filter for action recognition. In: IEEE Conference on Computer Vision and Pattern Recognition (CVPR) (2008)
5. Campbell, L.W., Bobick, A.F.: Recognition of human body motion using phase space constraints. In: IEEE International Conference Computer Vision, pp. 624–630 (1995)
6. Rao, C., Shah, M.: View-invariance in action recognition. In: IEEE Conferences on Computer Vision and Pattern Recognition (CVPR), vol. 2, p. II-316 (2001)
7. Sheikh, Y., Sheikh, M., Shah, M.: Exploring the space of a human action. In: IEEE International Conference on Computer Vision, vol. 1, pp. 144–149 (2005)
8. Chomat, O., Crowley, J.L.: Probabilistic recognition of activity using local appearance. In: IEEE Conference on Computer Vision and Pattern Recognition, vol. 2 (1999)
9. Zelnik-Manor, L., Irani, M.: Event-based analysis of video. In: IEEE Conference on Computer Vision and Pattern Recognition (CVPR), vol. 2, p. II-123 (2001)
10. Laptev, I.: On space-time interest points. International Journal of Computer Vision 64(2-3), 107–123 (2005)
11. Yilmaz, A., Shah, M.: Actions sketch: A novel action representation. In: IEEE Conference on Computer Vision and Pattern Recognition, vol. 1, pp. 984–989 (2005)
12. Blank, M., Gorelick, L., Shechtman, E., Irani, M., Basri, R.: Actions as space-time shapes. In: IEEE International Conference on Computer Vision (ICCV), vol. 2, pp. 1395–1402 (2005)

13. Yu, J., Tao, D., Wang, M., Rui, Y.: Learning to Rank Using User Clicks and Visual Features for Image Retrieval. IEEE Transactions on Cybernetics (2014), 10.1109/TCYB.2014.2336697

14. Niebles, J.C., Wang, H., Fei-Fei, L.: Unsupervised learning of human action categories using spatial-temporal words. International Journal of Computer Vision 79(3), 299–318 (2008)

15. Ryoo, M.S., Aggarwal, J.K.: Spatio-temporal relationship match: Video structure comparison for recognition of complex human activities. In: IEEE International Conference on Computer Vision (ICCV), pp. 1593–1600 (2009)

16. Hong, C., Yu, J., Chen, X.: Image-Based 3D Human Pose Recovery with Locality Sensitive Sparse Retrieval. In: 2013 IEEE International Conference on Systems, Man, and Cybernetics (SMC), pp. 2103–2108 (2013)

17. Olshausen, B.A.: Emergence of simple-cell receptive field properties by learning a sparse code for natural images. Nature 381(6583), 607–609 (1996)

18. Chen, S.S., Donoho, D.L., Saunders, M.A.: Atomic decomposition by basis pursuit. SIAM Journal on Scientific Computing 20(1), 33–61 (1998)

19. Mallat, S.G., Zhang, Z.: Matching pursuits with time-frequency dictionaries. IEEE Transactions on Signal Processing 41(12), 3397–3415 (1993)

20. Yu, J., Rui, Y., Tao, D.: Click Prediction for Web Image Reranking using Multimodal Sparse Coding. IEEE Transactions on Image Processing 23(5), 2019–2032 (2014)

21. Liu, B.-D., Wang, Y.-X., Zhang, Y.-J., Shen, B.: Learning dictionary on manifolds for image classification. Pattern Recognition 46(7), 1879–1890 (2013)

22. Liu, B.-D., Wang, Y.-X., Shen, B., Zhang, Y.-J., Hebert, M.: Self-explanatory sparse representation for image classification. In: Fleet, D., Pajdla, T., Schiele, B., Tuytelaars, T. (eds.) ECCV 2014, Part II. LNCS, vol. 8690, pp. 600–616. Springer, Heidelberg (2014)

23. Yuan, M., Lin, Y.: Model selection and estimation in regression with grouped variables. Journal of the Royal Statistical Society: Series B (Statistical Methodology) 68(1), 49–67 (2006)

24. Jenatton, R., Mairal, J., Bach, F.R., Obozinski, G.R.: Proximal methods for sparse hierarchical dictionary learning. In: The 27th International Conference on Machine Learning (ICML), pp. 487–494 (2010)

25. Jia, Y., Salzmann, M., Darrell, T.: Factorized latent spaces with structured sparsity. In: Advances in Neural Information Processing Systems, pp. 982–990 (2010)

26. Zheng, M., Bu, J., Chen, C., Wang, C., Zhang, L., Qiu, G., Cai, D.: Graph regularized sparse coding for image representation. IEEE Transactions on Image Processing 20(5), 1327–1336 (2011)

27. Gao, S., Tsang, I.W.-H., Chia, L.-T.: Laplacian sparse coding, hypergraph laplacian sparse coding, and applications. IEEE Transactions on Pattern Analysis and Machine Intelligence 35(1), 92–104 (2013)

28. Zheng, M., Bu, J., Chen, C.: Hessian sparse coding. Neurocomputing 123, 247–254 (2014)

29. Liu, W., Tao, D., Cheng, J., Tang, Y.: Multiview hessian discriminative sparse coding for image annotation. Computer Vision and Image Understanding 118, 50–60 (2014)

30. Yu, J., Wang, M., Tao, D.: Semisupervised multiview distance metric learning for cartoon synthesis. IEEE Transactions on Image Processing 21(11), 4636–4648 (2012)

31. Kim, K.I., Steinke, F., Hein, M.: Semi-supervised regression using hessian energy with an application to semi-supervised dimensionality reduction. In: Advances in Neural Information Processing Systems, pp. 979–987 (2009)

32. Beck, A., Teboulle, M.: A fast iterative shrinkage-thresholding algorithm for linear inverse problems. SIAM Journal on Imaging Sciences 2(1), 183–202 (2009)

33. Nemirovsky, A.S., Yudin, D.B.: Problem complexity and method efficiency in optimization (1983)

34. Kuehne, H., Jhuang, H., Garrote, E., Poggio, T., Serre, T.: HMDB: A large video database for human motion recognition. In: IEEE International Conference on Computer Vision (ICCV), pp. 2556–2563 (2011)
35. Belkin, M., Niyogi, P., Sindhwani, V.: Manifold regularization: A geometric framework for learning from labeled and unlabeled examples. The Journal of Machine Learning Research 7, 2399–2434 (2006)
36. Lee, H., Battle, A., Raina, R., Ng, A.Y.: Efficient sparse coding algorithms. In: Advances in Neural Information Processing Systems, pp. 801–808 (2006)

Robust Multi-label Image Classification
with Semi-Supervised Learning and Active Learning

Fuming Sun, Meixiang Xu, and Xiaojun Jiang

Liaoning University of Technology, Jinzhou, 121001, P.R. China
sunfm@mail.ustc.edu.cn

Abstract. Most existing work on multi-label learning focused on supervised learning which requires manual annotation samples that is labor-intensive, time-consuming and costly. To address such a problem, we present a novel method that incorporates active learning into the semi-supervised learning for multi-label image classification. What's more, aiming at the curse of dimensionality existing in high-dimensional data, we explore a dimensionality reduction technique with non-negative sparseness constraint to extract a group of features that can completely describe the data and hence make the learning model more efficiently. Experimental results on common data sets validate that the proposed algorithm is relatively effective to improve the performance of the learner in multi-label classification, and the obtained learner is with reliability and robustness after data dimensionality using NNS-DR (Non-Negative Sparseness for Dimensionality Reduction).

Keywords: multi-label learning, semi-supervised regression, non-negative sparseness, dimensionality reduction.

1 Introduction

Recently the research on multi-label learning has received more and more attention in the society of multimedia. A large number of methods on multi-label learning have been proposed and widely applied to image processing and pattern recognition field, such as text classification [1], scene classification [2], bio-informatics [3], image and video annotation [4], and so on. The typical multi-label learning methods include Multi-Label K-Nearest Neighbor (ML-KNN) algorithm [5], decision tree algorithm [6], and Kernel method [3], etc. However, most of these existing algorithms may achieve success to efficiently work in supervised settings where a larger training set with labeled examples is available. In other words, in order to obtain a learning model with a higher classification accuracy and stronger generalization performance, the labeled samples must be sufficient. Unfortunately, it is often in difficulty to meet this requirement since the insufficiency of training data is often encountered in multi-label classification. One way to solve the shortage of training data is to resort to manually annotating images, whereas manual annotating images is laboring, time-consuming and quite costly. An alternative way is to use abundant unlabeled data, which is often called Semi-Supervised Learning (SSL). In particular, co-training [7] as one of the

X. He et al. (Eds.): MMM 2015, Part II, LNCS 8936, pp. 512–523, 2015.

main paradigms of semi-supervised learning has been proved to be an efficient method in single label learning, and its efficiency of co-training is based on an assumption that unlabeled examples are much more than labeled ones. However, this assumption doesn't always hold because of the large variations of classification labels when massive unlabeled examples are utilized. To target this issue, Co-training style semi-supervised regression algorithm, named CO-training REGressors (COREG), is proposed in [8]. Unlike co-training, the advantage of COREG is that it doesn't require two views of the data satisfying sufficient and redundant features. Therefore, it is applicable to any regression problems without natural attribute partitions.

Note that COREG mainly focuses on binary classification problems where a sample has only one label and it has not been investigated well in SSL under the multi-label learning background. Here we take image classification as a multi-label COREG problem and raise Multi-Label Semi-Supervised Active Image Classification (ML-SSAIC) method which attempts to incorporate active learning into semi-supervised regression. The proposed method combines the merits of semi-supervised learning and active learning. In a nutshell, we first preprocess the multi-label data by a novel NNS-DR (Non-Negative Sparseness Dimensionality Reduction) method to alleviate the curse of dimensionality. Then, inspired by COREG, ML-KNN together with QUIRE [20] the ML-SSAIC approach is proposed to handle the problem of high labors costs caused by manual annotation in supervised multi-label learning. Experimental results demonstrate that the proposed method is more robust and has higher classification performance compared with the other algorithms.

The outline of this paper is given as follows. Some related works are reviewed in Section2. After that, the proposed method is detailed in Section 3 and Section 4. Section 5 reports experimental results followed by the conclusion in Section 6.

2   Related Work

There are some classical SSL methods, like self-training [9], co-training [29], transductive SVM [10], and graph-based [12, 13] SSL algorithms. Two aspects should be considered. One is how to exploit the unlabeled examples. The other is how to excavate the associations implied in multiple labels.   Some multi-label algorithms based on semi-supervised methods [11-14] have been proposed in the assignments of image/video content analysis. Among them, Liu et al. [11] brought up a semi-supervised multi-label classification approach based on Constrained Non-negative Matrix Factorization (CNMF) with constraints. It assumes that two samples tend to have larger overlap in their assigned class memberships if they shared high similarity in their input properties. Chen et al. [12] and Zha et al. [13] first built two graphs on feature and label level respectively, and then excavated the multi-label inter-similarity, which gives rise to the label smoothness over multiple labels for each sample. Moreover, Guo et al. [14] raised a multi-label SSL algorithm by combining the multi-label maximum margin method with the unsupervised subspace representation approach.

Just like semi-supervised learning, despite that the main efforts in active learning worked on single-label learning, multi-label active learning has drawn more and more attention in image classification for it can significantly reduce the effort in labeling training samples [15, 16]. For different settings, there are a variety of methods to address different learning problems, such as the approach for multi-view data [17, 24], the batch model form algorithm [18], and a two dimensional approach to deciding iteratively whether a label is positive on a specific example [19] etc. The primary restriction of existing active learning methods for multi-label learning lies in that they are limited to selecting the most informative examples and ignoring the correlations among multiple labels. To address such limitations, Huang et al [20] proposed the QUIRE method by combing label correlation with the measures of representativeness and informativeness for query selection.

In this paper, we address the semi-supervised co-training regression method with the active QUIRE algorithm. It has double advantages. (1) The higher labor caused by human annotation can be greatly reduced. (2) Incorporating active learning into semi-supervised learning can help the classifier obtain the required performance with training samples as few as possible.

3   Multi-label Dimensionality Reduction via Non-negative Sparseness

Sparse representations theory has its background of cognitive science and has been widely applied to compressed sensing [30], dictionary learning [26], image classification [27] and image re-ranking [25], etc. Theories and extensive experiments demonstrate that the sample similarity measure based on sparse representation has certain discrimination ability. Usually one sample can be obtained by its similar samples through the linear weighted method, in which case the similarity between samples can be better characterized. To this end, in this section we first get the sparse reconstruction coefficient between the samples, then transfer it to the similarity between samples and eventually add it into the semi-supervised dimension as a constraint to process multi-label data in order to improve the performance of the multi-label learner.

3.1   Basics of Sparseness

For the convenience of further discussion, we first introduce some basic theories about sparse representation. Assuming that $X = [x_1 \quad x_2 \quad \cdots \quad x_n] \left(X \in R^{d \times N}, N > d \right)$, N is the number of the examples, and d denotes the dimension of each example. For a new example x_{new}, we expect to seek for its sparse linear reconstruction coefficients β by the following formula:

$$\hat{\beta}_0 = \arg \min \|\beta\|_0 \quad \text{s.t. } X\beta = x_{new} \tag{1}$$

where $\|\beta\|_0$ is the zero order norm of β, and its value is the number of the non-zero elements in β. However, solving (1) is a NP hard problem. If β is sparse, its solution is usually converted to solve the following equation:

$$\hat{\beta}_1 = \arg\min \|\beta\|_1 \quad \text{s.t.} \quad \|X\beta - x_{new}\|_2 \le \varepsilon \tag{2}$$

In (2), the constraint term is slack because noises often exist in the data. So it is difficult to reconstruct x_{new} accurately. In fact, $\hat{\beta}_1$ is less affected by such slackness, and $\hat{\beta}_1$ can be solved by the virtue of the linear programming method within polynomial time range. Further, the problem in (2) can be turned into the minimum problem under the l_1 norm regularization [23]:

$$\hat{\beta}_1 = \arg\min_{\beta} \|X\beta - x_{new}\|_2 + \lambda\|\beta\|_1 \tag{3}$$

Here, λ is a trade-off parameter between the l_1 norm and the reconstruction error. To overcome the problem that the similarity among the examples is negative, (3) can be rewritten as follows:

$$\hat{\beta}_1 = \arg\min_{\beta \ge 0} \|X\beta - x_{new}\|_2 + \lambda\|\beta\|_1 \tag{4}$$

where $\hat{\beta}_1 = [\hat{\beta}_{1,1}, \hat{\beta}_{1,2}, \cdots \hat{\beta}_{1,n}]$ is the sparse reconstruction coefficient with minimizing reconstruction error. In this paper $\hat{\beta}_1$ is used to measure the similarity among the examples, where $\hat{\beta}_{1,i}$ is the similarity between x_{new} and x_i. x_{new} is usually reconstructed by its homogenous examples, while reconstructed coefficients are estimated by its heterogeneous examples. Thus the similarity measure upon such reconstruction coefficients has its natural discriminative capability.

3.2   NNS-DR

Let S denote the known constraint set of the same class and D represent the known constraint set of different classes. Let $\{x_i, x_j\} \in S$ if x_i and x_j belong to the same class; $\{x_i, x_j\} \in D$, otherwise. The linear semi-supervised dimensionality reduction taking the edge information into consideration aims to seek for the projection matrix $P \in \mathbb{R}^{d \times r}$ under the supervision of both S and D. Then $X = [x_1, x_2, ..., x_n]$ is embedded into the r-dimension $(r \ll d)$ sub-space through P, the example set after dimensionality reduction is $Z = [z_1, z_2, \cdots z_n]$ and $Z = P^T X$ ($Z \in \mathbb{R}^{r \times n}$). For analysis convenience, we first discuss the case when $P \in \mathbb{R}^{d \times r} (r = 1)$. As for S and D, let O_S stand for the between-class compactness on S, and O_D denote the within-class scatter matrix on D respectively. Then O_S can be defined as follows:

$$O_S = \sum_{i,j}^{n} (z_i - z_j)^2 C_{ij}^S = \sum_{i,j}^{n} (p^T x_i - p^T x_j)^2 C_{ij}^S = \sum_{i,j=1}^{n} (p^T x_i x_i^T p - 2p^T x_i x_j^T p + p^T x_j x_j^T p) C_{ij}^S p$$

$$= 2 \left(\sum_{i,j}^{n} (p^T x_i x_i^T p - p^T x_i x_j^T p) C_{ij}^S \right) = 2p^T X (\Lambda^S - C^S) X^T p = 2p^T X L^S X^T p \tag{5}$$

where C^S is defined as (6):

$$C_{ij}^S = \begin{cases} 1, & \text{if } (x_i, x_j) \in S \\ 0, & \text{oterwise} \end{cases} \tag{6}$$

Additionally, $\Lambda^S = \sum_{j=1}^{n} C_{ij}^S$ is a diagonal matrix, $L^S = \Lambda^S - C^S$ is a graph Laplacian matrix and it is symmetric and positive semi-definite. Similarly, the between-class scatter among the heterogeneous samples is defined as (7):

$$O_D = \sum_{i,j}^{n} (z_i - z_j)^2 C_{ij}^D = \sum_{i,j=1}^{n} (p^T x_i - p^T x_j)^2 C_{ij}^D = \sum_{i,j=1}^{n} (p^T x_i x_i^T p - 2p^T x_i x_j^T p + p^T x_j x_j^T p) C_{ij}^D$$

$$= 2 \left(\sum_{i,j=1}^{n} (p^T x_i x_i^T p - p^T x_i x_j^T p) C_{ij}^D \right) = 2p^T X (\Lambda^D - C^D) X^T p = 2p^T X L^D X^T p \tag{7}$$

where C^D satisfies (8):

$$C_{ij}^D = \begin{cases} 1, & \text{if } (x_i, x_j) \in D \\ 0, & \text{oterwise} \end{cases} \tag{8}$$

And $\Lambda^D = \sum_{j=q}^{n} C_{ij}^D$ is a diagonal matrix, $L^D = \Lambda^D - C^D$ is a Laplacian matrix.

On the unlabeled data, the sparseness retaining term O is defined as (9):

$$O = \sum_{i,j=1}^{n} (z_i - z_j)^2 C_{ij} = \sum (p^T x_i - p^T x_j)^2 C_{ij} = 2p^T X (D - S) X^T p = 2p^T X L X^T p \tag{9}$$

where C_{ij} is the similarity between x_i and x_j (both x_i and x_j are represented with sparseness), $\Lambda = \sum_{j=1} C_{ij}$ is a diagonal matrix and L is a graph Laplacian matrix with $L = D - C$.

On the basis of O_S, O_D and O, the objective equation of semi-supervised dimensionality reduction via sparseness can be obtained as (10):

$$\hat{p} = \arg\max_p \frac{O_D}{O_S + \alpha O} = \arg\max_p \frac{p^T X L^D X^T p}{p^T X (L^S + \alpha L) X^T p} \tag{10}$$

where α is a balancing parameter between the regularization term and the constraints among the same class. With (10), it is guaranteed that in the mapping space, the homogenous examples is as close as possible and the heterogeneous ones become estranged as far as possible, and simultaneously the structure relations with sparseness retaining can be maintained in the low-dimension space. The problem in (10) can be transferred to solving the generalized eigenvalue problem regarding p:

$$XL^D X^T p = \mu X \left(L^S + \alpha L \right) X^T p \tag{11}$$

4   ML-SSAIC

In this section we first introduce the initial co-training algorithm, and then provide a high-level description of the proposed method ML-SSAIC.

4.1   Co-training

The standard co-training algorithm in [7] assumes that: (1) the features of a dataset can be partitioned into two sets; (2) each sub-feature set is sufficient to train a good learner; (3) the two sets are conditionally independent given the label. At first, two separate learners are trained utilizing the labeled data on the two sub-feature sets separately. Then, each learner makes predictions on the unlabeled data, and 'teaches' the other classifier with the few unlabeled examples (and the predicted labels) they consider most confident. Besides, each learner is retrained with the additional training examples provided by the other learner, and the process repeats. The final prediction on an unlabeled example is decided by the results of the two learners trained in each view. See Algorithm 1 for the detailed process of co-training.

Algorithm 1. Process of Co-training

Given: two views V_1 and V_2, the sets T and U of labeled and unlabeled examples, the number t of iterations

When U is not empty D_o

use $V_1(T)$ and $V_2(T)$ to train two learners h_1 and h_2

For each class do

Let E_1 and E_2 be the unlabeled examples on which h_1 and h_2 make the most confident predictions for each class

Remove E_1 and E_2 from U, label them based on h_1 and h_2, respectively, and add them to T

End   when

   Combine the predictions of h_1 and h_2

4.2   ML-SSAIC

For convenience, we denote by $D = \{(x_1, y_1), \dots (x_{|L|}, y_{|L|}), (x_{|L|+1}, y_{|L|+1}), \dots, x_{|L|+|U|}\}$ the training data set including $|L|$ labeled examples and $|U|$ unlabeled examples, where each example $x_i = [x_{i1}, x_{i2}, ...x_{id}]$ is a vector of d dimensions and $y_i = [y_{i1}, y_{i2}, ...y_{iL}]$ is the class label of x_i. Let $L = \{(x_1, y_1), ..., (x_{|L|}, y_{|L|})\}$ represent the labeled sample set and $U = \{x_{|L|+1}, ..., x_{|L|+|U|}\}$ be the unlabeled sample set whose real-valued labels are unknown, where $|L|$ and $|U|$ are the number of labeled examples and unlabeled examples respectively.

Motivated by the co-training paradigm, ML-SSAIC tries to exploit U to improve the performance of classification in multi-label learning. In the process of co-training of semi-supervised regression, two classifiers with diversity are iteratively trained on the training set, and the predictions of either classifier on the unlabeled examples are used to augment the training set of the other. Here, the ML-KNN model is adopted as the base learner. In detail, ML-SSRIC uses two regressors with diversity, in each round of iterative regression, the two regressors are retrained by adding the new labeled data. The new labeled data are partly from the valuable examples by the other regressor and partially from the examples selected through the active QUIRE method that measures both the informativeness and representativeness of an example by its prediction uncertainty. Detailed information about the active QUIRE can be found in literature [20]. Note that the learners used by ML-SSAIC could be implemented in different ways, we utilize ML-KNN as the base classifier in this paper. Besides, to guarantee that the initial two classifiers are diverse, different distance metrics and different k values are adopted in ML-KNN.

It is worth mentioning that the standard co-training requires sufficient but redundant views, but some related works show that using two different supervised learners instead of two attribute sets can also work better, which is really the way ML-SSAIC goes. Note that other kinds of semi-supervised and active learning paradigms can also be used here to exploit the unlabeled data. ML-SSAIC uses the co-training scheme just because this enables semi-supervised learning and active learning be easily and gracefully integrated together. In summary, the pseudo-code of ML-SSAIC is presented as Algorithm 2.

5   Experiments and Analysis

5.1   Discriminative Capability of NNS-DR

In this section, we use ML-KNN and ML-NB [22] (Multi-Label Naive Bayes Classification) as the base classifier respectively to evaluate the discriminative capability of NNS-DR on 3 real-world tasks with 5 evaluation metrics, i.e., Average Precision (AP), Hamming Loss (HL), Ranking Loss (RL), One Error (OE), and Coverage. The test results are shown in Table 1 and Table 2. Details of three datasets are introduced as follows. (1) **Image annotation.** This task is automatic image annotation on Corel database containing 5000 images, each represented by a 500-dimensional feature vector with 374 words utilized for annotation. (2) **Text classification.** In this task, the experiment is done on the Rcvlv2 database. The used subset is represented by a 4000-dimensional feature vector, including 6000 documents with 50 labels. (3)

Video Annotation. This task is automatic video annotation on Kodak dataset containing 3590 instances, and each instance has 409 features and 20 possible labels.

In Table 1 and Table 2, Original means the dataset is with the initial feature dimension, NNS means the data is preprocessed by the NNS-DR method, and P&I (%) is the percentage of the performance improvement after dimensionality reduction. '+'conveys that the performance of the classifier is improved by NNS-DR, while '-' shows that the performance of the classifier is degraded on the data with dimensionality reduction. Symbol $\uparrow$ ($\downarrow$) means the bigger (smaller) the value is, the better the performance is. It can be seen from the above results that with NNS-DR the performance of ML-KNN and ML-NB is enhanced greatly on most evaluation metrics, although it is sometimes worsened on certain evaluation metrics. Overall, NNS-DR can help enhance the classification capability of the learner in most cases.

Algorithm 2. SSR-CT for Multi-label Classification

Input: data set D
Procedure:
1. Initialize
   Divide D into L, U ; create two pools P_1 and P_2 from U randomly, choose two labeled set L_1 and L_2 from L, empty containers C_1, C_2
2. Train classifier h_1 and h_2 on L_1 and L_2 respectively.
3. When U is not empty do
4. For h_1

   1) h_1 make prediction on P_1, and select the example from P_1 on which h_1 makes the most confident prediction on L_1 then add it to C_1, and remove the selected examples from P_1

   2) as for the remained examples in P_1, use the active QUIRE method to query the most informative and representative one and add it to C_1

5. For h_2

   1) h_2 make prediction on P_2, and select the example from P_2 on which h_2 makes the most confident prediction on L_2 then add it to C_2, and remove the selected examples from P_2

   2) as for the remained examples in P_2, use the active QUIRE method to query the most informative and representative one and add it to C_2

6. Update h_1 using C_1, Update h_2 utilizing C_2
7. Replenish P_1 and P_2 from U randomly, and clear C_1, C_2
8. End when
Output: h_1 and h_2. Make predictions on testing set combining h_1 and h_2.

Table 1. Results on three datasets when ML-KNN is utilized as the base classifier

| Data | | Evaluation Metrics | | | | |
		AP $\uparrow$	HL $\downarrow$	RL $\downarrow$	OE $\downarrow$	Coverage $\downarrow$
Corel	Original	0.4104	0.4002	0.5351	0.6957	10.3532
	NNS	0.6097	0.3412	0.3219	0.4228	8.3213
	P&I(%)	+48.5	+14.7	+38.0	+39.9	+19.6
Rcv1v2	Original	0.5219	0.2476	0.7061	0.4810	2.0251
	NNS	0.6757	0.3291	0.2878	0.4955	1.4078
	P&I(%)	+29.5	-32.9	+58.6	-3.0	+30.5
Kodak	Original	0.5742	0.3132	0.4097	0.5608	3.0464
	NNS	0.7705	0.2509	0.1896	0.3025	1.8738
	P&I(%)	+34.2	+19.9	+53.7	+46.1	+38.5

Table 2. Results on three datasets when ML-NB is utilized as the base classifier

| Data | | Evaluation Metrics | | | | |
		AP $\uparrow$	HL $\downarrow$	RL $\downarrow$	OE $\downarrow$	Coverage $\downarrow$
Corel	Original	0.6235	0.4163	0.3480	0.2771	1.293
	NNS	0.6968	0.3685	0.3260	0.6033	1.5271
	P&I(%)	+11.8%	+11.5%	+6.7%	-55.7%	-18.1%
Rcv1v2	Original	0.2830	0.2961	0.2973	0.676	138.233
	NNS	0.4498	0.1405	0.1885	0.134	132.4875
	P&I(%)	+58.9%	+52.5%	+36.6%	+80.1%	+4.2%
Kodak	Original	0.4479	0.5119	0.6705	0.6983	3.6814
	NNS	0.5577%	0.3133	0.4454	0.6373	3.0898
	P&I(%)	+24.5%	+38.8%	+33.6%	+8.7%	+16.1%

5.2   Classification Performance

In this part, experiments are conducted on two datasets (i.e. scene and yeast). Their relevant descriptions are detailed as follows. (1)   **Scene.** This data set contains 2000 scene images each represented by a 294-dimensional feature vector with all the 5 possible labels including desert, mountains, sea, sunset and trees, where multiple labels is manually assigned to each image. (2) **Yeast.** Yeast is a data set for predicting the gene functional classes of the Yeast Saccha-romyces cerevisiae, consisting of 2417 genes and 14 possible labels, and on average each gene is associated with 4.24 $\pm$ 1.57 labels.

Relevant experimental settings are described as follows: for each dataset, 15% of the examples in the data set are used as testing data set, 20% as training data set, and the rest of the examples as the unlabeled data set. According to the above settings, the performance comparison between ML-SSAIC, ML-Cotrain and TRAM (Transductive Multi-label classification) [21] on two data sets is shown as Fig.1 and Fig.2. It can be

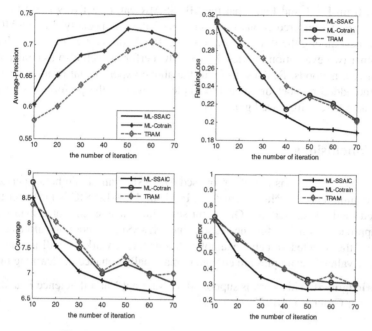

Fig. 1. Comparisons on Yeast with four evaluations

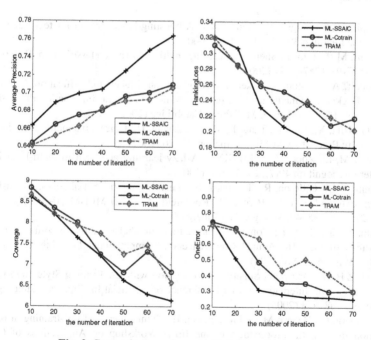

Fig. 2. Comparisons on Image with four evaluations

observed from Fig.1 and Fig.2 that the ML-SSAIR can effectively boost the performance of regression predictions using unlabeled data. Especially, ML-SSAIC is always the best among the three algorithms on AP, OE on the two datasets. Moreover, on the other two evaluation metrics, ML-SSAIC performs well also as compared with the other two methods. Besides, from all the above experimental results, we can find that as the added number of labeled examples increases, the performance of the approach taken here is improved gradually.

6   Conclusions

We can draw conclusions that the proposed ML-SSIC can obtain better performance on four common used evaluation metrics. Particularly, ML-SSIC is far better than the compared methods on AP and OE.   But sometimes it is even worse than the compared approaches on RL for the reason that ML-SSIC is possibly affected by the number of the selected valuable examples. In our future work we will further discuss how many valuable examples are enough to maximally enhance the learning model.

Acknowledgment. This work is supported by National Natural Science Foundation of China under Grant No. 61272214.

References

1. Schapire, R.E., Singer, Y.: BoosTexter: A boosting-based system for text categorization. Machine Learning 39(2-3), 135–168 (2000)
2. Boutell, M.R., Luo, J., Shen, X.: Learning multi-label scene classification. Pattern Recognition 37(9), 1757–1771 (2004)
3. Elisseeff, A., Weston, J.: A kernal method for multi-labeled classification. In: Dietterich, T.G., Becker, S., Ghahrarnani, Z. (eds.) Advances in Neural Information Processing Systems, vol. 14, pp. 681–687. MIT Press, Cambridge (2002)
4. Qi, G., Hua, X., Rui, Y., Tang, J., Mei, T., Zhang, H.: Correlative multi-label video annotation. In: Proc. of ACM Multimedia, pp. 17–26 (2007)
5. Zhang, M.-L., Zhou, Z.H.: ML-KNN: A lazy learning approach to multi-label learning. Pattern Recognition 40(7), 2038–2048 (2007)
6. Comite, F.D., Gilleron, R., Tommasi, M.: Learning multi-label alternating decision trees from texts and data. In: Perner, P., Rosenfeld, A. (eds.) MLDM 2003. LNCS (LNAI), vol. 2734, pp. 35–49. Springer, Heidelberg (2003)
7. Blum, A., Mitchell, T.: Combining labeled and unlabeled data with co-training. In: Proceedings of the 11th Annual Conference on Computational Learning Theory (COLT 1998), Wisconsin, Ml (1998)
8. Zhou, Z.H., Li, M.: Semi-Supervised Regression with Co-Training Style Algorithms. In: Pro. of the 19th International Joint Conference on Artificial Intelligence, Edinburgh, Scotland, pp. 908–913 (2005)
9. Rosenberg, C., Hebert, M., Schneiderman, H.: Semi-supervised self-training of object detection models. In: Proceedings of the IEEE Workshop on Applications of Computer Vision, pp. 29–36 (2005)

10. Zhang, T., Oles, F.J.: A probability analysis on the value of unlabeled data for classification problems. In: Proceedings of the International Conference on Machine Learning (2000)
11. Liu, Y., Jin, R., Yang, L.: Semi-supervised multi-label learning by constrained Non-Negative matrix factorization. In: Proc. of AAAL, pp. 421–426. AIAA, USA (2006)
12. Chen, G., Song, Y., Wang, F., Zhang, C.: Semi-supervised multi-label learning by solving a Sylvester equation. In: Proc. of SDM, pp. 410–429. AAAI, Boston (2008)
13. Zha, Z.J., Mei, T., Wang, J., Hua, A.S.: Graph-based semi-supervised learning with multi-label. In: Proc. of ICME, pp. 1321–1324. Computer Society, Hannover (2009)
14. Guo, Y., Schuurmans, D.: Semi-supervised multi-label classification. In: Flach, P.A., De Bie, T., Cristianini, N. (eds.) ECML PKDD 2012, Part II. LNCS, vol. 7524, pp. 355–370. Springer, Heidelberg (2012)
15. Tang, J.H., Zha, Z.J., et al.: Semantic-gap oriented active learning for multi-Label image annotation. IEEE Trans. Image Processing 21(4), 2354–2360 (2012)
16. Tian, X.M., Tao, D.C., et al.: Active reranking for web image search. IEEE Trans. Image Processing 19(1), 805–820 (2010)
17. Zhang, X., Cheng, J., Xu, C., Lu, H., Ma, S.: Multi-view multi-label active learning for image classification. In: Proceedings of the IEEE International Conference on Multimedia and Expo., pp. 258–261 (2009)
18. Chakraborty, S., Balasubramanian, V., Panchanathan, S.: Optimal batch selection for active learning in multi-label classification. In: Proceedings of the 19th ACM International Conference on Multimedia, pp. 1413–1416 (2011)
19. Qi, G.-J., Hua, X.S., Rui, Y., et al.: Two-dimensional active learning for image classification. In: CVPR, pp. 1–8 (2008)
20. Huang, S.J., Jin, R., Zhou, Z.H.: Active learning by querying informative and representative examples (2014)
21. Kong, X.N., Michael, K., Zhou, Z.H.: Transductive Multi-Label Learning via Label Set Propagation. IEEE Trans. Knowledge and Data Engineering 25(3), 704–719 (2013)
22. Zhang, M.L., Pena, J.M., Robles, V.: Feature Selection for Multi-Label Naive Bayes Classification. Information Sciences 179(19), 3218–3229 (2009)
23. Huang, K., Aviyente, S.: Sparsity representation and signal classification. In: The 19th Annual Conference on Nerual Information Processing System, Vancouver, pp. 609–616 (2006)
24. Yu, J., Wang, M., Tao, D.: Semi-supervised Multiview Distance Metric Learning for Cartoon Synthesis. IEEE Trans. Image Processing 21(11), 4636–4648 (2012)
25. Yu, J., Rui, Y., Tao, D.: Click Prediction for Web Image Reranking using Multimodal Sparse Coding. IEEE Trans. Image Processing 23(5), 2019–2032 (2014)
26. Liu, B.-D., Wang, Y.-X., Zhang, Y.-J., Shen, B.: learning dictionary on manifolds for image classification. Pattern Recognition 46(7), 1879–1890 (2013)
27. Liu, B.-D., Wang, Y.-X., Shen, B., Zhang, Y.-J., Hebert, M.: Self-explanatory sparse representation for image classification. In: Fleet, D., Pajdla, T., Schiele, B., Tuytelaars, T. (eds.) ECCV 2014, Part II. LNCS, vol. 8690, pp. 600–616. Springer, Heidelberg (2014)
28. Liu, W., Tao, D.: Multiview Hessian Regularization for Image Annotation. IEEE Trans. Image Processing 22(7), 2676–2687 (2013)
29. Liu, W., Li, Y., Lin, X., Tao, D., Wang, Y.: Hessian regularized co-training for social activity recognition. PLoS ONE 9(9), e108474, doi:10.1371/journal.pone.0108474
30. Doloho, D.L.: Compressed sensing. IEEE Trans. Information Theory 52(4), 1289–1306 (2006)

Photo Quality Assessment with DCNN that Understands Image Well

Zhe Dong, Xu Shen, Houqiang Li, and Xinmei Tian

University of Science and Technology of China, Hefei, Anhui, 230027, China
{ustcdz,shenxu}@mail.ustc.edu.cn, {lihq,xinmei}@ustc.edu.cn

Abstract. Photo quality assessment from the view of human aesthetics, which tries to classify images into the categories of good and bad, has drawn a lot of attention in computer vision field. Up to now, experts have proposed many methods to deal with this problem. Most of those methods are based on the design of hand-crafted features. However, due to the complexity and subjectivity of human's aesthetic activities, it is difficult to describe and model all the factors that affect the photo aesthetic quality. Therefore those methods just obtain limited success. On the other hand, deep convolutional neural network has been proved to be effective in many computer vision problems and it does not need human efforts in the design of features. In this paper, we try to adopt a deep convolutional neural network that "understands" images well to conduct the photo aesthetic quality assessment. Firstly, we implement a deep convolutional neural network which has eight layers and millions of parameters. Then to "teach" this network enough knowledge about images, we train it on the ImageNet which is one of the largest available image database. Next, for each given image, we take the activations of the last layer of the neural network as its aesthetic feature. The experimental results on two large and reliable image aesthetic quality assessment datasets prove the effectiveness of our method.

Keywords: Image aesthetics, photo quality assessment, deep convolutional neural network.

1 Introduction

With the fast development of Internet, the amount of online images has gained an explosive growth that easily breaks through the magnitude of billions. What's more, even general consumers can also easily take a lot of digital photos as the high-performance capture devices become cheaper and more popular. According to this, new methods are needed to manage photos online or offline better and more intelligently.

Photo quality assessment from the perspective of human aesthetics attempts to classify images into good and bad. Fig. 1 shows an example. Most people will prefer the left four images as they are more beautiful. With effective photo aesthetic quality assessment method, image retrieval system can return images not only related to the given queries but also with high quality. This will surely provide better

X. He et al. (Eds.): MMM 2015, Part II, LNCS 8936, pp. 524–535, 2015.

(a)   (b)

Fig. 1. It may be easy for most people to agree on that the left four images in (a) look more beautiful than the right four images in (b). In other words, they have higher aesthetic quality.

user experience according to the user investigation conducted in [1]. Offline users who have a large image collection can also benefit from this method as it can help to select a small number of representative and beautiful photos automatically [2]. In addition, photo aesthetic quality assessment also helps to develop new image enhancement tools to make images look better [3, 4].

Researchers have made a lot of efforts to conduct photo quality assessment [3, 5, 6, 7, 8, 9, 10, 14, 15, 16, 24, 25]. They try to find out image attributes which are related to image aesthetic quality and describe them with mathematical model that can be handled by computers to extract features. The way by which they select the image attributes is mainly based on the user intuition and photography knowledge, such as the rule of thirds [3, 5, 6], colorfulness [5, 6, 7, 8], simplicity [7, 8] and so on.

But this kind of method suffers from several drawbacks. Firstly, we cannot point out all attributes that are related to image aesthetic quality assessment. In fact, it is hard to discover attributes that can affect image aesthetic quality. Researchers just adopt a small number of them which are well-known and easy to be implemented. Secondly, it is hard to explain how those image attributes affect the image aesthetic quality. For example, we may agree on that image color will affect the image aesthetic quality a lot. Professional photographers carefully set the image color to gain better visual effect. But there is no fixed discipline that the color scheme of beautiful images must follow. Both images with simple color palette and images which are colorful can have high aesthetic quality. Thirdly, it is not easy to describe the image attribute accurately with mathematical model even we are sure that the attribute can affect the image aesthetic quality. For example, sharpness is one of the most popular image attributes used in the research of image aesthetic quality assessment [6, 7, 8, 9, 14]. But the measure of sharpness itself is a hard problem which has already been studied for a long time and there still exists no method that can assess sharpness properly on all kinds of photos. When turning to more abstract image attributes such as simplicity and image composition, the problem becomes more complex and challenging.

To avoid those problems, some experts propose to adopt general image descriptors which are used to describe image content to do image aesthetic quality assessment [10, 25]. Those general image descriptors have gained remarkable success in the general image classification problem which tries to classify images into different categories according to their content [11]. As images look alike may also have similar image aesthetic quality, general descriptors such as dense SIFT also gain acceptable performance in this higher level computer vision problem reported by Marchesotti et al. in [10].

Our method is similar to Marchesotti as we also want to avoid the problems listed above that exist in the hand-crafted features. We apply the deep convolutional neural networks [13], which achieve the best performance in general image classification problem, to do the image aesthetic quality assessment. The neural network is firstly trained on part of the ImageNet image database, which contains millions of images with various categories, to "understand" images well. Then this deep network can directly compress the image into a relatively lower dimensional feature vector and meanwhile reserve most information in the image. We speculate that those information may implicitly reflect the aesthetic quality of the image. The experimental results obtained on two large and reliable datasets further confirm our assumption. Our method achieves much better performance compared with state-of-the-art methods.

The rest of this paper is organized as follows. In Section 2, we review related methods. In Section 3, we describe the deep convolutional neural networks and our methods in details. Then the experimental settings, results and comparisons with state-of-the-art methods are presented in Section 4. Finally, we give a conclusion in Section 5.

2   Related Work

Photo aesthetic quality assessment has been studied for a long time. In [9], Tong et al. tried to classify images taken by photographers and home users. This was actually the same with photo aesthetic quality assessment. They adopted many low-level features such as color histogram, image energy which were proposed for previous content-based image retrieval system. Those features did not have close relationship with aesthetic quality and just achieved limited success [8].

Subsequently, Datta et al. and Ke et al. proposed different ways from Tong et al. to do photo quality assessment [5, 8]. Both of them tried to firstly analyze which attributes can affect the image aesthetic quality. After this, Datta et al. used many low-level features which were related to those attributes. Ke et al. designed seven new features which all had high-level semantics such as simplicity, colorfulness and so on. Both of their methods performed better in effectiveness and efficiency.

Luo et al. and Wong et al. adopted other image attributes [6, 14]. They found out that most of professional photos carefully set the contrast between the subject and background. Due to this, there existed big differences between the subject that the photographers wanted to capture and the background they chose. So Luo et al. and Wong et al. determined to separate the subject and background first and then dealt with them respectively. They extracted features describing those two regions and the relationship between them.

As we have mentioned above, hand-crafted features suffer from several problems. Though there exist other works which adopt new attributes or refine mathematical models to describe previous attributes [7, 15, 16], limited improvement is gotten. Most of them lack robustness thus they perform badly when handling with different kinds of images or the dataset changes.

Different from those works which try to analyze the image aesthetic quality in the high semantic level, Marchesotti et al. choose another innovative way that they adopt the generic image descriptors such as BOV to represent the image [10]. This kind of descriptors is used to describe the image content previously. We can infer that it is related to the image aesthetic quality in an implicit way.

In recent years, deep learning methods have brought breakthroughs in many traditional computer vision problems. They need no human ingenuity but enough training data. Among them, Krizhevsky et al. used the deep convolutional neural networks (DCNN) to conduct image classification on ImageNet [13]. Their method performed better than previous ones in which hand-crafted features and generic image descriptors were used to represent the image. This deep neural network could describe and encode the image content more effectively. Inspired by this, in this paper, we propose to extract image features with the help of the deep convolutional neural network well trained on the ImageNet database and apply those features to do image aesthetic quality assessment.

3   Features Extracted by the Deep Convolutional Neural Network

In general, when facing a new computer vision problem, experts are accustomed to analyze it in a high semantic level and design features to represent the data. Then those features are transferred into the learning algorithm with or without preprocessing such as PCA and so on. Though this scheme can make use of human intelligence inside, it is hard to figure out how the brain perceives the outside world sometimes especially when the given question is very abstract such as photo aesthetic quality assessment we try to deal with here.

Convolutional neural networks, firstly proposed by LeCun et al. and applied in handwritten digit recognition [17], are quite different from other machine learning methods. Fig. 2 shows the architecture of their neural network model. It can directly handle with the raw image instead of feature vector. The structure of convolutional neural network imitates the biological vision system and therefore obtains strong ability to understand the nature of images.

General convolutional neural network applies several architectural thoughts, including the local receptive fields, shared weights and subsampling [18]. Before fed to the network, the given image should be scaled to a fixed size. Then pixels are translated to the range -1~1 in order to decrease the effect of the image exposure. But the resized image still has many pixels. If a traditional full-connected network with enough discriminative power is applied, too many parameters should be learned. For convolutional network, it applies the local receptive field to avoid this problem. This idea is originated from the survey on

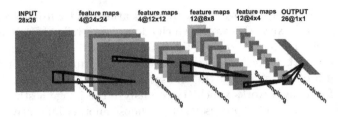

Fig. 2. The convolutional neural network designed for hand-written digit recognition proposed in [17]. It can directly deal with the raw image.

the visual system of cats [19]. Hubel et al. finds that visual neurons are locally-sensitive and orientation-selective. The local receptive field in convolutional neural network means for each unit in elementary layers, it only receives inputs from a small set of units in the neighborhood in the previous layer. This mathematical model tactfully imitates the locally-sensitive character of visual neurons and avoids to train too many parameters which may lead to over-fitting.

The setting of shared weights means receptive fields in different places of the image will have the identical weight vectors. This idea is proposed based on the knowledge that feature detector which is useful in one part of the image maybe also work across the whole image. It is equivalent to scan all over the image with a small convolutional kernel followed by a non-linear function. This is also why we call the method convolutional neural network.

Outputs of neurons in the same layer form a plane which we call feature map. We can obtain different feature maps by setting different weights of the local receptive field. Then local operations, such as averaging or maximum selection, and subsampling are conducted all over the feature map. Those operations form the pooling layer. In general, the convolutional layer and pooling layer alternately appear in the convolutional neural network. The output layer is fully connected with its previous layer and produces a feature vector which can further transferred to a logistic regression layer to accomplish the recognition task. All the weights in the network are learned with the back-propagation method.

Due to the application of those innovative and outstanding ideas, convolutional neural networks are adopted in many pattern recognition problems and achieve good performances [20, 21, 22]. But in the past, it is almost impossible to train an artificial neural network with many hidden layers and millions of neurons due to the limitation of computing power and lack of methods to avoid falling into the local optimization and over-fitting. Luckily, recently deep learning methods have gained a speedy development. Those methods improve the traditional artificial neural network in several aspects, such as the unsupervised and layer-wised pre-training, better activation functions and new training methods. Besides, large scale available datasets help to train more powerful models without over-fitting. The application of cheap and high-performance GPU and multi-core computers also greatly reduce the training time.

Due to this, Krizhevsky et al. designs a deep convolutional neural network with millions of parameters and applied it on the ImageNet classification [13]. Their

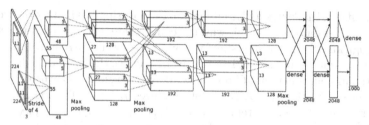

Fig. 3. The architecture of deep convolutional neural network used in [13]. It is similar with the neural network in [17], but has more layers and neurons. The first five layers of it are convolutional and the remaining three layers are full connected. It can take the 224*224*3 image patches as input.

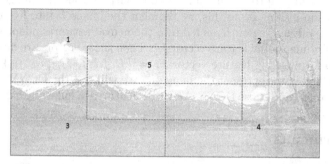

Fig. 4. In order to reduce the impact of image scaling and encode the information of image composition, we adopt the idea of spatial pyramid. We segment the image into five regions and each patch is fed into the deep convolutional neural network. All of their activations are concatenated with the activation gotten by the whole image.

model contains eight learned layers and adopts the Rectified Linear Units as the activation function [23]. The overall architecture of their network is shown in Fig. 3. It can handle with the RGB images instead of gray images only. Their neural network consists of five convolutional layers and three full connected layers. The output of the last layer is followed by a 1000-way soft-max that produces a distribution over the given 1000 image class labels. Then supervised learning process is conducted on the training set of the ImageNet database. Finally it achieves a breakthrough on this challenging dataset which proves its descriptive power on all kinds of images.

In view of this, we here adopt the deep convolutional neural network, trained on the ImageNet dataset to understand images well, to represent the image and challenge the problem of photo quality assessment. This task requires to classify images into high quality and low quality which seems to be easier in the question format but require more intelligence inside. But we can still trust that a model which know the image content well may also be powerful in aesthetic quality assessment.

Firstly, we implement a deep convolutional neural network that has the same architecture with [13]. Then this neural network is carefully trained on the ILSVRC-

2012 training set that extracts 1.2 million images covering 1000 categories from the whole ImageNet database. In this way, the convolutional neural network accumulates enough knowledge to understand various images well. We remove the external 1000-way soft-max in the network. For a given image I_i, the last hidden layer of the convolutional neural network C produces 4096-dimensional activations. We normalize those activations as DCNN_Aesth features f_i for I_i. The image aesthetic quality training set $\mathcal{L}$ will be represented as $\{(f_1, y_1),(f_2, y_2),...,(f_N, y_N)\}$, where the aesthetic quality label $y \in \{-1, 1\}$. Here $y = 1$ denotes high aesthetic quality and $y = -1$ denotes low aesthetic quality. Finally, we train a new classifier $\mathcal{S}$ on $\mathcal{L}$ to predict the aesthetic quality of new images.

As the convolutional neural network requires to scale the image into a fixed size which may lose some information, we further adopt the idea of spatial pyramid on the image. This improvement can also implicitly encode the image composition information inside. To do this, we segment the image I_i into five regions as shown in Fig. 4. Each region R_j is fed into the neural network independently to produce 4096-dimensional feature vector f_{ij}. Then those vectors are directly concatenated to form a long feature vector $F_i = \{f_i, f_{i1}, f_{i2}, f_{i3}, f_{i4}, f_{i5}\}$, which is 24576-dimensional, to represent the image I_i. We will test the DCNN_Aesth features talked above and those DCNN_Aesth_SP features in following experiments.

4   Experiments

In this section, we report the experimental results to verify the effectiveness of deep convolutional neural network when dealing with photo quality assessment. It will be compared with several state-of-the-art methods. Most of them are based on hand-crafted features [5, 6, 7, 8, 10]. All the experiments are conducted on two large scale and reliable public datasets, CUHKPQ and AVA, which are specifically designed for the research of photo quality assessment [24, 25].

4.1   Datasets

CUHKPQ. To explore the image aesthetics better, a large dataset which is carefully labeled and contains various kinds of images is needed. It is impossible for individual to form a dataset like this. Most experts turn to Internet for help. There exist many photo sharing websites which allow users to upload photos freely and score photos for each other. For example, DPChallenge.com contains hundreds of thousands images, most of which are scored by tens of different users. This makes it reliable to form a dataset used for the survey of image aesthetic quality assessment.

CUHKPQ is a public dataset contains 17690 images collected from the university students and professional photo sharing websites [24]. Each image is labeled as "good" or "bad" by ten independent volunteers and is reserved only when more than eight of them have the same opinions. Therefore, the image labels in CUHKPQ are more reliable and have less noise. Those make it easier to classify images in CUHKPQ into high quality and low quality. Another character of CUHKPQ is that the categories of images in this dataset can be divided into "animal", "architecture", "human", "landscape", "night", "static" and "plant". Those categories are common in natural images that occupy a large percentage of images we see in daily life. In our experiment, we randomly and evenly separate "good" and "bad" images in each category into training and test set. Finally both sets contain 8845 images.

AVA. Compared with CUHKPQ, Aesthetic Visual Analysis (AVA) is a larger dataset formed by more than 250 thousands of images [25]. This database is specifically constructed for the purpose of learning more about image aesthetics. All those images are directly downloaded from the DPChallenge.com. Though many other datasets make use of this website such as [6], they only contain a far less number of images. Besides, image categories in this dataset are more abundant including abstract images, advertisement images and so on.

For each image in AVA, there is an associated distribution of scores voted by different viewers. As reported in [25], the number of votes that per image gets is ranged in 78~549. Besides, users in the DPChallenge.com consist of all kinds of people, without restriction of age, gender and profession. Both professional photographers and amateur enthusiasts can enjoy images shared in this website. All of them have viewed a large amount of photos and can make the independent judgments. So it is reliable to take the average score of the image as its photo aesthetic quality.

Actually, the providers of AVA database only release the web links of those images. We successfully download 193077 images and other image links are invalid due to the update of the website. In our experiments, to increase the gap between images with high aesthetic quality and images with low aesthetic quality, we firstly sort all images by their mean scores. Then we pick out the top 10% images as good and the bottom 10% images as bad as done in other works [5, 8]. Then both good images and bad images are randomly and evenly separated into training set and test set. Due to this, each set contains 19308 images.

4.2 Experimental Settings

Here we implement several state-of-the-art methods as comparisons, including the 56-d features proposed by Datta et al. [5], 7-d features proposed by Ke et al. [8], 5-d features proposed by Luo et al. [6], 17-d original features proposed by Lo et al. [7], and bag-of-visual-words features of dense SIFT image descriptor with 1024 visual words proposed by Marchesotti et al. [10]. Among them, the method of Marchesotti is similar with ours which also try to apply features used to describe image content instead of hand-crafted features in aesthetic quality assessment.

Besides, the deep convolutional neural network used in our method is implemented with the help of the open source deep leaning framework called Caffe [27]. We inherit the architecture of DCNN model proposed by Krizhevsky et al. in [13]. The overall structure of this network has already been shown in Fig. 3.

Table 1. Classification accuracy comparisons between state-of-the-art methods and our proposed DCNN aesthetic features on CUHKPQ dataset

Aesthetic Features	Accuracy (%)
Luo[6]	76.91
Lo[7]	81.76
Datta[5]	85.27
Ke[8]	81.70
Marchesotti[10]	79.53
DCNN_Aesth	90.76
DCNN_Aesth_SP	**91.93**

ImageNet is a very large dataset which collects more than 14 million high-resolution images from the Internet [26]. This dataset is meaningful for the researches on image understanding algorithms and helpful to train powerful machine learning models. ILSVRC is an annual competition on computer vision problems, especially the image classification. The dataset provided by ILSVRC is a part of the ImageNet. We train our implemented deep convolutional neural network on the dataset used in ILSVRC-2012 [27]. It contains more than 1.2 million images come from 1000 categories. Each category provides 732 to 1300 images. With the help of those image data, this DCNN obtains enough power to describe various image content.

As reported in [13], the 4096-dimensional output of the last full connected layer reveals the networks visual knowledge. The images look alike may also have a small Euclidean distance between their 4096-d activations. We conduct a further normalization on this output and take it as DCNN_Aesth features.

Before fed to the network, the image have to be scaled into a fixed resolution of 224*224. This may lead to the loss of information and DCNN may overlook the whole layout of the image. Due to this, we also adopt the spatial pyramid method that has been introduced in the last section. Its performance is also shown below denoted as DCNN_Aesth_SP.

For each kind of features, we train a two-class SVM classifier and take the classification accuracy as the performance evaluation. We get the parameters of all those SVM classifiers by conducting the cross-validation processes on the training set.

4.3   Experimental Results on the CUHKPQ

The experimental results on the CUHKPQ dataset are summarized in Table. 1. It can be observed that the DCNN_Aesth_SP features achieve the best performance of 91.93%. Then the DCNN_Aesth features follow with the accuracy of 90.76%. Among above methods which are mainly based on the hand-crafted features, Datta et al. performs the best with the accuracy of 85.27%. Other methods just obtain accuracies around 80%. Though Marchesotti et al. hold the similar idea

Table 2. Classification accuracy comparisons between state-of-the-art methods and our proposed DCNN_Aesth features on AVA dataset

Aesthetic Features	Accuracy (%)
Luo[6]	61.49
Lo[7]	68.13
Datta[5]	68.67
Ke[8]	71.06
Marchesotti[10]	68.55
DCNN_Aesth	78.92
DCNN_Aesth_SP	**83.52**

with us, the image descriptive power of dense SIFT is not as good as the deep convolutional neural network. Finally, their method just submits an ordinary grade with the accuracy of 79.53%.

In conclusion, features extracted by the deep convolutional neural network beat the best hand-crafted features on CUHKPQ dataset that they improve the accuracy by almost 6%. This phenomenon proves the effectiveness of DCNN when applied to deal with the problem of photo aesthetic quality assessment.

4.4 Experimental Results on the AVA

We present the experimental results obtained on the AVA dataset in Table. 2. Again, it can be observed that DCNN_Aesth_SP features get the best performance with the accuracy of 83.52%. DCNN_Aesth also performs very well that it obtains an accuracy of 78.92%. The best performance obtained by the methods based on hand-crafted features is 71.06%, which is still worse than the DCNN features. Other hand-crafted features just get accuracies under 70%. In addition, the method of Marchesotti et al. only gets a performance of 68.55% which is closed with most hand-crafted features.

Another thing we have to explain here is that all the methods perform better on the CUHKPQ dataset than on the AVA dataset. This is firstly due to that there are more various image categories existing in the AVA dataset. Secondly, as images in the AVA dataset are directly downloaded from the website without the careful selection that conducted on the CUHKPQ, their labels inevitably have more noise inside. On such a challenging dataset, DCNN_Aesth and DCNN_Aesth_SP features still perform very well that they improve state-of-the-art by more than 6%. This result is quite promising.

5   Conclusion

In this paper, we adopt the deep convolutional neural network to conduct the photo quality assessment. The network is firstly trained on the ImageNet to gain strong image descriptive power. Then several experiments are conducted on two public dataset specifically designed for the research of image aesthetic quality assessment. The experimental results show that our method is considerably better than state-of-the-art methods which are mostly based on the hand-crafted features. To our best knowledge, this paper is the first to apply deep learning methods in image aesthetic quality assessment. The success of it further proves the potential power of those kinds of deep neural networks with millions of neurons. In the future, we will further apply multi-view methods to combine our DCNN features and other hand-crafted aesthetic features [28].

Acknowledgment. This work is supported by the 973 project under the contract No.2015CB351803, the NSFC under the contract No.61390514 and No.61201413, the Fundamental Research Funds for the Central Universities No. WK2100060007 and No. WK2100060011, the Specialized Research Fund for the Doctoral Program of Higher Education No. WJ2100060003.

References

1. Geng, B., Yang, L., Xu, C., Hua, X.-S., Li, S.: The role of attractiveness in web image search. In: ACM MM, pp. 63–72 (2011)
2. Li, C., Loui, A.C., Chen, T.: Towards aesthetics: A photo quality assessment and photo selection system. In: ACM MM, pp. 827–830 (2010)
3. Bhattacharya, S., Sukthankar, R., Shah, M.: A Framework for Photo-Quality Assessment and Enhancement based on Visual Aesthetics. In: MM, pp. 271–280 (2010)
4. Zhang, F.-L., Wang, M., Hu, S.-.M.: Aesthetic image enhancement by dependence-aware object recomposition. IEEE Trans. on Multimedia, 1480–1490 (2013)
5. Datta, R., Joshi, D., Li, J., Wang, J.Z.: Studying aesthetics in photographic images using a computational approach. In: Leonardis, A., Bischof, H., Pinz, A. (eds.) ECCV 2006. LNCS, vol. 3953, pp. 288–301. Springer, Heidelberg (2006)
6. Luo, Y., Tang, X.: Photo and video quality evaluation: Focusing on the subject. In: Forsyth, D., Torr, P., Zisserman, A. (eds.) ECCV 2008, Part III. LNCS, vol. 5304, pp. 386–399. Springer, Heidelberg (2008)
7. Lo, K.Y., Liu, K.H., Chen, C.S.: Assessment of photo aesthetics with efficiency. In: ICPR, pp. 2186–2189 (2012)
8. Ke, Y., Tang, X., Jing, F.: The design of high-level features for photo quality assessment. In: CVPR, pp. 419–426 (2006)
9. Tong, H., Li, M., Zhang, H.-J., He, J., Zhang, C.: Classification of digital photos taken by photographers or home users. In: Aizawa, K., Nakamura, Y., Satoh, S. (eds.) PCM 2004. LNCS, vol. 3331, pp. 198–205. Springer, Heidelberg (2004)
10. Marchesotti, L., Perronnin, F., Larlus, D., Csurka, G.: Assessing the aesthetic quality of photographs using generic image descriptors. In: ICCV, pp. 1784–1791 (2011)
11. Dance, C., Willamowski, J., Fan, L., Bray, C., Csurka, G.: Visual categorization with bags of keypoints. In: ECCV International Workshop on Statistical Learning in Computer Vision, Prague (2004)
12. Lazebnik, S., Schmid, C., Ponce, J.: Beyond bags of features: spatial pyramid matching for recognizing natural scene categories. In: CVPR (2006)

13. Krizhevsky, A., Sutskever, I., Hinton, G.E.: Imagenet classification with deep convolutional neural networks. In: NIPS (2012)
14. Wong, L.K., Low, K.L.: Saliency-enhanced image aesthetics class prediction. In: ICPR, pp. 997–1000 (2009)
15. Yin, W., Mei, T., Chen, C.W.: Assessing photo quality with geo-context and crowdsourced photos. In: Visual Communications and Image Processing, pp. 1–6 (2012)
16. Chu, W.T., Chen, Y.K., Chen, K.T.: Size does matter: how image size affects aesthetic perception? In: ACM MM, pp. 53–62 (2013)
17. LeCun, Y., Boser, B., Denker, J.S., Henderson, D., Howard, R.E., et al.: Handwritten digit recognition with a back-propagation network. In: Advances in Neural Information Processing Systems (1990)
18. LeCun, Y., Bengio, Y.: Convolutional networks for images, speech, and time series. In: The Handbook of Brain Science and Neural Networks. MIT Press (1995)
19. Hubel, D.H., Wiesel, T.: Receptive fields, binocular interaction, and functional architecture in the cat's visual cortex. Journal of Physiology (London) 160, 106–154 (1962)
20. Lawrence, S., Giles, C.L., Tsoi, A.C., Back, A.D.: Face recognition: A convolutional neural network approach. IEEE Trans. Neural Networks 8, 98–113 (1997)
21. Le Callet, P., Viard-Gaudin, C., Barba, D.: A convolutional neural network approach for objective video quality assessment. IEEE Trans. Neural Netw. 17(5), 1316–1327 (2006)
22. Szarvas, M., Yoshizawa, A., Yamamoto, M., Ogata, J.: Pedestrian detection with convolutional neural networks. In: Proc. IEEE Intell. Veh. Symp., pp. 224–229 (2005)
23. Nair, V., Hinton, G.E.: Rectified linear units improve restricted boltzmann machines. In: Proc. 27th International Conference on Machine Learning (2010)
24. Luo, W., Wang, X., Tang, X.: Content-based photo quality assessment. In: ICCV, pp. 2206–2213 (2011)
25. Murray, N., Marchesotti, L., Perronnin, F.: AVA: A largescale database for aesthetic visual analysis. In: CVPR, pp. 2408–2415 (2012)
26. Deng, J., Berg, A., Socher, R., Li, L.-J., Li, K., Fei-Fei, L.: ImageNet: A large-scale hierarchical image database. In: CVPR (2009)
27. Jia, Y.: Caffe: An open source convolutional architecture for fast feature embedding (2013), http://cafe.berkeleyvision.org/
28. Liu, W.F., Tao, D.: Multiview hessian regularization for image annotation. IEEE Trans. on Image Processing, 2676–2687 (2013)

Non-negative Low-Rank and Group-Sparse Matrix Factorization

Shuyi Wu[1], Xiang Zhang[1], Naiyang Guan[1], Dacheng Tao[3], Xuhui Huang[2],
and Zhigang Luo[1,*]

[1] Science and Technology on Parallel and Distributed Processing Laboratory
College of Computer
National University of Defense Technology, Changsha, Hunan, P.R. China, 410073
[2] Department of Computer Science and Technology, College of Computer
National University of Defense Technology, Changsha, Hunan, P.R. China, 410073
[3] Centre for Quantum Computation & Intelligent Systems and the Faculty of
Engineering and Information Technology
University of Technology, Sydney, 235 Jones Street, Ultimo, NSW 2007, Australia
wushuyi09@163.com, zhangxiang_43@aliyun.com,
{ny_guan,xuhuihuang,zgluo}@nudt.edu.cn, dacheng.tao@uts.edu.au

Abstract. Non-negative matrix factorization (NMF) has been a popular data analysis tool and has been widely applied in computer vision. However, conventional NMF methods cannot adaptively learn grouping structure from a dataset. This paper proposes a non-negative low-rank and group-sparse matrix factorization (NLRGS) method to overcome this deficiency. Particularly, NLRGS captures the relationships among examples by constraining rank of the coefficients meanwhile identifies the grouping structure via group sparsity regularization. By both constraints, NLRGS boosts NMF in both classification and clustering. However, NLRGS is difficult to be optimized because it needs to deal with the low-rank constraint. To relax such hard constraint, we approximate the low-rank constraint with the nuclear norm and then develop an optimization algorithm for NLRGS in the frame of augmented Lagrangian method(ALM). Experimental results of both face recognition and clustering on four popular face datasets demonstrate the effectiveness of NLRGS in quantities.

Keywords: Non-negative matrix factorization, low-rank representation, group sparse.

1 Introduction

Data representation reveals the intrinsic structure of data to boost subsequent data processing. Due to its effectiveness, it has been widely applied in computer vision, such as image annotation [23] and image classification [19]. Principal component analysis(PCA, [16]) and non-negative matrix factorization (NMF, [17][18])

---

* Corresponding author.

X. He et al. (Eds.): MMM 2015, Part II, LNCS 8936, pp. 536–547, 2015.

have been the state-of-the-art methods. Particularly, PCA learns the holistic representation which is not consistent with the human intuition, i.e., the parts forming the whole. In contrast, NMF learns the parts-based representation to relax this limitation.

Non-negative matrix factorization (NMF) learns two non-negative factors to approximate original data. Since it learns the parts-based representation, it has attracted much attention from computer vision community. However, NMF cannot capture the cluster memberships among examples meanwhile remain immune to the outliers. Recently, low-rank and sparse decomposition (LRSD) [37][33][6] recovers the low-rank part of the data from the corrupted observations and meanwhile captures the outliers, i.e., sparse component, within the dataset under certain conditions [29]. Besides, low-rank representation (LRR) [22] incorporates the low-rank constraint to represent each sample as a linear combination of the other examples. However, these low-rank methods cannot derive the projection subspace of original examples.

Both LRR and NMF have been successfully utilized in image classification [7][11]. However, for LRR, it does not identify the group structure and thus performs unsatisfactorily in classification. Zhang et al. [36] learns the structured low-rank representation (SLRR) by incorporating label constraint in LRR. SLRR often succeeds in the supervised settings but the labels of training samples are so expensive. However, it cannot maintain the non-negativity of the data meanwhile learning parts-based representation. On the other hand, for NMF, it neither identifies the group structure nor has the robustness to the outliers.

To defeat the aforementioned deficiencies, this paper proposes a non-negative low-rank group-sparse matrix factorization (NLRGS) method by simultaneously integrating low-rank and group-sparse constraints. Particularly, we can capture the relationship of the coefficients via the low-rank constraint meanwhile identifies the outliers by imposing the group-sparse constraint over the sparse component. These constraint strategies have been proven beneficial for various vision tasks such as classification [7]. Benefiting from such property, NLRGS enhances the performance of NMF in classification and clustering. However, it is difficult to optimize NLRGS because its objective is non-convex. To avoid such limitation, we relax the low-rank constraint via the equality constraint and then develop an optimization algorithm for NLRGS in the frame of augmented Lagrangian method (ALM, [15][28]). Experimental results of both face recognition and clustering on Yale [1], UMIST [8], ORL [31] and FERET [27] datasets suggest the effectiveness of NLRGS in quantities.

The rest of this paper is organized as follows: Section 2 briefly reviews related works. We present NLRGS and optimize it in Section 3. Section 4 verifies the effectiveness of NLRGS and Section 5 concludes this paper.

2   Related Works

2.1   Non-negative Matrix Factorization

Non-negative matrix factorization (NMF) [17] decomposes the data matrix, i.e., $X \in R_+^{m \times n}$ into two reduced-dimensional non-negative matrices, i.e., $W \in R_+^{m \times r}$

and $H \in R_+^{r \times n}$, where r denotes the reduced dimensionality such that $r \ll \min\{m, n\}$. The objective function of NMF is

$$\min_{W \geq 0, H \geq 0} \|X - WH\|_F^2.$$ (1)

Lee and Seung [18] proposed the multiplicative update rule (MUR) for NMF. Since NMF only allows additive, not subtractive, combination of the basis matrix, NMF and its variants has been widely applied in face recognition [13][10] and clustering [9]. Obviously, NMF cannot capture the relationships among examples meanwhile identifying the outliers.

2.2 Low-Rank Representation

Low-rank representation [22] learns the structural representation Z over the specific dictionary D via the low-rank constraint. Thus, the final objective function is

$$\min_Z rank(Z), s.t., X = DZ.$$ (2)

To optimize (2), Candès et al. [6] solve an equivalent problem with the tractable nuclear norm. LRR might performs unsatisfactorily in classification tasks. To address this issue, recent works [26][36] utilize discriminative information to improve the performance of LRR. However, they cannot maintain the non-negativity property of the data. Thus, Zhuang et al. [38] imposed the sparsity and low rank constraint as well as the non-negative constraint over the coefficient matrix to achieve the above purpose. But it still can not take the group relationships among samples into account.

2.3 Group Sparse Coding

Recently, group sparsity has been widely used in computer vision but it is unsuited for classification tasks. To address this issue, Samy et al. [32] proposed group sparse coding (GSC) via mixed-norm regularizers, i.e., the l_2/l_1 regularization, to learn the group sparse representation. GSC utilized the structure of bags of visual words to promote discriminative power. Similar to sparse coding [35][24][20], it can be integrated into various learning methods such as dictionary learning [21]. Saha et al. [30] applied such regularization over the coefficient matrix for sparse subspace clustering to capture the joint sparse and common semantic structure among clustering tasks. But they cannot guarantee the learned coefficient matrix to be low-rank.

3 Non-negative Low-Rank and Group Sparse Matrix Factorization

Here we give the important notations throughout the remainder of the paper in Table 1.

Table 1. Important Notations and Their Descriptions

Notations	Descriptions
A^T	The transpose of matrix A
$\|A\|_*$	The nuclear norm of matrix A
$\|A\|_F$	The Frobenius norm
$\|A\|_g$ & $\|A\|_{1,2}$	The $l_{1,2}$-norm of matrix A, i.e., the group sparse norm
$tr(A)$	The trace operator of matrix A
ε	The tolerance of the stopping criteria
$(A)_+ = \max(A, 0)$	It denotes the operator shrinks the negative value of A to zero.
$(A)_-$	It equals to $\max(-A, 0)$.

3.1 The Proposed Model

Benefitting from the low-rank and group sparse constraint, this section proposes a non-negative low-rank group-sparse matrix factorization (NLRGS) to enhance the performance of NMF both in clustering and in classification.

According to (2), LRR does not learn the projection subspace. To address this issue, we can rewrite it as follows:

$$\min_{Z,D} rank(Z), s.t., X = DZ. \tag{3}$$

To learn sparse representation, we impose the non-negativity constraint over the basis and the coefficients in (3) as follows:

$$\min_{Z \geq 0, D \geq 0} rank(Z), s.t., X = DZ. \tag{4}$$

The objective (4) assumes that data matrix has not any data corruption. This assumption is usually incorrect in real-world scenarios. To address this issue, we assume the outliers to be group sparse and further derive

$$\min_{Z \geq 0, D \geq 0} rank(Z) + \lambda \|E\|_g, s.t., X = DZ + E, \tag{5}$$

where λ tradeoffs the group sparse regularization term.

Similar to LRR [22], we utilize the nuclear norm to replace the low-rank constraint:

$$\min_{Z \geq 0, D \geq 0} \|Z\|_* + \lambda \|E\|_g, s.t., X = DZ + E. \tag{6}$$

According to (6), NLRGS learns the relationships among all the examples meanwhile capturing the outliers using the group-sparse constraint.

3.2 Optimization Algorithm

It is difficult to solve function (6) because the objective (6) is non-convex. To address this issue, we first introduce an auxiliary variable J and the equality constraint such that

$$\min_{Z,D \geq 0, E, J} \|J\|_* + \lambda \|E\|_g s.t., X = DZ + E, Z = J. \tag{7}$$

To solve (7), we utilize the augmented Lagrangian method (ALM, [15][28]) to remove the equality constraint into the unconstrained optimization problem. This is because ALM optimizes a constrained optimization problem with a series of unconstrained sub-problems. Thus, according to [2], ALM can rewrite (7) as the following augmented Lagrangian formulation:

$$
\begin{aligned}
L = \min_{Z,D\geq0,E,J} \|J\|_* + \lambda\|E\|_g &+ tr\left(Y_1^T(X - DZ - E)\right) + tr\left(Y_2^T(Z - J)\right) \\
&+ \tfrac{\mu}{2}\left(tr\left((Z - J)^T(Z - J)\right) + tr\left((X - DZ - E)^T(X - DZ - E)\right)\right),
\end{aligned}
\tag{8}
$$

where Y_1 and Y_2 represent two Lagrangian multipliers, respectively, μ is the penalty parameter. Thus, ALM recursively iterate the following procedures until convergence:

$$
(J_{k+1}, D_{k+1}, Z_{k+1}, E_{k+1}) = \arg\min_{D,Z\geq0,J,E} L(J_k, D_k, Z_k, E_k),
\tag{9}
$$

$$
Y_1 = Y_1 + \mu\left(X - D_k Z_k - E\right),
\tag{10}
$$

$$
Y_2 = Y_2 + \mu\left(Z_k - J_k\right).
\tag{11}
$$

Since the objective function (9) is non-convex, we alternately optimize one variable with the others fixed. Then we have the objective with respect to J as follows:

$$
J = \arg\min \tfrac{1}{\mu}\|J\|_* + \tfrac{1}{2}\left\|J - (Z + \tfrac{Y_2}{\mu})\right\|_F^2.
\tag{12}
$$

According to [5], we can employ the singular value thresholding algorithm to get the closed form solution of (12), i.e.

$$
J = J_{\frac{1}{\mu}}(T) = U J_{\frac{1}{\mu}}\left(\sum\right) V^T,
\tag{13}
$$

where $T = Z + \tfrac{Y_2}{\mu}$, $J_{\frac{1}{\mu}}(\sum) = diag\left\{\left(\sigma_i - \tfrac{1}{\mu}\right)_+\right\}$.

Accordingly, we need to optimize the following objective (14) with respect to E.

$$
E = \arg\min \tfrac{\lambda}{\mu}\|E\|_{1,2} + \tfrac{1}{2}\left\|E - \left(X - DZ + \tfrac{Y_1}{\mu}\right)\right\|_F^2.
\tag{14}
$$

According to **Lemma 1**, we can get the closed-form solution of (14).

Lemma 1: Let $Q = [Q_1, \ldots, Q_n]^T$ be a given matrix. If the optimal solution of $L_E = \min_E \lambda\|E\|_{1,2} + \tfrac{1}{2}\|E - Q\|_F^2$ is E^*, then the j-th row of E^* is

$$
E_{j.}^* = \begin{cases} \frac{\|Q_{j.}\| - \lambda}{\|Q_{j.}\|} Q_{j.}, & if\ \lambda < \|Q_{j.}\| \\ 0 & , otherwise. \end{cases}
\tag{15}
$$

Proof. We can compute the sub-gradient of L_E with E_{ji} as follows:

$$
\frac{\partial L_E}{\partial E_{ji}} = (E - Q)_{ji} + \lambda \frac{E_{ji}}{\sqrt{\sum_{j=1}^n \left(\sum_{i=1}^n |E_{ji}|\right)^2}}.
\tag{16}
$$

By setting $\frac{\partial L_E}{\partial E_{j\cdot}} = 0$, we can obtain

$$E_{j\cdot} - Q_{j\cdot} + \lambda \frac{E_{j\cdot}}{\sqrt{\sum_{j=1}^{n} \left(\sum_{i=1}^{n} |E_{ji}| \right)^2}} = 0. \tag{17}$$

By simple algebra, we further obtain

$$\|Q_{j\cdot}\| = \sqrt{\sum_{j=1}^{n} \left(\sum_{i=1}^{n} |E_{ji}| \right)^2} + \lambda, s.t., \|Q_{j\cdot}\| \geq \lambda. \tag{18}$$

Thus, substituting (18) into (17), we can get the closed-form solution of (14), i.e., Equation (15). This completes the proof.

The sub-gradient of (9) with Z is as follows:

$$\frac{\partial L}{\partial Z} = -D^T Y_1 + Y_2 + \mu \left[\left(I + D^T D \right) Z - J - D^T X + D^T E \right]. \tag{19}$$

According to (19) and [18], we obtain the multiplicative update rule (MUR) for Z:

$$Z_{i,j} = Z_{i,j} \frac{(Y_{2-} + D^T Y_{1+} + \mu D^T E_- + D^T X + \mu J_+)_{i,j}}{(Y_{2+} + \mu (I + D^T D) Z + \mu D^T E_+ + D^T Y_{1-} + \mu J_-)_{i,j}}. \tag{20}$$

Lemma 2: The sub-gradient of L over D is Lipschitz continuous, and its corresponding Lipschitz constant is $L_D = \mu \| ZZ^T \|_2$.

This proof can be referred to [12]. According to **Lemma 2** and [12], we apply the OGM method to optimize D. We summarized our algorithm for NLRGS in **Algorithm 1**, where the stopping criterion is as follows:

$$\|X - D_k Z_k - E\|_\infty < \varepsilon, \|Z_k - J_k\|_\infty < \varepsilon, \tag{21}$$

where the tolerance of ε is set to 10^{-7}.

The time complexity of **Algorithm 1** is mainly spent on updating D_k, Z_k, J_k and E. When updating D_k takes the time complexity in $O(mnr + nr^2) + K \times O(mr^2 + m^2 r)$, the time complexities for updating Z_k, J_k and E are $O(mnr^2)$, $O(2n^2 r)$ and $O(mn)$, respectively. For the reason that ALM converges very quickly, updating D_k takes few iteration loops, especially when the number of iteration meets $K < r$. Thus, the time complexity of one iteration in **Algorithm 1** is $O(mnr + nr^2 + mnr^2 + 2n^2 r + mn) + K \times O(mr^2 + m^2 r)$.

4   Experiments

We compare NLRGS with GNMF [3], NMF [17], K-means [14] in clustering and with GNMF [3], NMF [17] in face recognition on Yale [1], UMIST [8], ORL [31] and FERET [27]. The Yale dataset includes 165 grayscale images of 15 individuals. There are 11 images per individual with different facial expressions

Algorithm 1. NLRGS via ALM

Input: Data $X \in R^{m \times n}$, parameters λ, μ, ρ, and r
Output: D, Z and E
Initialize $Z_0 = J_0 = rand(r,n)$, $D_0 = rand(m,r)$, $Y_1 = Y_2 = E = \mathbf{0}$, $\mu_{\max} = 10^9$.
repeat
    Updating Z_k according to (20).
    $D_k = \mathbf{OGM}\,(X, D_k, Z_k, Y_1, E, \mu)$.
    Updating J_k according to (13).
    Updating E according to (15).
    Update two multipliers via (10) and (11), respectively.
    $\mu = \min(\mu_{\max}, \rho\mu)$.
    $k \leftarrow k + 1$.
until {stopping criterion (21) is satisfied}.
$D = D_k, Z = Z_k, E$.

---

or configurations. Each image is normalized to 32×32 pixel array. The UMIST dataset includes 575 images taken from 20 individuals. Each image is normalized to 40×40 pixel array. The Cambridge ORL database includes ten different images of each of 40 distinct subjects. Each image is normalized to 32×32 pixel array. The FERET database includes 13,539 face images taken from 1,565 subjects with varying pose, size, facial expression and age. Each image is normalized to 40×40 pixel array.

For each dataset, each image has been aligned according to the eye position and is reshaped into a long vector. In clustering, we select all images from different numbers of individuals to construct data subsets. For each subset, all compared methods learn the coefficients of all instances and then feed them to K-means method for clustering. For face recognition, we randomly choose different numbers of images as the training set and the remaining images consist of the testing set. Their clustering results are evaluated with normalized mutual information (NMI) [34] and clustering accuracy (AC) [4]. Meanwhile, We use the nearest neighbor classifier based on the subspace learned by all algorithms to calculate the percentage of correctly classified testing images as the accuracy of the face recognition [25]. To remove the influence of randomness, we conduct the trails 10 times for clustering and classification, respectively. For NMF and K-means, they do not involve any parameter tuning. For GNMF, we utilize its default parameters. Moreover, we set the maximum number of loops of NMF-based methods to 1500.

4.1   Face Clustering

For clustering, we set the parameters $\lambda = 0.1$, $\mu = 0.008$ and $\rho = 1.05$ on Yale and FERET datasets, $\lambda = 0.2$, $\mu = 0.08$ and $\rho = 1.02$ on UMIST dataset, and $\lambda = 0.2$, $\mu = 0.002$ and $\rho = 1.18$ on ORL dataset.

Figure 1 compares clustering performance of K-means, NMF, GNMF and NLRGS on four datasets. Figure 1(a)-(b) show that NLRGS consistently outperforms NMF, GNMF and K-means in terms of both the AC and NMI on Yale

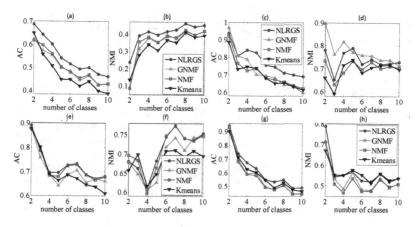

Fig. 1. Average clustering (a) AC and (b) NMI versus number of classes on the Yale dataset,(c) AC and (d) NMI on UMIST, (e) AC and (f) NMI on ORL, (g) AC and (h) NMI on FERET

dataset. Figure 1(c) shows that NLRGS outperforms NMF, GNMF and K-means in terms of clustering accuracy, and Figure 1(d) shows that NLRGS yields the slightly lower NMI than that of GNMF on UMIST dataset. Figure 1(e)-(f) show that NLRGS derives the competitive results with NMF on ORL dataset. Figure 1(g) shows that NLRGS performs better than K-means, NMF and GNMF in terms of AC on FERET dataset. Besides, Figure 1(h) also obtains the competitive results with K-means and performs better than both NMF and GNMF in terms of the NMI.

4.2   Face Recognition

For recognition, we set the parameters $\lambda = 0.2$, $\mu = 0.1$ and $\rho = 1.27$ on Yale dataset, $\lambda = 0.1$, $\mu = 0.08$ and $\rho = 1.02$ on UMIST dataset, $\lambda = 0.2$, $\mu = 0.015$ and $\rho = 1.18$ on ORL dataset, and $\lambda = 0.2$, $\mu = 0.08$ and $\rho = 1.02$ on FERET dataset.

Figure 2 compares recognition performance of NMF, GNMF and NLRGS on four datasets. Figure 2(a)-(d) report that NLRGS performs better than both NMF and GNMF in terms of recognition accuracy on Yale dataset. Figure 2(e)-(l) demonstrate that NLRGS outperforms both NMF and GNMF in terms of recognition accuracy on both UMIST and ORL datasets. Figure 2(m)-(p) show that NLRGS vastly outperforms NMF and GNMF in terms of recognition accuracy on FERET dataset.

For Yale dataset, NLRGS can remove the group-wise noise which illumination changes induce. Besides, the explicit low-rank constraint can further cluster different groups which is highly beneficial for both clustering and classification. Thus, NLRGS performs better than other methods on this dataset especially for clustering task. However, NLRGS achieves the competitive results of clustering with other methods on three datasets including UMIST, ORL and FERET.

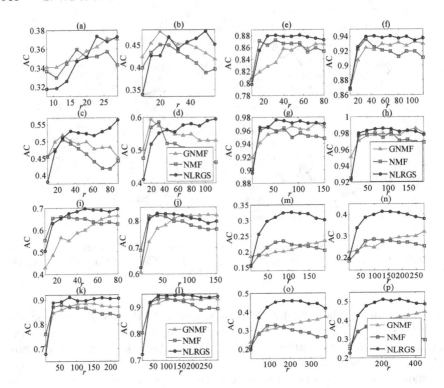

Fig. 2. Average face recognition accuracies versus reduced dimensionalities when (a) 2, (b) 4, (c) 6 and (d) 8 images of each subject were selected for training on the Yale dataset,(e) 4, (f) 6, (g) 8 and (h) 10 images on UMIST,(i) 2, (j) 4, (k) 6 and (l) 8 images on ORL, (m) 2, (n) 3, (o) 4 and (p) 5 images on FERET

This is because these datasets involve both pose variation and occlusion presence rather than illumination change. Although the group sparse constraint can remove the effect of some noises, the low-rank constraint influences little for performance. Hence, NLRGS can obtain better classification performance. In summary, NLRGS is well-suited for classification.

4.3   Parameter Selection

Empirical results show that the parameters of NLRGS make similar difference on both recognition and clustering performance, respectively. Therefore, this section only study the effect of them in NLRGS over the clustering performance on four datasets. Meanwhile, we select the reasonable parameter settings for $\lambda = 0.2$, $\mu = 0.001$ and $\rho = 1.02$ on Yale dataset, $\lambda = 0.2$, $\mu = 0.1$ and $\rho = 1.2$ on UMIST dataset, $\lambda = 0.2$, $\mu = 0.001$ and $\rho = 1.2$ on ORL dataset, and $\lambda = 0.4$, $\mu = 0.001$ and $\rho = 1.02$ on FERET dataset.

Figure 3 shows the best average clustering accuracy (AC) of NLRGS on four datasets. In detail, we fix one parameter with varying the other two ones ac-

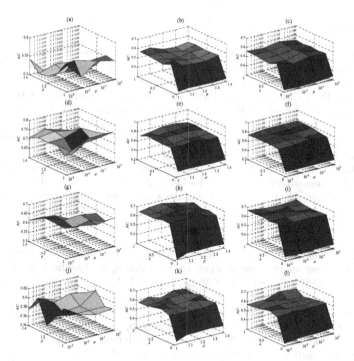

Fig. 3. Average clustering accuracy(AC) of parameter (a) λ, (b) μ and (c) ρ versus other two parameters on the Yale dataset, (d) λ, (e) μ and (f) ρ on UMIST, (g) λ, (h) μ and (i) ρ on ORL, (j) λ, (k) μ and (l) ρ on FERET

cording to specific range sizes. Obviously, the parameters μ and ρ highly affect the clustering performance when we select reasonable the parameter λ. Thus, we suggest selecting suitable parameters for both clustering and classification on different datasets.

5   Conclusion

This paper proposes a non-negative low-rank group sparse matrix factorization (NLRGS) method which incorporates both low-rank constraint and group sparse constraint to capture the cluster memberships of examples meanwhile identify the outliers in the data. To efficiently solve NLRGS, we developed the optimization method in the frame of the augmented Lagrangian method (ALM). Experiment results of both clustering and classification on the four face datasets suggest the effectiveness of NLRGS in quantities.

Acknowledgment. This work is partially supported by Plan for Innovative Graduate Student at National University of Defence Technology, Scientific Research Plan Project of NUDT (under grant No. JC13-06-01) and Australian Research Council Projects (under grant No. DP-120103730, FT-130101457, and LP-140100569).

References

1. Belhumeur, P.N., Hespanha, J.P., Kriegman, D.: Eigenfaces vs. fisherfaces: Recognition using class specific linear projection. IEEE Transactions on Pattern Analysis and Machine Intelligence 19(7), 711–720 (1997)
2. Bertsekas, D.P.: Constrained optimization and lagrange multiplier methods. Computer Science and Applied Mathematics, p. 1. Academic Press, Boston (1982)
3. Cai, D., He, X., Han, J., Huang, T.S.: Graph regularized nonnegative matrix factorization for data representation. IEEE Transactions on Pattern Analysis and Machine Intelligence 33(8), 1548–1560 (2011)
4. Cai, D., He, X., Wu, X., Han, J.: Non-negative matrix factorization on manifold. In: IEEE International Conference on Data Mining, pp. 63–72 (2008)
5. Cai, J.F., Candès, E.J., Shen, Z.: A singular value thresholding algorithm for matrix completion. SIAM Journal on Optimization 20(4), 1956–1982 (2010)
6. Candès, E.J., Li, X., Ma, Y., Wright, J.: Robust principal component analysis? Journal of the ACM 58(3), 11 (2011)
7. Cui, X., Huang, J., Zhang, S., Metaxas, D.N.: Background subtraction using low rank and group sparsity constraints. In: Fitzgibbon, A., Lazebnik, S., Perona, P., Sato, Y., Schmid, C. (eds.) ECCV 2012, Part I. LNCS, vol. 7572, pp. 612–625. Springer, Heidelberg (2012)
8. Graham, D.B., Allinson, N.M.: Characterising virtual eigensignatures for general purpose face recognition. In: Face Recognition, pp. 446–456 (1998)
9. Guan, N., Tao, D., Luo, Z., Shawe-Taylor, J.: Mahnmf: Manhattan non-negative matrix factorization. arXiv preprint arXiv:1207.3438 (2012)
10. Guan, N., Tao, D., Luo, Z., Yuan, B.: Manifold regularized discriminative nonnegative matrix factorization with fast gradient descent. IEEE Transactions on Image Processing (2011)
11. Guan, N., Tao, D., Luo, Z., Yuan, B.: Non-negative patch alignment framework. IEEE Transactions on Neural Networks (2011)
12. Guan, N., Tao, D., Luo, Z., Yuan, B.: Nenmf: an optimal gradient method for nonnegative matrix factorization. IEEE Transactions on Signal Processing 60(6), 2882–2898 (2012)
13. Guan, N., Tao, D., Luo, Z., Yuan, B.: Online nonnegative matrix factorization with robust stochastic approximation. IEEE Transactions on Neural Networks and Learning Systems (2012)
14. Hartigan, J.A., Wong, M.A.: Algorithm as 136: A k-means clustering algorithm. Applied statistics, 100–108 (1979)
15. Hestenes, M.R.: Multiplier and gradient methods. Journal of Optimization Theory and Applications 4(5), 303–320 (1969)
16. Jolliffe, I.: Principal component analysis. Wiley Online Library (2005)
17. Lee, D.D., Seung, H.S.: Learning the parts of objects by non-negative matrix factorization. Nature 401(6755), 788–791 (1999)
18. Lee, D.D., Seung, H.S.: Algorithms for non-negative matrix factorization. In: Advances in Neural Information Processing Systems, pp. 556–562 (2001)
19. Li, P., Bu, J., Yang, Y., Ji, R., Chen, C., Cai, D.: Discriminative orthogonal nonnegative matrix factorization with flexibility for data representation. Expert Systems With Applications 41(4), 1283–1293 (2014)
20. Liu, B.-D., Wang, Y.-X., Shen, B., Zhang, Y.-J., Hebert, M.: Self-explanatory sparse representation for image classification. In: Fleet, D., Pajdla, T., Schiele, B., Tuytelaars, T. (eds.) ECCV 2014, Part II. LNCS, vol. 8690, pp. 600–616. Springer, Heidelberg (2014)

21. Liu, B.-D., Wang, Y.-X., Zhang, Y.-J., Shen, B.: Learning dictionary on manifolds for image classification. Pattern Recognition 46(7), 1879–1890 (2013)
22. Liu, G., Lin, Z., Yu, Y.: Robust subspace segmentation by low-rank representation. In: International Conference on Machine Learning, pp. 663–670 (2010)
23. Liu, W., Tao, D.: Multiview hessian regularization for image annotation. IEEE Transactions on Image Processing 22(7), 2676–2687 (2013)
24. Liu, W., Tao, D., Cheng, J., Tang, Y.: Multiview hessian discriminative sparse coding for image annotation. Computer Vision and Image Understanding 118, 50–60 (2014)
25. Liu, W., Zheng, N., Lu, X.: Non-negative matrix factorization for visual coding. IEEE International Conference on Acoustics, Speech, and Signal Processing 3, 293–296 (2003)
26. Ma, L., Wang, C., Xiao, B., Zhou, W.: Sparse representation for face recognition based on discriminative low-rank dictionary learning. In: IEEE Conference on Computer Vision and Pattern Recognition, pp. 2586–2593 (2012)
27. Phillips, P.J., Moon, H., Rizvi, S.A., Rauss, P.J.: The feret evaluation methodology for face-recognition algorithms. IEEE Transactions on Pattern Analysis and Machine Intelligence 22(10), 1090–1104 (2000)
28. Powell, M.J.: A method for non-linear constraints in minimization problems (1967)
29. Recht, B., Fazel, M., Parrilo, P.A.: Guaranteed minimum-rank solutions of linear matrix equations via nuclear norm minimization. SIAM Review 52(3), 471–501 (2010)
30. Saha, B., Pham, D.S., Phung, D., Venkatesh, S.: Sparse subspace clustering via group sparse coding. In: Proceedings of the Thirteenth SIAM International Conference on Data Mining, pp. 130–138 (2013)
31. Samaria, F.S., Harter, A.C.: Parameterisation of a stochastic model for human face identification. In: IEEE Workshop on Applications of Computer Vision, pp. 138–142 (1994)
32. Samy, B., Fernando, P., Yoram, S., Dennis, S.: Group sparse coding. In: Advances in Neural Information Processing Systems (2012)
33. Wright, J., Ganesh, A., Rao, S., Peng, Y., Ma, Y.: Robust principal component analysis: Exact recovery of corrupted low-rank matrices via convex optimization. In: Advances in Neural Information Processing Systems, pp. 2080–2088 (2009)
34. Xu, W., Liu, X., Gong, Y.: Document clustering based on non-negative matrix factorization. In: ACM SIGIR Conference on Research and Development in Informaion Retrieval, pp. 267–273 (2003)
35. Yu, J., Rui, Y., Tao, D.: Click prediction for web image reranking using multimodal sparse coding. IEEE Transactions on Image Processing 23, 2019–2032 (2014)
36. Zhang, Y., Jiang, Z., Davis, L.S.: Learning structured low-rank representations for image classification. In: IEEE Conference on Computer Vision and Pattern Recognition, pp. 676–683 (2013)
37. Zhou, T., Tao, D.: Godec: Randomized low-rank & sparse matrix decomposition in noisy case. In: International Conference on Machine Learning, pp. 33–40 (2011)
38. Zhuang, L., Gao, H., Lin, Z., Ma, Y., Zhang, X., Yu, N.: Non-negative low rank and sparse graph for semi-supervised learning. In: IEEE Conference on Computer Vision and Pattern Recognition, pp. 2328–2335 (2012)

Two-Dimensional Euler PCA
for Face Recognition

Huibin Tan[1], Xiang Zhang[1], Naiyang Guan[1], Dacheng Tao[3], Xuhui Huang[2],
and Zhigang Luo[1,*]

[1] Science and Technology on Parallel and Distributed Processing Laboratory
College of Computer
National University of Defense Technology, Changsha, Hunan, P.R. China, 410073
[2] Department of Computer Science and Technology, College of Computer
National University of Defense Technology, Changsha, Hunan, P.R. China, 410073
[3] Centre for Quantum Computation & Intelligent Systems and the Faculty of
Engineering and Information Technology
University of Technology, Sydney, 235 Jones Street, Ultimo, NSW 2007, Australia
tanhuibin2815@163.com, zhangxiang_43@aliyun.com,
{ny_guan,xuhuihuang,zgluo}@nudt.edu.cn, dacheng.tao@uts.edu.au

Abstract. Principal component analysis (PCA) projects data on the
directions with maximal variances. Since PCA is quite effective in dimension reduction, it has been widely used in computer vision. However,
conventional PCA suffers from following deficiencies: 1) it spends much
computational costs to handle high-dimensional data, and 2) it cannot reveal the nonlinear relationship among different features of data. To overcome these deficiencies, this paper proposes an efficient two-dimensional
Euler PCA (2D-ePCA) algorithm. Particularly, 2D-ePCA learns projection matrix on the 2D pixel matrix of each image without reshaping it into 1D long vector, and uncovers nonlinear relationships among
features by mapping data onto complex representation. Since such 2D
complex representation induces much smaller kernel matrix and principal subspaces, 2D-ePCA costs much less computational overheads than
Euler PCA on large-scale dataset. Experimental results on popular face
datasets show that 2D-ePCA outperforms the representative algorithms
in terms of accuracy, computational overhead, and robustness.

Keywords: Principal component analysis (PCA), Euler PCA, face
recognition.

1 Introduction

In many fields such as multimedia and cybernetics, data consists of high-
dimensional features and thus confronts the so-called curse of dimensionality
problem [3]. Recently, many dimension reduction techniques [12][9][6][10] have
been proposed to address this issue. Dimension reduction uncovers the intrinsic

---

* Corresponding author.

X. He et al. (Eds.): MMM 2015, Part II, LNCS 8936, pp. 548–559, 2015.

structure of the datasets and boosts subsequent data processing. It has attracted much attention from the computer vision community. The most representative method including principal component analysis (PCA, [12]) has become greatly popularized due to its effectiveness. Recent years witness various PCA extensions such as sparse PCA (SPCA) [15] applied in image classification. PCA learns a group of orthogonal projection axis by maximizing the variance of the data. Due to its simplicity and efficacy, it has been widely applied in image classification such as face recognition. However, it is required to transform each 2D image into a 1D vector to derive the high-dimensional covariance matrix. This induces high computational overhead especially when the size of image is relatively large. Besides, PCA fails to unveil nonlinear relationships among the features of data.

To alleviate high computational cost of PCA, Yang *et al.* proposed two-dimensional principal component analysis (2DPCA [21]) which utilizes 2D matrices rather than 1D vectors. Since 2DPCA preserves the spatial structure of the pixels, it can obtain better data representation than PCA. Similar to PCA, 2DPCA still cannot discover the non-linear relationships among the features as well. To defeat this limitation, Scholkopf *et al.* [20] proposed kernel PCA (KPCA) which utilizes the kernel trick to project the original examples to a higher dimensional space. Benefit from the kernel trick, KPCA can reveal the nonlinear patterns in the data. Meanwhile, Liwicki *et al.* [17] developed Euler PCA (*e*-PCA), which employs an explicit mapping to overcome such limitation. More particularly, *e*-PCA suppresses the outliers via a dissimilarity measure, before which pixel intensities are first normalized and then mapped using the Euler representation of complex numbers. Then *e*-PCA commences standard l_2-norm measure. *e*-PCA appears to be robust to the data noise. However, it still induces the spatial information loss in data representation and thus performs unsatisfactorily in classification tasks. Besides, it would bear the expensive computational burden especially when the scale of image data is so large.

To address the above issues, this paper proposes a 2D Euler PCA (2D-*e*PCA) method designed for dimension reduction. Firstly, 2D-*e*PCA inherits the advantages of *e*-PCA in revealing the structure of features of data and robustness in eliminating noises by the dissimilarity metric. Moreover, 2D-*e*PCA is an algorithm that learns the projection matrix over 2D image matrices rather than 1D vector. Thus it preserves the spatial structure of the pixels and reduces the scale of total kernel matrix, which decreases computational overheads of decomposition and the dimensionality of feature subspace. Experimental results of face recognition on Yale [2], ORL [19], UMIST [5], and FERET [18] datasets show 2D-*e*PCA's superiority to the representative methods in quantities.

The remainder of this paper is organized as follows: Section 2 reviews related works. We propose 2D-*e*PCA and optimize it in Section 3, and then validate it in Section 4, respectively. Finally, we conclude this paper in Section 5.

2 Related Works

Prior to reviewing the related work, we introduce the following notations used throughout this paper shown in Table 1.

Table 1. Notations and definitions

N	the number of examples
p	the dimensionality of example vector such that $p = mn$
x_i	the i-th vector example
$\bar{x}$	the centroid of all examples
S_A, S_B	the covariance (scatter) matrix
k	the number of principal components of samples
U	the principal component of the samples
A_i	the i-th 2D image
m, n	the dimensionality of row and column of a 2D image, respectively
d	the number of principal components of 2D images
z_i	the complex representation of examples
K	the kernel matrix
B	the principal subspace

2.1 PCA

PCA [12] learns a group of orthogonal projection axis by maximizing the variance of the data. Due to its effectiveness and simplicity, PCA has been widely applied in computer vision. Given $N \times p$-dimensional examples $X = \{x_1, x_2, \cdots, x_N\}$, PCA projects data on the directions with maximal variances with the projective matrix U.

$$\max_U(tr(U^T S_X U)) = \max_U tr\left(U^T \sum_{i=1}^{N}(x_i - \bar{x})(x_i - \bar{x})^T U\right) \qquad (1)$$

where $\bar{x}$ denotes the centroid of all examples and $tr(\bullet)$ represents the trace operator. Maximizing (1) is to find the projection directions $U = \{u_1, u_2, \cdots, u_k\}$ under the orthonormal constraints such that

$$\begin{cases} \{u_1, u_2, \cdots, u_k\} = \arg\max tr\left(U^T S_X U\right) \\ u_i^T u_j = 0; i \neq j \\ u_i^T u_j = 1; i = j. \end{cases} \qquad (2)$$

However, PCA suffers from high computational cost. Therefore, it fails to uncover non-linear patterns among the features of the data.

2.2 2DPCA

Recently Yang *et al.* [21] proposed two-dimensional PCA (2DPCA) based on 2D intensity matrices rather than 1D vectors to preserve the spatial relationships among the pixels. Besides, it extremely reduces the computational cost. Given N images $\{A_1, A_2, \cdots, A_N\}$, where $A_i \in R^{m \times n}$ denotes the i-th 2D image and $p = mn$, 2DPCA also maximizes the total scatter of the projections of original examples.

$$\max_{U} tr \left(U^T S_A U \right) = \max_{U} tr \left(U^T \sum_{i=1}^{N} \left(A_i - \bar{A} \right) \left(A_i - \bar{A} \right)^T U \right) \tag{3}$$

Benefit from 2D data representation, it spends much less time cost in solving (3). By the eigenvalue decomposition (EVD), we can get the orthonormal vectors $U = \{u_1, u_2, \cdots, u_d\}$, corresponding to the first d largest eigenvalues of S_A. Similar to PCA, it still fails to unveil non-linear relationships among the features of the data.

2.3   e-PCA

Liwicki *et al.* [17] developed an Euler PCA (*e*-PCA) based on an explicit kernel on Euler representation of complex numbers using the following dissimilarity measure:

$$d\left(x_j, x_q \right) = \sum_{c=1}^{p} \{ 1 - \cos \left(\alpha \pi \left(x_j \left(c \right) - x_q \left(c \right) \right) \right) \}, \tag{4}$$

where $\alpha \in [0, 2)$. We map x_j onto the complex representation $z_j \in C^p$, where

$$Z_j = \frac{1}{\sqrt{2}} \begin{bmatrix} e^{i\alpha\pi x_j(1)} \\ \vdots \\ e^{i\alpha\pi x_j(p)} \end{bmatrix} = \frac{1}{\sqrt{2}} e^{i\alpha\pi x_j}. \tag{5}$$

By (5), we get $Z = \{z_1, z_2, \cdots, z_N\}^T \in C^{N \times p}$. Similar to PCA, *e*-PCA learns the orthonormal vectors in complex region by using EVD to decompose the following kernel matrix:

$$K = \sum_{i=1}^{n} Z_i Z_i^T = U \Lambda U^T \in C^{N \times N} \tag{6}$$

Then we can find the k-reduced sets $U_k \in C^{N \times k}$ and $\Lambda_k \in R^{k \times k}$ of U and Λ, subsequently compute the principal feature subspace B as follows:

$$B = Z^T U_k \Lambda_k^{-\frac{1}{2}}. \tag{7}$$

Based on (5), *e*-PCA can unveil the non-linear patterns among the features of the data. However, it inevitably costs high time overhead to yield the principal subspace (7) especially for massive images.

3   Two-dimensional Euler Principal Component Analysis

To address the above issues, this section proposes two-dimensional Euler PCA(2D-ePCA). Similar to 2DPCA, 2D-*e*PCA learns the projection matrices on 2D images. Additionally, it employs the same mapping as *e*-PCA to uncover the non-linear patterns within the data. Thus, 2D-*e*PCA can be so low both in both time and in space complexity meanwhile unveil the nonlinear patterns of the features of the data. We normalize each A_i and then map it onto the complex space (7) to obtain Z_i. All the transformed features can be put together as $Z_0 = \{Z_1, \cdots, Z_i, \cdots, Z_N\}$.

First, we need to calculate the kernel matrix based on the 2D complex matrix.

$$K = \sum_{i=1}^{N} Z_i Z_i^T \in C^{m \times m}. \tag{8}$$

Such kernel matrix is very different from that of e-PCA. That is because the kernel matrix of 2D-ePCA based on the 2D matrix maintains the spatial relationships among the pixels while e-PCA takes no account of such spatial structure. 2D-ePCA learns the linear subspace B in the high-dimensional subspace via EVD:

$$K = B^T \Sigma B, \tag{9}$$

where the diagonal entries of Σ denotes the eigenvalues of K.

Thus we select a set of projection axes from the kernel matrix, $\mu_1, \mu_2, \cdots, \mu_k$ corresponding to the largest k eigenvalues to meet the orthonormal constraints. Interestingly, the complex representation of each image can be transformed into the pixel space via the $\angle$-operator, which returns the angle of a complex number.

Algorithm 1. 2D-ePCA

Input: Given N images $\{A_1, \cdots, A_i, \cdots, A_N\}$, where $A_i \in R^{m \times n}$, the number d of principal components and the parameter α

Output: B

1. Represent A_i in the range $[0, 1]$ as I_i.
2. Calculate $Z_i = \frac{1}{\sqrt{2}} e^{i\alpha\pi I_i} \in C^{m \times n}$.
3. Calculate the kernel matrix$K = \sum_{i=1}^{N} Z_i Z_i^T \in C^{m \times m}$.
4. Obtain the eigenvalue decomposition of K via (9).
5. Select the k eigenvectors corresponding to the k largest eigenvalues as B.

To illustrate the efficiency of 2D-ePCA, we summarize time complexity of four representative PCA methods into Table 2. Table 2 shows that both 2D-ePCA and 2DPCA takes much less time than PCA because $p^3 >> m^3$. 2D-ePCA costs competitive time cost as e-PCA when the number of samples is relatively so little, but 2D-ePCA cost much less time than e-PCA on large-scale data. Similar to e-PCA, 2D-ePCA can benefit from a dissimilarity measure to handle noisy data. Besides, 2D-ePCA preserves the spatial structure among pixels and thus boosts e-PCA. In summary, 2D-ePCA enhances PCA both in efficacy and in efficiency.

Table 2. Time Complexity of 2D-ePCA, PCA, 2DPCA and e-PCA

PCA	$O\left(Np + Np^2 + p^3\right)$
2DPCA	$O\left(Nmn + Nm^2n + Nm^2 + m^3\right)$
e-PCA	$O\left(NP + N^2p + N^3 + Npm^2\right)$
2D-ePCA	$O\left(Nmn + Nm^2n + Nm^2 + m^3\right)$

4 Experiments

This section evaluates 2D-ePCA by comparing with PCA [12], 2DPCA [21] and e-PCA [17] on four face image datasets including Yale [2], ORL [19], UMIST [5], and FERET [18]. Figure 1 illustrates some instances of four face datasets. For fair comparison, we compare 2D-ePCA and e-PCA in terms of their highest average accuracy. All face images have been aligned according to the position of eyes. We adopt nearest neighbor classifier(NN) to calculate the recognition accuracy. To eliminate the influence of randomness, we repeat each trials five times and then report the final results with their average accuracy. For PCA, the maximum reduced dimensionality equals to the number of non-zero eigenvalues. The rank-deficient covariance induces them to be far less than the dimensional size of original examples. For e-PCA, the eigenvalues of the kernel matrix lie in complex region and are non-zero as well, and thus the reduced dimension maximum often equals to the number of samples. But the maximum reduced dimensionality of both 2DPCA and 2D-ePCA depends on the size of the covariance matrix and the kernel matrix, respectively. Henceforth, it is not surprise that they own different numbers of reduced dimensionalities in our experiments.

4.1 Face Recognition

Yale Dataset. The Yale face dataset contains 165 images of 15 individuals. Each person has 11 different images under various facial expressions and lighting conditions. All images are grayscale and cropped to 32×32 pixels. We select 3, 4, 5 and 6 images in each individual for training and the remainder for testing.

Figure 2 shows that 2D-ePCA is significantly superior to PCA, 2DPCA and e-PCA in terms of recognition accuracy. Specially, 2D-ePCA obtains the highest

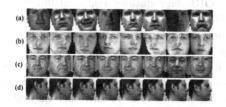

Fig. 1. Image instances of (a) Yale, (b) ORL, (c) UMIST and (d) FERET dataset

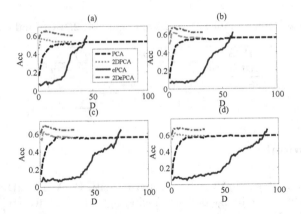

Fig. 2. Average accuracies versus different dimensionalities when (a) 3, (b) 4, (c) 5 and (d) 6 images per individual were selected for training on Yale dataset

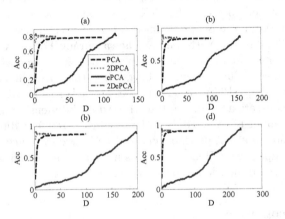

Fig. 3. Average accuracies versus different dimensionalities when (a) 3, (b) 4, (c) 5 and (d) 6 images per individual were selected for training on ORL dataset

recognition accuracy in lower dimension compared with the compared methods e-PCA. This also attributes to the spatial structure of pixels.

ORL Dataset. The ORL face dataset contains images from 40 individuals, each containing 10 different images. All images are grayscale and cropped to 32×32 pixels. We select 3, 4, 5 and 6 images each individual for training and the remaining samples for testing.

Figure 3 demonstrates that e-PCA and 2D-ePCA outperforms PCA, 2DPCA in terms of recognition accuracy when the dimensionalities D of low-dimensional space are identical. Besides, 2D-ePCA obtains much highest recognition accuracy than e-PCA when D is low. This observation reflects that 2D-ePCA benefits much from the relationship among pixels.

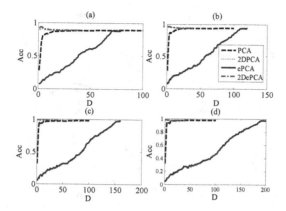

Fig. 4. Average accuracies versus different dimensionalities when (a) 4, (b) 6, (c) 8 and (d) 10 images per individual were selected for training on UMIST dataset

UMIST Dataset. The UMIST face dataset contains images from 20 individuals, totally 575 different images. All images are grayscale and cropped to 40×40 pixels. The fifteen right profile images each subject consist of a subset of 300 images for comparison. Based on this subset, we respectively select 4, 6, 8 and 10 images each individual for training and the remaining samples for testing.

Figure 4 suggests that 2D-ePCA outperforms the compared methods in the identical cofigurations. Likewise, 2D-ePCA achieves the highest recognition accuracy due to the spatial relationship among pixels.

FERET Dataset. The FERET dataset contains images from 40 individuals. Each subject has 10 different images from various facial expressions such as smile. All images are grayscale and cropped to 40×40 pixels. We respectively select 3, 4, 5 and 6 images each individual for training and the remaining samples for testing.

Figure 5 also shows that 2D-ePCA is superior to PCA, 2DPCA and e-PCA in terms of recognition accuracy when the dimensionalities D of low-dimensional space are identical. It is implies that 2D-ePCA always induces the superiority of recognition on FERET dataset.

4.2   The Effect of α

This section studies the effect of the parameter α in 2D-ePCA at average accuracy.

Figure 6 shows that the parameter α influences little to the face recognition accuracy, especially on both ORL and UMIST datasets. It also implies that 2D-ePCA can achieve the highest average accuracy on Yale, ORL and FERET datasets when $\alpha = 1.99$. In UMIST dataset, 2D-ePCA can achieve the highest average accuracy when $\alpha = 1.3$.

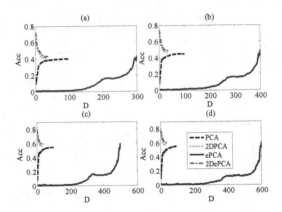

Fig. 5. Average accuracies versus different dimensionalities when (a) 4, (b) 6, (c) 8 and (d) 10 images per individual were selected for training on FERET dataset

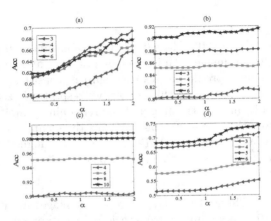

Fig. 6. Average accuracies of 2D-ePCA versus different α on(a) Yale, (b) ORL, (c) UMIST and (d) FERET dataset

4.3   Time Evaluation

Table 3 shows time cost of PCA, 2DPCA, e-PCA and 2D-ePCA when six images per individual are selected for training in four face datasets. It also shows that 2D-ePCA consistently spends the least CPU seconds on four face datasets.

Table 3. Time Overhead (CPU seconds)

Method	Yale	ORL	UMIST	FERET
PCA	2.359	2.426	7.532	7.902
2DPCA	0.0197	0.1251	0.0985	1.434
e-PCA	0.0125	0.1047	0.0529	1.213
2D-ePCA	0.0104	0.0278	0.0257	0.1089

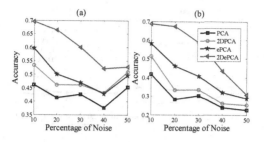

Fig. 7. Average accuracy versus different percentages of noises in pixels of (a) 70% images and (b) 90% images on Yale dataset

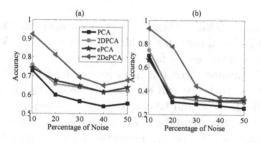

Fig. 8. Average accuracy versus different percentages of noises in pixels of (a) 70% images and (b) 90% images on ORL dataset

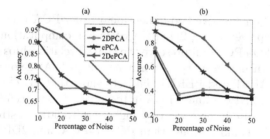

Fig. 9. Average accuracy versus different percentages of noises in pixels of (a) 70% images and (b) 90% images on UMIST dataset

4.4 Robustness Evaluation

Figure 7 shows that the higher the percentage of the noise is the lower the average accuracy of all compared methods becomes. It also demonstrates that 2D-ePCA still achieves the highest average accuracy on Yale dataset even if the ratio of the noise to the pixel is 50%. Figure 8 demonstrates that 2D-ePCA consistently achieves the highest average accuracy on ORL dataset with different percentages of the noise in the pixels. Figure 9 also shows 2D-ePCA still achieves the highest average accuracy on UMIST even if the ratio of the noise to the pixel is 50%. Figure 10 reports how the percentage of the noise affects the average accuracy of

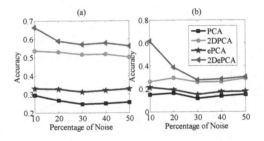

Fig. 10. Average accuracy versus different percentages of noises in pixels of (a) 70% images and 90% images on FERET dataset

all compared methods. It shows that 2D-ePCA outperforms the representative methods on FERET even if the ratio of the noise to the pixel is 50%. In summary, 2D-ePCA has achieved success in face recognition.

But 2D-ePCA is unsupervised but achieves unsatisfactory results in classification. In our future work, we will employ the geometric structure [14][7][23], popular features like LBP [4] and label data [16][8][11] to improve its discriminative power; meanwhile, to extend the applicability, we can borrow some strategies from [13][1][22] to apply it for other applications such as image retrieval.

5   Conclusion

Here we proposed an effective two-dimensional Euler PCA (2D-ePCA). Benefit from 2D image representation, 2D-ePCA costs less computation overhead than e-PCA. Besides, 2D-ePCA can preserve the spatial relationships among the pixels and thus it can enhance the performance of e-PCA. Experimental results on Yale, ORL, UMIST and FERET datasets show that 2D-ePCA consistently outperforms PCA, 2DPCA and e-PCA in terms of average accuracy. Even if the percentage of the noise in pixels is very high close to 50%, it still performs well and thus 2D-ePCA is more robust than PCA, e-PCA, and 2DPCA.

Acknowledgment. This work is partially supported by Research Fund for the Doctoral Program of Higher Education of China, SRFDP (under grant No. 20134307110017) and Scientific Research Plan Project of NUDT (under grant No. JC13-06-01), and Australian Research Council Projects (under grant No. DP-120103730, FT-130101457 and LP-140100569).

References

1. Hypergraph-based multi-example ranking with sparse representation for transductive learning image retrieval. Neurocomputing 101(0), 94 – 103 (2013)
2. Belhumeur, P.N., Hespanha, J.P., Kriegman, D.: Eigenfaces vs. fisherfaces: Recognition using class specific linear projection. IEEE Transactions on Pattern Analysis and Machine Intelligence 19(7), 711–720 (1997)

3. Bellman, R.E., Dreyfus, S.E.: Applied dynamic programming (1962)
4. Feng, X., Lv, B., Li, Z., Zhang, J.: A novel feature extraction method for facial expression recognition. In: JCIS (2006)
5. Graham, D.B., Allinson, N.M.: Characterising virtual eigensignatures for general purpose face recognition. In: Face Recognition, pp. 446–456. Springer (1998)
6. Guan, N., Tao, D., Luo, Z., Shawe-Taylor, J.: Mahnmf: Manhattan non-negative matrix factorization. arXiv preprint arXiv:1207.3438 (2012)
7. Guan, N., Tao, D., Luo, Z., Yuan, B.: Manifold regularized discriminative nonnegative matrix factorization with fast gradient descent. IEEE Transactions on Image Processing (2011)
8. Guan, N., Tao, D., Luo, Z., Yuan, B.: Non-negative patch alignment framework. IEEE Transactions on Neural Networks (2011)
9. Guan, N., Tao, D., Luo, Z., Yuan, B.: Nenmf: an optimal gradient method for nonnegative matrix factorization. IEEE Transactions on Signal Processing 60(6), 2882–2898 (2012)
10. Guan, N., Tao, D., Luo, Z., Yuan, B.: Online nonnegative matrix factorization with robust stochastic approximation. IEEE Transactions on Neural Networks and Learning Systems 23(7), 1087–1099 (2012)
11. Guan, N., Zhang, X., Luo, Z., Tao, D., Yang, X.: Discriminant projective nonnegative matrix factorization. PLoS ONE (2013)
12. Jolliffe, I.: Principal component analysis. Wiley Online Library (2005)
13. Liu, B.-D., Wang, Y.-X., Shen, B., Zhang, Y.-J., Hebert, M.: Self-explanatory sparse representation for image classification. In: Fleet, D., Pajdla, T., Schiele, B., Tuytelaars, T. (eds.) ECCV 2014, Part II. LNCS, vol. 8690, pp. 600–616. Springer, Heidelberg (2014)
14. Liu, B.-D., Wang, Y.-X., Zhang, Y.-J., Shen, B.: Learning dictionary on manifolds for image classification. Pattern Recognition 46(7), 1879–1890 (2013)
15. Liu, W., Zhang, H., Tao, D., Wang, Y., Lu, K.: Large-scale paralleled sparse principal component analysis. ArXiv e-prints (2013)
16. Liu, W., Song, C., Wang, Y.: Facial expression recognition based on discriminative dictionary learning. In: International Conference on Pattern Recognition, pp. 1839–1842 (2012)
17. Liwicki, S., Tzimiropoulos, G., Zafeiriou, S., Pantic, M.: Euler principal component analysis. International Journal of Computer Vision 101(3), 498–518 (2013)
18. Phillips, P.J., Moon, H., Rizvi, S.A., Rauss, P.J.: The feret evaluation methodology for face-recognition algorithms. IEEE Transactions on Pattern Analysis and Machine Intelligence 22(10), 1090–1104 (2000)
19. Samaria, F.S., Harter, A.C.: Parameterisation of a stochastic model for human face identification. In: Proceedings of the Second IEEE Workshop on Applications of Computer Vision, pp. 138–142 (1994)
20. Schölkopf, B., Smola, A., Müller, K.R.: Nonlinear component analysis as a kernel eigenvalue problem. Neural Computation 10(5), 1299–1319 (1998)
21. Yang, J., Zhang, D., Frangi, A.F., Yang, J.Y.: Two-dimensional pca: a new approach to appearance-based face representation and recognition. IEEE Transactions on Pattern Analysis and Machine Intelligence 26(1), 131–137 (2004)
22. Yu, J., Rui, Y., Tao, D.: Click prediction for web image reranking using multimodal sparse coding. IEEE Transactions on Image Processing (2014)
23. Yu, J., Tao, D., Wang, M.: Adaptive hypergraph learning and its application in image classification. IEEE Transactions on Image Processing 21(7), 3262–3272 (2012)

Multiclass Boosting Framework for Multimodal Data Analysis

Shixun Wang, Peng Pan*, Yansheng Lu, and Sheng Jiang

School of Computer Science and Technology
Huazhong University of Science and Technology
Wuhan 430074, China
{wsxun,panpeng,lys,jwt}@hust.edu.cn

Abstract. A large number of multimedia documents containing texts and images have appeared on the internet, hence cross-modal retrieval in which the modality of a query is different from that of the retrieved results is being an interesting search paradigm. In this paper, a multimodal multiclass boosting framework (MMB) is proposed to capture intra-modal semantic information and inter-modal semantic correlation. Unlike traditional boosting methods which are confined to two classes or single modality, MMB could simultaneously deal with multimodal data. The empirical risk, which takes both intra-modal and inter-modal losses into account, is designed and then minimized by gradient descent in the multidimensional functional spaces. More specifically, the optimization problem is solved in turn for each modality. Semantic space can be naturally attained by applying sigmoid function to the quasi-margins. Extensive experiments on the Wiki and NUS-WIDE datasets show that the performance of our method significantly outperforms those of existing approaches for cross-modal retrieval.

Keywords: multiclass boosting, loss function, intra-modal and inter-modal, cross-modal retrieval, semantic space.

1   Introduction

In the past few years, a tremendous amount of multimedia content had been generated on the web, such as Flickr for photographs and YouTube for videos. Many unimodal retrieval approaches in which a query and the retrieved results are of the same modality have been proposed, including image retrieval [5], music retrieval [17] and video retrieval [15]. Actually, these methods cannot effectively deal with the case where a user wants to retrieve texts or other modalities with a query image. Therefore, the cross-modal retrieval for solving such problem remains a big challenge.

In parallel, the solutions to various problems in machine learning have involved the design of a classifier. One reliable tool is boosting technique whose idea is to

---

* Corresponding author.

X. He et al. (Eds.): MMM 2015, Part II, LNCS 8936, pp. 560–571, 2015.

combine multiple weak learners into a strong ensemble. Existing boosting methods on single modality data can be basically categorized into two aspects: binary (e.g., AdaBoost [7] and TaylorBoost [13]) and multiclass (e.g., SAMME [19] and GD-MCBoost [14]). In general, two main strategies can be used to design multiclass boosting algorithm. The first is to decompose the multiclass problem into several independent binary sub-problems. Popular methods in this type include one-versus-all and one-versus-one. However, these schemes can have several disadvantages, such as imbalanced data distribute, increased complexity and lack of a joint optimal predictor. Alternatively, one can resort to the direct strategy in which a set of codewords usually plays an important role when multiclass weak learners are boosted. In [14], the vertices of multidimensional regular simplex which is centered at the origin are regarded as the optimal set of codewords. Nevertheless, there is few work about multiclass boosting method for analyzing multimodal data.

Although the automatic annotation of media [3,16] can be treated as the origin of cross-modal system, the annotations have few text data and only describe visible or auditory object. In cross-modal retrieval, the modality of a query is different from that of the returned results. For example, given a query image of lion, the response may be text or audio of lion. How to measure the similarity among objects of different modalities is a key challenge. A common space, such as correlative subspace [1,8], hash space [2,18,20] and semantic space [12], is usually learned so that the traditional measure could be directly applied.

As a solution to learn the correlative subspace, canonical correlation analysis (CCA) [8] is a linear dimensionality reduction method that maximizes the correlation between two sets of heterogeneous data. In [9], CCA is applied to localize visual events associated with sound sources. Kernel canonical correlation analysis (KCCA), an extension of CCA, is utilized to fuse text and image for spectral clustering [1]. The idea of hash learning is to convert different modalities into hash space such that the neighborhood relationships can be preserved. Cross-modal similarity sensitive hashing (CMSSH) [2] learns the projection function for each dimension of the hash space. In [18], a probabilistic latent factor model (MLBE) is used to learn hash functions from different modalities data. Linear cross-modal hashing (LCMH) [20] represents each object using its distance to cluster centroids, and maps the new representation into a latent space. However, these methods do not consider the similarity in a semantic space.

Semantic space [11] is a probability simplex, in which each data point is a vector of posterior category probabilities. In [12], multiclass logistic regression is utilized to attain the semantic vectors of texts and images, and the best experimental results can be obtained by combining semantic matching and CCA. When the feature vector has high dimensionality and sparse values, logistic regression cannot effectively work. To the best of our knowledge, the research which utilizes multimodal multiclass boosting to jointly mine intra-modal semantic information and inter-modal semantic correlation from multimodal data has not been found in previous papers.

In this paper, our tasks are the retrieval of images in response to a query text and vice-versa. To analyze multimodal data, the proposed multiclass boosting framework designs an objective function which considers both intra-modal and inter-modal losses. The intra-modal loss is useful for getting accurate semantic category information in each modality, while the inter-modal loss plays an important role in capturing the semantic correlation between different modalities. Both of them are significant and their combination may complement each other. In the multidimensional functional spaces, gradient descent is utilized to solve the optimization problem, which is minimized in turn for every modality. After learning the optimal predictor, posterior category probabilities can be attained by applying sigmoid function to the learned quasi-margins. The experimental results on two datasets show the effectiveness of the proposed approach, compared with some existing methods.

The rest of this paper will be organized as follows. In Section 2, we introduce our multimodal multiclass boosting framework. The algorithm for cross-modal retrieval is presented in Section 3. The experimental results and analyses are provided in Section 4. Finally, the concluding discussions and possible future works are listed in Section 5.

2   Risk Construction with Multiclass Loss Function

In this section, we present a multimodal multiclass boosting framework for analyzing multimodal data. Although the key idea of our framework can be applied to other modalities, the discussions are limited to documents containing images and texts. The boldface letters are used to denote matrices or vectors.

Let multimedia dataset be $(X, Y, S) = \{(\mathbf{x}_1, \mathbf{y}_1, s_1), \ldots, (\mathbf{x}_N, \mathbf{y}_N, s_N)\}$, in which X, Y, S and N denote images, texts, semantic vocabulary and the number of documents, respectively. Suppose the vocabulary includes K unique semantic categories, each image $\mathbf{x}_i \in \mathbb{R}^I$ or text $\mathbf{y}_i \in \mathbb{R}^T$ in the training set is assigned with a semantic category $s_i \in \{1, \ldots, K\}$, while images and texts in the testing set remain unlabeled. Given an image (text) query $\mathbf{x}_q$ ($\mathbf{y}_q$) in the test set, the goal of cross-modal retrieval is to search for the closest match in the text (image) space $\mathbb{R}^T$ ($\mathbb{R}^I$) of the retrieved set.

According to [6,14], a set of K distinct unit vectors, which are the vertices of $K - 1$ dimensional regular simplex centered at the origin, can form a codeword matrix $\mathbf{C} = [\mathbf{c}^1, \ldots, \mathbf{c}^K]$. Therefore, each semantic category k can be recoded with a codeword vector $\mathbf{c}^k \in \mathbb{R}^{K-1}$. If $\mathbf{f}(\mathbf{x})$ and $\mathbf{u}(\mathbf{y}) \in \mathbb{R}^{K-1}$ are the predictors for image and text, their quasi-margins with respect to category k can be respectively defined as $< \mathbf{f}(\mathbf{x}), \mathbf{c}^k >$ and $< \mathbf{u}(\mathbf{y}), \mathbf{c}^k >$, where $< \cdot, \cdot >$ denotes the standard inner product. To search for the optimal predictors for different modalities, the empirical risk is defined as follow

$$R[\mathbf{f}, \mathbf{u}] = R_1[\mathbf{f}(\mathbf{x})] + R_2[\mathbf{u}(\mathbf{y})] + R_3[\mathbf{f}(\mathbf{x}), \mathbf{u}(\mathbf{y})]$$

$$= \sum_{i=1}^{N} L_1[\mathbf{c}_i, \mathbf{f}(\mathbf{x}_i)] + \sum_{i=1}^{N} L_2[\mathbf{c}_i, \mathbf{u}(\mathbf{y}_i)] + \sum_{i=1}^{N} \left\| \mathbf{C}^T[\mathbf{f}(\mathbf{x}_i) - \mathbf{u}(\mathbf{y}_i)] \right\|_2^2 , (1)$$

where $L[\cdot,\cdot]$ stands for multiclass loss function. In Eq. 1, the first and second terms are the corresponding intra-modal losses for images and texts, and the third term is the inter-modal loss which may capture semantic correlation between different modalities. In general, the empirical risk can be minimized by solving the following optimization problem

$$\begin{cases} \min_{\mathbf{f},\mathbf{u}} R[\mathbf{f}(\mathbf{x}),\mathbf{u}(\mathbf{y})] \\ s.t \quad \mathbf{f}(\mathbf{x}) \in span(H),\ \mathbf{u}(\mathbf{y}) \in span(\bar{H}), \end{cases}$$

where $H = \{\mathbf{g}_i(\mathbf{x})\}$ and $\bar{H} = \{\mathbf{v}_i(\mathbf{y})\}$ are the sets of multiclass weak learners $\mathbf{g}_i(\mathbf{x}) : X \to \mathbb{R}^{K-1}$ and $\mathbf{v}_i(\mathbf{y}) : Y \to \mathbb{R}^{K-1}$, respectively; and $span(\cdot)$ denotes the functional space of linear combinations of weak learners.

2.1 Multiclass Exponential Loss

To guarantee large quasi-margin, the multiclass loss for image is usually defined as the following non-negative function of quasi-margin

$$L[\mathbf{c},\mathbf{f}(\mathbf{x})] = \sum_{k=1}^{K} \exp\left(- < \mathbf{f}(\mathbf{x}), \mathbf{c} - \mathbf{c}^k > \right). \tag{2}$$

Similarly, the loss for text can be defined. By adjusting only one modality at a time with the other fixed, the optimization problem is easy to solve. After t boosting iterations, let $\mathbf{f}^t(\mathbf{x})$ and $\mathbf{u}^t(\mathbf{y})$ are the predictors of image and text, respectively. Based on gradient descent, the update procedure for image predictor is firstly implemented. Around point $\mathbf{f}^t(\mathbf{x})$, the functional derivative of the objective function $R[\mathbf{f}(\mathbf{x}),\mathbf{u}^t(\mathbf{y})]$, along the direction of weak learner $\mathbf{g}(\mathbf{x})$, is

$$\delta R[\mathbf{f}^t;\mathbf{g}] = \left.\frac{\partial R[\mathbf{f}^t + \xi\mathbf{g},\mathbf{u}^t]}{\partial \xi}\right|_{\xi=0} = -\sum_{i=1}^{N} < \mathbf{g}(\mathbf{x}_i),\mathbf{P}_i(\mathbf{x}_i,\mathbf{y}_i) >, \tag{3}$$

where $\mathbf{P}_i(\mathbf{x}_i,\mathbf{y}_i) \in \mathbb{R}^{K-1}$

$$\mathbf{P}_i = \sum_{k=1}^{K} (\mathbf{c}_i - \mathbf{c}^k) \exp\left(- < \mathbf{f}^t(\mathbf{x}_i),\mathbf{c}_i - \mathbf{c}^k > \right) - 2\mathbf{C}\mathbf{C}^{\mathbf{T}}\left(\mathbf{f}^t(\mathbf{x}_i) - \mathbf{u}^t(\mathbf{y}_i) \right). \tag{4}$$

At $t+1$ image iteration, the direction of greatest risk decrease and the corresponding optimal step can be obtained by

$$\mathbf{g}^* = \arg\min_{\mathbf{g}\in H} \delta R[\mathbf{f}^t;\mathbf{g}] = \arg\max_{\mathbf{g}\in H} \sum_{i=1}^{N} < \mathbf{g}(\mathbf{x}_i),\mathbf{P}_i(\mathbf{x}_i,\mathbf{y}_i) >, \tag{5}$$

$$\alpha^* = \arg\min_{\alpha\in\mathbb{R}} R[\mathbf{f}^t + \alpha\mathbf{g}^*,\mathbf{u}^t]. \tag{6}$$

Therefore, the image predictor is updated to $\mathbf{f}^{t+1}(\mathbf{x}) = \mathbf{f}^t(\mathbf{x}) + \alpha^*\mathbf{g}^*(\mathbf{x})$. Next, the update of text predictor is similarly executed. At point $\mathbf{u}^t(\mathbf{y})$, the partial derivative of the risk $R[\mathbf{f}^{t+1}(x), \mathbf{u}(\mathbf{y})]$, along the direction of weak learner $\mathbf{v}(\mathbf{y})$, is calculated as follow

$$\delta R[\mathbf{u}^t; \mathbf{v}] = \left.\frac{\partial R[\mathbf{f}^{t+1}, \mathbf{u}^t + \varepsilon\mathbf{v}]}{\partial \varepsilon}\right|_{\varepsilon=0} = -\sum_{i=1}^{N} < \mathbf{v}(\mathbf{y}_i), \mathbf{Q}_i(\mathbf{x}_i, \mathbf{y}_i) >, \quad (7)$$

where

$$\mathbf{Q}_i = \sum_{k=1}^{K} \left(\mathbf{c}_i - \mathbf{c}^k\right) \exp\left(- < \mathbf{u}^t(\mathbf{y}_i), \mathbf{c}_i - \mathbf{c}^k >\right) + 2\mathbf{C}\mathbf{C}^{\mathbf{T}} \left(\mathbf{f}^{t+1}(\mathbf{x}_i) - \mathbf{u}^t(\mathbf{y}_i)\right) . \quad (8)$$

Similar to Eq. 5, the optimal direction $\mathbf{v}^*$ at $t+1$ text iteration is easily got. The optimal step size along this direction is slight different

$$\beta^* = \arg\min_{\beta \in \mathbb{R}} R[\mathbf{f}^{t+1}, \mathbf{u}^t + \beta\mathbf{v}^*] . \quad (9)$$

With exponential loss, our model is called as E_MMB.

2.2   Multiclass Logistic Loss

Alternatively, the logistic loss for image is defined as follow

$$L[\mathbf{c}, \mathbf{f}(\mathbf{x})] = \sum_{k=1}^{K} \log\left[1 + \exp\left(- < \mathbf{f}(\mathbf{x}), \mathbf{c} - \mathbf{c}^k >\right)\right] , \quad (10)$$

and the loss for text can be similarly attained. As before, around points $\mathbf{f}^t(\mathbf{x})$ and $\mathbf{u}^t(\mathbf{y})$, the first order partial derivatives of the corresponding risks along the directions of weak learners $\mathbf{g}(\mathbf{x})$ and $\mathbf{v}(\mathbf{y})$ are respectively computed as follow

$$\delta R[\mathbf{f}^t; \mathbf{g}] = -\sum_{i=1}^{N} < \mathbf{g}(\mathbf{x}_i), \mathbf{PP}_i(\mathbf{x}_i, \mathbf{y}_i) >, \quad (11)$$

$$\delta R[\mathbf{u}^t; \mathbf{v}] = -\sum_{i=1}^{N} < \mathbf{v}(\mathbf{y}_i), \mathbf{QQ}_i(\mathbf{x}_i, \mathbf{y}_i) >, \quad (12)$$

where $\mathbf{PP}_i(\mathbf{x}_i, \mathbf{y}_i), \mathbf{QQ}_i(\mathbf{x}_i, \mathbf{y}_i) \in \mathbb{R}^{K-1}$

$$\mathbf{PP}_i = \sum_{k=1}^{K} \frac{(\mathbf{c}_i - \mathbf{c}^k) \exp\left(- < \mathbf{f}^t(\mathbf{x}_i), \mathbf{c}_i - \mathbf{c}^k >\right)}{1 + \exp\left(- < \mathbf{f}^t(\mathbf{x}_i), \mathbf{c}_i - \mathbf{c}^k >\right)} - 2\mathbf{C}\mathbf{C}^{\mathbf{T}} \left(\mathbf{f}^t(\mathbf{x}_i) - \mathbf{u}^t(\mathbf{y}_i)\right) , \quad (13)$$

$$\mathbf{QQ}_i = \sum_{k=1}^{K} \frac{(\mathbf{c}_i - \mathbf{c}^k) \exp\left(- < \mathbf{u}^t(\mathbf{y}_i), \mathbf{c}_i - \mathbf{c}^k >\right)}{1 + \exp\left(- < \mathbf{u}^t(\mathbf{y}_i), \mathbf{c}_i - \mathbf{c}^k >\right)} + 2\mathbf{C}\mathbf{C}^{\mathbf{T}} \left(\mathbf{f}^{t+1}(\mathbf{x}_i) - \mathbf{u}^t(\mathbf{y}_i)\right) . \quad (14)$$

At $t+1$ iteration, the corresponding optimal parameters (directions and steps) and updated predictors for image and text are similarly derived from gradient descent. With logistic loss, our mode is named as L_MMB.

3   MMB Algorithm for Cross-Modal Retrieval

In this section, the algorithm for cross-modal retrieval is presented in detail, and then its extension is simply discussed.

In unimodal boosting, the intra-modal semantic information would not be captured well if the low-level features are of low quality. Moreover, inter-modal semantic correlation is not involved. To better preserve both intra-modal and inter-modal semantics, an effective mapping mechanism should be learnt by combining both modalities together. In multimodal boosting, the distance of quasi-margins of semantic relevant pairs is minimized in the inter-modal loss, which makes semantic relevant objects close in a semantic space. In addition, the intra-modal loss is also minimized to capture the intra-modal semantics, and some intra-modal semantics generated from the low quality objects may be enhanced by complementing their corresponding inter-modal semantic correlations.

For simplicity of the presentation, the multimodal multiclass boosting algorithm with exponential loss is summarized in Algorithm 1. However, the algorithm with logistic loss can be easily attained by replacing $\mathbf{P}_i$ and $\mathbf{Q}_i$ with $\mathbf{PP}_i$ and $\mathbf{QQ}_i$, as shown in lines 3 and 7.

Algorithm 1. MMB algorithm for cross-modal retireval

Input: Multimodal dataset $(X, Y, S) = \{(\mathbf{x}_i, \mathbf{y}_i, s_i)\}_{i=1}^{N}$, the number of iterations T,
  the number of categories K, and the generated codeword matrix $\mathbf{C} \in \mathbb{R}^{(K-1) \times K}$.
Output: The final predictors $\mathbf{f}^T(\mathbf{x})$ and $\mathbf{u}^T(\mathbf{y})$ for image and text.
 1: Initialization. Set $t = 0$, and $\mathbf{f}^t(\mathbf{x}) = \mathbf{u}^t(\mathbf{y}) = \mathbf{0} \in \mathbb{R}^{K-1}$.
 2: **While** $t < T$ **do**
 3: Compute $\mathbf{P}_i(\mathbf{x}_i, \mathbf{y}_i)$ with (4).
 4: Using (5), find the optimal direction $\mathbf{g}^*(\mathbf{x})$ for image.
 5: Using (6), search for the optimal step size α^* of image.
 6: Update $\mathbf{f}^{t+1}(\mathbf{x}) = \mathbf{f}^t(\mathbf{x}) + \alpha^* \mathbf{g}^*(\mathbf{x})$.
 7: Calculate $\mathbf{Q}_i(\mathbf{x}_i, \mathbf{y}_i)$ with (8).
 8: Find the optimal direction $\mathbf{v}^*(\mathbf{x})$ for text,

$$\mathbf{v}^* = \arg\max_{\mathbf{v} \in \tilde{H}} \sum_{i=1}^{N} < \mathbf{v}(\mathbf{y}_i), \mathbf{Q}_i(\mathbf{x}_i, \mathbf{y}_i) > .$$

 9: Using (9), find the optimal step size β^* for text.
10: Update $\mathbf{u}^{t+1}(\mathbf{y}) = \mathbf{u}^t(\mathbf{y}) + \beta^* \mathbf{v}^*(\mathbf{y})$.
11: $t = t + 1$.
12: **End while**

In each iteration, if $O(\rho)$ and $O(\eta)$ respectively denote the time costs of learning the optimal image and text predictors , then the time complexity of our proposed algorithm is $O(T \cdot \rho + T \cdot \eta)$. Next, the posterior category probabilities of an image can be computed

$$P\left(s = k | \mathbf{x}\right) = \sigma\left(< \mathbf{f}^T(\mathbf{x}), \mathbf{c}^k >\right) / \sum \sigma\left(< \mathbf{f}^T(\mathbf{x}), \mathbf{c}^k >\right), \qquad (15)$$

where $\sigma(\cdot)$ denotes sigmoid function. Similar to Eq. 15, the probabilities of text can be attained. Given any query q form one modality and the retrieved objects from the other modality, their semantic vectors are achieved by the learnt mapping mechanism. Therefore, the traditional distance measure can be used to accomplish cross-modal retrieval.

Finally, the MMB framework can naturally support more than two modalities. Let M be the number of modalities, the corresponding empirical risk $R[\mathbf{f}_1, \ldots, \mathbf{f}_M]$ is defined as follow

$$
R = \sum_{m=1}^{M} R_m[\mathbf{f}_m(\mathbf{z}^m)] + \sum_{m=1}^{M} \sum_{j>m}^{M} R_{mj}[\mathbf{f}_m(\mathbf{z}^m), \mathbf{f}_j(\mathbf{z}^j)]
$$

$$
= \sum_{m=1}^{M} \sum_{i=1}^{N} L_m[\mathbf{c}_i, \mathbf{f}_m(\mathbf{z}_i^m)] + \sum_{m=1}^{M} \sum_{j>m}^{M} \sum_{i=1}^{N} \left\| \mathbf{C}^T[\mathbf{f}_m(\mathbf{z}_i^m) - \mathbf{f}_j(\mathbf{z}_i^j)] \right\|_2^2, \quad (16)
$$

where $\mathbf{z}_i^m$ denotes the ith object of mth modality, and $\mathbf{f}_m$ is the predictor of mth modality. To minimize this risk, we can adjust only one modality with the others fixed, cycling through all the modalities in turn. For the extension, it is straightforward to add some appropriate lines in Algorithm 1 such that the final predictors of all modalities are attained. Hence, the new cross-modal retrieval problem is solved by transforming the quasi-margins into a semantic space.

4   Experiments

In this section, some experimental evaluations of the proposed framework are described, which are compared with those of the existing methods for cross-modal retrieval. The comparison methods include SM and SCM [12] where the semantic mapping mechanism is replaced by unimodal multiclass boosting [14]. As an example which does not consider semantics, LCMH [20] is also compared. One task of cross-modal retrieval is to use a text query to search for relevant images, and the other is to use an image query to retrieve relevant texts.

Two benchmark datasets, namely Wiki [12] and NUS-WIDE [4], are used in our experiments. The Wiki dataset is assembled from Wikipedia featured articles. Each article is split into several sections according to its section headings, and the images are assigned to the respective sections according to image position in the article. The final multimedia corpus contains 2,866 documents, which are image-text pairs and annotated with a label from the vocabulary of 10 semantic categories. Images are represented by 128 dimensional bag-of-visual-words vectors based on SIFI feature. Texts are represented by the probability distributions over 10 latent topics, which are derived from LDA model. Following the setting in [12], 2,173 documents form a training set and the remaining 693 documents constitute a testing set.

The NUS-WIDE dataset chosen from Flickr originally has 269,648 images associated with 81 ground truth concept categories. The image and its tags are regarded as image-text pair. Similar to the previous work [20], we randomly extract 4,800 image-text pairs by keeping that each pair only belongs to one of the most frequent 15 category concepts, such as animal, buildings, clouds and so on. The low-level features of images and texts are represented by 500-dimensional SIFI feature vectors and 1,000 dimensional tag occurrence vectors, respectively. The number of documents derived from each semantic category is 320, and this dataset is randomly split into a training set of 3,750 documents and a testing set of 1,050 documents.

To acquire a fair comparison, all the compared methods were trained and tested on the same training and testing sets, respectively. In Wiki dataset, the surrounding textual descriptions of each image are long and quality. On the contrary, there are smaller words per image in the NUS-WIDE dataset. Together, these two datasets explain the diversity of cross-modal multimedia retrieval. In accordance with the compared works, normalized correlation is used to measure distance in SM and SCM, hamming distance is used in LCMH. In our framework, the multiclass weak learner is decision tree whose depth is two, and the number of iterations T is set to be 100. Additionally, normalized correlation is utilized to rank the retrieved objects. A returned document is relevant to a query document if they belong to the same semantic category. Mean Average Precision (MAP) [10], one of the standard information retrieval metrics, is utlized as the major performance measures. The larger the MAP, the better the performance. Besides, the PR curve (11-point interpolated precision-recall curve) and recall curve are also reported on all the methods.

4.1   Results on Wiki Dataset

The MAP values of MMB are compared with those of other three methods in Table 1, where W denotes the number of returned documents. As can be seen, the proposed method consistently outperforms the compared approaches for the two retrieval tasks. For example, compared with the second best approach, L_MMB improves about 20.2% with $W = 50$ and 19.8% with $W = $ all, attaining the average MAP values of 0.31 and 0.23, respectively. The inter-modal semantics is not involved in the unimodal multiclass boosting, which is the reason why the MAP values of SM and SCM are smaller. On one hand, intra-modal semantic information reflects the internal regularities of each modality, while the inter-modal semantics captured by minimizing the inter-modal loss focuses on the correlation between different modalities, a combination of these two semantics is beneficial. On the other hand, some low quality intra-modal semantics can be enhanced by complementing the inter-modal semantic correlations. These may be the reasons behind the gain of our method.

For a more detailed analysis, the PR curves and recall curves of LCMH, SM, SCM, E_MMB and L_MMB are shown in Fig. 1. As can be seen from the PR curves, our methods again have improvements over their counterparts for text query and image query. These improvements are substantial and occur at all levels of recall, which indicates higher accuracy and better generalization. The recall curve can demonstrate the changes of recall when more objects in the returned rank list are inspected. For this performance measure, the curves of MMB

Table 1. The performance comparison (MAP Values) on Wiki dataset, the values shown in boldface are the best results

Experiment	Text Query		Image Query		Average	
	$W = 50$	$W = $ all	$W = 50$	$W = $ all	$W = 50$	$W = $ all
LCMH [20]	0.186	0.121	0.187	0.124	0.187	0.123
SM [12]	0.269	0.169	0.235	0.204	0.252	0.187
SCM [12]	0.280	0.179	0.235	0.204	0.258	0.192
E_MMB	0.329	**0.208**	**0.273**	**0.260**	0.301	**0.234**
L_MMB	**0.350**	0.207	0.270	0.252	**0.310**	0.230

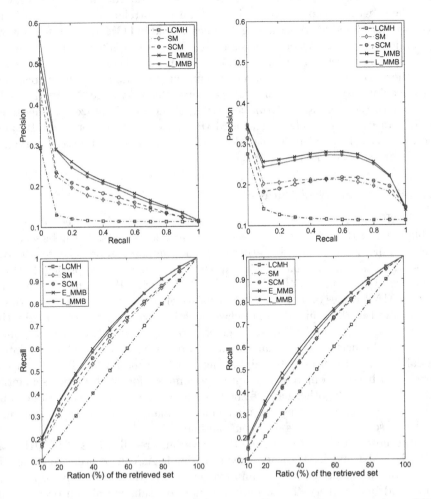

Fig. 1. Precision-recall curves (top row) and recall curves (bottom row) on Wiki dataset, for text query on the left and image query on the right

Table 2. The performance comparison (MAP Values) on NUS-WIDE dataset, the values shown in boldface are the best results

Experiment	Text Query		Image Query		Average	
	$W = 50$	$W = $ all	$W = 50$	$W = $ all	$W = 50$	$W = $ all
LCMH [20]	0.135	0.076	0.132	0.082	0.134	0.079
SM [12]	0.169	0.097	0.177	0.148	0.173	0.123
SCM [12]	0.164	0.097	0.166	0.100	0.165	0.099
E_MMB	0.241	0.139	0.220	0.198	0.231	0.169
L_MMB	**0.248**	**0.143**	**0.234**	**0.203**	**0.241**	**0.173**

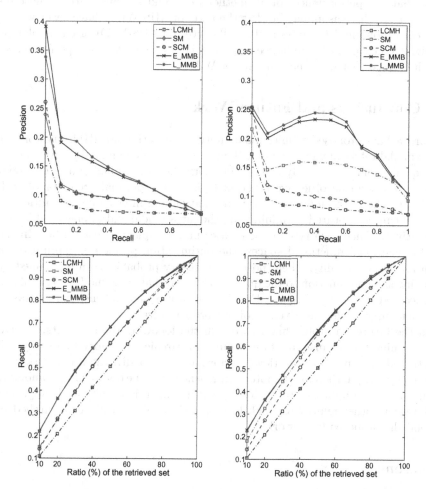

Fig. 2. Precision-recall curves (top row) and recall curves (bottom row) on NUS-WIDE dataset, for text query on the left and image query on the right

are always above those of the compared methods, which means a better recall can be attained when the same number of documents are inspected. In other words, our methods can rank more relevant documents in the front of returned list. In summary, these experimental results suggest that the combination of intra-modal semantic information and inter-modal semantic correlation can result in a significant benefit.

4.2 Results on NUS-WIDE Dataset

The similar evaluations are conducted on NUS-WIDE dataset. The MAP scores of all the methods are shown in Table 2. It is observed that our MMB again outperforms the compared methods by a large gain for $W = 50$ and $W = $ all. Note that the performances of all methods on Wiki are higher than those on NUS-WIDE, the reason may be that the texts of NUS-WIDE have fewer words. The corresponding PR curves and recall curves on NUS-WIDE are also plotted in Fig. 2. Once again, it can be seen that MMB has the best overall performance, which is consistent with that on dataset Wiki.

5   Conclusions and Future Work

So far, we have proposed a multimodal multiclass boosting (MMB) framework for cross-modal retrieval which enables the retrieval from one modality in response to a query from another. By considering both the intra-modal and inter-modal losses, we design an objective function to learn the optimal predictor of each modality. The intra-modal semantic information and inter-modal semantic correlation can be captured by minimizing the intra-modal and inter-modal losses, respectively. These two types of semantics may complement each other. In the multidimensional functional spaces, the optimization problem is minimized in turn for every modality. Then, posterior category probabilities are attained by applying sigmoid function to the quasi-margins so that cross-modal retrieval can be executed in a semantic space. Experiment results on two benchmark datasets well demonstrate the effectiveness of our proposed framework.

In the future, we would like to work in the following three aspects. Firstly, how to mine the intrinsic relations from a hierarchical category taxonomy will be studied to improve the retrieval performance. Secondly, the framework used here is supervised, therefore an interesting benefit may be obtained by combining an unsupervised approach or exploiting a semi-supervised method. Lastly, how to integrate other multiclass weak learns, such as SVM, will be researched to extend the proposed framework.

References

1. Blaschko, M.B., Lampert, C.H.: Correlational Spectral Clustering. In: Proceeding of IEEE Conference on Computer Vision and Pattern Recognition, pp. 1–8 (2008)
2. Bronstein, M.M., Bronstein, A.M., Michel, F., Paragios, N.: Data Fusion through Cross-modality Metric Learning Using Similarity-sensitive Hashing. In: Proceeding of IEEE Conference on Computer Vision and Pattern Recognition, pp. 3594–3601 (2010)

3. Carneiro, G., Chan, A.B., Moreno, P.J., Vasconcelos, N.: Supervised Learning of Semantic Classes for Image Annotation and Retrieval. IEEE Transactions on Pattern Analysis and Machine Intelligence 29(3), 394–410 (2007)
4. Chua, T.S., Tang, J., Hong, R., Li, H., Luo, Z., Zheng, Y.: NUS-WIDE: a real-world web image database from National University of Singapore. In: Proceedings of the ACM International Conference on Image and Video Retrieval, pp. 48–56 (2009)
5. Clinchant, S., Ah-Pine, J., Csurka, G.: Semantic Combination of Textual and Visual Information in Multimedia Retrieval. In: Proceeding of the 1st ACM International Conference on Multimedia Retrieval (2011)
6. Coxeter, H.S.M.: Regular polytopes. Courier Dover Publications (1973)
7. Freund, Y., Schapire, R.E.: A Decision-Theoretic Generalization of On-Line Learning and an Application to Boosting. Journal of Computer and System Sciences 55(1), 119–139 (1997)
8. Hotelling, H.: Relations Between Two Sets of Variates. Biometrika 28(3-4), 321–337 (1936)
9. Kidron, E., Schechner, Y.Y., Elad, M.: Pixels That Sound. In: Proceeding of IEEE Conference on Computer Vision and Pattern Recognition, pp. 88–95 (2005)
10. Manning, C.D., Raghavan, P., Schtze, H.: Introduction to information retrieval. Cambridge University Press, Cambridge (2008)
11. Rasiwasia, N., Moreno, P.J., Vasconcelos, N.: Bridging the gap: Query by semantic example. IEEE Transactions on Multimedia 9(5), 923–938 (2007)
12. Rasiwasia, N., Costa Pereira, J., Coviello, E., Doyle, G., Lanckriet, G.R., Levy, R., Vasconcelos, N.: A New Approach to Cross-modal Multimedia Retrieval. In: Proceedings of the ACM International Conference on Multimedia, pp. 251–260 (2010)
13. Saberian, M.J., Masnadi-Shirazi, H., Vasconcelos, N.: Taylorboost: First and Second-order Boosting Algorithms with Explicit Margin Control. In: Proceeding of IEEE Conference on Computer Vision and Pattern Recognition, pp. 2929–2934 (2011)
14. Saberian, M.J., Vasconcelos, N.: Multiclass Boosting: Theory and Algorithms. In: Advances in Neural Information Processing Systems, pp. 2124–2132 (2011)
15. Shen, J., Cheng, Z.: Personalized Video Similarity Measure. Multimedia Systems 17(5), 421–433 (2011)
16. Turnbull, D., Barrington, L., Torres, D., Lanckriet, G.: Semantic Annotation and Retrieval of Music and Sound Effects. IEEE Transactions on Audio, Speech, and Language Processing 16(2), 467–476 (2008)
17. Typke, R., Wiering, F., Veltkamp, R.C.: A Survey of Music Information Retrieval Systems. In: Proceeding of ISMIR, pp. 153–160 (2005)
18. Zhen, Y., Yeung, D.Y.: A Probabilistic Model for Multimodal Hash Function Learning. In: Proceedings of the 18th ACM SIGKDD International Conference on Knowledge Discovery and Data Mining, pp. 940–948 (2012)
19. Zhu, J., Zou, H., Rosset, S., Hastie, T.: Multi-class Adaboost. Statistics and Its Interface 2, 349–360 (2009)
20. Zhu, X., Huang, Z., Shen, H.T., Zhao, X.: Linear Cross-modal Hashing for Efficient Multimedia Search. In: Proceedings of the 21st ACM International Conference on Multimedia, pp. 143–152 (2013)

Author Index

Akai, Ryota   I-71
Al-Ani, Ahmed   I-1
Albatal, Rami   II-111, II-282, II-319, II-369
Al-Dmour, Hayat   I-1
Ali, Noman   I-1
Almeida, Jurandy   I-335
Amsaleg, Laurent   II-345
Andreadou, Katerina   I-372
Apostolidis, Evlampios   II-249
Apostolidis, Konstantinos   II-249
Atrey, Pradeep K.   I-396, I-430
Avgerinakis, Konstantinos   II-249

Babaguchi, Noboru   I-71
Bae, Changseok   II-447, II-458
Ballas, Nicolas   I-318
Baranauskas, M. Cecília C.   I-335
Barthel, Kai Uwe   II-237, II-287
Basu, Subhasree   II-135
Beecks, Christian   I-294
Belongie, Serge   II-357
Blažek, Adam   II-243
Böszörmenyi, Laszlo   I-203, II-87
Bu, Fenglin   II-491

Cai, Hongming   II-491
Cao, Qinglei   II-148
Carrive, Jean   I-140
Čech, Přemysl   II-315
Chang, Lu   I-224
Che, Xiaoyin   I-385
Chen, Fangdong   I-454
Chen, Hui   II-470
Chen, Jun   I-37, I-105, I-234, II-61
Chen, Liang   I-37
Chen, Liming   I-522
Chen, Xiaogang   I-246
Chen, Xiaoming   II-208
Chen, Zhicong   I-25
Chen, Zuo   I-180
Cheng, Shuying   I-25
Chua, Tat-Seng   I-269

Chung, Yuk Ying   II-208
Cobârzan, Claudiu   II-99, II-266

Dai, Manna   I-25
Danesh, Ali   II-1
Del Fabro, Manfred   II-266, II-291
Deng, Robert H.   II-38
Deng, Shihong   II-183
Ding, Lu   I-360
Dong, Jichao   II-311
Dong, ShaoLong   I-487
Dong, Wenyan   I-49
Dong, Zhe   II-524
Du, Youtian   II-427
Duane, Aaron   II-319
Duc, Thanh Ngo   II-50
Duong, Duc Anh   II-50, II-278
Dupont, Stéphane   II-255

Eiríksdóttir, Áslaug   II-345
El Saddik, Abdulmotaleb   II-1

Fang, Wenhua   I-234
Fang, Yuchun   I-224
Fang, Yuming   I-59
Fortmann, Jutta   II-323
Fu, Huiyuan   I-498, II-13

Gao, Li   I-487, I-534
Gao, Yue   I-269
Geng, Wenjing   I-546
Giangreco, Ivan   II-255
Gillhofer, Michael   II-380
Ginsca, Alexandru Lucian   I-318
Godbole, Suneeta   II-357
Gong, Yunfei   I-191
Gravier, Guillaume   I-140
Grošup, Tomáš   II-315
Grottel, Sebastian   II-159
Guan, Naiyang   II-536, II-548
Guggenberger, Mario   I-203, II-87
Gumhold, Stefan   II-159
Guntuku, Sharath Chandra   II-171

Guo, Zongming   II-307
Gurrin, Cathal   I-571, II-111, II-282,
    II-303, II-319, II-335, II-369

Han, Zhen   I-37
He, Can   II-227
He, Xiangjian   I-25, I-246, I-418, II-227
Hentschel, Christian   I-95
Heuten, Wilko   II-323
Hezel, Nico   II-237, II-287
Hinbarji, Zaher   II-111
Hirai, Tatsunori   II-299
Hoet, Miklas   II-123, II-261
Hu, Chunzhi   II-403
Hu, Pan   II-491
Hu, Qingmao   I-128
Hu, Ruimin   I-37, I-234, I-487, I-534,
    II-195
Huang, Bingyue   II-61, II-195
Huang, Di   I-522
Huang, Lei   II-148
Huang, Xiaodi   II-403
Huang, Xuhui   II-536, II-548
Hudelist, Marco A.   I-306, II-99, II-272
Hürst, Wolfgang   II-123, II-261

Ishikura, Kazumasa   I-408

Jia, Wenjing   II-227
Jiang, Junjun   I-37, II-195
Jiang, Sheng   I-257, II-560
Jiang, Shuhui   II-392
Jiang, Xiaojun   II-512
Johansen, Dag   II-319, II-335
Johansen, Håvard   II-335
Johnson, Eileen   II-357
Jónsson, Björn Þór   II-345
Ju, Ran   I-546
Ju, Zhen-fei   I-83

Kanellos, Ioannis   I-318
Kang, Kyuchang   II-447, II-458
Kansal, Kshitij   I-430
Kashino, Kunio   II-74
Katto, Jiro   I-408
Kawai, Masahide   I-155
Kerr, Jacqueline   II-357
Kimura, Akisato   II-74
K.L.E., Marissa   II-135

Kompatsiaris, Ioannis   I-372, II-249
Kuwahara, Daiki   II-295
Kwon, Yongjin   II-447, II-458

Lai, Songlin   I-25
Lang, Xufeng   I-559
Lárusdóttir, Marta Kristín   II-345
Le, Duy-Dinh   II-50, II-278
Le Borgne, Hervé   I-318
Leng, Qingming   I-105, II-195
Li, Feng   I-418
Li, Houqiang   I-454, I-466, II-524
Li, Jun   I-418
Li, Kan   I-510
Li, Ning   I-83
Li, Xuechen   I-128
Li, Yang   II-470
Li, Zhen   II-148
Lian, Zhouhui   I-49
Liang, Chao   I-105, I-234, II-61, II-195
Liao, Wenjuan   I-118, II-311
Lin, Peijie   I-25
Lin, Xiahong   I-442
Lin, Yang-Cheng   II-220
Lin, Yu-Hsun   I-13
Liu, Dawei   II-481
Liu, Liang   I-498
Liu, Mengmeng   II-470
Liu, Qiang   I-191
Liu, Weifeng   II-502
Liu, Zhihong   II-183
Lo, Felix   II-208
Lokoč, Jakub   II-243, II-291, II-315
Long, Wei   I-246
Lu, Dawei   I-498
Lu, Hanqing   I-214
Lu, Yansheng   I-257, II-560
Luo, Jun   I-214
Luo, Suhuai   I-128
Luo, Zhigang   II-536, II-548
Lux, Mathias   I-203, II-87
Lv, FuYu   I-510

Ma, Huadong   I-498, II-13
Mackowiak, Radek   II-237, II-287
Maddage, Namunu C.   I-396
Maejima, Akinobu   II-295
Mao, Xiao-jiao   I-83
Markatopoulou, Foteini   I-282
Markatopoulou, Fotini   II-249

Matzner, Filip   II-243
Mei, Tao   II-392
Meinel, Christoph   I-385
Meng, Shengbin   II-307
Meyer, Jochen   II-323
Mezaris, Vasileios   I-282, II-249
Mironidis, Theodoros   II-249
Moghimi, Mohammad   II-357
Mohanty, Manoranjan   I-430
Moon, Jinyoung   II-447, II-458
Morishima, Shigeo   I-155, II-295, II-299
Moumtzidou, Anastasia   I-372, II-249
Moynagh, Brian   II-303
Mu, Tingting   II-436

Nakajima, Jiro   II-74
Nan, Yuanyuan   I-234
Ngo, Thanh Duc   II-278
Nguyen, Ngoc-Bao   II-50
Nguyen, Vinh-Tiep   II-278
Nguyen, Vu-Hoang   II-278, II-50
Nie, Jie   II-148
Nitta, Naoko   I-71

O'Halloran, Kay L.   II-135
Ohya, Hayato   II-299
Ou, Yangxiao   I-477
Ouyang, Kun   II-38

Pan, Peng   I-257, II-560
Papadopoulos, Symeon   I-372
Papadopoulou, Olga   I-282
Patras, Ioannis   I-282, II-249
Paul, Manoranjan   I-167
Pei, JianMeng   I-510
Pinto-Cáceres, Sheila M.   I-335
Pittaras, Nikiforos   I-282
Popescu, Adrian   I-318
Primus, Manfred Jürgen   II-99

Qian, Xueming   II-392, II-436
Qu, Bingqing   I-140

Ren, Tongwei   I-546
Rossetto, Luca   II-255
Roy, Sujoy   II-171

Sack, Harald   I-95
Sahillioğlu, Yusuf   II-255
Saini, Mukesh   II-1
Sasaki, Shoto   II-299
Satoh, Shin'ichi   II-278

Schedl, Markus   II-380
Schinas, Emmanouil   I-372
Schoeffmann, Klaus   I-306, II-99, II-266, II-291
Schöffmann, Klaus   I-306
Schuldt, Heiko   II-255
Seddati, Omar   II-255
Seidl, Thomas   I-294
Sezgin, Metin   II-255
Sha, Feng   II-208
Shen, Jialie   II-38, II-392
Shen, Xu   II-524
Shepherd, John   II-403
Shui, Hanbing   II-427
SigurÞórsson, Hlynur   II-345
Skopal, Tomáš   II-243, II-315
Smeaton, Alan F.   I-571, II-282
Song, Dacheng   I-246
Song, Mofei   I-559
Spehr, Marcel   II-159
Su, Feng   I-348, II-26
Sudo, Kyoko   II-415
Sugimoto, Akihiro   II-74
Sun, Fuming   II-512
Sun, Jia-Hao   I-13
Sun, Jun   II-307
Sun, Kaimin   I-105, II-61, II-195
Sun, Lifeng   I-442
Sun, Yongqing   II-415
Sun, Zhengxing   I-559

Tan, Huibin   II-548
Tan, Sabine   II-135
Tang, Yingmin   I-49
Taniguchi, Yukinobu   II-415
Tao, Dacheng   II-536, II-548
Tao, Dapeng   II-502
Terziyski, Stefan   II-369
Tian, Xinmei   II-524
Tómasson, Grímur   II-345
Torres, Ricardo da S.   I-335
Tsikrika, Theodora   I-372
Tu, Weiping   I-534

Uemura, Aiko   I-408
Uysal, Merih Seran   I-294

Vallet, Félicien   I-140
van de Werken, Rob   II-261
Vo, Phong   I-318
Vrochidis, Stefanos   I-372, II-249

Wang, Cheng   I-385
Wang, Jinqiao   I-214
Wang, Peng   I-571
Wang, Shixun   II-560
Wang, Shizheng   I-118
Wang, Xiao   I-234
Wang, Xiaocheng   I-534
Wang, Xinpeng   II-13
Wang, Yanye   I-487
Wang, Yimin   II-61
Wang, Yongtao   II-183
Wang, Yunhong   I-522
Wang, Yuqi   I-191
Wang, Zhen   II-502
Wang, Zheng   I-105, II-61, II-195
Wang, Zhi   I-442
Wasmann, Merlin   II-323
Wei, Zhiqiang   II-148
Wei, Zhuo   II-38
Weisi, Lin   I-59, II-171
Wu, Gangshan   I-546
Wu, Ja-Ling   I-13
Wu, Lijun   I-25
Wu, Lin   II-403
Wu, Shuyi   II-536
Wu, Tingzhao   I-534
Wu, Yongdong   II-38

Xiao, Jianguo   I-49
Xiao, Rui   I-167
Xie, Liang   I-257
Xie, SongBo   I-487
Xu, Hang   I-510
Xu, Huazhong   I-214
Xu, Jiang   I-360
Xu, Meixiang   II-512
Xu, Min   I-418
Xu, Ming   I-360
Xu, Qing   I-306, II-272
Xu, Richard   I-418
Xu, Xiangyang   I-546
Xue, Hao   II-26
Xue, Jiao   II-427
Xue, Like   I-348, II-26

Yakubu, M. Abukari   I-396
Yan, Yan   II-148
Yang, Cheng   I-487
Yang, Haojin   I-385
Yang, Jie   I-246
Yang, Yang   I-269
Yang, Yu-bin   I-83
Yang, Yuhong   I-487, I-534
Yang, Zeying   I-180
Ye, Gensheng   II-311
Ye, Mang   I-105
Ye, TengQi   II-303
Yeh, Wei-Chang   II-208
Yoon, Wan Chul   II-447
Yu, Hongjiang   I-487
Yu, Jun   II-502
Yu, Zhihua   II-481
Yuan, Yuan   I-59

Zeng, Dingheng   I-118, II-311
Zhang, Feiqian   I-559
Zhang, Fengjun   II-470
Zhang, Fengli   I-418
Zhang, Jie   I-25
Zhang, Jinlei   I-466
Zhang, Liming   II-227
Zhang, Mengmeng   I-477
Zhang, Xiang   II-536, II-548
Zhang, Xixiang   I-180
Zhang, Yongtai   II-183
Zhang, Zheng   II-13
Zhang, Zhenxing   II-282
Zhao, Huan   I-180
Zhao, Sicheng   I-269
Zhao, Yisi   II-436
Zhen, Qingkai   I-522
Zheng, Ning   I-360
Zhong, Huicai   I-118, II-311
Zhou, Jiang   II-319
Zhou, Liguo   I-118
Zhou, Lijuan Marissa   II-303
Zhou, Liting   II-303
Zhu, Jixiang   II-38
Zimmermann, Roger   II-135